Advanced Chromatography: Methods and Industrial Applications

Advanced Chromatography: Methods and Industrial Applications

Editor: Carol Evans

New York

Published by NY Research Press
118-35 Queens Blvd., Suite 400,
Forest Hills, NY 11375, USA
www.nyresearchpress.com

Advanced Chromatography: Methods and Industrial Applications
Edited by Carol Evans

International Standard Book Number: 978-1-63238-624-3 (Hardback)

Cataloging-in-Publication Data

Advanced chromatography : methods and industrial applications / edited by Carol Evans.
p. cm.
Includes bibliographical references and index.
ISBN 978-1-63238-624-3
1. Chromatographic analysis. 2. Chromatographic analysis--Industrial applications.
I. Evans, Carol.
QD79.C4 A38 2019
543.8--dc23

Contents

Preface

The objective of this book is to give a general view of the different areas of chromatography and its applications across various industrial fields. Chromatography as a chemical technique uses a variety of methods and apparatus to separate mixtures into their constituent components. Some examples of the industries that use this technique are food industry, forensic science, pharmaceutical industry, etc. This book presents the upcoming techniques and modern applications of chromatography in a comprehensive manner for easy understanding of the reader. It strives to provide a fair idea about this discipline and to help develop a better understanding of the latest advances within this field. Extensive use of examples and student-friendly language makes this book a valuable source of knowledge. Those who wish to broaden their understanding of the subject will be greatly benefited by this book.

This book is the end result of constructive efforts and intensive research done by experts in this field. The aim of this book is to enlighten the readers with recent information in this area of research. The information provided in this profound book would serve as a valuable reference to students and researchers in this field.

At the end, I would like to thank all the authors for devoting their precious time and providing their valuable contribution to this book. I would also like to express my gratitude to my fellow colleagues who encouraged me throughout the process.

Editor

Activity Ratio of Caesium, Strontium and Uranium with Site Specific Distribution Coefficients in Contaminated Soil near Vicinity of Fukushima Daiichi Nuclear Power Plant

Mishra S[1,2], Sahoo SK[1*], Arae H[1], Watanabe Y[1] and Mietelski JW[3]

[1]*Project for Environmental Dynamics and Radiation Effects, National Institute of Radiological Sciences, 4-9-1 Anagawa, Inage-ku, Chiba 263-8555, Japan*
[2]*Environmental Monitoring and Assessment Section, Bhabha Atomic Research Centre, Mumbai, Trombay – 400 085, India*
[3]*Department of Nuclear Physical Chemistry, The Henryk Niewodniczanski Institute of Nuclear Physics, Polish Academy of Sciences, Krakow, Radzikowskiego 152, Poland*

Abstract

Activity concentrations of Cs radioisotopes (^{137}Cs and ^{134}Cs) were measured in four soil samples at the proximity of Fukushima daiichi nuclear power plant (FDNPP), located in and off the radioactive plume direction. Activity of Cs was higher than ^{90}Sr and ^{238}U in the plume direction. Lowest ^{137}Cs activity was found to be 950 ± 13 Bq/kg and highest was 62,200 ± 880 Bq/kg. In case of ^{90}Sr, the lowest and highest activity concentrations were found to be 8.4 ± 1.5 Bq/kg and 21.2 ± 2.6 Bq/kg respectively. Activity concentration of ^{238}U varied from 25.4 ± 0.1 Bq/kg to 45.2 ± 0.1 Bq/kg. Activity ratio of ^{134}Cs to ^{137}Cs was in range of 0.84-0.87 measured in 2012, indicates the origin from FDNPP accident. Activity ratio of ^{234}U/^{238}U varied from 0.996-1.029. To establish the fate and transfer of radionuclides, site specific distribution coefficient (K_d) were measured in the soil samples using standard method. K_d values for Cs, Sr and U were found to vary from (114 ± 3 to 404 ± 43), (65 ± 0.2 to 154 ± 7) and (1640 ± 22 to 8563 ± 458) L/kg respectively. Different soil parameters like particle size distribution, pH, organic content, cation exchange capacity, $CaCO_3$, elemental and oxide composition of soil has been carried out to understand the geochemical behaviour of these radionuclides. A good correlation was observed for K_d (Cs) and K_d (Sr) with cation exchange capacity and fine particle concentration and K_d (U) with Fe and organic content of soil.

Keywords: ^{137}Cs; ^{90}Sr; ^{238}U; Activity ratio; Distribution coefficient; FDNPP

Introduction

A catastrophic earthquake of magnitude (9.0) followed by tsunami on 11 March 2011, caused a major nuclear accident at the Fukushima Daiichi Nuclear Power Plant (FDNPP) about 250 km north to Tokyo, capital of Japan and inundated the nuclear site with seawater. The damage by the flooding of the site resulted in loss of cooling to the three reactor units. This led to eventual overheating, hydrogen explosions and a probable partial melting of the core of the three reactors [1]. This accident resulted in a substantial release of radioactive materials to the atmosphere, ocean, and has caused extensive contamination to the environment [2-4]. The nuclear accident was eventually classified at Level 7, the highest on the International Nuclear and Radiological Event Scale (INES) [5], the same as given as the Chernobyl Nuclear Power Plant accident (CNPP) in 1986.

Environmental monitoring with respect to different radionuclides is very important to understand the cause and severity of accident. Similarly, radionuclides with long half lives and their high solubility in aqueous solution under normal environmental conditions are of major concern from radiological safety point of view. During the accident significant deposition of ^{137}Cs occurred on surface soils of Fukushima prefecture [6,7] as fallout. Due to its relatively long half-life of 30.2 years and chemical behaviour similar to that of potassium, ^{137}Cs is considered as one of the most significant radionuclide in the environmental radioactivity monitoring. Measurements of Chernobyl fallout clearly indicate that soil is the main reservoir of ^{137}Cs but its migration behaviour and associated profile distribution are site-specific and depend on soil characteristics and environmental conditions [8]. High volatile fission products including ^{90}Sr, ^{129m}Te, ^{131}I, ^{134}Cs and ^{137}Cs released into the atmosphere and caused radioactive contamination of soils over a wide area [7,9]. ^{90}Sr is chemically similar to calcium and it has a high transfer rate to the skeletal system. It leads to internal irradiation which can cause bone cancer, cancer of soft tissues, leukemia inside the human bones [10] together with its decay product ^{90}Y. Therefore ^{90}Sr and ^{90}Y belong to the most hazardous fission products. From environmental monitoring point of view ^{90}Sr is also of major concern, due to its high solubility. The chemical toxicity of the uranium metal is the primary environmental health hazard, whereas radioactivity of uranium a secondary concern. Exposure to uranium can result in both chemical and radiological toxicity. The update of the toxicologic evidence on uranium adds to the established findings regarding nephrotoxicity, genotoxicity, and developmental defects. Additional novel toxicological findings include some at the molecular level that raise the biological plausibility of adverse effects on the brain, on reproduction, including estrogenic effects, on gene expression, and on uranium metabolism [11]. For a better radiological impact assessment due to radionuclide contamination in ecosystems apart from the information on radionuclide species, knowledge on their interactions in soil-water systems influencing mobility and biological uptake is essential. Uranium has significant importance in radio-ecology because of its high solubility as U (VI). For prediction of radionuclide-soil interaction and subsequently transportation of radionuclides in the soil column, site specific distribution coefficient, K_d plays an important role.

***Corresponding author:** Sahoo SK, Project for Environmental Dynamics and Radiation Effects, National Institute of Radiological Sciences, 4-9-1 Anagawa, Inage-ku, Chiba 263-8555, Japan, E-mail: sahoo@nirs.go.jp

Present work emphasises monitoring of different radionuclides like Cs, Sr and U in soil samples affected by FDNPP accident in order to know the extent of contamination. Further to understand their geochemical behaviour, the site specific distribution coefficients (K_d) values for the respective nuclides have been determined experimentally. The sorption and migration of these nuclides in soil could be influenced by many factors such as the nature of the radionuclides in solution as well as chemical and mineralogical nature and physical environment of the soil. Therefore in the present study the soil samples were also chemically characterized with respect to different soil parameters controlling sorption of these stable/radionuclides. Laboratory batch method was used for measurement of K_d in soil collected around FDNPP. Expecting contamination of soil samples with radio isotopes of Cs, Sr and U as a consequence of accident, stable isotopes are used for Cs and Sr whereas depleted uranium is used as tracer for uranium K_d estimation.

Materials and Methods

Sampling location

Sampling sites have been selected based on contamination level measured by gamma dose rate, which was influenced by radioactive plume direction during accident. Two contaminated soil samples were collected in the plume direction, FS1 (N:37.4140°, E:141.0020°) and FS2 (N:37.4240°, E:141.0004°) with measured dose rate 71 and 85 μSv/h respectively on 24th Nov, 2011. Apart from these, two more soil samples have been collected from less contaminated area in the opposite direction, FS3 (N:37.1982°, E:140.9598°) and FS4 (N:37.1683°: E:140.9894°) with measured dose rate 0.50 and 0.45 μSv/h respectively on 23rd Nov, 2011. All the four samples were within 20 km from FDNPP as shown in Figure 1.

Sample collection and processing

Surface soil samples were collected by a core sampler from a depth of 0-10cm.The samples were transferred to laboratory and air dried at room temperature to a constant weight followed by sieving using a 2 mm sieve for estimation of K_d and soil parameters. Sieved samples were further oven dried at 110°C for 24 hours and were packed in a plastic U-8 standard cylindrical containers (diameter=48 mm; h=58 mm) for γ-spectrometry measurement. Sieved and oven dried samples were further powdered (<150 μm) for chemical characterization. Powdered soil samples were chemically digested using microwave digestion system (Milestone, MLS 1200 Mega) for elemental analysis. Since K_d is highly dependent on ionic strength of water [12], instead of using distilled water or synthetic water, ground water samples were collected from the corresponding sites of soils to measure K_d values experimentally. The samples were filtered using 0.45 μm membrane filter and divided into two parts. One part was acidified for elemental analysis and non-acidified part water was used for K_d estimation.

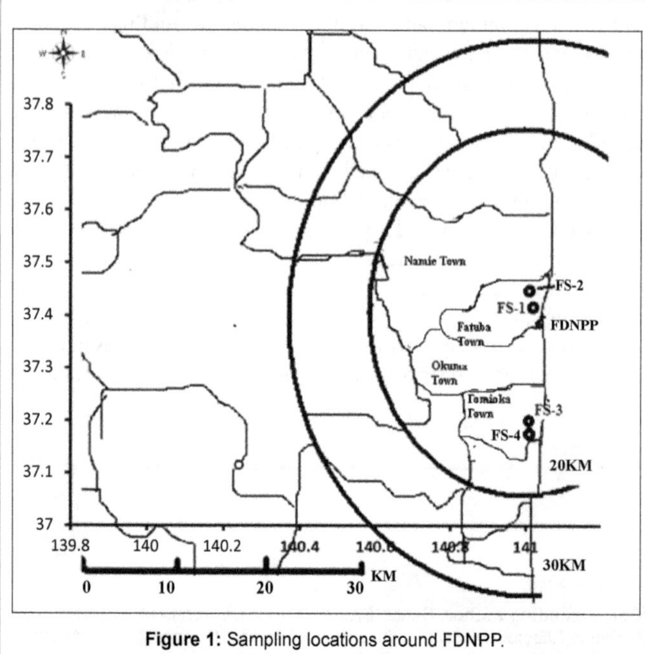

Figure 1: Sampling locations around FDNPP.

Instrumental analysis

Gamma spectrometry: A high-purity germanium detector (ORTEC GEM-100210) γ-spectroscopy coupled with a multi-channel analyzer (ORTEC-7700-010) with a range 0-4000 keV and gamma studio software (Seiko EG&G, 2000) was used for the measurement of ^{134}Cs and ^{137}Cs. The detector efficiency was determined using a 100 g multi-nuclide standard source supplied by Japan Radioisotope association with quoted gamma energies ranging from 60 to 1333 keV an overall uncertainty of less than 5%. The results were compared with a certified soil standard IAEA-375 to check the reproducibility of the method. The counting time was varied from 10,000 sec to 72,000 sec depending on the activity concentration. In the case of radioactivity concentration for ^{137}Cs, it was measured at energy of 662 keV whereas for ^{134}Cs concentration was derived from net peak areas due to 605 keV and 796 keV gamma rays [13].

Liquid scintillator: The ^{90}Sr activity was determined by a Wallac 1414-003 liquid scintillation spectrometer. Separation of strontium took place on a column filled with Sr-Spec (Eichrom Ltd). The resin was loaded on a column of 6 mm in diameter and 10 cm in height. The column was conditioned with 50 ml of 8 M HNO_3 and the sample was passed through the column. Strontium was later eluted afterwards together with some possible traces of lead from the column with 50 ml of double distilled water (DDW). Strontium was separated from lead by precipitation of lead iodide [14] and then filtration. Solution containing strontium was evaporated to dryness, iodine excess present in solution was destroyed by addition of concentrated HNO_3, evaporation and then addition of a few drops of peroxide followed by next evaporation to dryness. Dry residue (almost invisible) was dissolved in a mixture of 1.5 ml of 1 M HNO_3 and 1.5 ml of DDW and then transferred to a 15 ml scintillator plastic vial. Those samples were analyzed by an HPGe low background gamma spectrometer to determine the recovery of ^{85}Sr. Then, this solution was mixed with 10 ml of a liquid scintillation cocktail (HiSafe 3) and, after waiting two weeks for the equilibrium between ^{90}Sr and its daughter, ^{90}Y ($T_{1/2=}$64 h), was subject of LSC measurement. Typical counting time was 30,000 s.

Inductively coupled plasma mass spectrometry (ICP-MS): Inductively coupled plasma mass spectrometry (Agilent 7500, Agilent Technologies, USA) was used for the measurement of stable isotopes of Cs, Sr and U in both soil and water, which yielded detection limits of 0.03 μg/L for Cs, 1.17 μg/L for Sr and 0.001 μg/L for U. The ICP-MS detection limit was calculated as three times the standard deviation of the calibration blank measurements (1:1 v/v HNO_3: MilliQ water, n=10). The relative error of ICP-MS results for the reference sample (lake sediment JLK-1) for ^{238}U was 0.56 %. The parameters for data acquisition and optimization conditions are reported elsewhere [15].

Thermal ionisation mass spectrometry (TIMS): A VG Sector 54-30 thermal ionization mass spectrometer (VG Isotopes Ltd., UK), equipped with nine Faraday cup collectors and Daly ion detection system positioned behind axial Faraday and wide aperture retardation potential (WARP) energy filter, was used for the isotopic measurement of uranium. U was separated and pre-concentrated using a combination of anion exchange and extraction chromatography from digested soil [15]. First column was prepared by using pre-cleaned anion exchange resins (Dowex 1X-8, 200-400 mesh, Cl^- form) and packed into MUROMAC polypropylene column (10 x 120 mm size) up to a height of 8 cm. The second column was packed with 0.5 ml of pre-cleaned UTEVA resin (Eichrom industries Inc.) 100-150 μm particle size in MUROMAC polypropylene column (7 x 60 mm size). The UTEVA resin contains diamylamylphosphonate (DAAP) as a specific extractant. Anion exchange column was first conditioned by passing 50 ml of 8 M HNO_3. The digested soil sample in 8 M HNO_3 (5 ml) was transferred to anion exchange column followed by a washing of 80 ml 8 M HNO_3. Last 50 ml washing was collected after passing through anion exchange column, discarding the initial 30 ml washing of 8 M HNO_3. The UTEVA column was conditioned with 1.5 ml of 4 M HNO_3. The washing solution collected from the first column was evaporated completely and taken in 1 ml of 4 M HNO_3 to pass through UTEVA followed by a washing of 5 ml of 4 M HNO_3. Then column was washed with 6 ml of 5 M HCl to remove Th and finally uranium was eluted from UTEVA column with 6 ml of 0.1 M HNO_3. Complete organic destruction was carried out in the eluent collected from UTEVA column by treating with H_2O_2 and HNO_3 before loading onto TIMS filament. The recovery for uranium separation by this method was checked using ICP-MS and was found to be 80 ± 5 %.

Inductively coupled plasma optical emission spectrometry (ICP-OES): Major cations from soil and water samples were analysed using an inductively coupled plasma optical emission spectrometry (CCD simultaneous ICP-OES, VISTA-PRO, Seiko Instrument Inc.)

X-ray fluorescence spectrometer (XRF): Major oxide compositions in soil was analysed by X-ray fluorescence spectrometer (Rigaku ZSX 100e, Rigaku Corporation).

Estimation of distribution coefficient: Distribution coefficient, K_d for Cs, Sr and U nuclides were measured using laboratory batch method [16]. 1 g of air dried, homogenized soil sample (<2 mm size) was taken in triplicate in a 50 ml polypropylene centrifuge tube with screw cap and was equilibrated with 30 ml of water for 12 hours in a shaker with a shaking rate of 0.8-1.2, oscillations per second at 20°C. Solution mixture was centrifuged at 4500 rpm for 30 minutes and filtered through 0.45 μm filter paper and the supernatant solution was discarded and the procedure was repeated to ensure no change in the pH of the suspension as a result of the equilibration. 30 ml water, spiked with the radiotracer/stable nuclide, was then added to the soil and the suspension was kept in the shaker and equilibrated for 72 hours which have been determined earlier. 3.3 μg/ml of stable Cs ($CsNO_3$), 1.6 μg/ml of stable Sr ($Sr(NO_3)_2$) and 3.3 μg/ml of U ($UO_2(NO_3)_2$) were used as tracer. The amount of stable nuclides released during sorption already present in soil and water is negligible compared to the spike amount. Solution mixture was centrifuged at 4500 rpm for 30 minutes and filtered through a 0.45 μm filter paper. The amount of adsorbed Cs, Sr and U were estimated from their concentration in the aqueous phase before and after the adsorption. Three control solutions of ground water and river water containing tracers were prepared for the determination of initial activity (A_0). K_d has been calculated using the following equation:

$$K_d = \frac{(A_0 - A)/W_S}{A/V_W}$$

where, V_W is the volume of water in the suspension, W_S is the weight of soil, A_0 and A are initial activity and activity at equilibration respectively. Activity sorbed on the wall of polypropylene tube was found to be negligible in the present case for the three nuclides.

Physico-chemical characterization of soil and water: Some selected properties were analysed in soils of <2 mm size fraction. Particle size analysis of soil samples was carried out using different mesh sieves for sand (2-0.05 mm), silt and clay (<0.05 mm). Soil parameters like pH was measured using 1:2.5 soil: water suspension for 1 hour using a pH meter and concentrations of exchangeable base cations (Ca^{+2}, Mg^{+2}, Na^+ and K^+) extracted with 1 N ammonium acetate at pH 7 were determined using ICP-OES. Organic content (OC) has been determined by loss of ignization method [17]. Major and trace elements were analysed in water samples using ICP-OES and ICP-MS respectively.

Results and Discussion

Distribution of Caesium, Strontium and Uranium radio isotopes and their activity ratio in soil affected by FDNPP

Activity concentrations for different radionuclides (Cs, Sr and U) in four soil samples are given in Table 1. A large variation in ^{137}Cs activity was observed with lowest activity 950 ± 13 Bq/kg to highest activity 62,200 ± 880 Bq/kg. This is mainly because, the deposition of Cs released during nuclear power plant accident was local and strongly dependent on meteorological conditions-predominantly the radioactive plume direction during accident. For samples collected in the North West direction (FS-1 and FS-2), the caesium radioisotopes activity observed was much higher while compared to the samples collected from opposite direction (FS-3 and FS-4). However, ^{134}Cs and ^{137}Cs ratio was found to be same in all the samples (0.84-0.87 as measured in 8th May 2012), indicates that contamination comes from the accident. In case of ^{90}Sr activity, it varied from 8.4 ± 1.5 Bq/kg to 22.3 ± 1.5 Bq/kg, which was much lower than Cs activity. The relatively low levels of ^{90}Sr and a lack of any clear relation to caesium activity in samples (Table 1) suggest that predominantly strontium comes from global fallout, although one cannot exclude traces of FDNPP origin, which are masked by local variation of global fallout remains. It is very difficult to say something certain on ^{90}Sr origin in samples (after decay of any possible addition of short lived ^{89}Sr) since strontium generally is not conserved in soils, so its activity ratio to caesium as measured in soil samples even just a year after fallout can be much lower than it was in fallout originally. Activity concentration for ^{238}U was found to vary from 25.4 ± 0.11 Bq/kg to 45.2 ± 0.1 Bq/kg.

Generally $^{234}U/^{238}U$ activity ratio is found to vary considerably due to natural causes in water, soil, sediment and uranium ore of different geographical origin. The mechanism of such variation is preferential leaching of ^{234}U compared with ^{238}U from solid phase, caused by

Sample	^{134}Cs (Bq/kg)	^{137}Cs (Bq/kg)	^{90}Sr (Bq/kg)	^{238}U (Bq/kg)	$^{134}Cs/^{137}Cs$	$^{234}U/^{238}U$
FS-1	53,400 ± 770	62,200 ± 880	18.4 ± 3.4	31.7 ± 0.1	0.86 ± 0.02	1.015 ± 0.002
FS-2	53,200±770	61,830 ± 880	8.4 ± 1.5	49.2 ± 0.2	0.86 ± 0.02	1.029 ± 0.001
FS-3	850 ± 20	1,010 ± 20	21.2 ± 2.6	25.4 ± 0.1	0.84 ± 0.03	0.996 ± 0.011
FS-4	830 ± 20	950 ± 13	22.3 ±1.5	45.2 ± 0.1	0.87 ± 0.03	1.006 ± 0.003

Table 1: Activity concentrations and activity ratio of Cs, Sr and U in soil around FDNPP.

radiation damage of crystal lattice upon alpha decay of ^{238}U, oxidation of insoluble tetravalent ^{234}U to soluble hexavalent ^{234}U during decay, and alpha recoil of ^{234}Th (and its daughter ^{234}U) into solution phase [18]. $^{234}U/^{238}U$, activity ratio, in soil typically ranges from 0.5 to 1.2 and also varies due to anthropogenic discharge. The activity ratio found to vary from 0.996-1.029 in the studied samples and did not show much disequilibrium. The distributions of $^{234}U/^{238}U$ are relatively uniform in soil with activity ratio not much different from 1.0 (indicating radiological equilibrium between ^{234}U and ^{238}U). This probably is because parent materials of these soils are relatively new volcanic ejecta, which initially had $^{234}U/^{238}U$ activity ratio near 1.0 and has not been subjected to leaching of ^{234}U since its deposition.

Correlation of distribution coefficient and soil parameters

In order to investigate the fate of these nuclides under site specific conditions the distribution coefficients for the nuclides have been determined experimentally. The K_d values of three radionuclides at four sampling points are presented in Figure 2. K_d values, measured on triplicate analyses, for Cs, Sr and U were found to vary from (114 ± 3 to 404 ± 43), (65 ± 0.2 to 154 ± 7) and (1640 ± 22 to 8563 ± 458) L/kg respectively. In log scale, K_d values for Cs, Sr and U were found to be $\log K_d(U) \sim 3 > \log K_d(Cs) \sim 2 > \log K_d(Sr) \sim 1\text{-}2$. K_d values are highly dependent on site specific characteristics and the methodology adopted for analysis. Therefore in literature we can find 3 to 4 order of magnitude difference in the K_d values for these nuclides [19,20]. K_d being an important parameter, used as an input in contaminant transport model for long term prediction of radionuclide migration, needs to be accurately determined under specific conditions. During Fukushima accident a large area is contaminated, so these data can be used for prediction of long term transport of these radionuclides in soil water system. K_d values are site specific, therefore chemical characterization of the soil as well as water from the respective site has been carried out. The major oxide composition in soil sample and some important soil parameters are given in Tables 2 and 3.

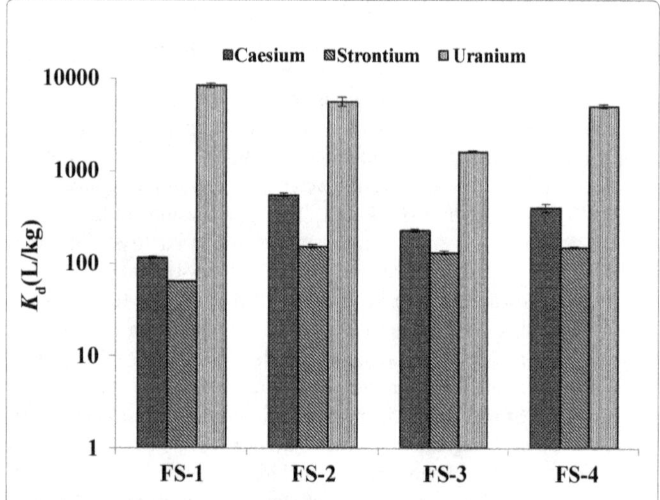

Figure 2: Distribution coefficient (K_d) for Cs, Sr and U radionuclides around FDNPP.

Soils are found to be richer in iron oxide content with minimum for FS-3 (3.0 %) and maximum for FS-1 (6.3 %). In FS-1 most of the major oxides are found to be at higher concentration than other soils. From the particle size distribution of soil, it is observed that FS-1 is mostly sandy having 96 % sand composition, whereas FS-2 is loamy sand with 27 % of silt and clay content. All four soil samples are acidic in nature having pH ranging from 4.97 to 6.24. A wide range for $CaCO_3$ content have been observed in these four samples with a minimum of 0.54 % (FS-1) to a maximum 1.99 % (FS-3). In all the soil samples stable Cs, Sr and U varies from 3.11 to 5.03 μg/g, 107.5 to 154.5 μg/g and 2.04 to 3.95 μg/g respectively.

The pH of the water samples used for K_d determination found to vary from 6.61-8.03 at 20°C. The concentration of major ions e.g., Ca, Mg, Na and K in water vary from 9.2-15, 2.7-5.4, 3.7-7.4 and 1.4 to 20.6 mg/L respectively. Stable concentrations in water for Cs, Sr and U found to vary from 0.03-0.36, 37.7-43.0 and 0.004-0.026 μg/L respectively.

In order to understand the sorption behaviour, a correlation of K_d with respect to different soil parameters were carried out and is given in Table 4. Soils contain different cations with adsorbing components especially in the silt and clay fractions [21]. Important soil components affecting the sorption of radionuclides are minerals like smectite, illite and chlorite as well as oxides and hydroxides of iron, aluminum and manganese. In the present study we could find K_d (Cs) shows significantly good correlation with fine particles (silt + clay) (r=0.91, p=0.08), CEC (r=0.86, p=0.14) and exchangeable cations (K, Na, Mg, Ca) shown in Table 4, where r is represented by Pearson correlation coefficient and p is for significance level. Cs is readily adsorbed on the surfaces of soil mineral components as well as on those of colloids and suspended particles in soil solution. Ion exchange is the adsorption mechanism and the sorption behavior of Cs is dominated by highly selective interlamellar exchange sites between the 2:1 layers of clay and mica minerals called frayed edge sites (FES) on which adsorption has been reported to be virtually irreversible [22]. Sr shows similar behavior to that of Cs. K_d (Sr) shows significant correlation with Ex Ca (r=0.97, p=0.02) and can be attributed due to similar chemical properties. There is a good correlation of K_d (U) with organic content (OC), (r=0.8,

Sample	SiO_2	TiO_2	Al_2O_3	Fe_2O_3	MgO	CaO	Na_2O	K_2O	P_2O_5	MnO	Total
FS-1	56.6	0.67	19.1	6.3	1.3	0.7	1.5	1.9	0.09	0.11	88.3
FS-2	60.7	0.53	15.6	4.1	1.1	1.6	1.9	1.5	0.47	0.06	87.7
FS-3	68.1	0.43	13.5	3.0	0.8	1.1	1.7	2.3	0.09	0.04	91.2
FS-4	66.1	0.44	14.4	3.2	0.8	1.1	1.9	2.2	0.13	0.04	90.4

Table 2: Major oxide composition (mass %) in soil.

Sample	pH	OC %	CEC mgeq/100g	$CaCO_3$ %	Sand %	Silt+Clay %
FS-1	5.87	6.53	1.9	0.54	96.3	3.7
FS-2	6.24	6.34	9.9	1.02	73	27
FS-3	4.97	3.59	7.2	1.99	88.1	11.9
FS-4	5.29	3.74	11.2	0.89	88.5	11.5

Table 3: Soil parameters affecting radionuclide sorption.

K_d (U)	pH	OC	SiO_2	TiO_2	Al_2O_3	Fe_2O_3	MgO	MnO	$CaCO_3$	P_2O_5
r	0.72	0.8	-0.92	0.87	0.92	0.88	0.88	0.85	-0.96	0.10
p	0.28	0.2	0.07	0.12	0.07	0.12	0.12	0.15	0.03	0.89
K_d (Cs)	Ex. Ca	Ex. K	Ex. Na	Ex. Mg	CEC	Sand	Silt +Clay	CaO	Na_2O	P_2O_5
r	0.87	0.4	0.98	0.87	0.86	-0.9	0.91	0.9	0.91	0.85
p	0.13	0.59	0.01	0.13	0.14	0.08	0.08	0.1	0.09	0.14
K_d (Sr)	Ex. Ca	Ex. K	Ex. Na	Ex. Mg	CEC	Sand	Silt +Clay	CaO	Na_2O	P_2O_5
r	0.97	0.78	0.79	0.98	0.97	-0.75	0.75	0.82	0.91	0.52
p	0.02	0.21	0.21	0.02	0.02	0.24	0.25	0.18	0.03	0.47

Table 4: Pearson correlation coefficient (r) and significance levels (p) for K_d with soil parameters.

p=0.2) and major oxides of soil components such as Fe, Al, Ti and Mn oxides and a significant reverse correlation is observed with soil $CaCO_3$ (r=-0.96, p=0.03). The uranyl ion and its complexes adsorb onto clays, organics and oxides [20]. Organic complexes play an important role in uranium aqueous chemistry. The non-complex uranyl ion has a greater tendency to form complexes with fulvic and humic acids than many other metals with a +2 valence. This has been attributed mainly to greater "effective charge" of the uranyl ion compared to other divalent metals. As the ionic strength of an oxidized solution increases, other ions, notably Ca^{2+}, Mg^{2+}, and K^+, will displace the uranyl ion from soil exchange sites, forcing it into solution. For this reason, the uranyl ion is particularly mobile in high ionic-strength solutions [20]. Other cations will not only dominate over the uranyl ion in competition for exchange sites, but also carbonate ions will form strong soluble complexes with the uranyl ion, thereby reducing the uranium sorption [20].

However, for a better correlation, sample size has to be increased which is in progress. Site specific K_d values could be used for contaminant transport model to predict the radionuclide migration and also the relationship between soil parameters and sorption behaviour of radionuclides will be helpful for soil remediation and long term dose assessment.

Conclusion

As a consequence of FDNPP accident a large area is contaminated with respect to Cs activity and is much higher than the global fallout. Cs activity ratio (^{134}Cs/^{137}Cs) indicates contamination from the accident. ^{90}Sr contamination was also observed in the studied samples. When compared with Cs activity, about 3 order lower than Cs activity is observed in contaminated sites and 1 order lower in less contaminated site and likely coming at least in some part from global fallout. Uranium activity ratio (^{234}U/^{238}U) found to vary from 0.996 to 1.029. The K_d values for Cs, Sr and U found in the order of log K_d(U) ~3 > log K_d(Cs) ~2 > : logK_d(Sr) ~1-2. Based on this study, K_d (Cs) and K_d (Sr) show a good correlation with CEC, exchangeable cations (K, Na, Mg, Ca) and with fine particles (silt+clay). K_d (Cs) is mostly associated with clay particles and the sorption is mainly controlled by ion exchange mechanism. K_d (Sr) shows significant correlation with Ex. Ca, Ex. Mg. and CEC, indicating the sorption may be controlled by cation exchange. K_d (U) shows a good correlation with organic content (OC) and major oxides of soil components such as Fe, Al, Ti and Mn oxides and a negative correlation was observed with $CaCO_3$. The results obtained are preliminary in nature and systematic studies are in progress to ascertain a better correlation among distribution coefficient and soil parameters.

Acknowledgement

This research was supported by a Grant-in-Aid for Scientific Research (P-11053) from the Japan Society for the Promotion of Science and Ministry of the Environment, Japan for sample collection. SM is thankful to the Japan Society for the Promotion of Science for the award of a research fellowship.

References

1. WHO (2012) Preliminary dose estimation from the nuclear accident after the 2011 Great East Japan earthquake and tsunami. WHO Press, Geneva, Switzerland. ISBN 978 92 4 150366 2.
2. NSC (2011) A testing estimates the amount of ^{131}I and ^{137}Cs released from the Fukushima Dai-ichi nuclear power plant into the atmosphere.
3. MEXT (2011) Preparation of distribution map of radiation doses, etc. (Map of radioactive caesium concentration in soil) by MEXT.
4. Masatoshoi Y (2012) A brief review of environmental impacts and health effects from the accidents at the Three Mile Island, Chernobyl and Fukushima Daiichi Nuclear power plants. Radiation Emergency Medicine 1: 33-39.
5. INES (2008) The International Nuclear and Radiological Event Scale (leaflet). Vienna, International Atomic Energy Agency.
6. Honda MC, Aono T, Aoyama M, Hamajima Y, Kawakami H, et al. (2012) Dispersion of artificial cesium- 134 and -137 (^{134}Cs and ^{137}Cs) in the western North Pacific one month after the Fukushima accident. Geochem J 46: el-e9.
7. Watanabe T, Tsuchiya N, Oura Y, Ebihara M, Inoue C, et al. (2012) Distribution of artificial radionuclides (^{110m}Ag, ^{129m}Te, ^{134}Cs, ^{137}Cs) in surface soils from Miyagi prefecture, northeast Japan, following the 2011 Fukushima Dai-ichi Nuclear Power Plant accident. Geochem J 46: 279-285.
8. Dragović S, Gajić B, Dragović R, Janković-Mandić L, Slavković-Beškoski L, et al. (2012) Edapic factors affecting the vertical distribution of radionuclides in the different soil types of Belgrade, Serbia. J Environ Monit 14: 127-137.
9. Yoshida N, Takahashi Y (2012) Land-surface contamination by radionuclides from the Fukushima Daiichi Nuclear Power Plant Accident. Elements 8: 201-206.
10. U.S. Department of Health and Human Services (2004) Toxicological profile for strontium (April 2004).
11. Brugge D, Buchner V (2011) Health effects of uranium: new research findings. Rev Environ Health 26: 231-249.
12. Mishra S, Maity S, Pandit GG (2012) Estimation of distribution coefficient of natural radionuclides in soil around uranium mines and its effect with ionic strength of water. Radiat Prot Dosim 152: 229-233.
13. Lee CJ, Chung C (1991) Determination of ^{134}Cs in environmental samples using a coincidence gamma ray spectrometer. Appl Radiat Isot 42: 783-788.
14. Gaca P, Skwarzec B, Mietelski JW (2006) Geographical distribution of ^{90}Sr contamination in Poland. Radiochim Acta 94: 175-179.
15. Sahoo SK, Parami VK, Quirit LL, Yonehara H, Ishikawa T, et al. (2011) Determination of uranium concentrations and its activity ratios in coal and fly ash from Philippine coal-fired thermal power plants using ICP-MS and TIMS. Proc Radiochim Acta 1: 257-261.
16. Mishra S, Arae H, Zamostyan PV, Ishikawa T, Yonehara H, et al. (2012) Sorption-desorption characteristics of uranium, cesium and strontium in typical podzol soils from Ukraine. Radiat Prot Dosim 152: 238-242.
17. Konare H, Yost RS, Doumbia M, McCarty GW, Jarju A, et al. (2010) Loss on ignition: Measuring soil organic carbon in soils of the Sahel. West Africa. Afr J Agric Res 5: 3088-3095.
18. Ivanovich M, Harmon RS (1992) Uranium series disequilibrium: Applications to earth, marine and environmental sciences. (2nd edn) Clarendon Press, Oxford.

19. Vandenhove H, Gil-Garćia C, Rigol A, Vidal M (2009) New best estimates for radionuclide solid-liquid distribution coefficients in soils. Part 2. Naturally occurring radionuclides. J Environ Radioactiv 100: 697-703.

20. EPA (1999) Understanding variation in partitioning coefficients, K_d, Values. Volume II USA EPA 402-R-99-004B.

21. Lusa M, Lempinen J, Ahola H, Söderlund M, Lahdenperä AM, et al. (2014) Sorption of cesium in young till soils. Radiochim Acta 102: 645-658.

22. Koch-Steindl H, Prohl G (2001) Considerations on the behavior of long-lived radionuclides in the soil. Radiat Environ Biophys 40: 93-104.

2

Evaluation of Gas Injection in the Horizontal Wells and Optimizing Oil Recovery Factor by Eclipse Software

Afshin Davarpanah*

Department of Petroleum, Science and Research Branch, Islamic Azad University, Tehran, Iran

Abstract

During the life of a producing oil field, several production stages are encountered. Initially, when a field is brought into production, oil flows naturally to the surface due to existing reservoir pressure in the primary phase. As reservoir pressure drops, water is typically injected to boost the pressure to displace the oil in the secondary phase. Lastly, the remaining oil can be recovered by a variety of means such as CO_2 injection, natural gas miscible injection, and steam recovery in the final tertiary or enhanced oil recovery (EOR) phase. Nowadays, despite of new technologies in petroleum industries, the volume of oil and gas production in many oil fields are plummeted because most of the wells in Iran and other countries were being produced for several years. Hence, recovery factor was a reduction trend. Petroleum engineers due to their field studies and experimental evaluation wrapped up and showed that drilling wells in forms of horizontal, the contact of reservoir formation and production facilities were merely increased. Therefore, the recovery factor was rose gradually. In most of scenario's injection, CO_2 is the best gas compound that should be injected to the model. Thereby, the results that was received from the injection methods for taking appropriate results were be compromised. This comparison included gas continuous injection and using horizontal wells impact.

Keywords: Gas continuous injection; Horizontal wells; CO_2 injection; Scenario's injection

Abbreviations: C1: Methane gas (Condensate form); N_2: Nitrogen gas; TB: Combination of some gases such as methane, ethane.

Introduction

Horizontal drilling

Horizontal drilling is a drilling process in which the well is turned horizontally at depth. It is normally used to extract energy from a source that it runs horizontally, such as a layer of shale rock. Horizontal drilling is a common was of extracting gas from the Marcellus Shale Formation.

Since the horizontal section of a well is at great depth, it must include a vertical part as well. Thus, a horizontal well resembles. When examining the differences between vertical wells and horizontal wells, it is easy to see that a horizontal well is able to reach a much wider area of rock and the natural gas that is trapped within the rock. Thus, a drilling company using the horizontal technique can reach more energy with fewer wells [1].

In this Figure 1, well B represents a vertically drilled well and well A represents a horizontally drilled well. Vertically drilled wells are only able to access the natural gas that immediately surrounds the end of the well. Horizontal wells are able to access the natural gas surrounding the entire portion of the horizontally drilled section.

As you can imagine, drilling a horizontal well is a more complicated process that drilling a conventional vertical well. The driller must first determine the depth of the energy-rich layer. That is done by drilling a conventional vertical well, and analyzing the rock fragments that appear at the surface from each depth.

Once the depth of the shale is determined, the driller withdraws the drilling assembly, and then inserts a special bit assembly into the ground that allows the driller to keep track of its vertical and horizontal location.

The driller calculates an appropriate spot above the shale in which the drill must start to turn horizontally. That spot is known as the 'kickoff point." From there, the drill bit is progressively angles so that it creates a borehole that curves horizontally. If done properly, the well reaches the 'entry point' and makes its way into the rock where the natural gas is trapped. The horizontal portion of the well is drilled, and provides much more contact with the rock than a vertical well.

Technology of drilling the well in a Horizontal way to achieve more volume of oil and gas widespread dramatically. In this method the contact between the reservoir and the well will be increased and the production engineers and companies try to drill most of those wells that produced for many years and their recovery factor are far less by horizontal wells to produce more volume of oil and gas. Drilling horizontal wells require much more facilities and equipment than drilling the vertical wells but this method economically due to the recovery factors increases and optimize the well efficiency will be preferred. It can be seen a sample of drilling horizontal well in Figure 2.

The use of horizontal drilling technology in oil exploration, development, and production operations has grown rapidly over the past 5 years. This report reviews the technology, its history, and its current domestic application [2]. It also considers related technologies that will increasingly affect horizontal drilling's future.

Drilling horizontal wells were more difficult than vertical wells. It needed more proper equipment and the procedure of drilling must be controlled in every hole change angle due to its high sensitivity of wellbore inefficiencies and in some cases it can be possible that facilities and equipment stocked in the borehole and provided fishing operation

***Corresponding author:** Afshin Davarpanah, Department of Petroleum, Science and Research Branch, Islamic Azad University, Tehran, Iran
E-mail: Afshindpe@gmail.com

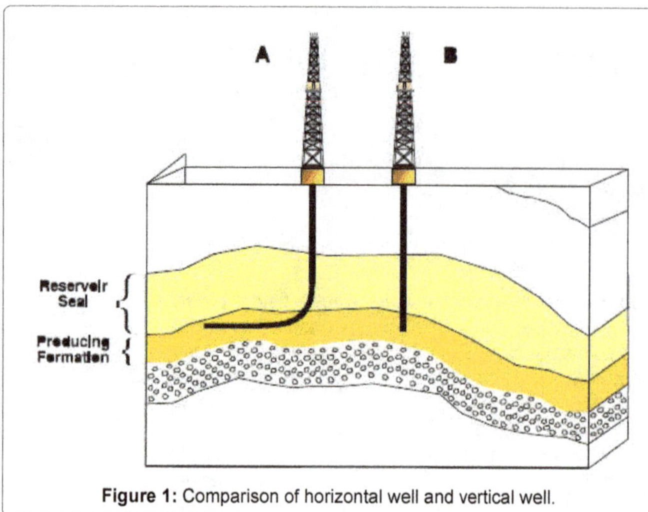

Figure 1: Comparison of horizontal well and vertical well.

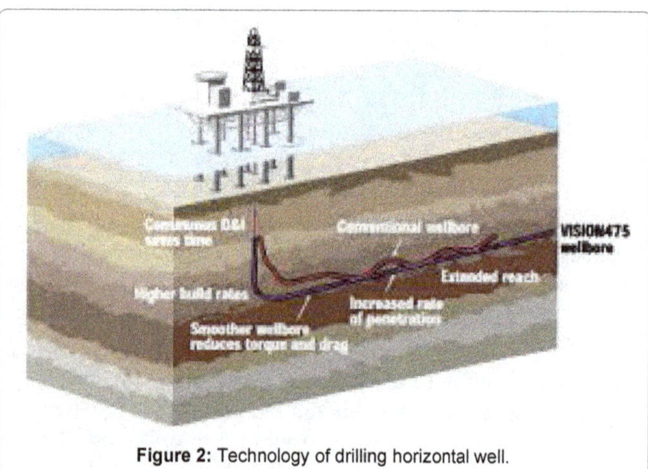

Figure 2: Technology of drilling horizontal well.

to recover the fallen facilities and reopen the way of drilling. If the drilling operation with less accuracy, it may be imposed extravagant expenditures to the government or in the worth cases we will lose the well in terms of high spending.

Some of the applications of drilled well in a horizontal well listed below. So, we can identify the procedure of enhanced oil recovery and optimizes its efficiencies [3-5].

- When the production oil and gas decreased, horizontal well helped us to increase the production rate.
- when we lost a well caused by many occasions like fish, horizontal wells will be drilled to assess the reservoir.
- If we have a reservoir beneath the sea or shallow water, we used this method of drilling.

Types of horizontal wells and their application favorability

Short-Radius horizontal wells: Short-radius horizontal wells are commonly used when re-entering existing vertical wells in order to use the latter as the physical base for the drilling of add-on arc and horizontal whole sections [6].

Medium-Radius horizontal wells: Medium-radius horizontal wells allow the use of larger hole diameters, near-conventional bottom hole (production) assemblies, and more sophisticated and complex completion methods [7].

Long-Radius horizontal wells: Long-radius holes can be drilled using either conventional drilling tools and methods, or the newer steerable systems. Long-radius wells, in the form of deviated wells (not, however, deviated to the horizontal), have been around quite a while. They are not suited to leases of less than 160 acres due to their low build rates [7].

Recovery mechanisms

When a reservoir were being drilled, firstly it was produced by the natural mechanisms. Natural mechanisms provided the substantial energy to push the fluid mainly included oil and gas to the surface. Oil expansion is a very important part among those mechanisms if without availability of other artificial introduced energy. The rock and fluids expand due to their individual compressibility [8]. Since the fluid was expanded and the matrix pore volume was imbibed by the surrounding fluid, the reservoir pressure was plunged. As a result, the crude oil and water will be forced out of the pore space to the wellbore [9]. If the natural energies couldn't provide appropriate power to transfer the oil and gas to the surface, we should use enhanced oil recovery methods like gas injection, water injection and etc. to alter the dead well to a productive well with high efficiency.

Injection of organic and inorganic gases into the reservoir is used during secondary and tertiary recovery of oil in order to maintain the balance of the reservoir energy. Gas methods involve blocking the action of capillary forces due to partial or complete mixing of the gas with oil. Process stability is achieved by alternating injection of bursts of gas, gas and water or a gas - water mixture. Pure gas treatment is applied in the case of vertical oil displacement or formations with low permeability where flooding is not applicable. Gas methods enable to increase the production of oil by 5-19% comparing to ordinary flooding applied during secondary recovery [10,11]. There are following gas methods:

Hydrocarbon gas drive, CO_2 drive, Inert and flue gas drive, High pressure gas drive, Water-gas drive.

Application of CO_2 injection: Carbon dioxide (CO_2) flooding is a process whereby carbon dioxide is injected into an oil reservoir in order to increase output when extracting oil.

When a reservoir's pressure is depleted through primary and secondary production, Carbon Dioxide flooding can be an ideal tertiary recovery method. It is particularly effective in reservoirs deeper than 2,500 ft., where CO_2 will be in a supercritical state, with API oil gravity greater than 22–25°C and remaining oil saturations greater than 20%. It should also be noted that Carbon dioxide flooding is not affected by the lithology of the reservoir area, but simply by the reservoir porosity and permeability, so that it is viable in both sandstone and carbonate reservoirs. Carbon dioxide flooding works on the physical phenomenon that by injecting CO_2 into the reservoir, the viscosity of any hydrocarbon will be reduced and hence will be easier to sweep to the production well [12,13].

As an oil field matures and production rates decline, there is growing incentive to intervene and attempt to increase oil output, via tertiary recovery techniques (also called improved or enhanced oil recovery). Petroleum Engineers will assess the available options-generally chemical injection, thermal/steam injection, or CO_2 injection. After gathering information and running simulations, the engineers will determine whether CO_2 is the optimal solution to increase oil production rates. To increase the rate of oil production, engineers must increase the amount of pressure within the reservoir.

In the case of CO_2 flooding, the first step is to inject water into the reservoir, which will cause the reservoir pressure to increase. Once the reservoir has sufficient pressure, the next step is to pump the CO_2 down through the same injection wells. The CO_2 gas is forced into the reservoir and is required to come into contact with the oil. This creates this miscible zone that can be moved more easily to the production well. Normally the CO_2 injection is alternated with more water injection and the water acts to sweep the oil towards the production zone [14].

Field Evaluation

Introduce eclipse software

Eclipse software that was used for this evaluation was an integrated development environment (IDE) used in computer programming, and is the most widely used Java IDE. It contains a base workspace and an extensible plug-in system for customizing the environment. Eclipse is written mostly in Java and its primary use is for developing Java applications, but it may also be used to develop applications in other programming languages through the use of plugins. Eclipse software is a powerful software for simulating the stages above.

Evaluation

Successful management of enhanced oil recovery projects was depended on proper planning. Even though, the primary planning was completed it was prevented from the poor project assessment. Appropriate planning has been divided as below:

- Determine proper process of enhanced oil recovery
- Determine reservoir properties include rock and fluid properties
- Determine engineering design parameters
- Doing field tests or experimental operations
- Assess to the comprehensive model for managing the project to what will be expected.

For developing a reservoir model, five-stage procedures must be considered as below:

- Select appropriate simulator
- Collect variable data
- History match
- Anticipate project operation
- Comprehensive Reservoir studies

It should be considered noticeably that before starting a project, economic studies and reservoir simulation had been analyzed. Therefore, the extra costs of lacking knowledge about the reservoir could be decreased dramatically. For investigating the effects of horizontal wells on the recovery processes, horizontal sample was only be preplanned for injection wells. One model from horizontal wells with network length of 10 and network width was received. For this model two injection rate of 500 MSCFD and 1000 MSCFD was used. In the second model, horizontal well was drilled in the middle of area for comprising the results with the first sample. The results are being shown in the Figures 3 and 4.

As it can be seen in Figure 4, 4 types of gases like CO_2, C1 (methane gas), N_2 (nitrogen gas) and TB gas are injected to the bottom and middle layer of the reservoir in the rate of 500 MSCFD and 1000 MSCFD. The recovery factor in the rate of 1000 MSCFD for all gases are reached the peak. In Figure 5, 4 types of gas are compared together. The Figure 3 demonstrated that CO_2 gas has the largest impact on the ultimate recovery factor. After that C1 (methane gas) is the second gas that has more effect on ultimate recovery factor. The results of this evaluation are illustrated in Tables 1 and 2.

As it can be seen in the Figure 6 pressure differences are being decreased by the time of production. The Figure 6 shows that CO_2 injection has the least pressure loss during the period of time until it approximately reached plateau during the 3000 to 4000 days after production.

Bottom horizontal wells			
Scenario	Gas type	Inj. rate (SCFD)	RF (%)
GI	CO_2	500	60.69
GI	C1	500	61.22
GI	N2	500	18.97
GI	TB	500	55.92
GI	CO_2	1000	63.79
GI	C1	1000	62.33
GI	N2	1000	26.06
GI	TB	1000	59.03

Table 1: Recovery factor on the Bottom horizontal wells.

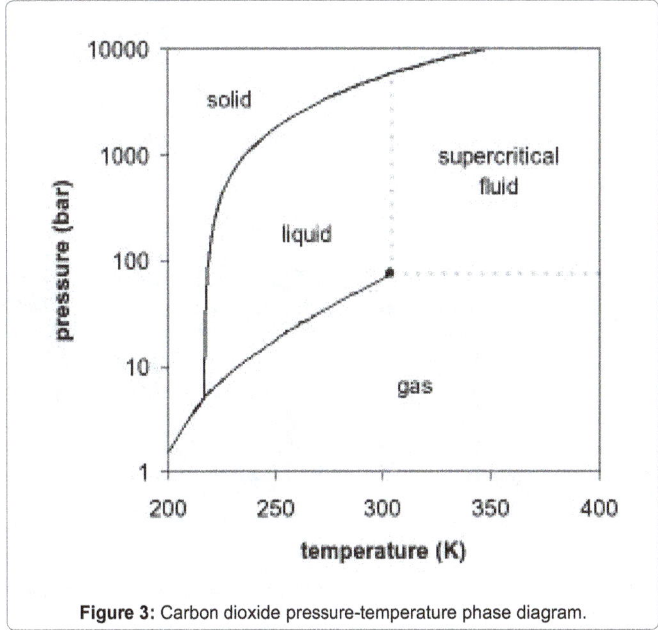

Figure 3: Carbon dioxide pressure-temperature phase diagram.

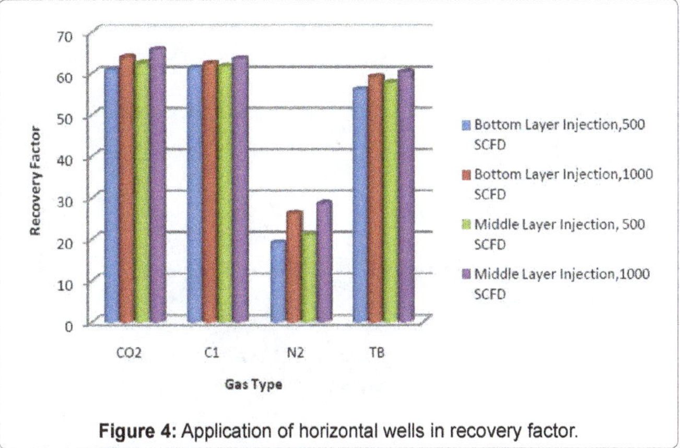

Figure 4: Application of horizontal wells in recovery factor.

Results

As you can see in the Tables 1 and 2, CO_2 injection is the most effective method in optimizing the recovery factor.

Conclusions

Drilling horizontal wells instead of vertical wells due to the high contact of reservoir formation helped the production engineer to produce more volume of oil from the reservoir. Thereby, the oil recovery factor will be dramatically increased. It also considers related technologies that will increasingly affect horizontal drilling's future. In most of scenario's injection, CO_2 is the best gas compound that should be injected to the model. Thereby, the results that was received from the injection methods for taking appropriate results were be compromised. This comparison included gas continuous injection and using horizontal wells impact. Among the oil compounds, CO_2 has the most recovery Factor. Thereby, it has the best efficiency up to all. The most completion that was used in the horizontal wells that fluid should be injected through the middle layers. Furthermore, the Figure 6 shows that CO_2 injection has the least pressure loss during the period of time.

Middle horizontal wells			
Scenario	Gas type	Inj. rate (SCFD)	RF (%)
GI	CO_2	500	62.43
GI	C1	500	61.68
GI	N2	500	20.83
GI	TB	500	57.6
GI	CO_2	1000	65.62
GI	C1	1000	63.42
GI	N2	1000	28.53
GI	TB	1000	60.22

Table 2: Recovery factor on the Middle horizontal wells.

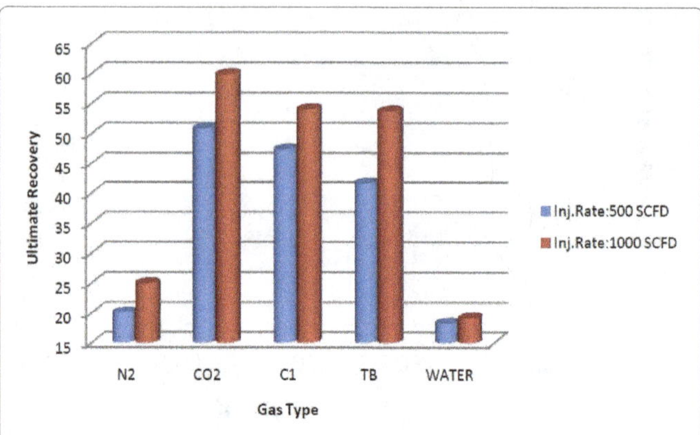

Figure 5: Comparison of different injection gases and its effect on ultimate recovery factor.

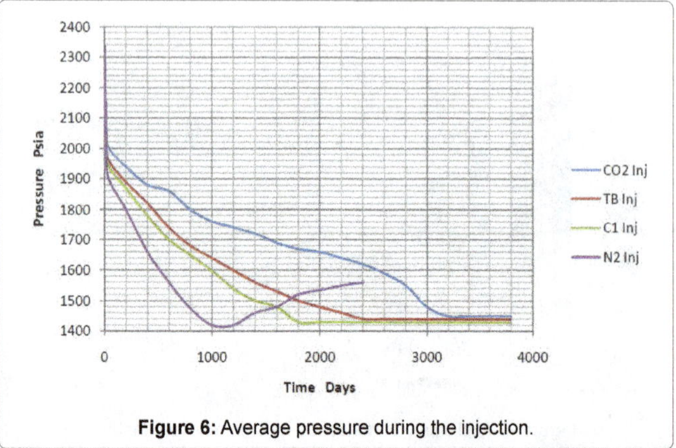

Figure 6: Average pressure during the injection.

References

1. Curtis R (2011) What is horizontal drilling, and how does it differ from vertical drilling. Institute for Energy and Environmental Research of Northeastern Pennsylvania Clearinghouse.
2. Mc Carthy K, Niemann M, Palmowski D, Peters K, Stankiewicz A (2011) Basic Petroleum Geochemistry for Source Rock Evaluation. Schlumberger.
3. Directed Technologies Drilling Inc., Horizontal Directional Drilling Services for Environmental remediation, Company Information.
4. Hazardous Waste Remedial Actions Program (HAZWRAP) (1995) In Situ Bioremediation Using Horizontal wells, Innovative Technology Summary Report, prepared for US Department of Energy.
5. US Department of Energy (1994) Volatile Organic Compounds in Non-Arid Soils Integrated Demonstration. DOE/EM-0135P, Chapter 2.0, Directional Drilling.
6. Rainer J (1991) Horizontal Drilling and Completions: A Review of Available Technology, Petroleum Engineer International, p: 18.
7. Lynn W (1992) Horizontal Drilling Is Feasible in Kansas. The American Oil & Gas Reporter, p: 84.
8. Ahmed T (2010) Reservoir Engineering Handbook. Gulf Professional Publishing, pp: 711-750.
9. Ceragioli ES (2008) Gas Injection: Rigorous Black-Oil or Fast Compositional Model. Paper SPE 12867 presented at the International Petroleum Technology Conference, Kuala Lumpur, Malaysia.
10. Kryaev DY (2013) Theory and practice of the use of EOR methods. Moscow.
11. Lions U, Plisg G (2009) Great Compendium of oil and gas engineer. Mining Equipment and mining technology. Moscow, Professiya.
12. Hebert M (2015) New technologies for EOR offer multifaceted solutions to energy, environmental, and economic challenges. Oil & Gas Financial Journal.
13. Brown K, Jazrawi W, Moberg R, Wilson M (2001) Role of Enhanced Oil Recovery in Carbon Sequestration. The Weyburn Monitoring Project, a case study (PDF) (Report). US Department of Energy, National Energy Technology Laboratory.
14. Weyburn-Midale CO_2 Project (2010) IEA GHG Weyburn- Midale CO_2 Storage and Monitoring Project. Regional Carbon Sequestration Partnerships Annual Review.

3

Detection and Quantification of Inorganic and Organic Anions in Natural, Potable, and Wastewaters in Northern New York Using Capillary Zone Electrophoresis and Indirect UV Detection

Lara Varden, Britannia Smith and Fadi Bou-Abdallah*

Department of Chemistry, State University of New York (SUNY) at Potsdam, 44 Pierrepont Avenue, Potsdam, NY, USA.

Abstract

Capillary zone electrophoresis (CZE) is a sensitive and rapid technique used for determining traces of inorganic and organic anions in potable, natural, and wastewaters. Here, CZE with indirect UV-diode array detection (CZE-DAD) was employed with a background electrolyte system comprising of an Agilent Technologies proprietary basic anion buffer at pH 12.0 and a forensic anion detection method. The limits of detection (LOD) for this method ranged between 3 and 5 ppm and involved hydrodynamic injection of 50 mbar for 6 s with a negative polarity separation voltage of −30 kV at 30°C, a detection wavelength of 350 nm and indirect reference of 275 nm. Fourteen different anions were checked for in the water samples that were examined and included bromide, chloride, thiosulfate, nitrate, nitrite, sulfate, azide, carbonate, fluoride, arsenate, phosphate, acetate, lactate, and silicate. The water samples were collected from Northern New York towns and the Raquette River water system, the third longest river in New York State and the largest watershed of the central and western Adirondacks. The concentrations detected for these anions ranged from <5.0 ppm to 260 ppm.

Keywords: Capillary zone electrophoresis; Indirect UV detection; Organic and inorganic anions; Northern New York Raquette River; Adirondack watershed

Introduction

Clean water is vital for sustaining life. As demands on limited water resources continue to increase, new and efficient purification and detection methods are needed for water treatment, recovery and reuse. This is particularly important in the case of a natural disaster or chemical spills where untapped water sources including saltwater might become the only source of clean drinking water for millions of people. Our life is sustained by complex interactions of inorganic and organic substances. Nine essential inorganic ions play a substantial role in supporting and sustaining health and life including Na^+, K^+, Ca^{2+}, Mg^{2+}, H^+, Cl^-, HCO_3^-, PO_4^{3-}, and OH^-, most of which are naturally found in our drinking water. Other non-essential ions with no known biological functions can be harmful to humans and wildlife if present above certain concentrations including lead (Pb), aluminum (Al), arsenic (As), barium (Ba), beryllium (Be), cadmium (Cd), gold (Au), lithium (Li), mercury (Hg), nickel (Ni), silver (Ag), and strontium (Sr) [1].

Although azide and lactate are not normally found in water, these compounds were added to the standards for additional possible detection. Azides are used in chemical industry as propellants and detonators and are precursors to amines [2]. Lactate is the conjugate base of lactic acid, which is used in topical preparations, cosmetics and detergents, found in milk products, and used as a food additive, among other uses [3]. Lactate could possibly be present in wastewater influent due to the usage and disposal by the general population as well as industry.

Separation of an aqueous mixture of analytes by capillary electrophoresis (CE) is achieved *via* the electrophoretic mobilities of each analyte, which depend on the charge-to-size ratio. Typically, for successful anions separation, the normal cathode to anode direction of the electroosmotic flow (EOF) must be reversed [4]. A variety of cationic surfactants such as tetradecyltrimethylammonium bromide, tetraethylenepentaamine,1,6-bis-(trimethylammonium)hexane, and 1,3,5-benzenetricarboxylic acid have also been used to enhance separation [5,6]. The cationic surfactants bind to the negatively charged wall of the capillary effectively creating a positively charged surfactant bi-layer and reversing the EOF direction. A negative voltage is then applied for analysis as both the EOF and the anions migrate in the same direction [4].

Because many anions do not have chromophores, detection must be achieved using indirect UV detection methods whereby compounds that absorb in the UV region and have high mobility are added to a buffer solution referred to as the background electrolyte (BGE). Several examples of indirect UV detection being implemented in the detection of anions exist in literature [7-13]. In brief, a UV-absorbing solute of the same charge is added to the BGE and serves as the visualizing reagent. When solute ions migrate past the detector window, they are measured as negative peaks relative to the absorbing solute. By reversing the signal and reference wavelengths of the diode-array detector, a positive signal is obtained whereby the area of the peak generated is linearly related to the sample concentration [4]. The migration speed of the UV-absorbing ion must closely match that of the ion being detected otherwise the analyte peak shape is severely distorted resulting in poor sensitivity. Depending on the size of the anion, different low concentration additives generating low currents (i.e. 5–10 μA) have been used for optimal peak shape and sensitivity [4,5,14]. Alternatively, zwitterionic buffers such as TRIS are often used to prevent excessively noisy baselines due to increased joule heating in the capillary [4].

***Corresponding author:** Bou-Abdallah F, Department of Chemistry, State University of New York (SUNY) at Potsdam, 44 Pierrepont Avenue, Potsdam, NY, USA, E-mail: bouabdf@potsdam.edu

The U.S. Environmental Protection Agency (EPA) has set regulations and parameters for various ions and compounds found in potable water [15]. Constituents existing in all natural waters include bicarbonate/carbonate, chloride, silicate and sulfate, but of these only chloride and sulfate are regulated by the EPA. Other regulated constituents found in water include arsenic/arsenate, bromide/bromate, fluoride, nitrite, nitrate, and phosphate, which arise from erosion and dissolution of minerals, ores and natural deposits, oxidative or other chemical reactions, agricultural and industrial run-off, sewage, or purposeful additions. Presence of regulated constituents in excess of the maximum contaminant levels (MCLs) can cause health issues including damage to major organs, cancer and/or mutations in DNA, diarrhea and dehydration, and various other effects depending on the constituent. Table 1 summarizes the tested anions/constituents that could be found in potable and natural water systems, EPAs regulation standard (if applicable), the possible sources or causes of the constituent, and the significance towards health of the constituent.

Capillary electrophoresis has been employed in recent years to examine ions in water and other media [7-13,16-20]. Particular interest is in capillary zone electrophoresis (CZE), an attractive separation method based on mass to charge ratios and the differences in mobilities among the various ions. In the present work, CZE with indirect UV detection is used to determine the presence and concentration of various anions in water samples taken in the Northern New York area, primarily of the Raquette River water system, which is the third longest river in New York State and the largest watershed of the central and Western Adirondacks. The river system begins at Blue Mountain Lake, flows south to Raquette Lake, turns Northeast to Long Lake, heads North to Tupper Lake through the Adirondack Park before it finally empties into the St. Lawrence River, Northeast of Massena and borderline with Canada (Figure 1). The Raquette River is one of the most popular, recreationally traveled and fished rivers of the Adirondack Park. Historically, every community along the River drew their drinking water from this river system; but due to changes the water quality of this watershed, only a handful of communities still draw their drinking water out of this source. The rest have been forced by the New York State Department of Health (NYSDOH) regulations

Constituent	Standard	Source or Cause	Significance
Acetate	---	Used as carbon source in denitrification process.	Health effects vary depending on what it is compounded with.
Arsenic/Arsenate	10 µg/L MCL	Dissolution of minerals and ores, from industrial effluents, and from atmospheric deposition.	Toxic to humans and animals. A cumulative poison that is slowly excreted. Can cause nasal ulcers; damage to the kidneys, liver, and intestinal walls; and death; carcinogenic.
Bicarbonate/Carbonate	---	Carbon dioxide dissolved by naturally circulating waters; represents linkage between carbon cycle and hydrologic cycle.	In combination with calcium and magnesium forms carbonate hardness.
Bromide/Bromate	10 µg/L MCL	Bromate occurs when bromide ions present in water are oxidized by ozone and some other oxidizing agents (including, it is believed, chlorine)	Bromate is both carcinogenic and mutagenic.
Chloride	250 mg/L SMCL	Exists in all natural waters; In soil and rock formations, sea spray, waste discharges, and industrial brine. Sewage contains large amounts of chloride, as do some industrial effluents.	Does not pose a health hazard to humans; principal consideration is related to palatability. Large concentrations increase the corrosiveness of water and, in combination with sodium, give water a salty taste.
Fluoride	4.0 mg/L MCL 2.0 mg/L SMCL	Occurs naturally in rare instances; arises almost exclusively from fluoridation of public water supplies and from industrial discharges.	Potential health effects of long-term exposure to elevated fluoride concentrations include dental and skeletal fluorosis, nephrotoxicity, and may affect neurodevelopment in children.
Nitrite (mg/L, as N)	1 mg/L MCL	Commonly formed as an intermediate product in bacterially mediated nitrification and denitrification of ammonia and other organic nitrogen compounds. Typically from untreated or partially treated wastes.	An acute health concern at certain levels of exposure. Concentrations greater than 1.0 mg/L, as nitrogen, may be injurious to pregnant women, children, and the elderly. Excess exposure can cause methemoglobinemia, or blue-baby syndrome, anemia and preeclampsia in pregnant women.
Nitrate (mg/L, as N)	10 mg/L MCL	Oxidation of ammonia: agricultural fertilizer run-off; leaking from septic tanks, sewage; erosion of natural deposits.	Concentrations greater than local background levels may indicate pollution by feedlot run-off, sewage, or fertilizers. Concentrations greater than 10 mg/L, as nitrogen, may be injurious to pregnant women, children, and the elderly. See nitrite for health effects.
Phosphate (mg/L, as P)	16 mg/L	Added to finished water to inhibit corrosion in distribution piping and residential plumbing. Phosphorus occurs widely in nature in plants, micro-organisms, and animal wastes. Widely used as agricultural fertilizer, major constituent of detergents, Run-off and sewage discharges.	No known adverse health effects.
Silicate	---	Always present in natural waters. Rocks and geologic formations.	No definite health implications in water.
Sulfate	250 mg/L SMCL	Rocks, geological formations, discharges and so on. Exists in all natural waters, concentrations vary according to terrain.	Sulfates of calcium and magnesium form hard scale. Large concentrations of sulfate have a laxative effect on some people and, in combination with other ions, give water a bitter taste.
Thiosulfate	---	Rapidly dechlorinates water, is notable for its use to halt bleaching in paper-making industry, used in smelting silver ore, producing leather goods, and in textile industry.	No known adverse health effects.

MCL: Maximum Contaminant Level; SMCL: Secondary Maximum Contaminant Level; mg/L: Milligrams per Liter (=ppm, parts per million); µg/L: Micrograms per Liter (=ppb, parts per billion); ---: No Limit Established.

Table 1: Water-quality criteria, standards, or limits for selected properties and constituents. All regulated standards are from U.S. Environmental Protection Agency, EPA [15].

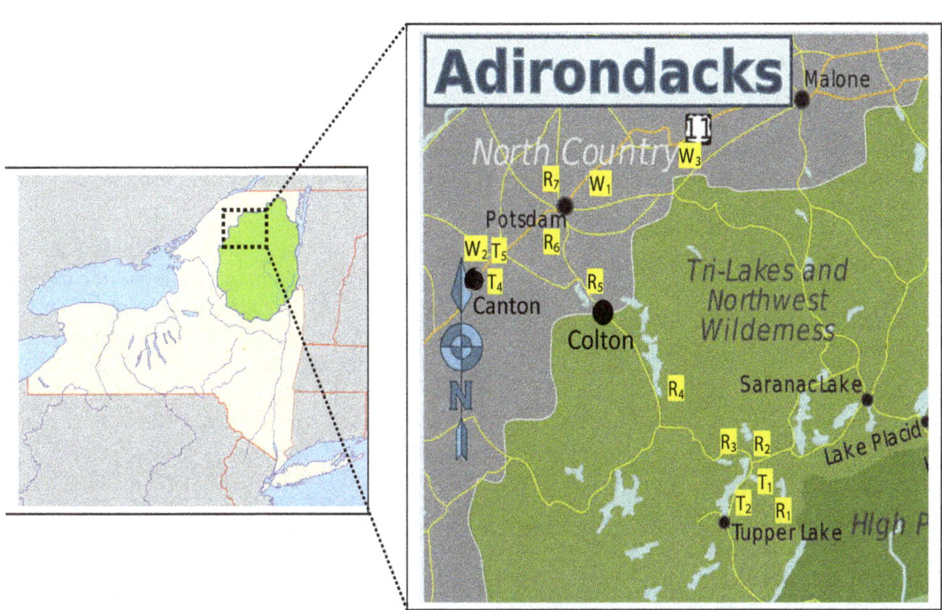

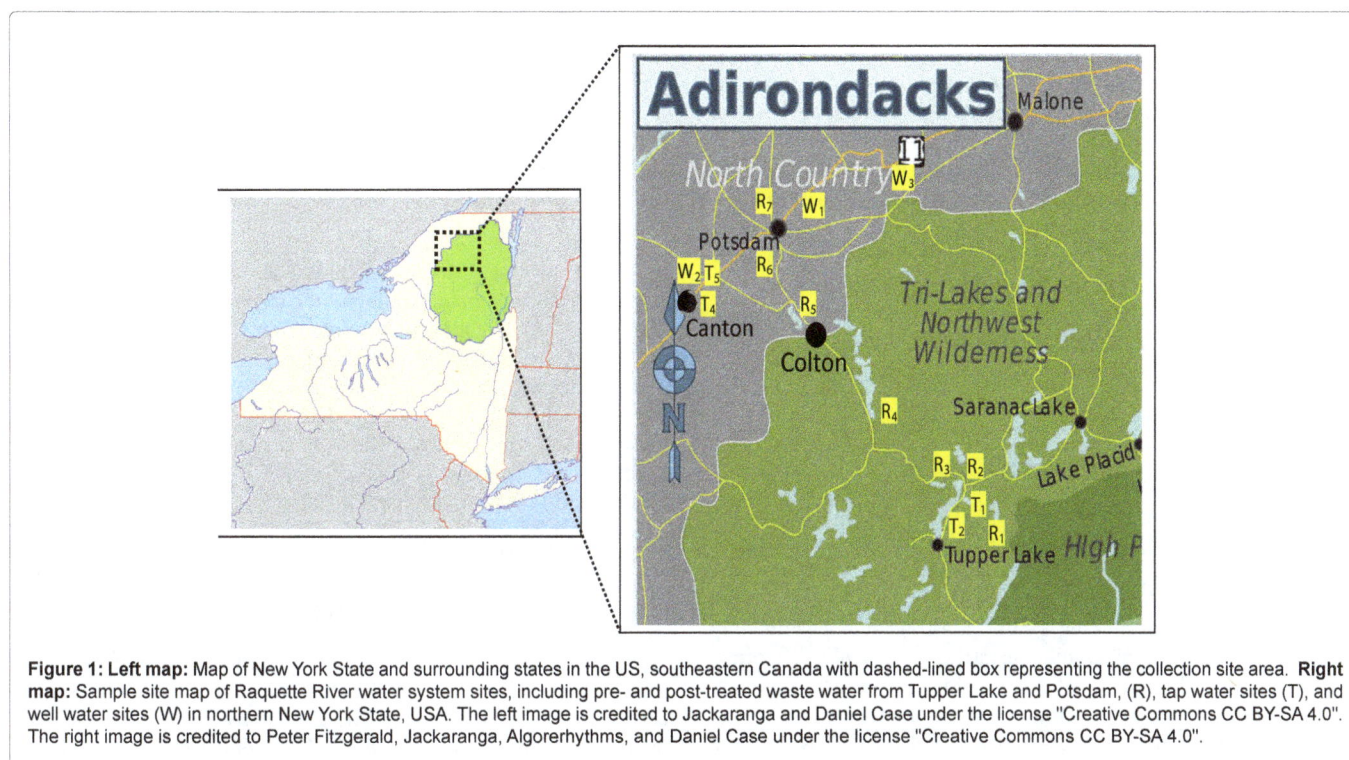

Figure 1: Left map: Map of New York State and surrounding states in the US, southeastern Canada with dashed-lined box representing the collection site area. **Right map:** Sample site map of Raquette River water system sites, including pre- and post-treated waste water from Tupper Lake and Potsdam, (R), tap water sites (T), and well water sites (W) in northern New York State, USA. The left image is credited to Jackaranga and Daniel Case under the license "Creative Commons CC BY-SA 4.0". The right image is credited to Peter Fitzgerald, Jackaranga, Algorerhythms, and Daniel Case under the license "Creative Commons CC BY-SA 4.0".

on water quality to obtain their drinking water from deep-water wells. The village of Potsdam in Upstate New York resides on the banks of the Raquette River and is the largest user of drinking water drawn off the River in addition to two local universities, State University of New York at Potsdam and Clarkson University.

Materials and Methods

Chemicals

Sodium azide, sodium bromide, sodium chloride, potassium fluoride, potassium nitrate, sodium phosphate dibasic, sodium silicate, sodium sulfate, and sodium thiosulfate were obtained from Fisher Scientific (Fairlawn, NJ, USA). Sodium acetate, sodium arsenate, sodium carbonate, and potassium nitrite were purchased from J.T. Baker (Phillipsburg, NJ, USA). Lactic acid was acquired from Eastman (Kingsport, TN, USA). All reagents were of analytical grade. Ultra-pure CE water procured from Agilent Technologies (Santa Clara, CA, USA) from their Forensic Anion Solution Kit (PN 5064-8208).

Standards and BGE

Agilent's Inorganic Anion Test Mixture stock (1000 ppm fluoride, chloride, bromide, nitrite, nitrate, sulfate, and 2000 ppm phosphate) was diluted 1:10 with deionized (DI) water obtained from EMD Millipore Elix10 Water Purification System (Darmstadt, Germany) and used as a standard. Stocks of the following inorganic anions (from their sodium or potassium salts) and organic acids (i.e., lactic acid) were made at 1000 ppm: acetate, arsenate, azide, bicarbonate/carbonate, bromide, chloride, fluoride, lactate, nitrate, nitrite, phosphate, silicate, sulfate, and thiosulfate. Each anion and organic acid was diluted from its stock solution to make a 20 ppm standard mixture and was filtered using a 0.22 µm PES syringe filter (Argos Technologies, Elgin, IL, USA). The running buffer/BGE was Agilent's proprietary Basic Anion Buffer (pH 12.0) obtained from their Forensic Anion Solution Kit. The BGE was degassed for 10 min in a TA Instruments degassing station (New Castle, DE, USA) using 2 mL glass CE vials from Agilent Technologies prior to runs.

Instrumentation

The capillary electrophoresis system used in this study is the Agilent G7100 with a UV-DAD (diode array detector) interfaced with ChemStation software for data acquisition (Agilent Technologies, Santa Clara, CA, USA). The separation capillary used was a standard bare fused silica from Agilent Technologies, measuring 112.5 cm total length, 104 cm effective length, and an internal diameter (id) of 50 µm. Baseline noise is markedly increased if a wider internal diameter capillary is used due to the high UV absorptivity of the buffer.

Procedures

Prior to first use, the capillary was flushed with running buffer/BGE for 30 min. Daily conditioning of capillary included 20 min flush of BGE. The capillary was conditioned between runs for 2 min using BGE from outlet Home vial and another 2 min of BGE from inlet Home vial. No sodium hydroxide was used in the conditioning of the capillary due to determined degradation of separation quality, reproducibility and performance with this application method [20]. Conditioning of the capillary at the end of the day involved 15 min ultra-pure water flush followed by 3 min air flush. The applied voltage was 30 kV negative polarity and the capillary cassette was kept at 30°C for all experiments. Negative polarity is used since the EOF is reversed. Sample injection was performed under Hydrodynamic injections of 50 mbar for 6 s followed by buffer injections of 50 mbar for 4 s. UV-vis signal detection was performed at 350 nm (20 nm bandwidth) with an indirect reference signal of 275 nm (10 nm bandwidth) and a runtime of 15 min. This procedure was based on Agilent's Forensic Anion Solution Kit procedure manual [21].

Samples

All water samples (natural river waters, potable tap and well waters, and pre- and post-treatment wastewaters) were collected in BPA-free plastic gallon water containers and filtered through a 0.22 μm PES syringe filter. No further manipulation of water samples (i.e., dilution, concentration, pH adjustment, complexation, etc.) was performed so as to mimic natural water conditions during testing. A set of samples were stored at 4°C and another set of same samples were kept at approximately 20°C. Samples were tested within 2 weeks of collection. Water samples were obtained from water sources feeding areas of interest, mainly Raquette River system from Tupper Lake (TL), NY, to Potsdam, NY, as well as pre- and post- wastewater treatment (WWT) from the two villages. Tap and well waters from other local areas including Canton and Winthrop, NY, were also tested. Figure 1 depicts Northern New York map of sample site location.

Results and Discussion

Anion analysis of standard mixtures

Figure 2 shows the electropherograms of (a) fourteen-anion standard mixture using the sodium or potassium salts of each inorganic anion, and lactic acid for the organic anion, lactate and (b) seven inorganic anion standard mixture from Agilent Technologies. The standard mix from Agilent consisted of inorganic anions commonly monitored in water (bromide, chloride, nitrite, nitrate, sulfate, fluoride, and phosphate). The additional seven anions in our standard mix include azide, arsenate, acetate, lactate, silicate and thiosulfate. As can be seen in Figure 2, the electropherograms of eluted anions from the two different sources are identical suggesting high reproducibility and robustness of the separation method. On the basis of CZE separation, bromide was the first ion out of the seven commonly monitored inorganic anions to elute followed by chloride, nitrite, nitrate, sulfate, fluoride, and then phosphate. Resolution between bromide and chloride can be improved by decreasing operating temperature to 20 °C, but was not necessary for detecting and quantifying the two anions in our tested samples. All seven ions eluted under 9 min. The detection of silicate found in the Agilent Anion Standard mix electropherogram (Figure 2b) was the result of silicate being present in our DI water which was used to dilute the concentrated Agilent standard mix. The anions from our fourteen-anion standard mix migrated in the following order: bromide, chloride, thiosulfate, nitrite, nitrate, sulfate, azide, bicarbonate/carbonate, fluoride, arsenate, phosphate, acetate, lactate, and silicate, all eluting under 13 min.

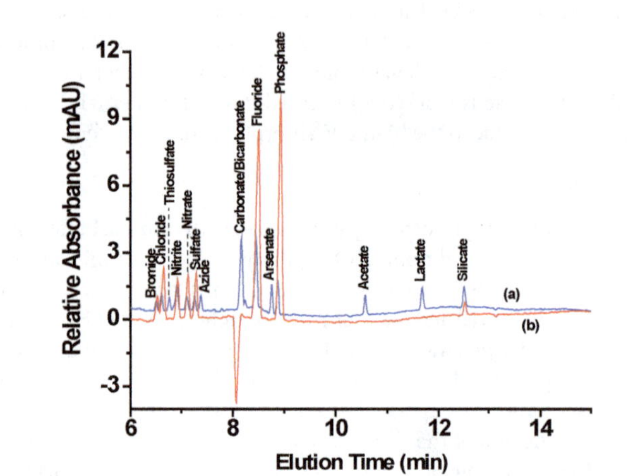

Figure 2: Electropherograms of two separate reference anion standard mixtures. **(a)** The blue electropherogram is that of fourteen-anion standard mixture prepared by us in the lab. The concentration of all anions (bromide, chloride, thiosulfate, nitrite, nitrate, sulfate, azide, fluoride, arsenate, phosphate, acetate, lactate, and silicate) was 20 ppm each whereas that of bicarbonate/carbonate was at 40 ppm. **(b)** The red electropherogram is that of seven-anion standard mixture from Agilent Technologies. The concentration of the anions bromide, chloride, nitrite, nitrate, sulfate, and fluoride was at 100 ppm each and that of phosphate was at 200 ppm. The background electrolyte (BGE) was Agilent's proprietary Basic Anion Buffer (pH 12.0). Separation conditions: 30 kV negative polarity, 30°C, hydrodynamic injection of 50 mbar for 6 s with post injection of buffer at 50 mbar for 4 s, UV-vis signal detection at 350 nm with indirect reference of 275 nm.

To quantify the amount of detected anions in the water samples, a series of anion standards calibration curves were created (Figure 3). The concentration of the standard anion solutions of bromide, chloride, nitrite, nitrate, sulfate, fluoride, acetate, silicate, phosphate, and bicarbonate/carbonate varied between 5 and 200 ppm. Only calibration curves for the ten anions detected in the water samples are being presented in Figure 3. The goodness of the fit is determined from the coefficients of determination, R-squared (R^2), which ranged between 0.99817 and 1.00000 as displayed in Table 2.

Table 2 displays the values for the propagation of the detected anion standards calibration curves (Figure 3). The curves were generated via amount in parts per million (ppm) by corrected area in milli-Absorbance Units (mAU). The corrected area was calculated by the area, which is absorbance x time (mAU × min), divided by migration time in min. All values for the calibration table, linear regression equations and subsequent calibration curves were auto-generated using the ChemStation program. The limits of detection (LOD) for all analytes were in the range of 3–5 ppm with 300 mbar pressure injection with a signal to noise ratio of 3.

Anions analysis from natural and wastewater sources

Natural water samples were obtained along the Raquette River watershed route, starting with Bog River Falls, the site closest upstream to the source (Blue Mountain Lake), following downstream as far north, heading toward the St. Lawrence River, as Potsdam, NY. The seven sites for which the electropherograms in Figure 4 are displayed (R1–R7) represent the watershed and wastewater sites analyzed in this study. All tested samples contained chloride, bicarbonate/carbonate, and silicate, and most samples contained sulfate. Arsenate, azide, bromide, lactate, and thiosulfate were not detected in any of the samples. The only anions detected in the natural water samples from Raquette River sites depicted in Figure 4 and located at the Crusher (R1), Carry Falls Reservoir (R4), and South Colton Reservoir (R5) were chloride,

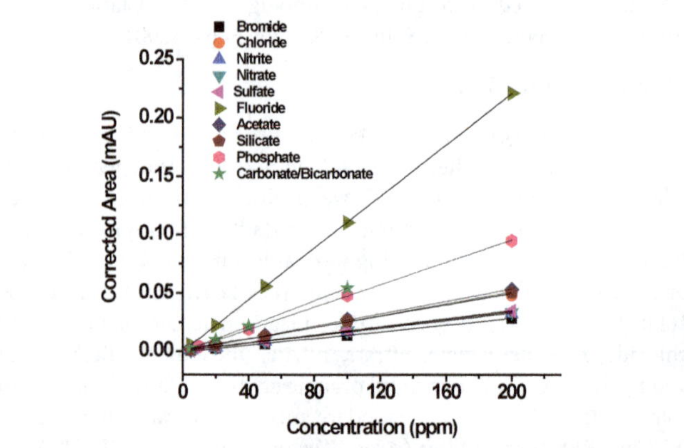

Figure 3: Calibration curves for ten anion standards with concentrations of 5, 20, 50, 100, and 200 ppm for bromide, chloride, nitrite, nitrate, sulfate, fluoride, acetate, and silicate; 10, 40, 100, and 200 ppm for phosphate; and 5, 20, 40, and 100 ppm for carbonate/bicarbonate.

Anion	Migration time (min)	Amount (ppm)	Corrected Area (mAU)	Correlation factor, R^2	Linear regression equation
Bromide	6.376	5.00	4.2676E-4	0.99941	y=1.4558E-4x−4.578E-4
		20.00	2.3861E-3		
		50.00	6.3833E-3		
		100.00	1.3658E-2		
		200.00	2.8992E-2		
Chloride	6.449	5.00	1.5469E-3	0.99817	y=2.4561E-4x+8.679E-4
		20.00	5.7247E-3		
		50.00	1.3783E-2		
		100.00	2.7383E-2		
		200.00	4.8873E-2		
Nitrite	6.700	5.00	9.1049E-4	0.99963	y=1.6453E-4x+2.465E-4
		20.00	3.4472E-3		
		50.00	8.7465E-3		
		100.00	1.7255E-2		
		200.00	3.2821E-2		
Nitrate	6.894	5.00	1.0516E-3	0.99951	y=1.6954E-4x+3.851E-4
		20.00	4.0557E-3		
		50.00	8.9077E-3		
		100.00	1.7889E-2		
		200.00	3.3982E-2		
Sulfate	7.044	5.00	9.4979E-4	0.99967	y=1.7493E-4x+2.592E-4
		20.00	3.8510E-3		
		50.00	9.0739E-3		
		100.00	1.8359E-2		
		200.00	3.4920E-2		
Fluoride	8.189	5.00	5.4907E-3	1.00000	y=1.10725E-3x−2.64E-5
		20.00	2.2093E-2		
		50.00	5.5623E-2		
		100.00	1.1030E-1		
		200.00	2.2155E-1		
Acetate	9.965	5.00	1.4667E-3	0.99990	y=2.6727E-4x−2.565E-4
		20.00	5.5550E-3		
		50.00	1.3915E-2		
		100.00	2.7385E-2		
		200.00	5.3444E-2		
Silicate	11.649	5.00	1.3897E-3	0.99992	y=2.5482E-4x+1.766E-4
		20.00	5.2543E-3		
		50.00	1.2947E-2		
		100.00	2.6121E-2		
		200.00	5.0906E-2		
Phosphate	8.581	10.00	3.8703E-3	0.99999	y=4.7854E-4x−2.218E-4
		40.00	1.9228E-2		
		100.00	4.7902E-2		
		200.00	9.5435E-2		
Carbonate/ bicarbonate	7.726	5.00	0.0254E-2	0.99878	y=5.6373E-4x−1.612E-3
		20.00	1.0159E-2		
		40.00	2.1784E-2		
		100.00	5.4371E-2		

Table 2: Raw data for the anion standards calibration curves with migration times.

sulfate, bicarbonate/carbonate, and silicate. Acetate was detected in the wastewater in Tupper Lake (Figure 4, R2) but not in the wastewater in Potsdam (Figure 4, R6) although it is used in the denitrification process in wastewater treatment solutions. Interestingly, nitrite and nitrate were both present in the wastewater effluent in Tupper Lake (R3), but not in Tupper Lake's wastewater influent (R2), whereas Potsdam's wastewater influent (R6) does contain nitrite and nitrate, but Potsdam's wastewater effluent (R7) shows no nitrite and an increase of nitrate. The peaks between acetate (8) and silicate (9) in sample R2 and phosphate (7) and silicate (9) in sample R5 in Figure 4 remained unidentified.

Potable water analysis

Potable water samples from taps and wells were obtained from various homes and institutions in and around Canton, Potsdam, Tupper Lake, and Brasher Falls-Winthrop, NY. All tap and well water samples contained chloride, sulfate, nitrate, bicarbonate/carbonate, and silicate (Figures 5 and 6). Table 3 displays the sites, locations, and concentrations in ppm of all the anions detected in the water samples. An expected small amount of fluoride, <5.0 ppm (estimated value to be <1 ppm based on linear extrapolation), was detected in the tap water

Site	Location	Concentration Levels (ppm)													
		Br^-	Cl^-	$S_2O_3^{2-}$	NO_2^-	NO_3^-	SO_4^{2-}	N_3^-	CO_3^{2-}/ HCO_3^-	F^-	AsO_4^{3-}	PO_4^{3-}	$C_3H_3O_2^-$	Lactate	SiO_3^{2-}
R1	Raquette River, at Crusher	---	<5*	---	---	---	<5*	---	5.5	---	---	---	---	---	6.8
R2	Pre-WWT in Tupper Lake	---	95	---	---	---	---	---	47	---	---	<10*	56	---	8.4
R3	Post-WWT in Tupper Lake	---	98	---	<5*	10	14	---	59	---	---	<10*	---	---	10
R4	Carry Falls Reservoir	---	<5*	---	---	---	<5*	---	7.3	---	---	---	---	---	5.0
R5	South Colton Reservoir	---	<5*	---	---	---	<5*	---	9.4	---	---	---	---	---	<5*
R6	Pre-WWT in Potsdam	---	198	---	54	6.1	38	---	133	<5*	---	<10*	---	---	10
R7	Post-WWT in Potsdam	---	223	---	---	5.8	36	---	136	<5*	---	<10*	---	---	11
T1	Tap, village of Tupper Lake	---	<5*	---	---	---	<5*	---	5.4	---	---	---	---	---	6.3
T2	Tap, Mt. Morris in Tupper Lake	---	<5*	---	---	---	<5*	---	6.1	---	---	---	---	---	5.3
T3	Tap, village of Potsdam	---	<5*	---	---	---	18	---	14	<5*	---	---	---	---	6.1
T4	Tap, St. Lawrence Univ. in Canton	---	9.2	---	---	<5*	6.9	---	135	---	---	---	---	---	9.2
T5	Tap, village of Canton	---	19	---	---	<5*	9.1	---	141	---	---	<10*	---	---	8.6
W1	Well, Town of Potsdam	---	147	---	---	---	56	---	103	<5*	---	---	---	---	10
W2	Well, Town of Canton	---	143	---	---	11	42	---	217	<5*	---	---	---	---	14
W3	Well, Winthrop	---	8.4	---	---	---	51	---	260	---	---	---	---	---	20

*Reported values are based on the lowest concentration run of standards used in the calibration curves.

Table 3: Concentration of anions (ppm) detected in the water samples tested in this study.

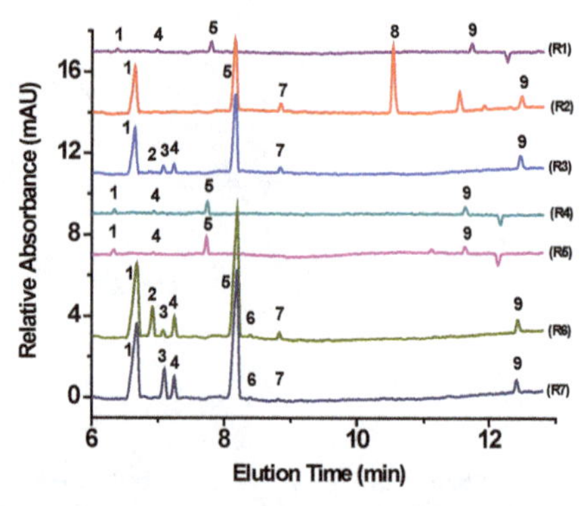

Figure 4: Electropherograms of water samples from (R1) Raquette River at the Crusher, (R2) Pre-WWT in Tupper Lake, (R3) Post-WWT in Tupper Lake, (R4) Carry Falls Reservoir, (R5) South Colton Reservoir, (R6) Pre-WWT in Potsdam, and (R7) Post-WWT in Potsdam. Conditions as in Figure 2 Peaks: 1-Chloride; 2-Nitrite; 3-Nitrate; 4-Sulfate; 5-Bicarbonate/carbonate; 6-Fluoride; 7-Phosphate; 8-Acetate; 9-Silicate.

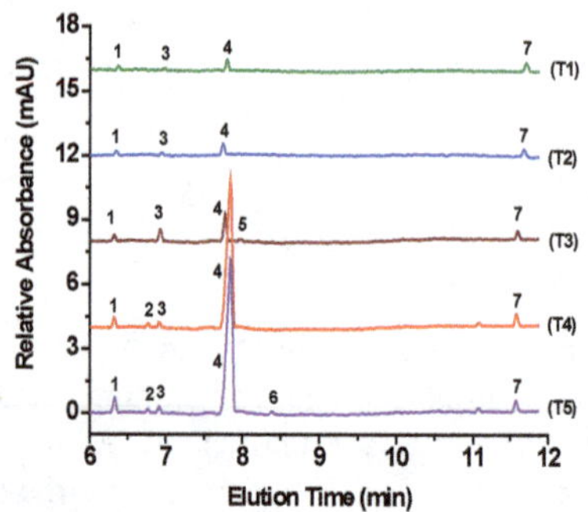

Figure 5: Electropherograms of tap water samples from (T1) Co-op located in village of Tupper Lake, (T2) a home located on Mt. Morris in Tupper Lake, (T3) State University of New York at Potsdam, (T4) St. Lawrence University, and (T5) home in village of Canton. Conditions as in Figure 2 Peaks: 1-Chloride; 2-Nitrite; 3-Nitrate; 4-Bicarbonate/carbonate; 5-Fluoride; 6-Phosphate; 7-Silicate.

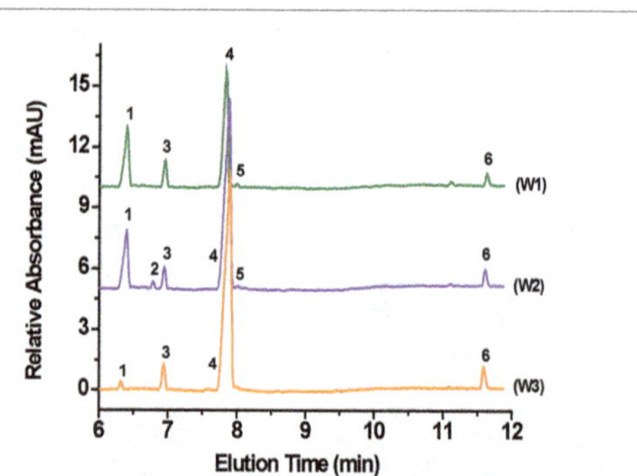

Figure 6: Electropherograms of well water samples from (W1) home in town of Potsdam, (W2) home in town of Canton, and (W3) home in village of Brasher Falls-Winthrop. Conditions as in Figure 2 Peaks: 1-Chloride; 2- Nitrate; 3-Sulfate; 4-Bicarbonate/carbonate; 5-Fluoride; 6-Silicate.

obtained from SUNY Potsdam (Figure 4 and Table 3, T3) since the water treatment facility for the Village of Potsdam fluoridates the water supply. Unexpectedly, fluoride was also detected at <5.0 ppm (estimated value to be <1 ppm based on linear extrapolation), in the well water samples of the Town of Potsdam (Figure 6, W1) and the Town of Canton (Figure 6, W2). However, the estimated concentrations of fluoride (based on linear extrapolation) found in the analyzed samples fall under the EPA Maximum Contaminant Level of 4.0 ppm (Table 1). All tap water samples represented in Figure 5 (T1–T5 samples) contained chloride, sulfate, silicate, and bicarbonate/carbonate with the two tap from Canton (T4 and T5) having the highest amount of bicarbonate/carbonate at 135 and 141 ppm, respectively. Furthermore, the two tap samples from Canton (T4 and T5) contained small amounts of nitrate (<5.0 ppm, Table 1) with sample T5 showing a small amount of phosphate (<5.0 ppm).

The well water samples obtained from homes in the Town of Canton (Figure 6, W2) and village of Brasher-Falls/Winthrop (W3) areas had the highest bicarbonate/carbonate concentrations of all samples tested,

217 and 260 ppm, respectively. This is not remarkably unusual since the geologic region is known to produce hard water for those who have wells as their potable water source. Sulfate and carbonate, in combination with calcium and magnesium cause water hardness and is typically found in natural waters. Even these high concentrations of anions fall within normal levels as dictated by the EPA (Table 1). Nitrate concentration at 11 ppm was detected in one well water sample in the Town of Canton, a level just over the EPA's MCL value. This rather high nitrate concentration could be due to either agricultural fertilizer run-off, given that many Northern New York communities (including Canton, NY) are dedicated to farming and agriculture, leaking from a septic tank/sewer, natural deposits erosion, or any combination of the three. No bromide, thiosulfate, azide, arsenate, and lactate were detected in any of the water samples tested. However, a more systematic study involving hundreds of samples from older homes is needed to rule out the presence of any of these anions in the water supply.

Samples tested from the same sources stored at different temperatures, 4°C and 20°C, were not found to have significant differences in findings regarding detection or quantification.

Conclusions

Our study showed that capillary zone electrophoresis (CZE) with indirect UV detection is a sensitive, reliable, and suitable method for the determination of many inorganic and organic anions. The simple, fast and cheap sample preparation method makes CZE an attractive tool for the detection and quantification of several anions present in wastewaters, natural waters, and potable waters including tap and well waters. Under our experimental conditions, all tested anions eluted within 12 min with excellent linearity and reproducibility. There were no significant differences in results from the water samples stored in 4°C versus same samples stored at 20°C. The concentrations of tested EPA regulated anions were found to be within acceptable limits although the presence of fluoride in some of the water samples tested is a concern and high levels of bicarbonate/carbonate or sulfate in combination with calcium and magnesium contribute to water hardness.

Acknowledgement

This work was supported by the T. Urling and Mabel Walker Research Fellowship Program, and the Collegiate Science and Technology Entry Program (CSTEP) at SUNY Potsdam. The Capillary Electrophoresis (CE) system used in this study was purchased with NIH Award Number R15GM104879 of the National Institute of Health (NIH). The authors would like to acknowledge Scott Varden for his assistance with sample collections.

References

1. Chang LW, Magos L, Suzuki T (1996) Toxicology of Metals. CRC Press: Boca Raton, FL.
2. Badgujar DM, Talawar MB, Asthana SN, Mahulikar PP (2008) Advances in science and technology of modern energetic materials: An overview. J Hazard Mater 151: 289-305.
3. Datta R, Henry M (2006) Lactic acid: recent advances in products, processes and technologies – a review. J Chem Technol. Biotechnol 81: 1119-1129.
4. Altria K (2011) Analysis of inorganic anions by capillary electrophoresis. LCGC Europe 24: 32-36.
5. Guo WP, Lau KM, Fung YS (2010) Microfluidic chip-capillary electrophoresis for two orders extension of adjustable upper working range for profiling of inorganic and organic anions in urine. Electrophoresis 31: 3044–3052.
6. Rovio S, Kalliola A, Siren H, Tamminen T (2010) Determination of the carboxylic acids in acidic and basic process samples by capillary zone electrophoresis. J Chromatogr A 1217: 1407–1413.
7. Padarauskas A, Olšauskaitė V, Paliulionyte Y (1998) Simultaneous Separation of Inorganic Anions and Cations by Capillary Zone Electrophoresis. J Chromatogr A 829: 359-365.
8. Padarauskas A, Olšauskaitė V, Paliulionytė V (1998) Simultaneous Determination of Inorganic Anions and Cations in Waters by Capillary Zone Electrophoresis. J Chromatogr A 829: 359-365.
9. Haumann I, Boden J, Mainka A, Jegle U (2000) Simultaneous Determination of Inorganic Anions and Cations by Capillary Electrophoresis with Indirect UV Detection. J Chromatogr A 895: 269-277.
10. Hiissa T, Sirén H, Kotiaho T, Snellman M, Hautojärvi A (1999) Quantification of anions and cations in environmental water samples: Measurements with capillary electrophoresis and indirect-UV detection. J Chromatogr A 853: 403-411.
11. Gries T, Sitorius E, Giesecke A, Schlegel V (2008) Feasibility of using capillary zone electrophoresis with photometric detection for the trace level detection of bromate in drinking water. Food Addit Contam 25: 1318-1327.
12. Oehrle SA (1996) Analysis of Anions in Drinking Water by Capillary Ion Electrophoresis. J Chromatogr A 733: 101-104.
13. Romano J, Krol J (1993) Capillary ion electrophoresis, an environmental method for the determination of anions in water. J Chromatogr A 640: 403-412.
14. Altria KD, Rogan MM, Goodall DM (1994) Quantitative determination of drug counter-ion stoichiometry by capillary electrophoresis. Chromatographia 38: 637–642.
15. Environmental Protection Agency. Table of Regulated Drinking Water Contaminants.
16. Pursell C, Chandler B, Bushey M (2004) Capillary Electrophoresis Analysis of Cations in Water Samples. An Experiment for the Introductory Laboratory. J Chem Educ 81: 1783-1786.
17. Environmental Protection Agency. Method 6500 (2007) Dissolved Inorganic Anions in Aqueous Matrices by Capillary Electrophoresis, pp: 1-34.
18. Shi M, Gao Q, Feng J, Lu Y (2012) Analysis of inorganic cations in honey by capillary zone electrophoresis with indirect UV detection. J Chromatogr Sci 50: 547-552.
19. Nair J, Izzo C (1993) Anion screening for drugs and intermediates by capillary ion electrophoresis. J Chromatogr A 640: 445-461.
20. Varden L, Bou-Abdallah F (2017) Detection and separation of inorganic cations in natural, potable, and wastewater samples using capillary zone electrophoresis with indirect UV detection. Am J Anal Chem 8 : 81-94.
21. Capillary Electrophoresis Forensic Kit Manual (2000) Agilent Technologies: Netherlands.

Development of a Method for Regioisomer Impurity Detection and Quantitation within the Raw Material 3-Chloro-5-Fluorophenol by Gas Chromatography

Justin R. Denton[1]*, Yun Chen[2] and Thomas Loughlin[1]

[1]*Manufacturing Division: Supply Analytical Services, Rahway, NJ, USA*
[2]*Research Laboratories: Analytical Research and Development, Rahway, NJ, USA*

Abstract

Control of regioisomer impurities within pharmaceutical raw materials, intermediates, and active pharmaceutical ingredients are of major concern for the pharmaceutical industry. If regioisomer impurities are possible, their detection and quantitation should be established as early as possible within the process. This work describes a gas chromatography area% method for the detection and quantitation of all regioisomer impurities associated with 3-chloro-5-fluorophenol, a raw material used within the pharmaceutical industry along with other related impurities found within 3-chloro-5-fluorophenol. Several method development aspects, as well as a general regioisomer impurity control strategy related to 3-chloro-5-fluorophenol are discussed.

Keywords: Gas chromatography; Regioisomer; API; Raw material

Introduction

Related compounds (impurities) are always of concern when it comes to quality of raw materials and intermediates used for the production of an active pharmaceutical ingredient (API) [1-3]. One subclass of related compounds which tends to be more challenging to detect and quantify are regioisomer impurities (also known as positional isomers) [4]. Figure 1 shows two sets of regioisomer impurity examples for 1-propanol and *para*-anisidine.

A considerable amount of investment must be committed to ensure appropriate analytical methods are developed to ensure these impurities are properly controlled in raw materials, intermediates, and active pharmaceutical ingredients. Tactically, the earlier regioisomer impurities are controlled to acceptable levels within the process; analysis of further downstream intermediates and ultimately the API become simplified to typical related impurities (non-regioisomer impurities). Therefore, control of regioisomer impurities should take place within raw materials or when these impurities are found to be formed within a given process [5].

The raw material 3-chloro-5-fluorophenol poses a unique separation challenge since all three substituents are not the same functional group. Therefore, there are nine total regioisomer impurities possible within this raw material (Figure 2). Since 3-chloro-5-fluorophenol has the potential of being a raw material used in the synthesis of an API, development of an analytical method to ensure proper detection and quantitation of all possible regioisomer impurities was deemed to be an important research objective.

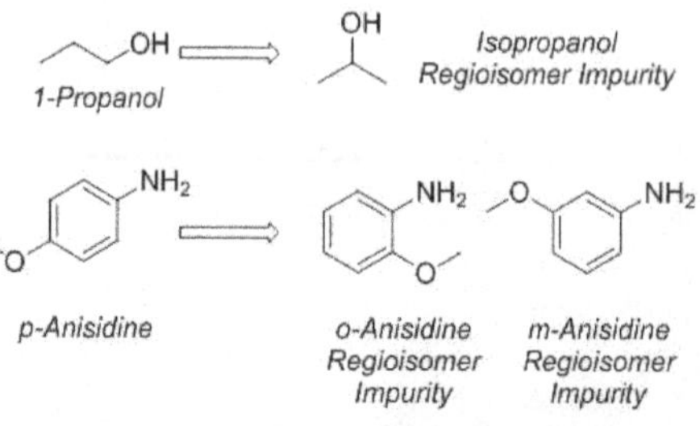

Figure 1: Regioisomer Impurity Examples.

Experimental

Chemicals and reagents

The following chemicals and reagents were utilized in this communication: Methanol (MeOH) Fisher Chemical Lot#161609; Acetonitrile (MeCN) Fisher Chemical Lot#162576; N,N-Dimethylacetamide (DMAc) Sigma-Aldrich Lot#SHBG3632V; N-Methyl-2-pyrrolidone (NMP) Acros Organics Lot#1338572; 2-Chloro-5-fluorophenol Acros Organics Lot#A020210301; 2-Chloro-3-fluorophenol Matrix Scientific Lot#N17P; 2-Chloro-4-fluorophenol Acros Organics Lot#A010381301; 2-Chloro-6-fluorophenol Aldrich Lot# 04724KH; 3-Chloro-4-fluorophenol Aldrich Lot#00419CS; 4-Chloro-2-fluorophenol Aldrich Lot#05418PU; 5-Chloro-2-fluorophenol BePharm Limited Lot#WZG091010-001; 3-Chloro-2-fluorophenol Matrix Scientific Lot#0091; 4-Chloro-3-fluorophenol Acros Organics Lot#A018902701; Phenol Sigma-Aldrich Lot#BCBK8781V, 3-Fluorophenol Acros Organics Lot#A003610401; 3-Chlorophenol Acros Organics Lot#A013855901; 3-Bromo-5-fluorophenol Aldrich Lot#MKBF3319V; and 3-Chloro-5-fluorophenol Combi-Blocks Inc. Lot#L42454 and Lot#L54033, BePharm Limited Lot#0032554-16070101, Ark Pharm Lot#WG0032554-140708001, Oakwood Chemical Lot#013283I13F.

Chromatographic conditions

A gas chromatography (GC) system was setup with the following:

Column: Rtx®-35 30-m × 0.25-mm, 1.0 μm film, S/N 1402372 cat. #10453 or Rtx®-65 30-m × 0.25-mm, 1.0 μm film, S/N 1394868 cat. #17053.

***Corresponding author:** Justin R Denton, Associate Principal Scientist, Manufacturing Division: Supply Analytical Services, Rahway, NJ, USA
E-mail: justin.denton@merck.com

Oven Program: 100°C to 210°C @10°C/min to 210°C (hold 14 min).

Inlet: Split Ratio 25:1, Temp. 240°C.

Detector: FID, Temp. 240°C.

Carrier Gas: Helium @ 1.0 mL/min.

Injection Volume: 1 μL.

Sample preparation

5.0 mg/mL Sample Solution: Transferred 50 mg of 3-chloro-5-fluorophenol to a 10-mL volumetric flask and diluted to volume with MeOH. Alternatively, acetonitrile (MeCN) may be used as a diluent.

Sensitivity Solution: Performed a 2000x dilution of the 5.0 mg/mL sample solution. For example, transferred 1.0 mL to a 100-mL volumetric flask then diluted to volume with MeOH and mixed thoroughly. Transferred 1.0 mL of the intermediate solution to a 20-mL volumetric flask then diluted to volume with MeOH and mixed.

Qualitative mixture solutions: Transferred 1 drop of each desired component to a 100-mL volumetric flask containing some MeOH, then diluted the volume with MeOH (For components in each mixture see Figure 3).

Equipment and software

The GC-FID system employed during these experiments was an Agilent Technologies 7890B GC System equipped with an Agilent 7693 Auto sample and an Agilent G4513A Injector. The GC columns discussed were purchased from the Restek® Corporation. The acquisition software utilized was Empower 3 licensed from the Waters Corporation.

Results and Discussion

Method Development: Due to the physical properties of 3-chloro-5-fluorophenol [6] GC-FID analysis was deemed the most appropriate analytical technique for development of a method which separates all of the regioisomer impurities as well as some previously known related impurities found within this raw material. One related compound mixture, two regioisomer mixtures, and a sample solution were prepared for GC column stationary screening (Figure 3). After a general GC column screen employing oven temperature gradient, it was found that all possible regioisomer impurities could be separated from each other and 3-chloro-5-fluorophenol utilizing an Rtx®-35 column (Figure 4).

However, the Rtx®-35 column was not able to cleanly separate the related compound mixture from regioisomer mixture #1 (see the region in boxed area of Figure 4). The Rtx®-35 stationary phase is composed of 35% diphenyl/ 65% dimethylpolysiloxane therefore to obtain better

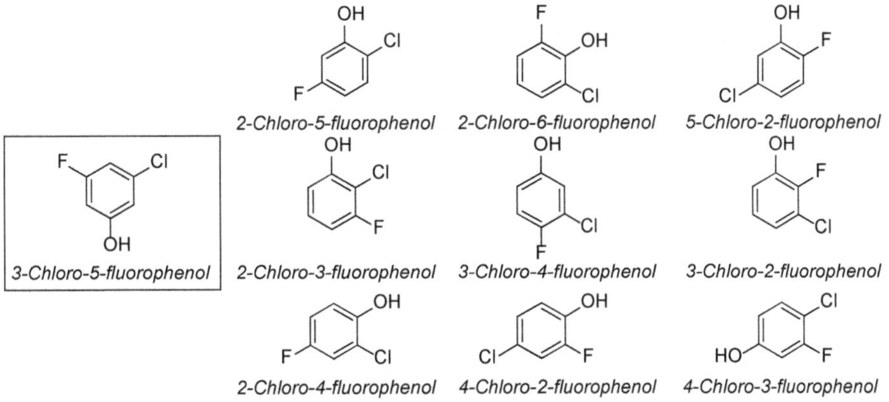

Figure 2: All Possible Regioisomer Impurities of 3-Chloro-5-fluorophenol.

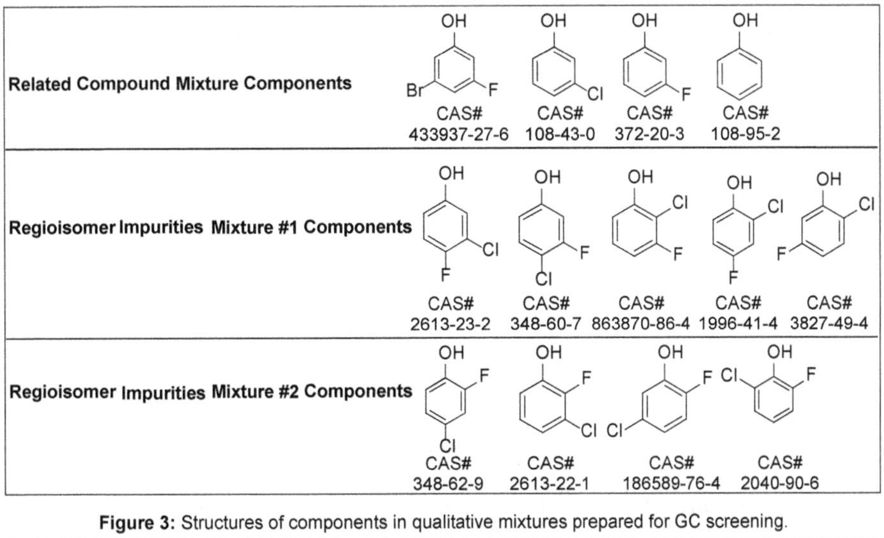

Figure 3: Structures of components in qualitative mixtures prepared for GC screening.

selectivity between the related compound mixture and regioisomer mixture #1 the stationary phase Rtx®-65 which contains 65% diphenyl to 35% dimethyl polysiloxane was attempted. Gratifyingly, this stationary phase with increased diphenyl content separated all related compounds as well as the nine possible regioisomer impurities (see the region in boxed area of Figures 5 and 6).

Method attributes: The method attributes (limit of detection, limit of quantitation, linearity, and carryover) for the impurity profile GC-FID method of 3-chloro-5-fluorophenol are shown in Table 1. Adequate detectability was achieved for the 0.001 mg/mL (0.02% with respect to (WRT) sample concentration) solution of 3-chloro-5-fluorophenol which gave greater than 3:1 signal to noise (S/N) for the 3-chloro-5-

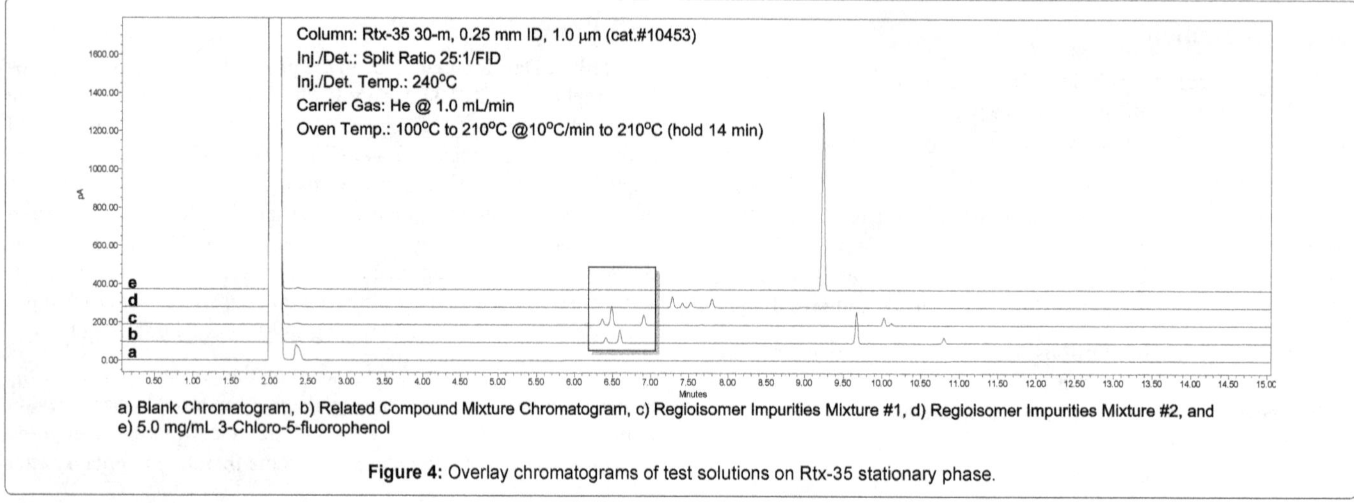

a) Blank Chromatogram, b) Related Compound Mixture Chromatogram, c) Regioisomer Impurities Mixture #1, d) Regioisomer Impurities Mixture #2, and e) 5.0 mg/mL 3-Chloro-5-fluorophenol

Figure 4: Overlay chromatograms of test solutions on Rtx-35 stationary phase.

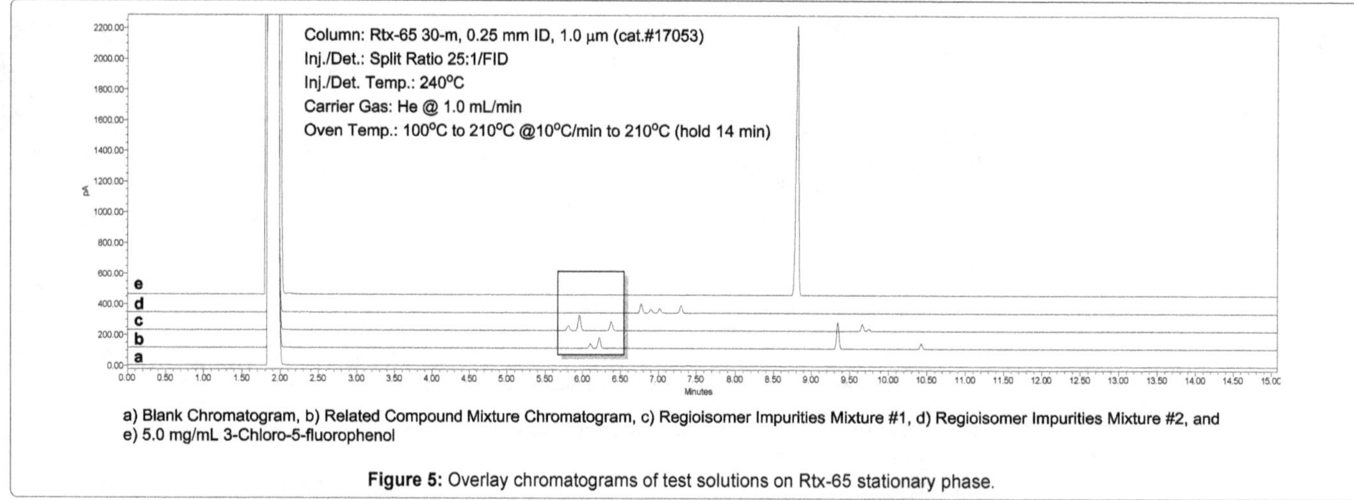

a) Blank Chromatogram, b) Related Compound Mixture Chromatogram, c) Regioisomer Impurities Mixture #1, d) Regioisomer Impurities Mixture #2, and e) 5.0 mg/mL 3-Chloro-5-fluorophenol

Figure 5: Overlay chromatograms of test solutions on Rtx-65 stationary phase.

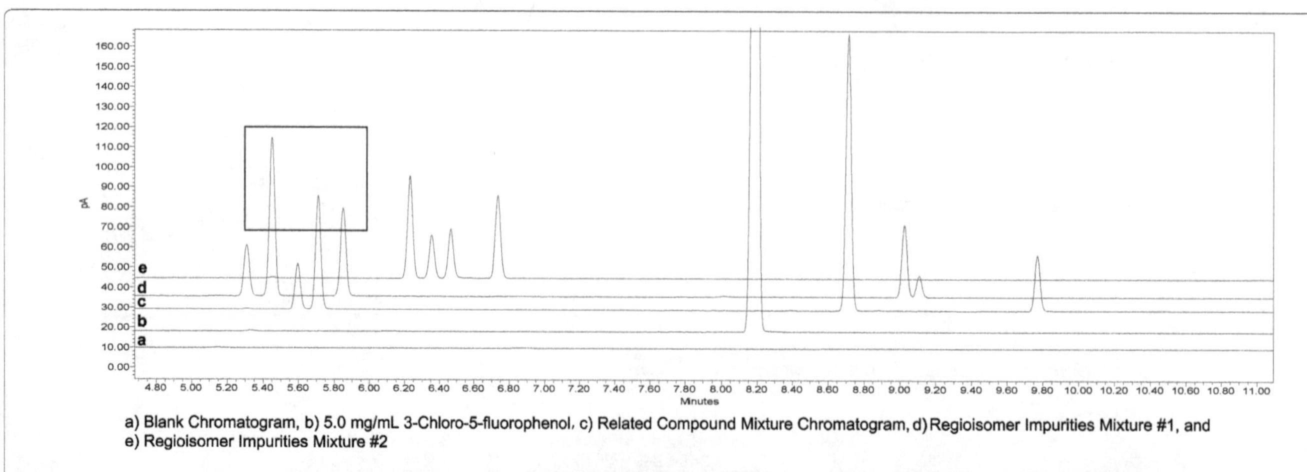

a) Blank Chromatogram, b) 5.0 mg/mL 3-Chloro-5-fluorophenol, c) Related Compound Mixture Chromatogram, d) Regioisomer Impurities Mixture #1, and e) Regioisomer Impurities Mixture #2

Figure 6: Expanded overlay chromatograms of test solutions on Rtx-65 stationary phase using H2 as the carrier **gas.**

fluorophenol (found 11 S/N) and is denoted as the reporting limit of detection for the method. The reporting limit of quantitation for this method was set to 0.05% sample solution of 3-chloro-5-fluorophenol since the S/N was greater than 15:1 ([7], found 29 S/N). Linearity assessment of 3-chloro-5-fluorophenol was conducted with six points over the concentration range of 0.001 mg/mL to 10.0 mg/mL which is equivalent to 0.02% to 200% of the target sample concentration of 5 mg/mL. Carryover of 3-chloro-5-fluorophenol was evaluated by performing a sample injection followed by a diluent injection and integration of the signal at the retention time of 3-chloro-5-fluorophenol. As shown in Table 1 all criteria set forth for limit of detection, limit of quantitation, linearity, and carryover method attributes were met.

Control Strategy: In order to develop a robust control strategy of the possible regioisomer impurities within the raw material 3-chloro-5-fluorophenol, samples were ordered from several commercially available manufacturers and their products were subjected to the developed GC-FID method. The area% results for these GC-FID analyses are in Table 2. Of the possible nine regioisomer impurities only three regioisomer impurities (2-chloro-5-fluorophenol, 2-chloro-6-fluorophenol, and 4-chloro-3-fluorophenol) were observed by retention time conformation from the four manufactures of 3-chloro-5-fluorophenol analysed. All area% values for the regioisomer impurities observed were below 0.10% area%. Therefore, we were confident that if we implemented an internal specification of no more than 0.15 area% for each of these regioisomer impurities within 3-chloro-5-fluorophenol the resulting impurities would not likely be of concern in downstream processing steps. Since regioisomers and their corresponding downstream intermediates may have different physical and chemical properties, their corresponding regioisomer intermediate levels may increase due to their inability to be rejected from the process but this scenario is typically a rare occurrence.

Method Attribute	Criteria	Result
Limit of detection 0.0010 mg/mL (0.02%)	S/N (n=3) 3:1, % RSD none	11, 3.4%
Limit of quantitation 0.0025 mg/mL (0.05%)	S/N (n=3) ≥ 15:1, % RSD ≤ 15.0%	29, 0.4%
Linearity	Correlation Coefficient ≤ 0.99	1.000
Carryover	Less than the area of the LOD injection (0.02%)	0.007%

Table 1: Summary of the method attributes limit of detection, limit of quantitation, linearity, and carryover.

As for the other known impurities (phenol, 3-fluorophenol, 3-chlorophenol, and 3-bromo-5-fluorophenol) within the raw material 3-chloro-5-fluorophenol, all vendors contained at least one of these impurities. Since each vendor's manufacturing process of this raw material was not disclosed, we would recommend the identification of any impurity above 0.15 area% before use. Identification would allow the customer the ability to track the fate and purge of any unknown impurity above the 0.15 area% threshold and should be easily achieved by coupling the GC-FID method with MS technology [8-9]. In the case for unknown impurities RRT 0.24 and 0.26 in Table 2 at impurity levels above 0.15 area%, these impurities could be simply the processing solvents and/or starting materials for this manufacturer. We chose not to identify these peaks.

Robustness evaluation: To demonstrate the robustness of the analytical method the following method parameters were altered: carrier gas, diluent, and site location. Replacement of the carrier gas from

Manufacture			Combi-blocks, inc.	Combi-blocks, inc.	BePharm Limited	Oakwood Chemical	Ark Pharma
Lot/batch			L42454	L54033	0032554-16070101	13283	WG0032554-140708001
Name	CAS#	RRT	%Area	%Area	%Area	%Area	%Area
		0.24	0.75				
		0.26		0.39			
		0.30			0.03		0.03
		0.32			0.04		0.04
2-chloro-5-fluorophen	3827-49-4	0.66	0.05		0.03		0.02
Phenol	108-95-2	0.69				0.09	
3-fluorophenol	372-20-3	0.71				0.05	
		0.75	0.03				
2-Chloro-6-fluorophenol	2040-90-6	0.80				0.06	
		0.86				0.06	
		0.90			0.11		
		0.99			0.02		
3-Chloro-5-fluorophenol	202982-70-5	1.00	99.11	99.37	99.70	99.58	99.67
3-Chlorophenol	108-43-0	1.06				0.04	
		1.09				0.03	
4-Chloro-3-fluorophenol	348-60-7	1.11				0.04	
3-Bromo-5-fluorophenol	433937-27-6	1.19		0.22			
		1.21			0.04		0.04
		1.56				0.02	
		1.65			0.04		0.04
		1.72	0.06				
		1.92		0.02			
		2.45				0.02	

Table 2: Impurity Profile Data from Several Manufacturers of 3-Chloro-5 fluorophenol.

helium to hydrogen resulted in an overall decrease in retention time of all compounds with similar relative retention times (RRT) in respect to the retention time of 3-chloro-5-fluorophenol (Table 3). The RRT were determined by individual identification injections of each impurity onto the GC-FID system. Diluent evaluation was either performed by preparation of a sample or by a single blank injection. Acetonitrile as diluent appears acceptable since the same area% impurity profile was obtained for the 3-chloro-5-fluorophenol from Oakwood Chemical Lot 013283I13F (Table 4). High boiling solvents like DMAc and NMP eluted in regions of interest and would interfere with the area% analysis if they were used as the diluent. To emphasis the robustness of the method, the method has been successful utilized externally as well as internationally for the purpose of detection and quantitation of regioisomer impurities within the raw material 3-chloro-5-fluorophenol used in the synthesis of a potential API.

Proposed System Suitability Criteria

After evaluation of the collected data, the following minimal system suitability criteria are proposed to ensure proper detection and quantitation of the regioisomer impurities within 3-chloro-5-fluorophenol utilizing the described GC-FID method

(1) The blank injection is free of significant interference at the retention times of 3-chloro-5-fluorophenol and known impurities.

(2) The sensitivity solution (2000x dilution of the sample solution) should provide at least 10:1 S/N.

	He Carrier Gas		H_2 Carrier Gas	
CAS#	RT (min)	RRT	RT (min)	RRT
3827-49-4	5.809	0.66	5.309	0.65
1996-41-4	5.959	0.68	5.452	0.67
108-95-2	6.103	0.69	5.596	0.68
372-20-3	6.222	0.71	5.711	0.70
863870-86-4	6.381	0.73	5.851	0.71
348-62-9	6.776	0.77	6.231	0.76
185689-76-4	6.902	0.79	6.355	0.78
2040-90-6	7.019	0.80	6.466	0.79
2613-22-1	7.302	0.83	6.736	0.82
202982-70-5(main)	8.784	1.00	80188	1.00
108-43-0	9.354	1.10	9.025	1.06
2613-23-2	9.664	1.10	9.025	1.10
348-60-7	9.744	1.11	9.107	1.11
433937-27-6	10.426	1.19	9.769	1.19

Table 3: Retention Time and Relative Retention Time Comparison Between the Carrier Gasses Helium and Hydrogen.

Oakwood Chemical Lot 013283I13F		MeOH	MeCN
Name	RRT	%Area	%Area
108-95-2	0.69	0.09	0.09
372-20-3	0.71	0.05	0.05
2040-90-6	0.80	0.06	0.06
	0.86	0.06	0.06
202982-70-5	1.00	99.58	99.57
108-43-0	1.06	0.04	0.04
	1.09	0.03	0.03
348-60-7	1.11	0.04	0.05
	1.56	0.02	0.02
	2.45	0.02	0.02

Table 4: Area% Comparison Between the Diluents MeOH and MeCN.

Conclusion

A general GC-FID area% method employing an Rtx®-65 column has been developed for the detection and quantitation of all possible regioisomer impurities of 3-chloro-5-fluorophenol. The method as described has been shown to have a reporting quantitation limit of 0.05% and a reporting detection limit of 0.02% WRT a sample concentration of 5.0 mg/mL.

Notes and References

1. The International Council for Harmonisation (ICH) (2006) Q3A (R2): Impurities in New Drug Substances (ICH, October 2006).

2. The United States Pharmacopeia and The National Formulary (USP-NF), General Chapter: <1086> Impurities in Drug Substances and Drug Products.

3. European Pharmacopoeia (EP), General Texts: 5.10. Control of Impurities in Substances for Pharmaceutical Use and references within text (04/2012:51000).

4. Regioisomer impurities are impurities that have the same carbon skeleton connectivity but differ in the connectivity of the functional groups onto the carbon skeleton.

5. ICH, Q11: Development and Manufacture of Drug Substances (ICH, may 2011)

6. SciFinder® A CAS Solution search of 3-Chloro-5-fluorophenol CAS# 202982-70-5, MW 146.55 g/mole, B Pt (predicted) 206 ± 20°C at 760 Torr, physical state liquid at room temperature.

7. A goal to have the S/N for the reporting limit of quantitation solution of at least 15:1 can aid in transferring the method to other laboratories.

8. Ecker J, Scherer M, Schmitz G, Liebisch G (2012) A rapid GC-MS method for quantification of positional and geometric isomers of fatty acid methyl esters. Journal of Chromatography B 897: 98-104.

9. Fariña L, Boido E, Carrau F, Dellacassa E (2007) Determination of volatile phenols in red wines by dispersive liquid-liquid microextraction and gas chromatography-mass spectrometry detection. Journal of Chromatography A 1157: 46-50.

Enhanced Removal of Some Cationic Dyes from Environmental Samples Using Sulphuric Acid Modified Pistachio Shells Derived Activated Carbon

Akl MA*, Mostafa MM and Mohammed SA Bashanaini

Chemistry Department, Faculty of Science, Mansoura University, Mansoura, Egypt

Abstract

In the present work, pistachio shell, local agriculture waste, has been successfully used for the preparation activated carbon (AC) by sulphuric acid. This sulphuric acid based activated carbon (ACS) was used for the removal of two cationic dyes viz., methylene blue (MB) and Brilliant green (BG) from aqueous solutions. The structure and physical and chemical properties of ACS are investigated using FTIR and SEM analysis and pH surface and Boehm titration. The effect of the differential experimental parameters controlling the adsorption of dyes onto ACS, was thoroughly investigated, such as the effect of pH, initial dye concentration, contact time and adsorbent dosage in batch mode. Employment of equilibrium isotherm models for the description of adsorption capacities of ACS explores the good efficiency of Langmuir model for the best presentation of experimental data with maximum adsorption capacity of 208.333 and 151.515 mg/g for MB and BG dyes. The kinetic data were fitted to the pseudo-second-order kinetic model with cooperation with interparticle diffusion model. Thermodynamic parameters were evaluated to predict the nature of adsorption. These results point out the endothermic and spontaneous nature of the sorption process. The results demonstrate that ACS is effective in the removal of MB and BG dyes from aqueous solutions and can be used as an alternative to the high-cost commercial adsorbents.

Keywords: Pistachio shell; Activated carbon; Cationic dyes; Adsorption

Introduction

The dyes and pigments usage in last two decades (more than 107 kg/year) in human life and industries like textile dyeing, dermatological agent, a biological stain, paper, carpet and printing wastewater widely was increased [1]. Dyes, which usually have a synthetic origin, are characterized by complex aromatic molecular structures that supply stabilities of thermal, physicochemical and optical [2]. Dyes can be classified as cationic (basic dyes), anionic (direct, acid, and reactive dyes) or non-ionic (disperse dyes) [3]. Cationic dyes such as Methylene Blue (MB) can be applied to leather, silk, paper wool, plastics, in addition to for the production of ink, copying paper and cotton mordant with tannin [4]. In spite of low toxicity of MB, it can cause harmful effects such as vomiting enhance in heart rate, cyanosis, diarrhea, shock, jaundice, quadriplegia and human tissue necrosis [5] Brilliant green (BG) is applied as dermatological agent, veterinary medicine and as inhibitor of mold propagation and following contact with skin and eye, inhalation and ingestion generates toxic to the lungs and other tissues and lead to target-organ damage [6].

A vast number of physicochemical, chemical and biological methods have been used for removing dyes from wastewater e.g., ozonation [7], adsorption [8], electrochemical techniques [9], coagulation and flocculation [10] biological treatment [11]. Among these methods, adsorption has gained favour in recent years due to proven efficiency in the removal of pollutants from effluents. Activated carbon (AC) is known to be a very efficient adsorbent because of its large surface area, highly developed porosity, changeable characteristics of surface chemistry, and the high degree of surface reactivity [12]. Activated carbon, as an adsorbent has been widely investigated for the adsorption of dyes, but its high cost limits its commercial application. In the last two decades, there has been growing interest in finding inexpensive and effective alternatives to carbon, such as jute sticks [13], acorn [14], Lantana camara [8], cocoa shell [15], aegle marmelos [16], Pea shells [17], sugarcane bagasse [18], Rice Husk [19], Waste Weed [20].

The objective of the present study is the preparation of pistachio shell based-activated carbon (ACS). Pistachio shell is considered as waste matter in the environment and has low cost. Similar to other agro-residues, pistachio shell is chiefly comprised of lignin, hemicellulose, and cellulose. Such composition makes pistachio shell a good raw material for the output of new adsorbents for processes of water treatment.

In this investigation, the pistachio shell is simultaneously carbonized and activated chemically by sulphuric acid [21] to the removal of methylene blue (MB) and Brilliant Green (BG) from aqueous solution and some water samples. The various analytical factors affecting the removal of dyes are investigated viz. pH, the dose of activated carbon, the initial concentration of dye, the temperature and the contact time. Also, the adsorption kinetic, thermodynamic and isotherm studies were determined to foresee the sorption behavior.

Experimental

Adsorbates and materials

Methylene Blue (MB), basic blue 9, C.I. 52015; chemical formula, $C_{16}H_{18}N_3ClS$, and molecular weight 319.85 g/mol, λ_{max}: 662 nm, (Figure 1a), supplied by Merck. A 1000 mg/L stock solution of MB dye was prepared by dissolving the required amount of dye powder in bi-distilled water.

Brilliant green dye, also called Basic Green 1 (C.I.: 42040, chemical formula, $C_{27}H_{34}N_2O_4S$, FW: 482.62 g/mol, λ_{max}: 623 nm), supplied by Titan Biotech Limited, Bhiwadi, India (75% dye content) (Figure 1b), to prepare a 1000 mg/L stock solution appropriate amount of the dye was dissolved in DDW.

***Corresponding author:** Akl MA, Chemistry Department, Faculty of Science, Mansoura University, Mansoura, Egypt, E-mail: magdaakl@yahoo.com

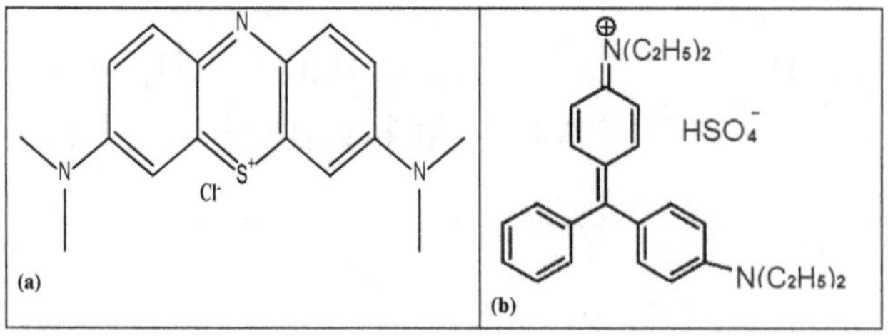

Figure 1: Molecular structure of (a) Methylene Blue dye (MB) and (b) Brilliant Green dye (BG).

Calibration curves were constructed in the concentration ranges 1.0-20.0 mg/L and the concentrations of the investigated dyes were determined spectrophotomterically at λ_{max}=662 and λ_{max}=623 nm for MB and BG, respectively.

Apparatus

The laboratory measurements of pH were performed using HANNA pH-meter model Hi 931401 (Portugal), The concentrations of MB and BG dyes were determined using UV-Vis spectrophotometer (Chrom Tech., Ltd., USA).

Preparations of adsorbents

Pistachio shells were collected from local markets, cleaned with tap water hardly and rinsed with distilled water for several times, then dried in at 373 K for 12 h. Sulphuric acid based activated carbon (ACS) was prepared by mixing 150 mL of 13 mol/L sulphuric acid with 30 g of pistachio shell. The mixture was put in furnace whose temperature was kept about 423-430 K for 90 min with occasional stirring. After cooling, the resulting black residue was filtered using a Buchner funnel under vacuum. Activated carbon was washed for some time with distilled water until pH value became 5-6 and dried at 373 K.

Adsorption studies of dye

To perform batch adsorption experiments, 0.025 grams of ACS with 25 mL aqueous solution of dye were introduced into 250 mL Erlenmeyer flasks, shake well in a temperature controlled water bath shaker using variable concentrations of dye between 10 and 400 mg/L of BG and between 50 and 500 mg/L of MB, pHs (between 3-10 for BG and 2-12 for MB), temperatures (between 32°C and 50°C), doses of ACS (between 0.005 and 0.0625 g) and ionic strength (between 0.005 and 0.2 mole/L) and shaking at a constant rate of 200 rpm. The concentrations of the non-adsorbed in the solution, was determined spectrophotometrically at 623 and 662 nm for BG and MB, respectively.

The capacity of adsorption of dye removed by adsorbent (q), and the removal percentage (R %) of cationic dyes are determined by the following equations. (1) and (2):

$$q = \frac{C_o - C_e}{m} \times V \quad (1)$$

$$R\% = \frac{C_o - C_e}{C_o} \times 100 \quad (2)$$

Where q is the adsorption capacity of dye (mg/g), C_o and C_e are the initial and equilibrium state concentrations of dye (mg/L), respectively. The term m is the mass of the adsorbent used (g), and V is the solution volume of the dye (mL).

Desorption of dyes

100 mg of ACS adsorbents were shaken separately with 0.1 L of 10 mg/L of the dye solution. The adsorption capacity of dyes was determined. Thereafter, the dye-loaded ACS was desorbed using 0.05, 0.1 and 0.2 mol/L HCl solutions.

The quantity of the desorbed dye was calculated from the following dependence:

$$q_d=(C_d-C_a)/m$$

Where: q_d – the quantity of dye desorbed, C_d – concentration of dye after desorption, C_a – concentration of dye before desorption (after adsorption), m – ACS mass

The percentage of desorption of dye % was calculated by using the following equation:

$$\text{Desorption of dye \%} = \frac{q_{d}}{q_a} \times 100 \quad (3)$$

Where q_d and q_a are the amount of adsorption capacity of dye (mg/g) of desorption and adsorption, respectively.

Results and Discussion

Characterization of adsorbents

A measurement of specific surface area of the activated carbons produced from the Pistachio shell (ACS) was made by N_2 adsorption (at 77 K), using a surface analyzer (QUANTACHROME – NOVA 2000 Series). The value by this method is 58.03 m^2/g and the total pore volume with diameter less than 25.8 nm at P/P_o=0.30019 is 1.735E-02 cc/g. While specific surface area value of ACS by DH method cumulative adsorption and desorption are 77.37 and 77.58 m^2/g, respectively. Table 1 shows a comparison of the results of the present study by the previously reported studies. Marked enhancement in the surface area and adsorption capacity was obtained upon using the proposed ACS in ftThe presenft sftudy [21-27].

Scanning electron microscopy (SEM) analyses showed that the sulphuric acid-treated activated carbon (ACS) had rough areas containing various irregular-shaped particles and macropores (Figures 2a and 2b).

The FTIR spectrum (Figure 3a) of the activated carbon ACS shows several bands at 3566, 3447 and 3422 cm^{-1} assigned to γ(OH) phenolic [23], $\nu_{as}(NH_2)$ and $\nu_s(NH_2)$ vibrations [28], respectively. Also, the bands observed in the 1797-2022 and 2929-3065 cm^{-1} ranges are attributed to the existence of hydrogen bonding in the activated carbon [29]. Moreover, the bands at 1703, 1683, 1649 and 1627 cm^{-1} are attributed to the free ν(C=O), ν(C=O; hydrogen bonded), ν(C=N) and ν(C=C) phenolic vibrations [30], respectively.

On other hands, Figure 3b shows FTIR spectrum of the activated carbon ACS loaded with methylene blue dye (MB). The results show new bands in the 1648-1465 cm^{-1} which are assigned to the C=C and C=N bands which are absent in the spectrum of the activated carbon ACS alone (Figure 3a). Also, the bands in the 1319-1165 cm^{-1} region are assigned to the C-O bands which have existed in the spectrum of ACS. The band at 3424 cm^{-1} in the case of AC attributed to ν(OH) vibration is shifted to lower wave number and observed at 3449 cm^{-1} (Figure 3b). This suggests the involvement of this group in bonding between ACS and MB.

Figure 3c shows the FTIR spectrum of the activated carbon ACS loaded with Brilliant Green dye (BG) several bands in the 1757-1266 cm^{-1}. These bands are obscured in the spectrum of the activated carbon

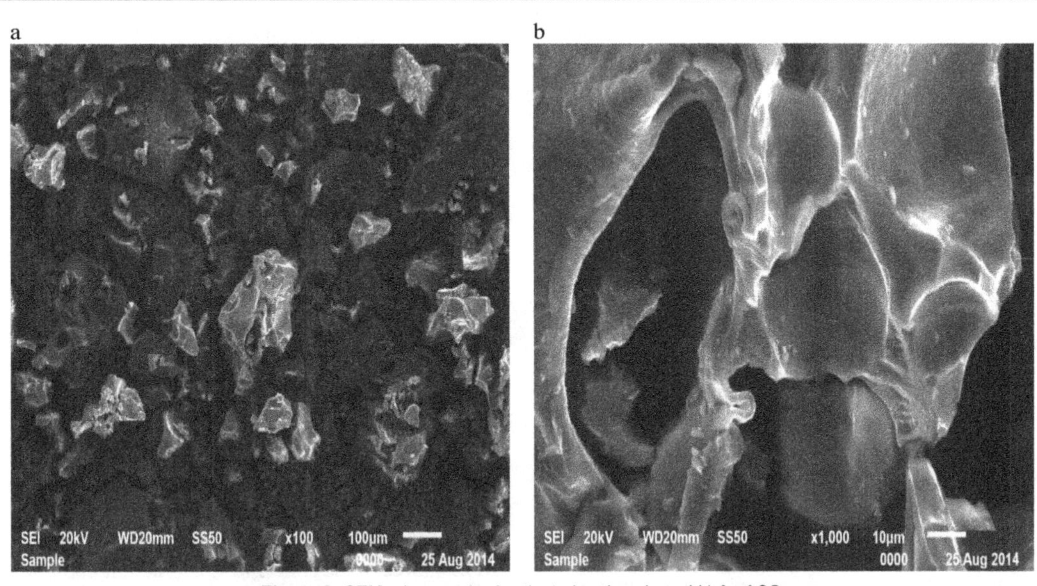

Figure 2: SEM micrograph of activated carbon (a and b) for ACS.

Parameters								Present study
S_{BET} (m²/g)	[a]SSH1 (at 25°C) 10.35	[b]PNS1 (at 303K) 25.34	[c]AC1 8.8	-	-	[d]BC400 2.8	72.00	[e]ACS 58.03
	SSH2 (at 80°C) 18.19	PNS2 (at 333K) 25.34	AC2 240.02			BC700 367		
	SSH3 (at 100°C) 21.06	PNS2 (at 353K) 25.34	AC3 114.77			-		
q_e (mg/g)	SSH1 (100 ppm) 34.00	PNS1 78.74	AC1 (at 25°C) 10.21	277.80	279.72	-	32.30	205.42 (149.00)
	SSH1 (100 ppm) 66	PNS2 93.45	AC2 (at 25°C) 16.43					
	SSH1 (100 ppm) 52	PNS3 125.00	AC3 (at 25°C) 15.80					
Sources	Sunflower Seeds hull	Pistachio Nut Shell	Sunflower Oil Cake	Lapsi seeds	Biomass plant	Bamboo Charcoal	Peanut Shell	Pistachio Shell
Adsorbate	Acid Violet 17	Acid Violet 17	Methylene Blue	Pb^{2+}	Pb^{2+}	NH_3	Se(IV)	MB (BG)

[a]SSH=Sunflower Seeds Hull; [b]PNS=Pistachio Nut Shell; [c]AC=Activated Carbon; [d]BC=Bamboo Charcoal and [e]ACS=Sulphuric acid based-activated carbon

Table 1: Comparison of surface area values and adsorption capacities of the proposed ACS with different adsorbents previously reported.

ACS alone (Figure 3a). The bands observed at 1757 cm^{-1} is assigned to the quinoid structure. Also, the band at 1473 cm^{-1} is attributed to the C=C groups. Moreover, the two bands at 1347 and 1266 cm^{-1} are mainly due to the ν(C-O) vibration of the lactonic and/or the phenolic groups existed in the ACS. All these observations suggest that BG is attached to activated carbon ACS *via* the C-O group.

The pH of the slurry of the ACS substance in aqueous solution provides a favorable pointer of the surface groups on carbon. The pH of the carbon slurry reflects the surface acidic groups which, in this case, for ACS are mainly carboxylic groups with only a very slight addition from sulfonic groups due to the low concentration of sulfur in the sorbent [31]. As the concentration of such functional groups increases on the carbon surface, the pH of the carbon decreases. The large quantities of the acidic functional groups on the surface of ACS result in increased cation exchange capacity and more adsorption of the dyes onto ACS surface. Boehm titration and surface acidity and basicity results are shown in Table 2. The synoptic number of the basic sites of the surface was calculated in all cases is smaller than the synoptic number of the acidic sites of the surface. The pH_{SUS} is in agreement with this result, which is also acidic, and point of zero charge pH_{PZC}.

Adsorption of MB and BG

The effect of pH on the dye adsorption: The pH of the dye solution plays a significant role in the whole sorption process and specifically on the adsorption capacity. The effects of initial pH of sample solution on the removal percentage of MB and BG dyes using ACS adsorbent were evaluated within the pH range 2–12. Figure 4 shows that the adsorption process of BG dye onto ACS adsorbent does not depend on the change in the pH values of the solution of the dye. This finding agrees with the FTIR results discussed above, which suggests that the sorption mechanism should not be electrostatic Otherwise, the adsorption should take place by the interaction of the dye with the aromatic rings

Acidic groups (mmol/g)				Basic groups (mmol/g)	Surface pH	Point of zero charge	Moisture content %	Ash content %
Carboxylic	Lactonic	Phenolic	Total					
3.670	1.940	2.400	8.010	0.330	2.93	3.02	2.62	0.56

Table 2: Chemical and physical Parameter of ACS adsorbent.

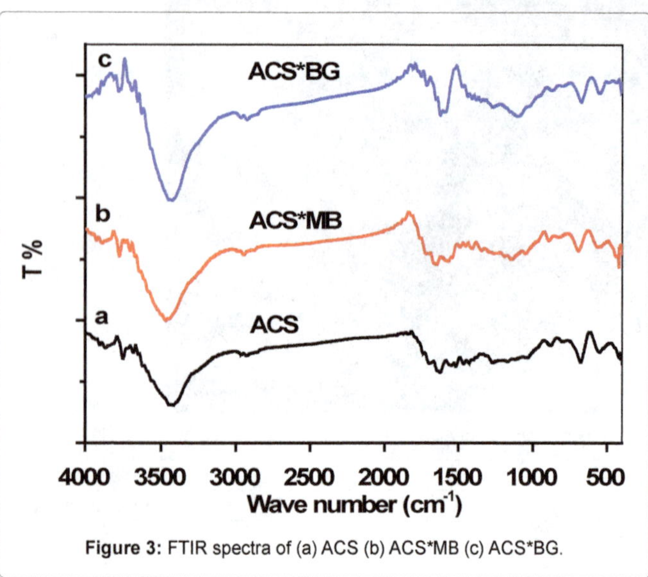

Figure 3: FTIR spectra of (a) ACS (b) ACS*MB (c) ACS*BG.

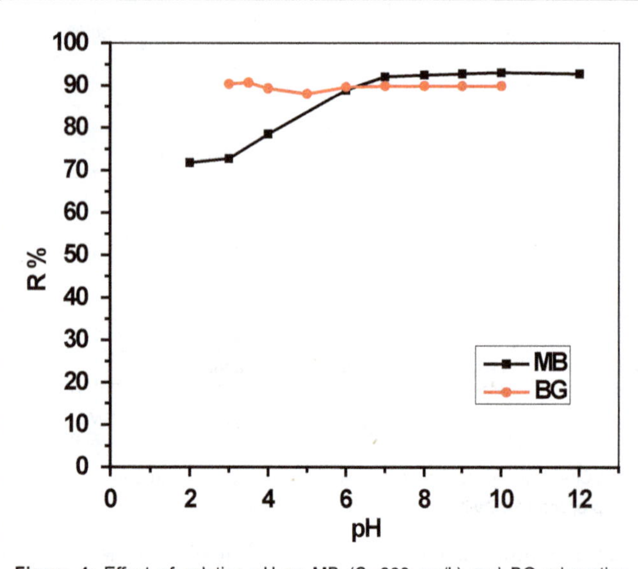

Figure 4: Effect of solution pH on MB (C_i=200 mg/L) and BG adsorption (C_i=100 mg/L) on ACS.

of the activated carbon [32]. This behavior is very significant from the analytical point of view as there is no need for pH adjustments for removing the BG dye using ACS adsorbent.

For MB, on the other hand, the removal percentage increases gradually upon increasing the pH from 2 to7. Then, the removal % of MB remained nearly constant over the pH range 7-12. The decreased removal % noticed at low pH values, may be due to competition between the protonated H^+ ions and the cationic dye molecules for the active sites on the adsorbents When the solution pH increased, the surface of ACS may get negatively charged due to adsorption of hydroxyl groups, and the functional groups got deprotonated producing negatively charged adsorption sites, which improved the interaction between the adsorbent and the cationic dye molecules. For both MB and BG, the subsequent experiments were carried out at ambient pH.

The effect the initial concentration of dyes: The effect of the initial concentration of MB and BG dyes on the capacity of adsorption of these dyes using ACS at normal ambient pH had been studied. Figure 5 shows that the capacity of adsorption of the ACS adsorbent increase sharply when the initial concentration of the dyes increased. Then the capacity of adsorption gradually increases with further increase of the concentration of the dyes. At higher concentrations of the dyes, the equilibrium is reached. At higher MB and BG dyes concentrations, adsorption capacity reached a plateau indicating saturation of the available binding sites on the adsorbent. The sharp increase in the adsorption capacity in the early phase can be attributed to the great driving force of the concentration gradient at the solid-liquid interface causing an increase of the amount of MB and BG dyes adsorbed on the adsorbent [33]. When the initial concentration of MB and BG dyes increases from 50 to 500 mg/L and 10 to 400 mg/L at 25°C, the amount of dye adsorbed at equilibrium (qe), the increase from 49.94 to 205.42 mg/g and from 9.998 to 125.00 mg/g, for MB and BG, respectively.

Effect of adsorbent dosage: The effect of the amount of adsorbent ACS on the removal % of the investigated dyes is illustrated in Figure 6. As it can be noticed the removal % of each dye increased gradually upon increasing the dose of the adsorbent. Increasing the dose of adsorbent would provide more functional sites on the adsorbent capable of binding more dye molecules and thus increasing the removal %. The removal % of MB and BG dyes adsorbed increased from 18.55% to 99.83% and 60.63% to 98.93% as the adsorbent dose increased from 0.005 to 0.0625 g, respectively.

The effect of interferents: The interference of some foreign cations such as Ca^{2+}, Mg^{2+}, Na^+, K^+ and Na(I) as well as anions (F^-, Cl^-, acetate and oxalate) on the process of MB and BG dyes adsorption from aqueous solutions was studied and estimated by ACS. Estimation of the possible interference of foreign ions was performed on the basis of 10 mg/L of dye versus other interfering species. The results presented in Table 3, show that the presence of anions and cations do not affect the adsorption percentage of MB and BG dyes on the ACS adsorbent, especially Ca^{2+}, Mg^{2+}, Na^+, K^+ at the concentration of 200 mg/L.

The effect of contact time: The contact time between the dye and the ACS adsorbent is important in the dye removal from the solution by the adsorption process. The effect of contact time is studied by batch adsorption processes. Figure 7, shows that the adsorption capacity of MB and BG dyes increased with time and reached equilibrium at 540 minutes for MB dye and 480 minutes for BG dye.

Adsorption kinetics: The linearized form of the pseudo-first-order kinetic model [34] can be expressed by equation 4:

$$\log(q_e - q_t) = \log q_e - \frac{K_1}{2.303}t \qquad (4)$$

Where q_e and q_t correspond to the amount of dye adsorbed per adsorbent unit mass (mg/g) at equilibrium and at time t (min), respectively, k_1 are the constant of equilibrium (min^{-1}) and q_e which were obtained from the slopes and intercept of the linear plots of log (q_e-q_t) versus t (Figure 8a).

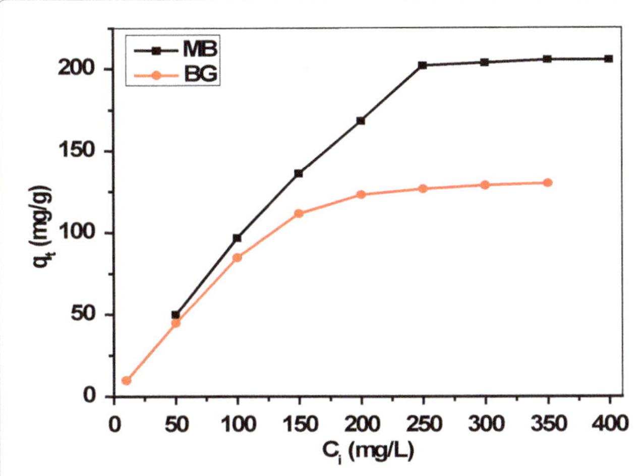

Figure 5: Effect of initial concentration of MB and BG on the adsorption capacity of ACS adsorbent.

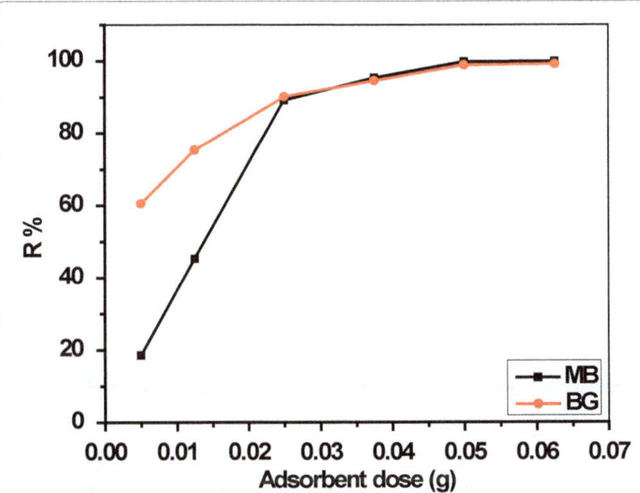

Figure 6: Effect of adsorbent dose on adsorption MB (C_i=200 mg/L) and BG (C_i=150 mg/L) on ACS.

Foreign ion (200 mg/L)	Removal percentage, R%	
	Methylene Blue	Brilliant Green
F^-	98.43	93.66
Cl^-	98.72	94.78
Acetate	98.74	98.14
Oxalate	94.83	96.27
Mg^{2+}	94.84	94.41
Ca^{2+}	93.71	92.17
Na^+	98.43	97.60
K^+	99.86	98.85

Table 3: Effect of interferents on the removal percentage of Methylene Blue and Brilliant Green.

The pseudo-second-order model of kinetic studies can be expressed by the following equation:

$$\frac{t}{q_e} = \frac{1}{K_2 q_e^2} + \frac{1}{q_e} t \tag{5}$$

Where, q_e and q_t represent the amount of dye adsorbed per adsorbent unit mass (mg/g) at equilibrium and at time t (min), respectively. q_e and k_2 can be calculated from the slope and intercept of plot t/qt versus t (Figure 8b) [35].

To determine the best-fit model for adsorption of MB and BG on the ACS, the linear correlation coefficient (R^2) was investigated. The results are shown in Table 4. The correlation coefficient (R^2) of pseudo-second-order ($R_2^2 \geq 0.9977$) which is close to 1 was much higher than that of pseudo-first-order ($R_1^2 \leq 0.9847$). The theoretical q_2 value calculated from pseudo-second-order was much closed to the experimental value q_{exp} (Table 4) for both dyes MB and BG indicating that the pseudo-second-order model fits well for the system.

The intra-particle diffusion model is an empirical found functional relationship, supposing that the adsorption capacity varies almost proportionally with $t^{0.5}$ [36,37] this model is expressed by the following equation:

$$q_t = K_{int} t^{0.5} + C \tag{6}$$

Where k_{int} (g/mg min$^{1/2}$) is that the adsorption constant, the intercept is C, both C and k_{int} are calculated from a plot q_t versus $t^{0.5}$ (Figure 8c). Two distinct linear trends (MB1 and MB2) and (BG1 and BG2) are present in intra-particle diffusion model. The first trend (MB1) and (BG1) represents rapid surface adsorption (<240 min) of MB and (<180 min) of BG molecules via boundary layer diffusion, in which dye molecules move from bulk solution to the external surface of ACS particles and tend to cover mesopores of ACS surface. The MB2 and BG2 trend shows the attainment of equilibrium stage (>240 min) and (>180 min), respectively. The plot in Figure 8c does not pass through the origin implying that dye adsorption is partly controlled by intraparticle diffusion. The large value of boundary layer (C_{int} of BG >85.592 mg/g) implies that adsorption is mainly controlled by boundary layer adsorption. The similar mechanism has been reported for BG adsorption on activated carbon [38].

The kinetic data were further investigated using the Boyd [39] equation kinetic expressed by equation (7):

Dye	q_{max} (mg/g)		
	32°C	40°C	50°C
MB	205.42	226.44	257.16
BG	149.00	154.59	162.11

Table 4: Parameters of kinetics for the adsorption of Methylene Blue and Brilliant Green onto ACS.

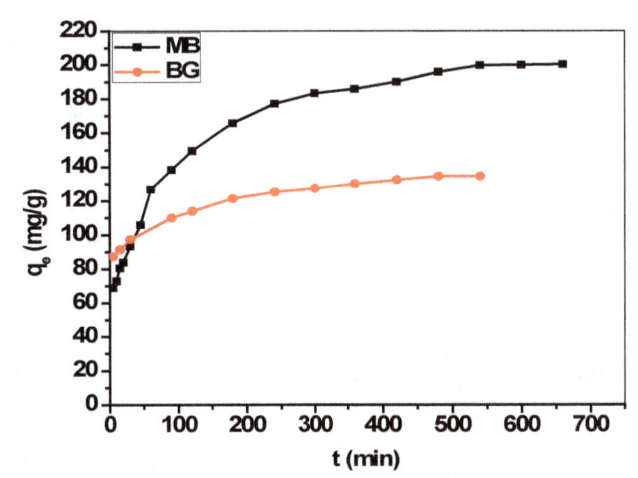

Figure 7: Effect of contact time on adsorption of MB (C_i=240 mg/L) and BG (C_i=150 mg/L) onto ACs.

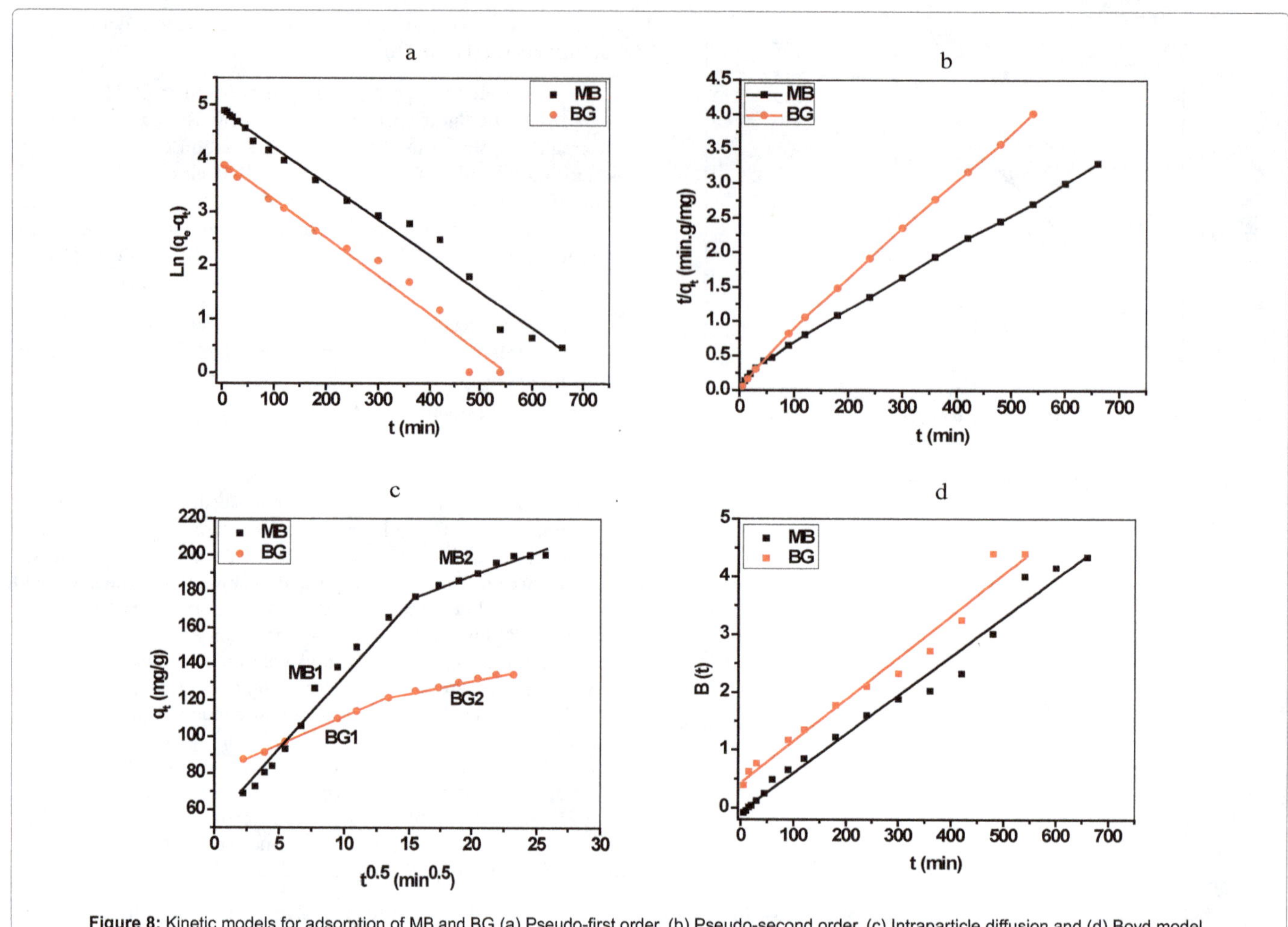

Figure 8: Kinetic models for adsorption of MB and BG (a) Pseudo-first order, (b) Pseudo-second order, (c) Intraparticle diffusion and (d) Boyd model.

$$F(t) = 1 - \frac{6}{\Pi^2} \sum_{1-n}^{\infty} \frac{1}{n^2} \exp(-n^2 Bt) \tag{7}$$

Where F(t) is the equilibrium fractional at different times t, and B(t) is mathematical function of F, n is an integer that defines the infinite series solution and F(t) is the fractional achievement of equilibrium at time (t) and is determined by the following equation:

$$F = \frac{q(t)}{q(e)} \tag{8}$$

Where $q_{(e)}$ and $q_{(t)}$ are the adsorbent capacity of dye adsorbed at equilibrium and the time t, respectively. Reichenberg [40] succeeded to obtain the following approximations:

For F values>0.85;

$$B(t) = -0.4977 - \ln(1-F) \tag{9}$$

And for F values<0.85

$$B(t) = \left\{ \sqrt{\Pi} - \sqrt{\Pi - \left\{ \frac{n^2 F(t)}{3} \right\}} \right\}^2 \tag{10}$$

The plots of Boyd's equation (Figure 8d) didn't go through the origin denoting that film diffusion is that the rate-limiting process of sorption for the sorption of dyes onto ACS.

The ionic strength effect: The effect of ionic strength on the removal of MB and BG dye as shown in Figure 9. The capacity of adsorption is reduced as the the ionic strength increased. Similarly, the concentration of NaCl ions increased from 0.01 to 0.20 mol/L, the capacities of adsorption is reduced from 95.23 to 93.01 mg/g, from 90.75 to 86.34 mg/g for MB and BG, respectively.

Effect of solution temperature and thermodynamic studies: The effect of temperature on the adsorption capacities of dyes on ACS was investigated at various temperatures (32, 40 and 50°C). The capacities of adsorption of dyes were represented in Table 5. Figures 10a and 10b show that the capacity of adsorption of MB and BG increases from 205.42 mg/g to 257.16 mg/g and 149.00 mg/g to 162.11 mg/g as the temperature increases from 32 to 50°C, respectively. Increasing temperature could also enhance the rate of diffusion of the dye molecules across the external boundary layer and internal pores of peat which reduces the viscosity of BG solution [6].

The thermodynamics studies of MB and BG dye adsorption onto ACS were performed to find the energy dependent mechanism of the adsorption process. The standard free energy change (ΔG°), enthalpy change (ΔH°) and entropy change (ΔS°) associated with the adsorption processes are calculated using the following equations:

$$\Delta G^o = -RTL_n k_d \tag{11}$$

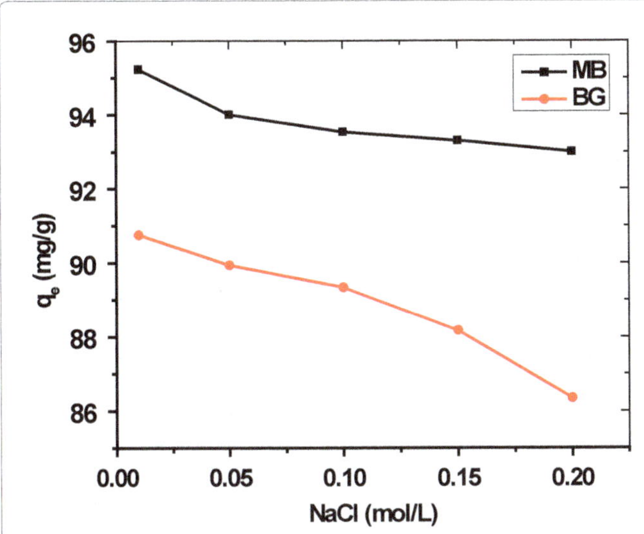

Figure 9: Effect of ionic strength on the removal of MB and BG by ACS. (C_o=10 mg/L).

Model	Kinetic parameter	Methylene Blue	Brilliant Green
	q_e, exp (mg/g)	205.42	149.00
Pseudo-First-order Equation	q_{e1} (mg/g)	131.079	51.676
	k_1 (min^{-1})	0.00673	0.0071
	R^2_1	0.9847	0.9746
Pseudo-second-order Equation	q_{e2} (mg/g)	208.3333	136.99
	K_2 (g/mg.min)	0.00014	0.00074
	R^2_2	0.9977	0.9985
Intra-particle diffusion equation	K_{int} ($mg/g.min^{1/2}$)	0.03479	2.3227
	C	4.8821	85.592
	R^2_{int}	0.9606	0.9663
Boyd equation	Intercept	-0.0657	0.4321
	R^2	0.9847	0.9749

Table 5: Effect of temperature on adsorption capacities of Methylene Blue and Brilliant Green by ACS activated carbon.

$$L_n k_d = \frac{-\Delta H^o}{RT} + \frac{\Delta S}{R} \quad (12)$$

Where, K_d is the distribution coefficient (Langmuir constant), R is gas constant (8.314 J/mol K) and T is the solution temperature (K), respectively. The (ΔS°) and (ΔH°) values are obtained from the intercept and slope of ln K_d versus 1/T plot (Van't Hoff) for the adsorption of MB and BG onto ACS at different temperature (Figure 11). The parameters of thermodynamic values are given in Table 6. The negative (ΔG°) values suggested that the adsorption process is spontaneous and more favorable at low temperature [41]. Moreover, the endothermic nature of adsorption process confirmed by the positive value of (ÄH°) and positive (ΔS°) value point out the increased randomness at the solid/liquid interface during the adsorption of dyes onto ACS [42].

Adsorption Isotherms: In the present investigation, the isotherm data were analyzed using the Langmuir and Freundlich isotherm equations (Figure 12).

Langmuir model is represented by the following equations [43]:

$$\frac{C_e}{q_e} = \frac{1}{K_L q_m} + \frac{C_e}{q_m} \quad (13)$$

Where, q_e the amount was adsorbed at equilibrium (mg/g), q_m is the theoretical maximum adsorption capacity (mg/g), C_e is the equilibrium concentration of dye (mg/L) and k_L is Langmuir adsorption constant (L/mg) related to the energy of adsorption. A plot of C_e/q_e against C_e (Figure 12a) gave a straight line graph with a slope $1/q_m$ and intercept of $1/k_L q_m$. Values of q_m and k_L are calculated from the graph and reported in Table 7. The fundamental characteristics of the Langmuir isotherm can be expressed by a dimensionless separation factor, R_L, defined by:

$$R_L = \frac{1}{1+K_L C_o} \quad (14)$$

Where K_L is Langmuir constant and C_0 is the highest initial dye concentration (mg/L). The R_L parameter indicates the type of the isotherm as follows: (R_L>1), unfavourable; (R_L=1), linear; (0<R_L<1), favourable; (RL=0), irreversible. The R_L value was found to be (0<R_L<1) in the present study which indicated that ACS adsorbent exhibit favourable adsorption for both MB and BG dyes (Table 7).

Isotherm model of Freundlich is expressed as equation [44]:

$$\log q_e = \log K_F + \frac{1}{n}\log C_e \quad (15)$$

Where k_F and n are Freundlich constants, the characteristic of the system, k_F and n are the indicator of adsorption capacity and adsorption intensity, respectively (Figure 12b).

The parameters of Langmuir and of Freundlich isotherm models are given in Table 7. Examination of Table 7 reveals that the values of R^2 acquired from Langmuir model are closer to unit than that of the model of Freundlich. A finding that denotes that the isotherm of Langmuir fits better with the adsorption both MB and BG dyes onto ACS.

Effect of desorption: The regeneration studies were conducted using three different concentrations of hydrochloric acid (0.05, 0.10 and 0.20 mol/L) revealed a maximum ability of desorption of 99.30 and 98.10% with 0.20 mol/L of HCl for MB and BG, respectively. At lower pH the greater numbers of H^+ ions present will compete with dye cations for the same binding sites, resulting in greater desorption of dye cations; the data are presented in Table 8.

Analytical applications

By spiking known concentrations of MB dye the ability of application of the ACS for uptake of the MB dye from different samples of water was investigated. The Table 9 shows the recovery percentage (R %) is more than 99.18% with less than 1% of relative standard deviation (RSD %).

The removal of MB and BG dye by the proposed ACS activated carbons is in good comparison with the results of the previously

Dye	ΔH° (KJ/mol)	ΔS° (KJ/mol)	ΔG° (KJ/mol)		
			305°K	313°K	323°K
Methylene Blue	27.8594	0.0924	-0.3159	-1.0681	-1.8706
Brilliant Green	6.7885	0.0252	-0.8832	-1.0922	-1.2634

Table 6: Parameters of thermodynamics for the adsorption of Methylene Blue and Brilliant Green on ACS.

Dye	Langmuir parameters				Freundlich parameters		
	q_m (mg/g)	K_L (L/mg)	R_L	R^2	n	K_F	R^2
MB	208.333	0.2449	0.06372	0.9986	5.166	83.994	0.9830
BG	151.515	0.2734	0.0103	0.9993	4.039	47.196	0.9609

Table 7: Langmuir and Freundlich isotherm Parameters.

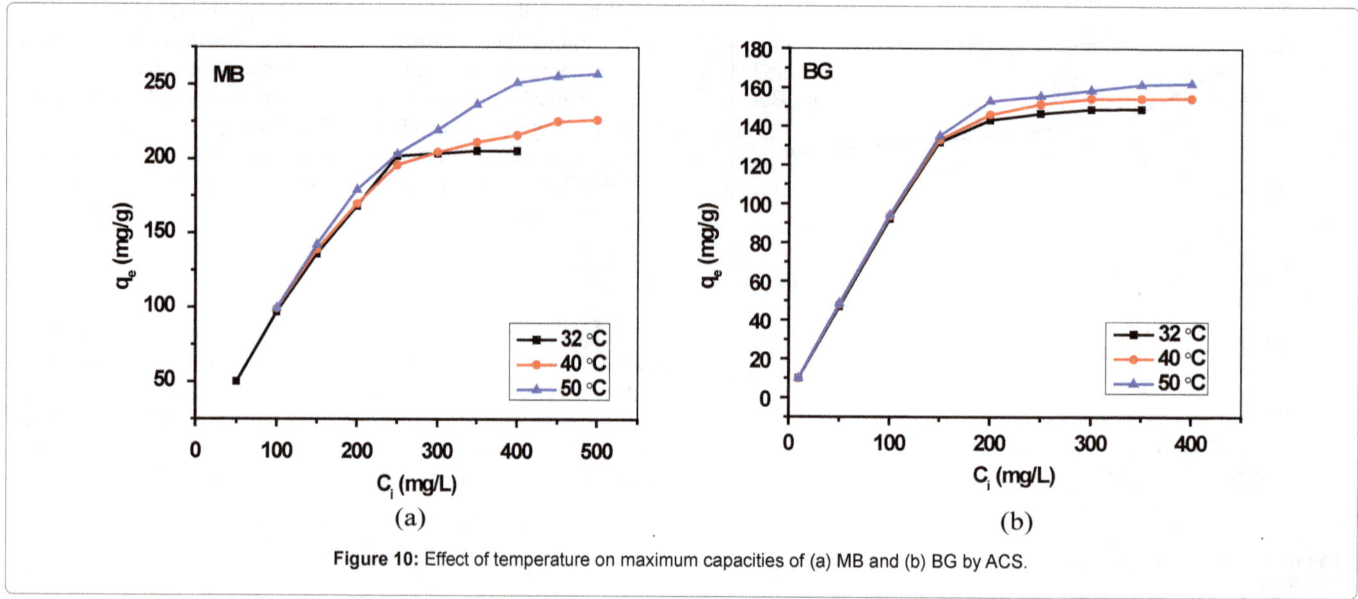

Figure 10: Effect of temperature on maximum capacities of (a) MB and (b) BG by ACS.

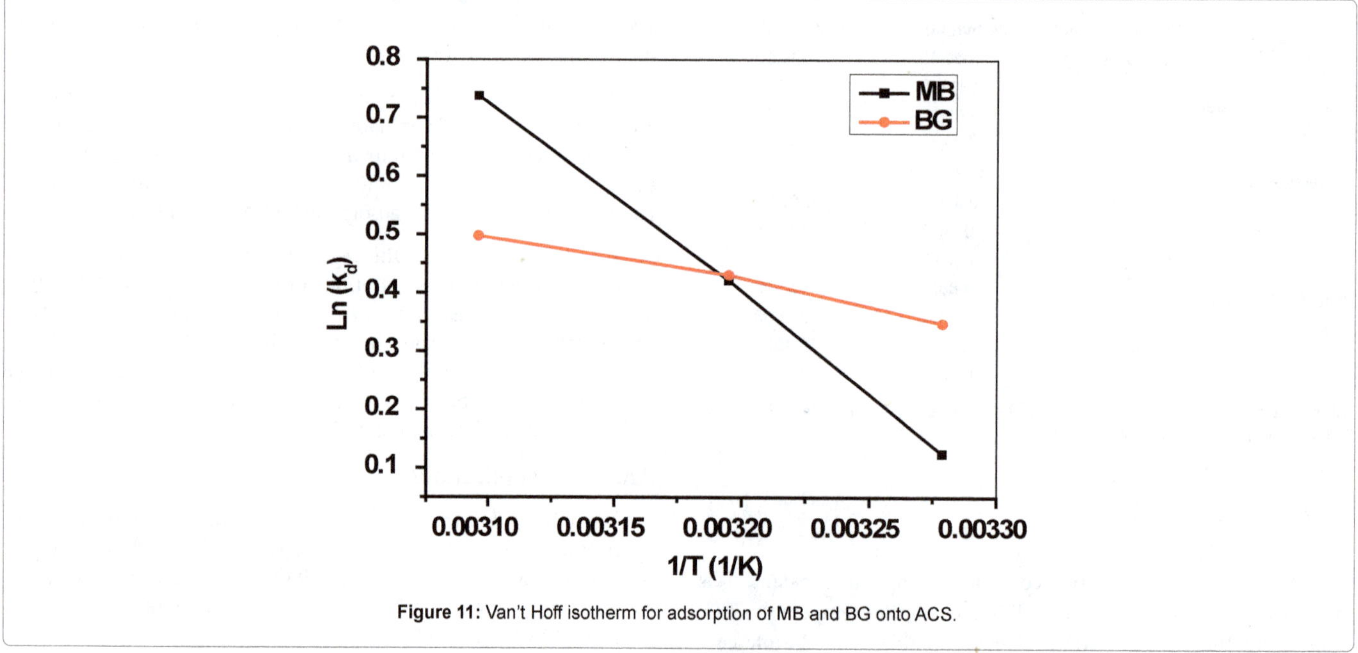

Figure 11: Van't Hoff isotherm for adsorption of MB and BG onto ACS.

(Dye)	q_e Adsorbed (mg/g)	Desorption of dye using 0.05 mol/L HCl		Desorption of dye using 0.1 mol/L HCl		Desorption of dye using 0.2 mol/L HCl	
		q_e desorbed (mg/g)	Desorption, %	q_e desorbed (mg/g)	Desorption, %	q_e desorbed (mg/g)	Desorption %
MB	10	8.97	89.70	9.475	94.75	9.93	99.30
BG	10	8.387	83.87	9.054	90.54	9.81	98.10

Table 8: Desorption % of Methylene Blue and Brilliant Green from ACS.

reported studies [13,14,45-52] (Table 10). The prepared low cost material such as pistachio shell activated carbon ACS is beneficial to the one high-cost commercial activated carbon.

Conclusion

Activated carbon prepared from pistachio shell (ACS) is identified to be an effective adsorbent for the removal of two dyes methylene blue and brilliant green from aqueous solution. The adsorption is highly dependent on various operating parameters, like; temperature, adsorbent dose, pH, initial dye concentration, contact time, ionic strength. The kinetic studies indicated that follows pseudo-second-order beside intra-particle diffusion and equilibrium in the adsorption of MB and BG on ACS were attained within 540 and 480 min, respectively. The equilibrium data were studied by Langmuir and Freundlich models, the adsorption equilibrium can be best represented by the Langmuir isotherm model, with maximum monolayer adsorption capacity of 208.333 and 151.515 mg/g for MB and BG on the adsorbent at 32°C. The results of thermodynamic indicated that the (- ÄG°) as expected for a process spontaneously under the optimum conditions. The synthesized ACS was successfully applied for the

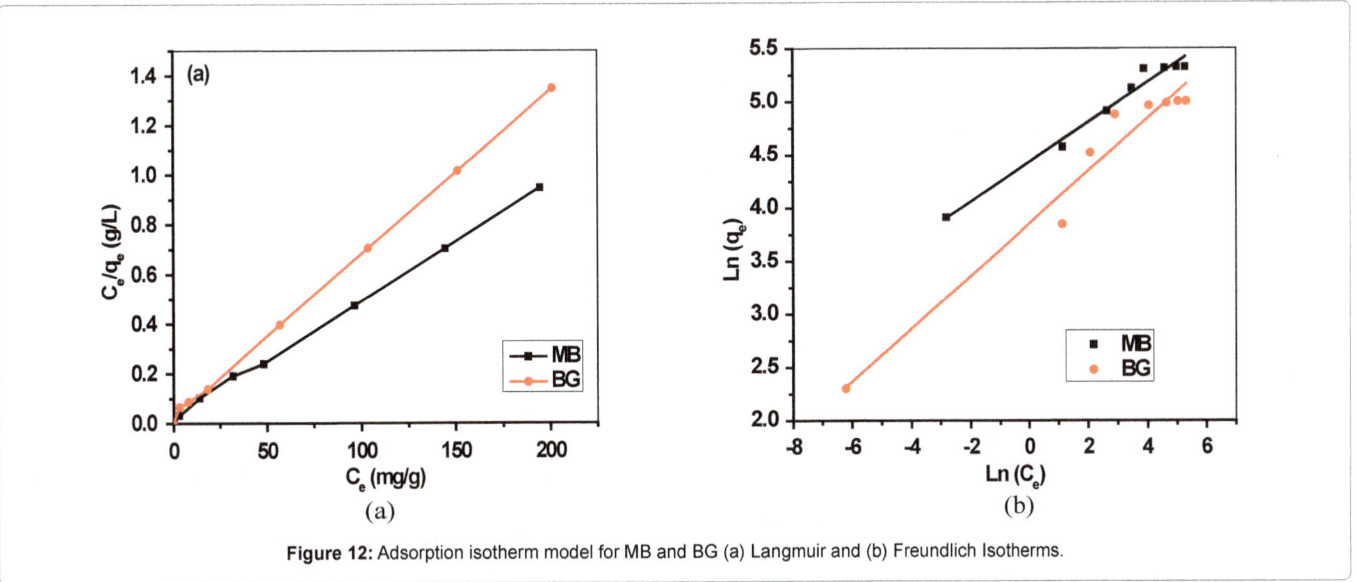

Figure 12: Adsorption isotherm model for MB and BG (a) Langmuir and (b) Freundlich Isotherms.

Sample	MB dye added (mg/L)	MB dye removed (mg/L)	R, %[a]	RSD, %[b]
Bi distilled water	5	4.99875	99.98	0.05
	10	9.99153	99.92	0.0245
	15	14.98058	99.87	0.026435
Underground water (Mit Gammr)	5	4.96602	99.32	0.0797
	10	9.91869	99.19	0.04686
	15	14.98762	99.18	0.04214
Domiat sea water	5	4.984224	99.69	0.002427
	10	9.963592	99.64	0.002802
	15	14.94782	99.65	0.00243
Alexandria sea water	5	4.979952	99.54	0.093388
	10	9.959952	99.60	0.024368
	15	14.94053	99.60	0.031115

[a]R, %=Removal % of methylene blue;
[b]RSD, %=Relative standard deviation

Table 9: Recovery (R %) of Methylene Blue from different samples of water using ACS adsorbent (n=4).

Adsorbent	Dye	Adsorption capacity (mg/g)	Reference
Pistachio shell Activated carbon (H_2SO_4) (ACS)	MB	205.42	Present study
Pistachio shell Activated carbon (H_2SO_4) (ACS)	BG	149.20	Present study
Strychnos Potatorum Seed (AC)	MB	100.00	[45]
Jute fiber-based activated carbon	MB	225.64	[46]
Coconut husk-based activated carbon	MB	434.78	[47]
E. strobilacea char (ESC)	MB	31.152	[48]
ESC impregnated with 95% H_3PO_4 (ESP)	MB	21.929	[48]
ESC impregnated with 85% $ZnCl_2$ (ESZ)	MB	37.037	[48]
Wood waste activated carbon	MB	4.937	[49]
Loofa activated carbon (Zn Cl_2/ H_3PO_4) (AC1)	MB	33.7496	[50]
Loofa activated carbon (HNO_3) (AC2)	MB	32.9992	[50]
Jute sticks charcoal	BG	52.00	[13]
Jute sticks steam activated carbon (ACS)	BG	150.00	[13]
Jute sticks chemical activated carbon (ACC)	BG	286.00	[13]
ACORN (AC)	BG	02.01	[14]
Bagasse fly ash	BG	116.00	[51]

Table 10: Adsorption capacities of different adsorbents previously reported for the removal of MB and BG compared with ACS.

removal of the MB dye from samples of natural water. Desorption of the both dyes from ACS could easily be achieved by using 0.2 mol/L HCl. Finally, this work shows the possibility of using this technique widely in different applications.

References

1. Ghaedi M, Zeinali Z, Ghaedi AM, Teimuori M, Tashkhourian J (2014) Artificial neural network-genetic algorithm based optimization for the adsorption of methylene blue and brilliant green from aqueous solution by graphite oxide nanoparticle. Spectrochim Acta Part A Mol Biomol Spectrosc 125: 264-277.
2. Silva MMF, Oliveira MM, Avelino MC, Fonseca MG, Almeida RKS, et al. (2012) Adsorption of an industrial anionic dye by modified-KSF-montmorillonite: Evaluation of the kinetic, thermodynamic and equilibrium data. Chem Eng J 203: 259-268.
3. Ak Mayk (2014) Solid Phase Extraction and Spectrophotometric Determination Activated Bentonite from Egypt. J Anal Bioanal Tech 5: 1-8.
4. Eren E (2009) Investigation of a basic dye removal from aqueous solution onto chemically modified Unye bentonite. J Hazard Mater 166: 88-93.
5. Chen L, Ramadan A, Lu L, Shao W, Luo F, et al. (2011) Biosorption of methylene blue from aqueous solution using lawny grass modified with citric acid. J Chem Eng Data 56: 3392-3399.
6. Kismir Y, Aroguz AZ (2011) Adsorption characteristics of the hazardous dye Brilliant Green on Sakli{dotless}kent mud. Chem Eng J 172: 199-206.
7. Al jibouri AKH, Wu J, Upreti SR (2015) Continuous ozonation of methylene blue in water. J Water Process Eng 8: 142-150.
8. Amuda OS, Olayiwola AO, Alade AO (2014) Adsorption of Methylene Blue from Aqueous Solution Using Steam-Activated Carbon Produced from Lantana camara Stem. Earth and Environmental Sciences 5: 1352-1363.
9. Bennajaha M, Darmaneb Y, Touhamic M, Maalmia M (2011) A variable order kinetic model to predict defluoridation of drinking water by electrocoagulation-electroflotation. Int J Eng Sci Technol 2: 12.
10. Petzol G, Schwarz S, Mende M, Jaeger W (2007) Dye flocculation using polyampholytes and polyelectrolyte-surfactant nanoparticles. J Appl Polym Sci 104: 1342-1349.
11. Junnarka N, Murty DS, Bhatt NS, Madamwar D (2005) Decolorization of diazo dye Direct Red 81 by a novel bacterial consortium. World J Microbiol Biotechnol 22: 163-168.
12. Rodriguez Reinoso F, Heintz EA, Marsh H (1997) Introduction to carbon technologies. Publicaciones de la Universidad de Alicante, Spain.
13. Asadullah M, Asaduzzaman M, Kabir MS, Mostofa MG, Miyazawa T (2010) Chemical and structural evaluation of activated carbon prepared from jute sticks for Brilliant Green dye removal from aqueous solution. J Hazard Mater 174: 437-443.

14. Ghaedi M, Hossainian H, Montazerozohori M, Shokrollahi A, Shojaipour F, et al. (2011) A novel acorn based adsorbent for the removal of brilliant green. Desalination 281: 226-233.

15. Obuge MA, Evbuomwan OB (2014) Adsorption of Methylene Blue onto Activated Carbon Impregnated With KOH Using Cocoa Shell. International Journal of Engineering and Technical Research 2: 11-18.

16. Yadav LS, Kumar A (2012) Adsorptive Removal Brilliant Green Onto Carbonised Aegle Marmelos. World Academy of Science, Engineering and Technology, pp: 491-493.

17. Gecgel U, Ozcan G, Gizem GC (2013) Removal of Methylene Blue from Aqueous Solution by Activated Carbon Prepared from Pea Shells (Pisum sativum). J Chem 614083: 9.

18. Vijaya Kumar G, Agrawal P, Hiremath L (2012) Utility of plant cellulose waste for the bioadsorption of brilliant green dye pollutant. Int J ChemTech Res 4: 319-323.

19. Rahman MA, Amin SMR, Alam MS (2012) Removal of Methylene Blue from Waste Water Using Activated Carbon Prepared from Rice Husk. Dhaka Univ J Sci 60: 185-189.

20. Shendkar CD, Torane RC, Mundhe KS, Lavate SM, Pawar AB (2013) Characterization and Application of Activated Carbon Prepared From Waste. JSIR 2: 5-7.

21. El-Shafey El (2007) Removal of Se (IV) from aqueous solution using sulphuric acid-treated peanut shell. J Environ Manage 84: 620-627.

22. Thinakaran N, Baskaralingam P, Pulikesi M, Panneerselvam P, Sivanesan S (2008) Removal of Acid Violet 17 from aqueous solutions by adsorption onto activated carbon prepared from sunflower seed hull. J Hazard Mater 151: 316-322.

23. Vijayalakshmi P, Bala VSS, Thiruvengadaravi KV, Panneerselvam P, Palanichamy M, et al. (2010) Removal of Acid Violet 17 from Aqueous Solutions by Adsorption onto Activated Carbon Prepared from Pistachio Nut Shell. Sep Sci Technol 46: 155-163.

24. Karag S, Tay T, Ucar S, Erdem M (2008) Activated carbons from waste biomass by sulfuric acid activation and their use on methylene blue adsorption. Bioresour Technol 99: 6214-6222.

25. Gercel O, Gercel HF (2007) Adsorption of lead(II) ions from aqueous solutions by activated carbon prepared from biomass plant material of Euphorbia rigida. Chem Eng J 132: 1-3.

26. Shrestha RM, Pradhananga RR, Varga M, Varga I (2011) Preparation of Activated Carbon for the Removal of Pb (II) from Aqueous Solutions. J Nepal Chem Soc 28: 94-101.

27. Asada T, Ohkubo T, Kawata K, Oikawa K (2006) Ammonia Adsorption on Bamboo Charcoal with Acid Treatment. J Heal Sci 52: 585-589.

28. Bellamy LJ, Williams RL (1957) Infrared spectra and polar effects. Part VII. Dipolar effects in α-halogenated carbonyl compounds. J Chem Soc, pp: 4294-4304.

29. Nakamoto K (2016) Infrared and Raman Spectra of Inorganic and Coordination Compounds. John Wiley & Sons. 2nd edn. New York, USA.

30. Bellamy LJ (1980) The Infrared Spectra of Complex Molecules. Dordrecht: Springer Netherlands.

31. Cox M, El Shafey E, Pichugin AA, Appleton Q (1999) Preparation and characterisation of a carbon adsorbent from flax shive by dehydration with sulfuric acid. J Chem Technol Biotechnol 74: 1019-1029.

32. Ren JL, Sun RC, Liu CF, Lin L, He BH (2007) Synthesis and characterization of novel cationic SCB hemicelluloses with a low degree of substitution. Carbohydr Polym 67: 347-357.

33. Li WH, Yue QY, Gao BY, Ma ZH, Li YJ, et al. (2011) Preparation and utilization of sludge-based activated carbon for the adsorption of dyes from aqueous solutions. Chem Eng J 171: 320-327.

34. Demirbas A (2009) Oil from Tea Seed by Supercritical Fluid Extraction Energy Sources. Part A Recover Util Environ Eff 31: 217-222.

35. Hameed BH, Ahmad AL, Latiff KNA (2006) Adsorption of basic dye (methylene blue) onto activated carbon prepared from rattan sawdust. J Hazard Mater 7: 5143-5149.

36. Franca AS, Oliveira LS, Ferreira ME (2009) Kinetics and equilibrium studies of methylene blue adsorption by spent coffee grounds. Desalination 249: 267-272.

37. Annadurai G, Juang RS, Lee DL (2002) Use of cellulose-based wastes for adsorption of dyes from aqueous solutions. J Hazard Mater 92: 263-274.

38. Wu FC, Tseng RL, Juang RS (2001) Adsorption of dyes and phenols from water on the activated carbons prepared from corncob wastes. Environ Technol 22: 205-213.

39. Calvete T, Lima EC, Cardoso NF, Dias SLP, Ribeiro ES (2010) Removal of brilliant green dye from aqueous solutions using home made activated carbons. Clean Soil Air Water 38: 521-532.

40. Boyd GE, Adamson W, Myers LS (1947) The exchange adsorption of ions from aqueous solutions by organic zeolites; kinetics. J Am Chem Soc 69: 2836-2848.

41. Aguedach A, Brosillon S, Morvan J, Lhadi EK (2008) Influence of ionic strength in the adsorption and during photocatalysis of reactive black 5 azo dye on TiO2 coated on non woven paper with SiO2 as a binder. J Hazard Mater 150: 250-256.

42. Youssef AM, Dawy AM, Akl MB, Abou-Elanwar MA (2013) EDTA Versus Nitric Acid Modified Activated Carbon For Adsorption Studies of Lead. J Appl Sci Res 9: 897-912.

43. Ak MAA, Dawy MB, Serage AA (2014) Efficient Removal of Phenol from Water Samples Using Sugarcane Bagasse Based Activated Carbon. J Anal Bioanal Tech 5: 2-12.

44. Langmuir I (1918) The Adsorption Of Gases On Plane Surfaces Of Glass, Mica And Platinum. J Am Chem Soc 40: 1361-1403.

45. Freundlich HMF (1906) Over the Adsorption in Solution. Zeitschrift fur Phys Chemie 57: 385-470.

46. Joseph MJA, Xavier N, Joseph S (2012) Equilibrium Isotherm Studies Of Methylene Blue From Aqueous Solution Unto Activated Carbon Prepared Form Strychnos Potatorum Seed. Int J Appl Biol Pharm Technol, pp: 27-31.

47. Senthilkumaar S, Varadarajan PR, Porkodi K, Subbhuraam CV (2005) Adsorption of methylene blue onto jute fiber carbon: kinetics and equilibrium studies. J Colloid Interface Sci 284: 78-82.

48. Tan IAW, Ahmad AL, Hameed BH (2008) Adsorption of basic dye on high-surface-area activated carbon prepared from coconut husk: Equilibrium, kinetic and thermodynamic studies. J Hazard Mater 154: 1-3.

49. Agarwal S, Tyagi I, Gupta VK, Ghasemi N, Shahivand M, et al. (2016) Kinetics, equilibrium studies and thermodynamics of methylene blue adsorption on Ephedra strobilacea saw dust and modified using phosphoric acid and zinc chloride. J Mol Liq 218: 208-218.

50. Ghaedi M, Kokhdan SN (2015) Removal of methylene blue from aqueous solution by wood millet carbon optimization using response surface methodology. Spectrochim Acta Part A Mol Biomol Spectrosc 136: 141-148.

51. Cherifi H, Fatiha, Salah H (2013) Kinetic studies on the adsorption of methylene blue onto vegetal fiber activated carbons. Appl Surf Sci 282: 52-59.

52. Mane VS, Mall ID, Srivastava VC (2007) Use of bagasse fly ash as an adsorbent for the removal of brilliant green dye from aqueous solution Dye. Pigment 73: 269-278.

A Rapid Liquid Chromatography Method for Determination of Gylphosate in Crude Palm Oil with Fluorescence Detection

Fariq Fitri MSM[1*], Halimah Muhamad[2], Dzolkhifli Omar[1] and Norhayu Asib[1]

[1]*Department of Plant Protection, Faculty of Agriculture, Universiti Putra Malaysia, 43400 Serdang, Malaysia*
[2]*Analytical and Quality Development Unit, Malaysian Palm Oil Board, No. 6, Persiaran Institusi, Bandar Baru Bangi, 43000 Selangor, Malaysia*
China

Abstract

A rapid and simple method for the determination of glyphosate in crude palm oil (CPO) was developed and validated using high performance liquid chromatography with fluorescence detector. Glyphosate was derivatized with 9-fluorenylmethylchloroformate (FMOC-Cl) and then separated using a C_{18} reverse phase column with potassium dihydrogen phosphate and acetonitrile as the mobile phase. A linear correlation was obtained for the concentration of glyphosate from 0.05-1.5 μg mL^{-1} with a correlation coefficient of 0.9998. The average recovery obtained for glyphosate ranged between 80% and 100% at five fortification levels with the relative standard deviation (RSD) of less than 3% of all cases. The limit of detection and limit of quantification for glyphosate were 0.05 and 0.1 μg/g, respectively. The method will facilitate palm oil trade through quality assurance in terms of glyphosate residues in palm oil products and also to counter any issues related to food safety for palm based products.

Keywords: Palm oil; Glyphosate; Herbicide; Recoveries; HPLC; Fluorescence

Abbreviations: CPO: Crude palm oil; FLD: Fluorescence detector; FMOC-Cl: 9-fluorenylmethyl chloroformate; RSD: Relative standard deviation.

Introduction

Nowadays, the use of pesticides is one of the methods to control pests effectively with low management cost. However, its impact towards the environment should be taken into account as the use of pesticides poses risk on the safety of food products. It is indeed a concern if the residues still remain in the crops for in a long-term period, besides affecting food products. Thus, the determination of pesticide residue in palm oil should be used as an important parameter in ensuring the quality of vegetable oil produced. Moreover, monitoring the levels of pesticide residue content in palm oil is among the important consensus from the study on methods development for the determination of pesticide residues. This should be a priority to ensure that palm oil is free from any chemical residue for human safety and to meet the standard of waste management for a significant palm oil importer [1].

According to Gibon et al. [2] crude palm oil (CPO), which is rich in minor components, has high value nutrients, such as tocopherol and tocotrienol (vitamin E), as well as carotenoid (α and β-carotene). Meanwhile, Nuzul Amri et al. [3] reported that the properties of crude palm kernel oil (CPKO) are unchanged over the last 17 years based on a survey carried out for a year. In addition, several analytical methods for the determination of pesticide in palm oil matrices had been reported. Halimah et al. [4] developed an analytical method for the detection of chlorpyrifos in pure olein oil sample. The developed method was a modification of the method reported by Cloborn et al. [5] regarding the detection of chlorpyrifos residues in tissues and cow's milk. In this study, the researchers used liquid-liquid extraction involving n-hexane and acetonitrile solvents before the clean-up step was preceded with sililic acid chromatography column. The gas chromatography with electron capture detector (GC-ECD) was selected as the quantification method as the percentage of the recovery study was 97%.

Although a lot of researches on pesticide residue determination in palm oil had been carried out, the analytical methods for determining the residue of glyphosate in palm oil matrices via pre-column derivatization with FMOC-Cl have limited reports. The use of pre-column derivatization provides numerous advantages in terms of the use of non-complicated instruments, rapid, fewer restrictions, and efficient. The method developed, thus, should be able to facilitate palm oil trade through quality assurance in terms of the absence of glyphosate residues in palm oil products, as well as to counter any issue related to food safety for palm-based products.

Experimental

Materials and methods

Glyphosate standard with purity at >97.5% was purchased from Dr. Ehrenstorfer (Augsburg, Germany). Meanwhile, HPLC grade acetone, 9-fluorenylmethylcholoroformate, potassium dihydrogen phosphate, disodium tetraborate, dichloromethane, and acetonitrile were obtained from MERCK (Darmstadt, Germany). Micro liter pipettes, adjustable between 100 and 1000 μL, and pipette tips were obtained from Eppendorf (Hamburg, Germany). Microvials were purchased from Agilent (Palo Alto, CA, USA) and vortex mix from Barnstead/Thermolyne Inc (Dubuque, IA, USA). Besides, a sonicator (Bransonic, USA) was also used. The extracts were filtered by using syringe filter (nylon, 0.45 μm) and both were purchased from Whatman (Maidstone, Kent, UK). Blank CPO that was used as a control had been obtained from MPOB Labu refinery, while blank CPKO was obtained from Felda Pandamaran in Pelabuhan Klang refinery.

***Corresponding author:** Fariq Fitri MSM, Department of Plant Protection, Faculty of Agriculture, Universiti Putra Malaysia, 43400 Serdang, Malaysia
E-mail: fariqfitri@gmail.com

Instrumentation

The sample extracts were analyzed on an Agilent 1100 HPLC system equipped with a quaternary pump (model G1311A), an auto sampler (model G1313A), and a degasser (model G1322A). The temperature of the column heater was maintained at 40°C with a column heater temperature control module. Meanwhile, the glyphosate derivatives were detected with a fluorescence detector with excitation at 370 nm and emission at 415 nm. The analytical column was Waters C_{18} reverse phase column (250 mm × 4.60 mm i.d., 5 µm, XBridge Waters). The system was controlled by HP ChemStation (Agilent Technologies), which also performed functions, such as 1453 data collection from the FLD detector and quantitative measurements. The mobile phase used was potassium dehydrogen phosphate (50 mM, pH 2.5) and acetonitrile in gradient mode. The flow rate was 1.00 mL min^{-1} and the volume injected was 15 µL. The analytical column was set at 40°C and the samples were run for 30 min.

Preparation of stock standard solution

A stock standard solution of glyphosate was prepared at a concentration of 2000 µg/mL by dissolving 0.1 g of glyphosate in 15 mL deionized water and 35 mL of acetone in a 50 ml volumetric flask. Then, intermediate working standard solutions were prepared by diluting the stock solutions in deionized water to obtain glyphosate standards of 100 and 10 µg/mL. Finally, serial dilutions of the working standard solutions were prepared to obtain seven calibration solutions (1.0, 0.8, 0.5, 0.3, 0.1, 0.08, and 0.05 µg/mL) in deionized water. All the standard solutions were kept in scintillation vials at 4°C in the refrigerator.

Preparation of mobile phase solution

The mobile system consisted of 0.05 M KH_2PO_4 buffer phase for pre-column. A 6.8 g amount of KH_2PO_4 was dissolved in 1 L of deionized water and the pH was adjusted to 2.5 with H_3PO_4. The solution was filtered through a 0.45 mm membrane and degassed. The flow rate of this mobile phase was optimized and maintained at 1.0 mL/min.

Crude Palm Oil (CPO) and Crude Palm Kernel Oil (CPKO) samples for fortification

CPO and CPKO which were free from glyphosate, used as control were melted at 60°C in an oven. After homogenization by shaking the samples, recoveries of glyphosate were determined at fortification levels of 0.05, 0.08, 0.3, 0.5, and 1.0 µg/g by using oil samples. Then, an appropriate amount of the fortification solution was pipetted into a screw cap centrifuge tube containing 5.0 g of the CPO and CPKO sample. The mixture was then vortexed for 5 min. The extraction was carried out without any clean-up process, as described below, prior to HPLC analysis.

Derivatization with FMOC-Cl analysis

A total of 1 mL of the upper layer from the 50 mL screw cap centrifuge tube was taken after extraction and prior to derivatization before injecting into HPLC. A molar ratio of disodium tetraborate, acetone, and 0.01 M FMOC-Cl was studied before added to complete the derivatization process, followed by 5 min of vortex, and it was left for few hours for optimum derivatization period study before analysis (0.5, 1, 1.5, 2, 2.5, 3, 3.5 and 4 hours).

Extraction optimization and Liquid-Liquid Extraction (LLE) analysis

Five g of CPO and CPKO samples were transferred into each 50-mL screw cap centrifuge tubes. Each sample was fortified with an appropriate volume of working standard solution in acetone: water (70:30, v/v) for the recovery experiment (based on fortification levels of 0.05, 0.08, 0.3, 0.5, and 1.0 µg/g). The spiked samples were mixed well by using a vortex mixer. Dichloromethane (10 mL) was added to the spiked sample in each tube and the mixtures were shaken for 3 min by using a vortex mixer before 5 mL of water was added. Then the mixture was shaken on the vortex mixer for about 5 min, and then, centrifuged for 30 min at 3000 rpm. The aqueous layer was separated from the oil layer for derivatization.

Method validation analysis

Typically, an analytical instrument must be determined of its efficiency level, linearity, and repeatability of injections before the instruments can be used for analysis. This is to ensure that the instruments are in good condition [6,7] Level of efficiency, linearity, and reproducibility for HPLC-FLD injections were determined by injecting a series of standard solution of glyphosate from 0.01, 0.05, 0.08, 0.1, 0.3, 0.5, to 1.0 µg/mL to sketch a calibration curve. Glyphosate standard calibration curve was prepared by plotting the peak area of the chromatogram as the y-axis against the concentration injected into the HPLC-FLD as the x-axis.

Repeatability and precision of the method developed are important criteria, in which the methods should be tested in the same condition, but by different operators and laboratories, known as inter-laboratory test. Intra-laboratory test was by two different operators/analysts at two different days. The selectivity of the analytical method in this work was determined by comparing the chromatograms of a blank matrix solution with the fortified matrix solution.

LOD and LOQ of instrument were determined by comparing the peak height of the chromatogram obtained with the height of instrument noise level (S/N). For LOD, peak height is three times higher than the level of noise ($S/N \geq 3$) while the LOQ, the height of the peak is 10 times higher than the level of noise ($S/N \geq 10$) [8-10].

Monitoring of CPO and CPKO samples analysis

As for the monitoring study, samples of crude palm oil were collected from 30 different refineries and producers (100 mL each), then obtained from Registration and Licensing Department, Wisma Sawit MPOB Kelana Jaya. Meanwhile, samples of crude palm kernel oil were obtained from 10 palm kernel oil refineries (100 mL each) throughout Malaysia. All samples were kept in a 100 mL brown bottle and stored in room temperature. All the samples were ready to be analyzed to determine the level of glyphosate residue by using the method that had been developed and optimized. Each sample was analyzed by 3 replicates.

Results and Discussion

Optimization of derivatization with FMOC-Cl

The derivatization process requires some reagents to react with the analyte to improve the physical and the chemical properties of the analyte. The main functions of the derivatization process are to change the molecular structure and the polarity of the analyte, to improve and to stabilize the separation of analyte, in addition to increase the detectability of the analyte. In some previous studies conducted by Bo et al. [11], Hanke et al. [12], as well as Nedelkoska and Low [13] several parameters, such as the optimization period of derivatization, temperature, concentration of reagents, and study on molar ratio, were carried out. This was to ensure that the process had been rapid and qualitative in order to minimize excess noise from FMOC-Cl.

In addition, a study on the optimization of the molar ratio of borate buffer, acetone and FMOC-Cl was successfully performed and gave a consistent, as well as good symmetry peak, as shown in Figures 1 and 2. On the other hand, the effect of derivatization period was studied and shown in Table 1. This study demonstrated that the optimum time was 30 min since there was no significant difference in intensity, retention time and the area for derivatization period longer than that. Hence, the relative standard deviation (RSD) values for peak areas, retention times and intensity by HPLC-FLD for standard solutions (0.5 µg/mL) were 0.4485%, 0.056% and 2.9453%, respectively. As the percentages of RSD were less than 3% the method developed had been satisfactory and could be repeated.

Extraction optimization and Liquid-Liquid Extraction (LLE)

In fact, various methods have been developed for determination of pesticides in environmental samples, either through the process of extraction or purification (separation and characterization). Some of the common extraction methods used are LLE and low temperature precipitation, while some of the common clean-up processes are SPE, MSPD, SPME, and GPC. However, 70% of the preparation methods that are often used to treat pesticide residues in fatty vegetable matrix are LLE, SPE, and GPC [14].

Liquid-liquid extraction (LLE) is a traditional method that has long been applied. LLE is also known as liquid phase extraction (LPE) and solid-liquid extraction (LSE), as LLE extraction techniques typically use one or more types of solvents. LLE is applied either by shaking manually or by using high-speed homogenization for separating the analyte from the solid samples or semi-solid by using a suitable solvent. Besides, solvents that are commonly used for extraction of pesticides in samples are acetonitrile, ethyl acetate, acetone, and n-hexane. The extraction method that employs glyphosate in palm oil samples was carried out by testing several types of solvents, the optimum volume of solvents, and the optimum period of extraction.

The extraction of trace residues from fatty foods or high lipid samples is problematic when the extracts contain a large amount of lipids that need to be removed. Since glyphosate is highly polar compound with high solubility in water (12.0 g/L) and very low solubility in oil matrix, the standard solution for fortification was prepared in the acetone: water solution because of high solubility of acetone towards oil. Based on solubility of glyphosate in non-polar solvent, acetone have the highest solubility (0.078 g/L) compare to ethyl acetate (0.012 g/L) and toluene (0.036 g/L). Besides, several ratios of dilution of acetone with water had been tested to determine the most suitable ratio that can meet the requirement to mix well with glyphosate. In this study, the ratio of 70:30 (acetone:water) had been identified as the most suitable.

To obtain a high percentage of recovery, the appropriate solvents selected to enhance of extraction is necessary. Extraction of glyphosate residues in palm oil was carried out by comparing the usage of organic solvents, such as dichloromethane and chloroform. For each selected solvents, some concentrations of 0.1, 0.5, and 1.0 ug/g were used to

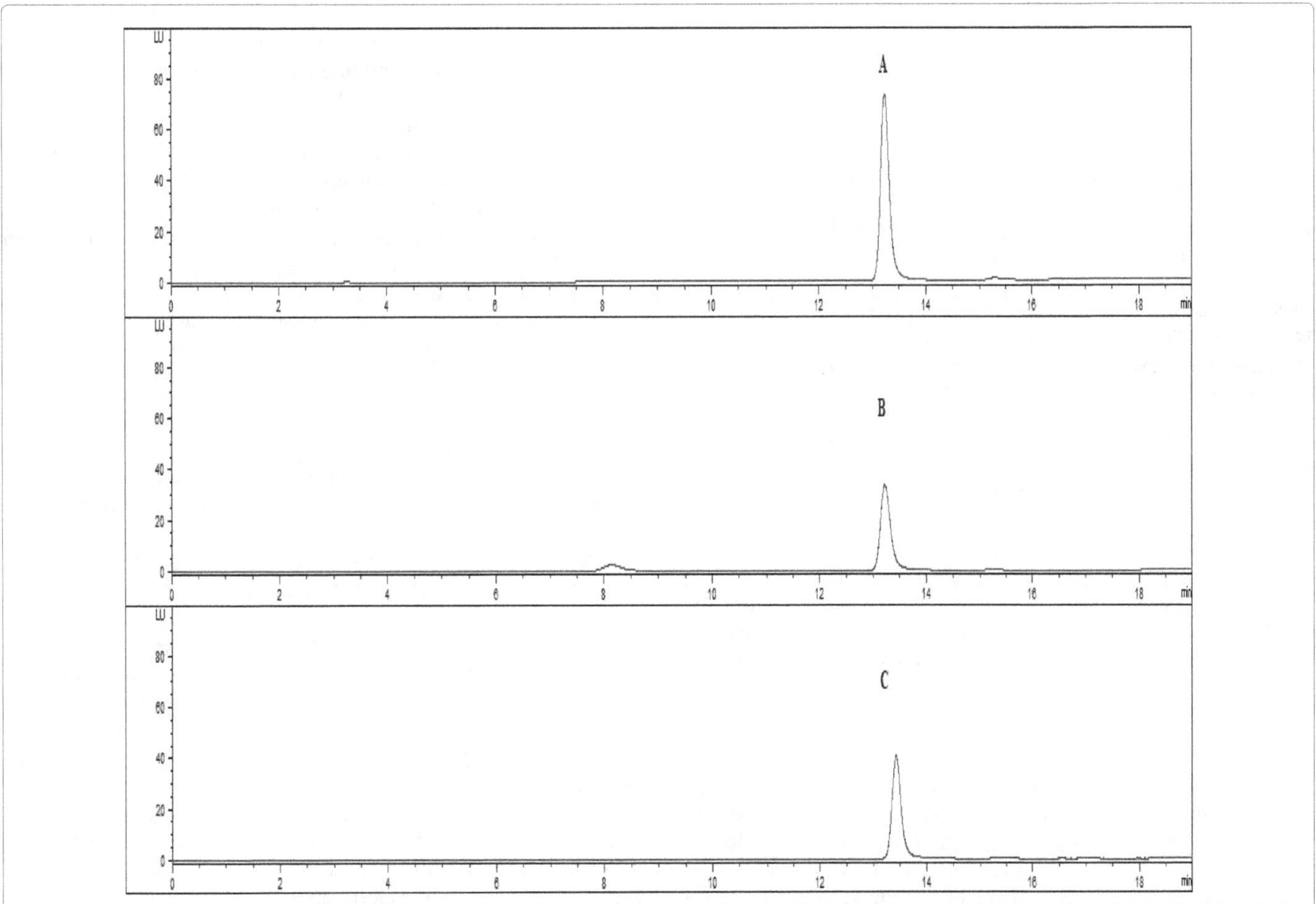

Figure 1: Glyphosate chromatogram (1.0 µg/ mL) at flow rate of 1.0 mL with molar ratio of borate buffer, acetone and FMOC-Cl (A) 1:1:0.5 (B), 1:0.5:0.5 and (C) 0.5:1:0.5.

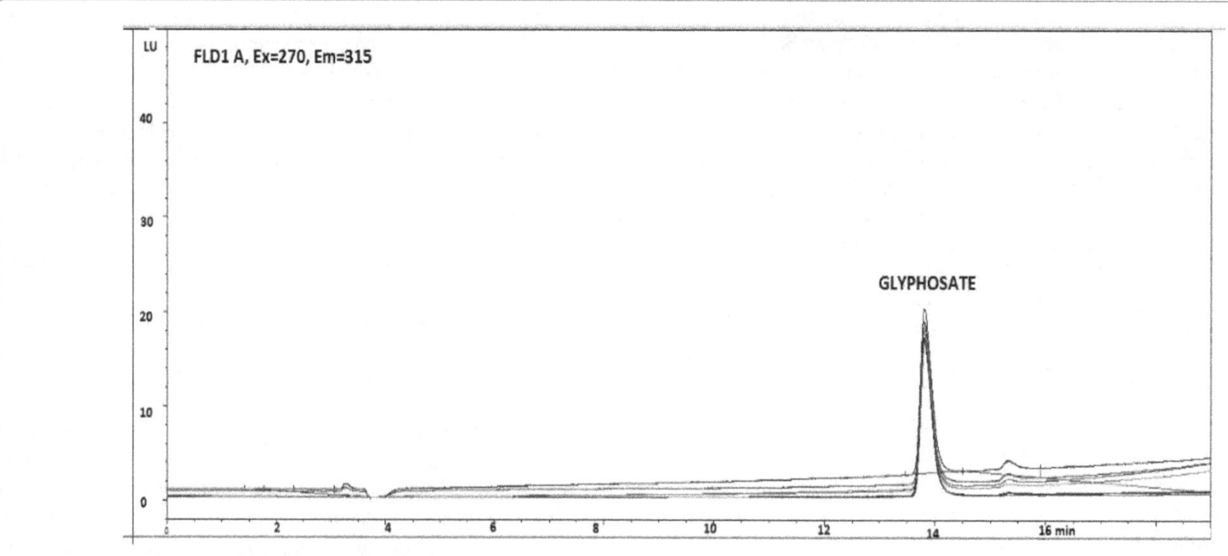

Figure 2: Overlay chromatogram of eight glyphosate standards at same concentration (0.5 µg/mL), with molar ratio of borate buffer, acetone and FMOC-Cl at 1:1:0.5.

Time (hour)	Area	Retention Time (min)	Intensity of peak
0.5	256.5399	13.82	17.50582
1	255.9935	13.821	17.16232
1.5	253.7324	13.823	17.07647
2	255.5514	13.825	16.17608
2.5	253.4689	13.814	16.58685
3	254.0001	13.81	16.36814
3.5	254.0706	13.819	16.63254
4	254.5403	13.836	16.16225
RSD (%)	0.4485	0.056	2.9453

Table 1: Optimum derivatization period for pre-column analysis (n=3).

compare the percentage of recovery. The result from the Table 2 shows that both solvents can be used in the extraction of glyphosate in palm oil samples, in which the percentage of recoveries ranged from 70-100%. The use of chloroform solvent gave a percentage of recovery at 74-83% with RSD range between 1.34% and 7.8% in contrast to the use of dichloromethane, which gave the percentage recoveries between 87 and 100% with RSD ranged from 1.11%-1.87%.

In addition, Chen et al. [15] agreed that the use of dichloromethane to replace the chloroform usage can reduce carcinogenic risk. Thus, dichloromethane was selected as the extraction solvent for this extraction study. According to Gelsomino et al. [16], the use of dichloromethane and acetone in extraction can completely remove the co-extractive hydrophilic that interferes with the analysis because of its properties. Dichloromethane was also used to separate and trap the lipid from the extract by separating the lipids into the dichloromethane layer. Without dichloromethane, oil droplets were still observed in the aqueous layer. Hence, centrifugation for 30 min after vortex for 5 min was necessary to ensure that all analytes were partitioned into the aqueous layer.

Furthermore, the method for extracting glyphosate in food matrices with high fat content was carried out by using DCM and water. Glyphosate with its polar nature dissolves in water, while the long chain fatty component dissolves in DCM solvent. This is because; the fat component with high hydrocarbon chains is likely to dissolve in non-polar solvents, such as DCM [14]. Therefore, based on the results and observations obtained, a simple and efficient method of extraction for determining glyphosate residues in CPO and CPKO was successfully developed.

Method validation

Linearity: Typically, an analytical instrument must be determined of its efficiency level, linearity, and repeatability of injections before the instruments can be used for analysis. This is to ensure that the instruments are in good condition [6,7]. Level of efficiency, linearity, and reproducibility for HPLC-FLD injections were determined by injecting a series of standard solution of glyphosate from 0.01, 0.05, 0.08, 0.1, 0.3, 0.5, to 1.0 µg/mL to sketch a calibration curve. Glyphosate standard calibration curve was prepared by plotting the peak area of the chromatogram as the y-axis against the concentration injected into the HPLC-FLD as the x-axis (Figure 3). The equation derived from the standard calibration curve was y=572.03x+3, and the regression coefficient (R^2) was=0.9998. This showed that the response of the HPLC-FLD detector to glyphosate residues had been very good where the linearity factor was at R^2>0.999.

Limit of Detection (LOD) and Limit of Quantification (LOQ): Anon [17] reported that limit of detection (LOD) is necessary to verify the analytical method that had been developed. LOD is also defined as minimum concentration at which the analyte can be identified and reported at the level of 99%, where the analyte concentration exceeds zero and it is determined from the analysis of samples containing the analyte [18]. Moreover, LOD statistically determines or explains the measurement of compound analyte carried out by using analytical protocols to distinguish the measurement control (blank) with interference (background noise) [19]. On the other hand, limit of quantification (LOQ) is the result obtained at a certain confidence level that is greater than the quantitative results. Therefore, LOD and LOQ were determined for glyphosate in this study. The method used in this study to determine the LOD and the LOQ was based on the methodology proposed by the EPA, as reported by Corley [20] and EPA [21] For the determination of LOD for all the methods developed, three lowest concentrations expected, which were 0.01, 0.03, and 0.05 ppm, had been added into all the crude palm oil and crude palm kernel oil samples. Each concentration was then analyzed with seven replicates with a control (blank sample) to ensure the consistency of the instrument. The analysis of both CPO and CPKO showed that the

Glyphosate	0.1 µg/g		0.5 µg/g		1.0 µg/g	
	Recovery (%)	RSD (%)	Recovery (%)	RSD (%)	Recovery (%)	RSD (%)
Dicholoromethane (DCM)	92	1.47	87	1.12	90	1.87
Choloroform	79	1.34	83	5.76	74	7.80

Table 2: Recoveries of glyphosate with two different solvents (n=3) at three different concentrations (0.1 µg/g, 0.5 µg/g and 1.0 µg/g).

values of LOD and LOQ were 0.01 µg/g and 0.05 µg/g, respectively. Based on these results, it can be concluded that the developed method had been suitable for detecting residues of glyphosate in palm oil.

Recovery and precision: According to Anon [22], the percentage of recoveries value accepted at the global stage is between 70% and 110% with the RSD value <20% in certain conditions. APVMA [23], on the other hand, outlined several criteria that need to be considered in determining the percentage of recovery, such as sample matrix, sample processing procedure, and analyte concentration. The accuracy and the precision of the methods were tested by adding five series of standard concentration of glyphosate into CPO and CPKO. Both CPO and CPKO were then prepared by using the optimized method before they were ready to be analyzed with HPLC-FLD. The percentage of recovery obtained from the five replicates for each concentration in CPO and CPKO are shown in Tables 3 and 4. The percentage recovery of glyphosate in CPO added with concentration of 0.05 - 1.0 ug/g were 85-90%, while for CPKO sample, it was 87-92%. Meanwhile, the percentage of RSD obtained for CPO was 1.13 - 2.47%, whereas for CPKO, it was 0.04-0.90%. As the percentages of recovery achieved for both CPO and CPKO had been above 80% with % RSD less than 3%; the method developed had been satisfactory and could be repeated.

Level of Spiking (µg/g)	Mean Recovery (%)	RSD (%)
0.05	**86**	2.47
0.08	**97**	1.18
0.3	**85**	1.81
0.5	**87**	1.13
1.0	**90**	1.87

Table 3: Recovery and statistical data obtained from analysis of glyphosate in CPO samples (n=5).

Level of Spiking (µg/g)	Mean Recovery (%)	RSD (%)
0.05	**87**	0.44
0.08	**91**	0.45
0.3	**90**	0.90
0.5	**88**	0.61
1.0	**92**	

Table 4: Recovery and statistical data obtained from analysis of glyphosate in CPKO samples (n=5).

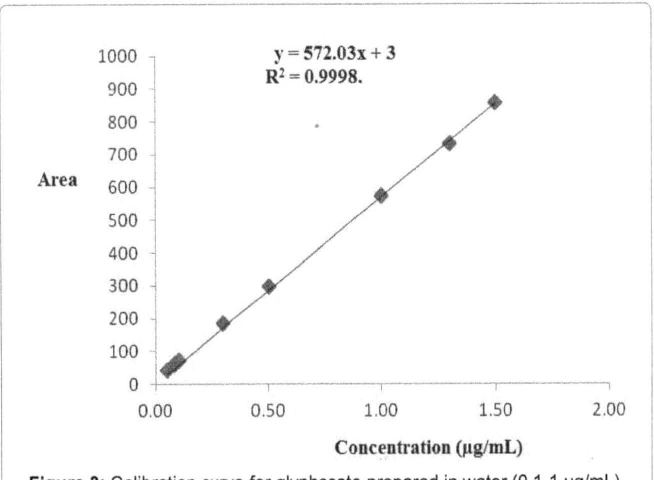

Figure 3: Calibration curve for glyphosate prepared in water (0.1-1 µg/mL).

Conclusion

Method development for determination of glyphosate in palm oil matrices had been conducted via HPLC-FLD optimization with three parameters, which were selection of mobile phase, FLD wavelength, and flow rate of mobile phase. From the results obtained, it had been discovered that the use of acetonitrile and KH_2PO_4 with a flow rate at 1.0 mL/min was the most suitable, while the selected FLD wavelength was at excitation of 370 nm and emission of 415 nm. Besides, optimization of HPLC-FLD was carried out to determine the best operation condition to achieve the most apt glyphosate chromatogram in the analysis of glyphosate residue.

Moreover, the accuracy and the precision of the method had been based on the percentage of recovery and the percentage of relative standard deviation. For CPO, the recovery and % RSD obtained were in the range of 85-97% and 1.1-2.5%, meanwhile 87-92% and 0.04-0.9% for CPKO, respectively. Besides, repeatability of the method was measured with percentage of recovery and % RSD via intra-laboratory test. The percentage of recovery and % RSD obtained from both the operators were in the range of 91-96% and 3.3-4.5% for CPO, while 90-95% and 0.5- 1.3% for CPKO.

Meanwhile, the values of limit of detection (LOD) and limit of quantification (LOQ) obtained for CPO and CPKO were 0.01 and 0.05 ug/g for both, respectively. This proved that the analytical method developed had been precise, simple, accurate, and efficient. Therefore, this method had been determined as suitable to be used to analyze and monitor glyphosate residue in CPO and CPKO.

Acknowledgements

The authors would like to thank the Malaysian Palm Oil Board (MPOB) for financial support and the Director General of the MPOB for permission to publish this paper. The authors are also indebted to Puan Rosilawati and Puan Rosliza for their excellent technical assistance and discussion.

References

1. Ainie K, Tan YA, Norman K, Yeoh CB (2007) Pesticide Application in oil palm plantation. Oil Palm Bulletin 54: 52-67.
2. Gibon V, De Greyt, Kellens M (2007) Palm Oil Refining. European Journal of Lipid Science and Technology 109: 315-335.
3. Nuzul Amri I, Ainie K, Tang TS, Siew WL (2003) Current status of Malaysian crude palm kernel oil characteristics. Oil Palm Bulletin 47: 15-27.
4. Halimah M, Osman H, Ainie K, Tan YA, Md Pauzi A (1999) Determination of chlorpyrifos in refined palm olein by GC-FPD and GC-ECD. Journal of Oil Palm Research 11: 89-97.
5. Cloborn VH, Mann HD, Oehler DD (1968) Dursban determination in milk and body tissues of cattle. J AOAC int 51: 1243-1245.
6. Halimah BM (2000) Pembangunan kaedah analisis residu klopirifos di dalam minyak sawit. Tesis Sarjana, Pusat Pengajian Sains Kimia & Teknologi Makanan, Fakulti Sains & Teknologi, Universiti Kebangsaan Malaysia.
7. Halimah BM (2006) The environmental fate of of fluroxypyr and chlorpyrifos in oil palm agroecosystem. Tesis Ijazah Doktor Falsafah, Pusat Pengajian Sains Kimia & Teknologi Makanan, Fakulti Sains & Teknologi, Universiti Kebangsaan Malaysia.

8. Cesnik BH, Gregorcic A (2006) Validation of the method for the determination of dithiocarbamates and thiram disulphide on apples, lettuce, potato, strawberry and tomato matrix. Acta Chemica Slovenica 53: 100-104.

9. Blasco C, Font G, Pico Y (2004) Determination of dithiocarbamate and metabolites in plants by liquid chromatography-mass spectrometry. Journal of Chromatography A 1028: 267-276.

10. Walia S, Sharma RK, Parmar BS (2009) Isolation and simultaneous LC analysis of thiram and its less toxic transformation product in DS formulation. Bulletin Environmental Contamination Toxicology 83: 363-368.

11. Bo L, Xiaojun D, Dehua G, Shuping J (2007) Determination of glyphosate and aminomethylphosphonic Acid Residues in Foods Using High Performance Liquid Chromatography-Mass Spectrometry/Mass Spectrometry. Chinese Journal of Chromatography 25: 486.

12. Hanke I, Singer H, Hollender J (2008) Ultratrace-level determination of glyphosate, aminomethylphosphonic acid and glufosinate in natural water by solid-phase extraction followed by liquid chromatography-tandem mass spectrometry: performance tuning of derivatization, enrichment and detection. Analytical and Bioanalytical Chemistry 391: 2265-2276.

13. Nedelkoska TV, Low KCG (2004) High performance liquid chromatography determination of glyphosate in water and plant material after pre-column derivatisation with 9-fluorenylmethyl chloroformate. Analytica Chimica Acta 511: 145.

14. Gilbert-Lopez B, Garcia-Reyes JF, Molina-Diaz A (2009) Sample treatment and determination of pesticides residues in fatty vegetable matrices. Talanta 79: 109-128.

15. Chen IS, Shen CSJ, Sheppard AJ (1981) Comparison of methylene chloride and chloroform for the extraction of fats from food products. JAOCS, pp: 599-601.

16. Gelsomino A, Petrovicova B, Tiburtini S, Magnani E, Felici M (1997) Multiresidue analysis of pesticide in fruits and vegetables by gel permeation chromatography followed by gas chromatography with electron-captured and mass spectrometry detection. Journal of chromatography A 782: 105-122.

17. Anon (2002) Method detection limit (MDL) and reporting limit (RL/DLR) requirements for ELAP certification.

18. USEPA (1994) Method detection limit-Definition in in 40 CFR. Part 136, Appendix B, Washington: USEPA.

19. Anon (1996) Analytical detection limit guidance and laboratory guide for determining method detection limits.

20. Corley J (2003) Best practices in establishing detection and quantification limit for pesticide residues in food. Philip WL, Aldos CB, John JM (eds.), Handbook of Residue Analytical Methods for Agrochemicals. USA: John Wiley & Sons Ltd.

21. EPA (Environmental Protection Agency) (1996) Results of the EPA method 1631 validation study.

22. Anon (2008) European Commission. Guidance document on residue analytical methods.

23. APVMA (Australian Pesticides and Veterinary Medicines Authority) (2004) Guidelines for the validation and analytical methods for active constituent, agricultural and veterinary chemical products.

Gas Chromatographic Assessment of Residual Solvents Present in Excipient-Benzyl Alcohol

Anumolu PD*, Krishna VL, Rajesh CH, Alekya V, Priyanka B and Sunitha G

Department of Pharmaceutical Analysis, Gokaraju Rangaraju College of Pharmacy, Osmania University, Hyderabad, Telangana, India

Abstract

Purpose: This article describes a simple and rapid gas chromatographic method for identification of residual solvents present in benzyl alcohol. Benzyl alcohol is used in foods and pharmaceutical products. The organic solvents such as benzene, chlorobenzene and toluene are frequently used in manufacturing of benzyl alcohol. Even after such manufacturing process, some solvents still remain in small quantities.

Methods: Method for the quantification of residual solvents present in benzyl alcohol was done by gas chromatography with flame ionization detector and utilizes the Agilent 7700 with FID (DB-624, 30 m × 0.53 mm, 3 µ) capillary column, nitrogen as carrier gas with a flow rate of 2.5 ml mn^{-1}. The critical experimental parameters such as oven temperature, zero air, make up flow; injection volume, split ratio and the selection of diluent were studied and optimized.

Results: The retention time of various residual solvents taken individually and in spiked standard solutions were determined as 8.824, 13.467 and 11.461 min for benzene, chlorobenzene and toluene respectively. The proposed method was applied for the quantification of residual solvents present in marketed benzyl alcohol and was statistically validated as per standard guidelines.

Conclusion: The results obtained from validation proved that the proposed method was scientifically sound. The proposed method was fruitfully applied for the quantification of residual solvents (which are involved in the manufacturing process) present in benzyl alcohol.

Keywords: Benzyl alcohol; GC; FID; Residual solvents

Introduction

Organic volatile chemicals may be present in pharmaceutical products (or) excipients, which are involved in the manufacturing/ synthesis process, but they are not desirable in the final product due to their toxicity, non therapeutic effect and may interfere with the quality of the pharmaceutical substance. These organic solvents still remain in pharmaceutical products even after subjecting to various manufacturing process (or) techniques. These types of solvents are called as residual solvents. As per standard guidelines, quantification of residual solvents in pharmaceutical substances is enforced for their release testing. Now days, there is an increase in the stipulate for analytical techniques, which are helpful to control the minimum quantity of organic solvents in final pharmaceutical products/ excipients by identifying /quantifying them. Benzyl alcohol is a colorless liquid with a mild pleasant aromatic odor. It is partially soluble in water and completely miscible in alcohols and diethyl ether. Benzyl alcohol is used in foods and pharmaceutical products it is also a precursor to a variety of esters, used in the soap, perfume and flavor industries. It has been used as a dielectric solvent for the dielectrophoretic reconfiguration of nano wires and as a bacteriostatic preservative at low concentration in intravenous medications, cosmetics and topical drugs. The organic solvents such as benzene, chlorobenzene and toluene are frequently used in manufacturing of benzyl alcohol [1,2]. Even after such manufacturing process of benzyl alcohol, some solvents still remain in small quantities, which are potential toxic to humans. All these points signify that the need for some efforts for quantification of residual solvents present in benzyl alcohol. For this investigation purpose, we made several trials and finally optimized the gas chromatographic conditions, which is the most popular technique to quantify the residual solvents in benzyl alcohol.

Extensive literature review revealed that few analytical methods are available for quantification of residual solvents present in some pharmaceutical drug substances and excipients [3-14] by gas chromatography which is the most useful technique to quantify residual solvents. It also accomplished that, there is no analytical method available for the quantification of residual solvents present in benzyl alcohol.

Keeping these points into deliberation, we contemplated a method with an objective of development and validation of the simple gas chromatographic method for the quantification of residual solvents present in benzyl alcohol.

Materials and Methods

Materials

Benzyl alcohol, benzene, chlorobenzene, methanol and toluene (HPLC grade) were purchased from Merck chemicals Co (Mumbai). DMSO (AR grade) was purchased from Merck chemicals Co (Mumbai).

Instrumentation and operating gas chromatographic condition

Gas chromatograph Azilent technologies 7700 equipped with flame ionization detection, standard oven for temperature ramping, and

***Corresponding author:** Panikumars D Anumolu, Department of Pharmaceutical Analysis, Gokaraju Rangaraju College of Pharmacy, Osmania University, Hyderabad-500 090, Telangana, India, E-mail: panindrapharma@yahoo.co.in

split/split less injection port. The analytes of interest were separated on a DB-624 capillary column (30 m × 0.53 mm, 3.0 µm film thickness) with nitrogen as carrier gas. An analytical balance (AVW 220D from Sartorious) and ultra-sonicator (RK-106 from Bandeliasonorex) were used for this investigation purpose.

Methods

Preparation of standard stock solution: Benzene 50 mg, chloro benzene 50 mg and toluene 50 mg were accurately weighed and transferred into 100 mL volumetric flask containing 50 mL of diluent, dissolved and diluted up to the mark with diluent.

Preparation of standard solution: Accurately transferred 1 mL of standard stock solution into 10 mL volumetric flask containing 5 mL of diluent, dissolved and diluted up to the mark with diluent.

Preparation of sample solution (assay of commercially available benzyl alcohol): The proposed method was evaluated by the assay of commercially available Benzyl alcohol (Hi-media) for the quantification of residual solvents present in it. The results obtained were compared with the corresponding specification limits of standard guidelines. Accurately weighed 5.0 g of sample was transferred into 10 mL volumetric flask containing 2 mL of diluent, dissolved and diluted up to the mark with diluent.

Procedure for calibration plot

Condition the column for 2 hours at 200°C column oven temperature before starting the analysis. The linearity of analytical method was established at LOQ to 150% of specification level concentration. A series of standard solutions were prepared of different concentrations at LOQ, 50%, 80%, 100%, 120% and 150% from standard solution to 10 mL with diluent to obtain the required concentration. Inject the blank as diluent once, six replicate injections of standard solution and then inject the sample solution. Inject these solutions into the GC system and record the area of solvent peak. Contrive the graph of concentration (at X-axis) versus average peak of solvent (at Y-axis) and assess the correlation coefficient (r), slope and Y-intercept.

Method validation

The method was validated according to the ICH guidelines and all validation parameters useful to evaluate the overall performance of analytical method were investigated including specificity, linearity, precision, accuracy, and robustness, limit of detection and limit of quantification [15].

Results and Discussion

Residual solvents, which may be present in the final pharmaceutical products /excipients, are not desired due to their probable toxicity and their quantification in pharmaceutical products is mandatory as per standard guidelines to release into the market. Benzyl alcohol is used in foods and pharmaceutical products. It may contain some residual solvents even after the last part of manufacturing process. These facts stand for the need of some efforts to quantify the residual solvents in benzyl alcohol. Hence present inference was undertaken with an objective of developing GC- analytical method for identification of residual solvents present in benzyl alcohol. Several trials were performed to optimize the most suitable chromatographic conditions which are having the ability of quantifying the residual solvents. Optimized chromatographic conditions includes nitrogen as a mobile phase with flow rate of 2.5 ml min^{-1}, DB-wax (300 m × 0.53 mm, 3 µm) as a column and other operating gas chromatographic conditions are specified in Table 1. The retention times with good resolution were observed as 8.824, 13.467 and 11.461 min for benzene, chloro benzene and toluene respectively. The chromatographs are shown in Figure 1.

Method validation

The method was validated for all the residual solvents under the study undertaken for specificity, linearity, precision, accuracy, LOD and LOQ, robustness and system suitability as per ICH guidelines.

The method was validated for all the residual solvents under the study for their specificity, linearity, precision, accuracy, LOD and LOQ, robustness and system suitability as per ICH guidelines.

Specificity: Specificity of the method was determined by injecting blank solution under the same experimental conditions and no peak interference at the retention time of the each solvent in the standard and sample solution, indicating that the method is specific.

Linearity: Linearity of the method was determined over the concentration range of LOQ, 50% 80% 100% 120% and 150%. Correlation coefficient (R^2), slope and Y-intercept were calculated from Linearity data and shown in Table 2.

Limit of detection (LOD) and limit of quantification (LOQ): The LOD and LOQ for the proposed method were determined using calibration standards and calculated using 3.3 σ /s and 10 σ/s formulae respectively, where s is the slope of the calibration curve and σ is the standard deviation of y- intercept of the regression equation. Results are shown in Table 2.

Precision: System precision and method precision was determined by injecting six replicate injections of standard solution and sample solution respectively and analyzed as per ICH guidelines. Intermediate precision was carried out to demonstrate the reproducibility of sample results obtained by the analytical method for the variability

Parameters	Condition
Column	DB-624, 30 m × 0.53 mm, 3.0 µm
Column Flow	2.5 ml min^{-1}
Carrier Gas	Nitrogen
Split Ratio	1:2
Injector Temperature	180°C
Injection Volume	2.0 µL
Detector	FID
Detector Temperature	240°C
Zero Air	300 ml min^{-1}
Hydrogen gas	30 ml min^{-1}
Makeup Flow	25 ml min^{-1}
Diluent	DMSO

Table 1: Optimized gas chromatographic conditions.

Parameter	Benzene	Toluene	Chloro benzene
Linearity range (ppm)	0.9-3.2	3.7-162.7	2.5-153.3
Regression equation	Y=4.525x+0.142	Y=46.34x+33.60	Y=3.5661x+3.7082
Correlation coefficient	0.996	0.9993	0.9997
Slope	4.5257	46.34	3.5661
Intercept	0.1424	33.60	3.7082
LOD (µg ml^{-1})	0.3	1.2	0.8
LOQ (µg mL^{-1})	0.9	3.7	2.5

Table 2: Linearity plot details of three residual solvents.

of instrument, column (serial number/ lot number), analyst and day. Peak responses for each solvent peak were measured and %RSD was calculated. The %RSD of each solvent from the six preparations of system precision should not be more than 15.0 and obtained results were within the acceptable range. The data of this study were summarized in Table 3.

Accuracy: Accuracy of the methods was assured by applying the standard addition technique. The % recovery was calculated. The mean % recovery of each solvent at LOQ level should be not less than 70.0 and not more than 130.0, at 50%, 100% and 150% level should not be less than 80.0 and not more than 120.0. Results obtained were within the limits indicating the method as accurate and are shown in Table 4.

Robustness: This study was performed by making small and deliberate variations in the method parameters. The variation in the column flow ± 0.1 ml min-1, column oven temperature ± 5°C was done and the results obtained were within the acceptance criteria indicating the method is robust within the specified range. % RSD values were less than 15% as shown in Table 5.

System suitability: System suitability was evaluated by injecting six replicates of standard solution into the chromatographic system as per the test method and solvents peak area was measured.% RSD was calculated and the results revealed that the system meets the required system suitability. Results are summarized in Table 6.

Application of the proposed method (Analysis of commercially available benzyl alcohol): The proposed method was evaluated by the assay of commercially available benzyl alcohol (Hi-media) for quantification of residual solvents present in it. The results obtained for residual solvents were compared with the corresponding specification limits of standard guidelines and reported in Table 7. This revealed that concentration of residual solvents present in benzyl alcohol in ppm levels which were less than the specified limits.

Conclusion

As per standard guidelines, the quantification of residual solvents is mandatory for all the pharmaceutical products and excipients before their release into the market because of its potential toxicity and may be the interference with the quality. Based on this point, we developed a method with an objective of development and validation of simple analytical method for simultaneous quantification of residual solvents present in excipients- benzyl alcohol by gas chromatography with flame ionization detector (which is powerful tool to quantify the residual solvents). The residual solvents benzene, chlorobenzene and toluene were well separated from each other and quantified by the proposed method. This method was also applied for the quantification of residual solvents in the marketed benzyl alcohol, which were present in ppm specification limits as per standard guidelines. The proposed method was validated as per the standard guidelines and the results revealed that the method was scientifically sound. Finally we conclude that the proposed method can be effectively applied for the quantification of residual solvents present in benzyl alcohol. This investigation may be helpful to the manufacturers for controlling and minimization of the residual solvents.

Acknowledgements

The authors are thankful to Gokaraju Rangaraju College of Pharmacy and Startech Labs Private Limited, Hyderabad for providing the facilities for this research work.

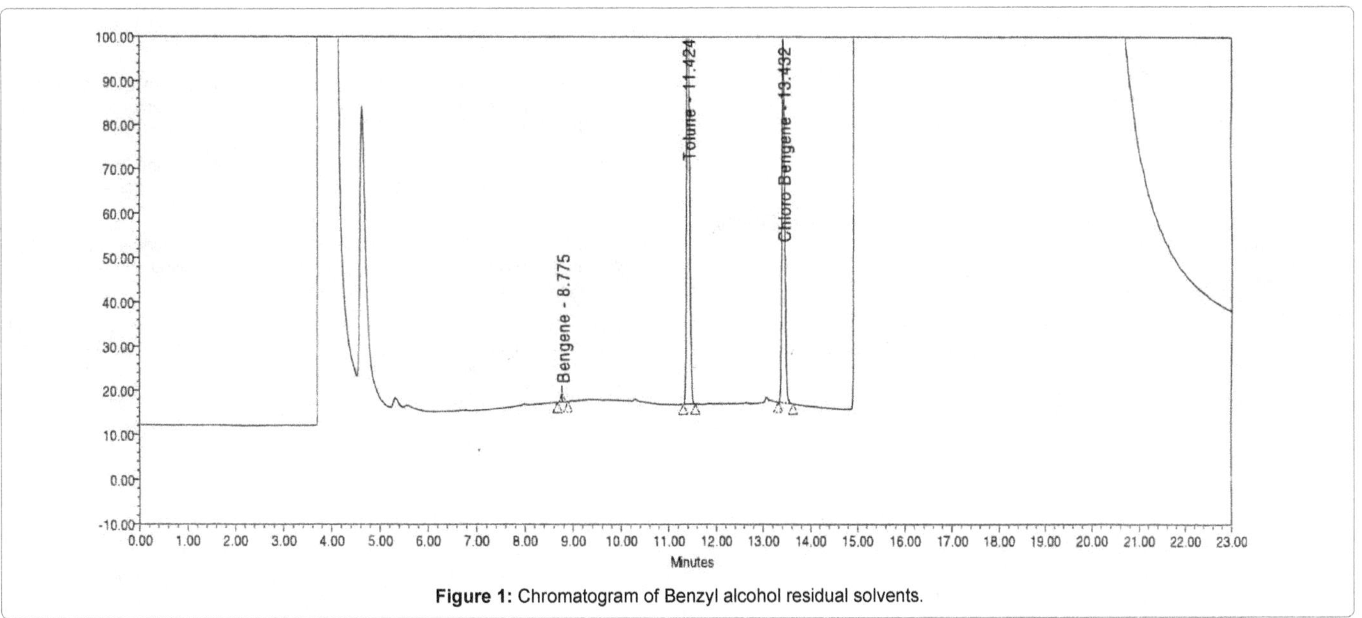

Figure 1: Chromatogram of Benzyl alcohol residual solvents.

Solvent name	System Precision		Method Precision		Intermediate Precision		
	Cocentration Mean[a] (ppm)	%RSD[b]	Concentration Mean[a] (ppm)	%RSD[b]	Mean[a] (ppm)		%RSD[b]
					Analyst 1	Analyst 2	
Benzene	12.55	1.5	2.0	4.9	2.0	2.1	5.9
Toulene	624.04	0.8	103.8	1.5	103.8	102.3	1.4
Chloro benzene	434.02	0.7	96.5	1.5	96.5	101.5	2.9
[a]Mean of 6 determinations							
[b]Relative standard deviation							

Table 3: Precision data of proposed method.

Solvent name	Parameter	Spike levels			
		LOQ	50s%	100%	150%
Benzene	Conc. Of solvent (ppm) Conc. Of spiked solvent (ppm) % Mean Recovery[a]	0.9 0.9 100.0	1.1 1.0 107.0	2.3 2.1 109.5	3.5 3.1 109.6
Toluene	Conc. Of solvent (ppm) Conc. Of spiked solvent (ppm) % Mean Recovery[a]	3.5 5.0 122.1	51.8 55.3 104.4	103.7 107.7 102.8	155.5 166.1 104.8
Chloro benzene	Conc. Of solvent (ppm) Conc. Of spiked solvent (ppm) % Mean Recovery[a]	2.6 2.7 103.8	51.6 52.2 101.6	103.2 103.2 100.0	154.7 159.3 104.0

Mean of 6 determinations

Table 4: Accuracy data of proposed method.

Parameter	% RSD[a]		
	Benzene	Toluene	Chlorobenzene
Effect of Variation Flow			
2.4 ml/min	3.6	0.6	0.7
2.5 ml/min	1.5	0.8	0.7
2.6 ml/min	2.6	1.7	1.7
Effect of Variation in Column oven temperature			
55°C	4.2	2.7	1.9
60°C	1.5	0.8	0.7
65°C	3.2	2.2	1.7
[a]Relative standard deviation (n=6)			

Table 5: Robustness data.

Parameter	Benzene	Toluene	Chloro benzene
% RSD	2.6	1.7	1.9
Acceptance criteria	Not more than 15.0		

Table 6: System suitability.

Solvent name	R_t	Area	Amount found (ppm)
Benzene	8.775	13	2.8
Toluene	11.424	434	134
Chloro benzene	13.432	624	120

Table 7: Assay results of commercially available Benzyl alcohol.

References

1. Maryadele J, O'Neil O (2006) The Merck index, an encyclopedia of chemicals, drugs, and biological. 13th edition. Merck research laboratories, USA: Division of Merck & Co Inc., p: 1138.
2. ICH Q3C (R5) (2005) ICH Harmonized tripartite guideline, Impurities in New Drug Products, International Conference on Harmonization ICH, Geneva.
3. Pugazhendhy S, Sivasaikiran B, Chowdary YN, Sreelakshmi V, Shrivastava SK (2014) Development and validation of a headspace gas chromatographic method for determination of residual solvents in bosentan monohydrate. Int J Pharm Tech Resm 6: 421-427.
4. Mishra G, Saxena V, Jawla S, Kumar VS (2014) Method development and validation for the determination of residual solvents in omeprazole by using head space gas chromatography. Asian J Pharm Clin Res 7: 54-56.
5. Yadav MR, Patil K (2013) Development and validation of gas chromatographic analytical method for estimation of residual solvents in excipients. Int J Res Pharm Bio Sci 4: 618-623.
6. Pugazhendhy S (2013) Determination by head space gas chromatography with flame ionization detector in Ropinirole API. Int J Res Pharm Bio Sci 4: 227-230.
7. Paul SD, Mazumder R, Bhattacharya S, Jha AK (2013) Method development and validation for the determination of residual solvents in ophthalmic nanoparticle suspension. WJPPS 2: 5802-5810.
8. Ramos CS (2013) Development and validation of a headspace gas chromatographic method for determination of residual solvents in five drug substances. Int J Pharm Sci Inve 2: 36-41.
9. Kumar AV, Aravind G, Srikanth I, Srinivasarao A, Dharma Raju Ch (2012) Novel analytical method development and validation for the determination of residual solvents in amlodipine besylate by gas chromatography. Der Pharma Chemica 4: 2228-2238.
10. Jahnavi N, Saravanan VS (2012) Method development and validation for the determination of residual solvents in methocarbamol pure drug by HS-GC. Int J Res Pharm Chem 2: 456-466.
11. Reddy BP, Reddy MS (2009) Residual solvents determination by HS-GC with flame ionization detector in omeprazole pharmaceutical formulations. Int J PharmTech Research 1: 230-234.
12. Saurabh P, Preeti P, Kumar R, Pai NS (2011) Residual solvent determination by head space gas chromatography with flame ionization detector in omeprazole API. Brazillian J Pharma Sci 47: 380-384.
13. Manish K, Suman L (2013) A Review on residual solvents and various effective gas chromatographic techniques in the analysis of Residual solvent. IJPRR 2: 25-40.
14. Kumar PB, Singh RP, Saurabh A (2009) Simultaneous estimation of Residual solvents (Isopropyl alcohol and Dichloromethane) in dosage form by GC-HS-FID. Asian Journal of Chemistry 21: 1739-1746.
15. Revathi R, Ethiraj T, Thenmozhi P, Saravanan VS, Ganesan V (2011) High performance liquid chromatographic method development for simultaneous analysis of doxofylline and montelukast sodium in a combined form. Pharm Methods 2: 223-228.

8

Direct Determination of N-Nitrosodiethanolamine (NDELA) in Ethanolamines by LC-MS-MS

George Kuriakose[1]*, Raghunandana KS[2] and Saeed Al-Shahrani[1]

[1]*SABIC Plastic Application Development Center, King Saud University Campus, Riyadh, KSA 12373, Saudi Arabia*
[2]*SABIC Technology Center-Jubail, P.O. Box 11669, Al-Jubail, 31961 Saudi Arabia*

Abstract

A sensitive and direct method for quantification of N-nitrosodiethanolamine (NDELA) in ethanolamine has been developed. The sample was dissolved in water and directly injected to high performance liquid chromatography with mass spectrometry to quantify NDELA. The eluent was transferred to mass spectrometer and ionized by heated electrospray positive ionization. The analyte was quantified in parts per billion level by using multiple reaction monitoring (MRM) mode. The MRM transition was *m/z* 135.0 to 74.1 and confirmed by *m/z* 104.0. To avoid ion suppression and better sensitivity for quantification, cation exchange column was used. A linearity coefficient of R^2 >0.995 was observed in calibration over a wide range of concentration 1 ppb to 10 ppb of NDELA. The method was validated by linearity, accuracy and precision studies. The results show that recoveries were in between 97% to 106%.

Keywords: N-nitrosodiethanolamine (NDELA); LCMS; MRM; Method validation; Ethanolamine

Introduction

Ethanolamines (monoethanolamine (MEA), diethanolamine (DEA) and triethaolamine (TEA)) are a group of chemicals combine the properties of both amine and alcohol. These ethanolamines are bifunctional compounds allowing for wide spread use in the chemical industry as a components in detergents, pharmaceuticals and cosmetics [1,2]. Ethanolamines are hydrophilic in nature and most of the alcohols making them ideal for use as surfactants, either alone or esterified. N-nitrosodiethanolamine (NDELA, Figure 1) is formed by the action of nitrosating agents such as nitrites and nitrogen oxides on diethanolamine and triethanolamine [3]. The rate of formation of NDELA in ethanolamines is pH, temperature and time-dependent. Tertiary amines nitrosate at a slower rate than secondary amines because the reaction involves a nitrosative dealkylation rate-limiting step to yield a secondary amine which is then available for further nitrosation [4].

Ethanolamines are mainly used in personal-care products, detergents and one of the potential health contaminant is NDELA. N-nitrosodiethanolamine has been recognized as a class of hazardous compounds and considered as more toxic in more animal species than any other category of chemical carcinogen. It has been proven that, NDELA can induce cancer in experimental animals [5,6]. As considering the toxicity of NDELA, in parts per billion level quantitation is very critical to qualify ethanoloamines used for personal-care products. There are a great number of scientific papers reported in the literature for qualitative and quantitative estimation of NDELA in different matrixes [7-9]. In general, these quantification methods recommend the multi-step extraction of nitrosamines from the matrix by extraction methods, including distillation (steam, vacuum, or atmospheric), solvent extraction, solid-phase extraction, autoclave extraction, and supercritical fluid extraction [10]. Most of the methods like colorimetry, spectrophotometry, polarography, capillary electro-chromatography, micellar electro-kinetic capillary chromatography, gas chromatography with flame ionization detection, nitrogen phosphorous detection, thermal energy detection, nitrogen chemiluminescence detection, mass spectrometry detection, high-performance liquid chromatography with thermal energy analyzer, mass spectrometry and fluorescence detection, comprise two or more clean-up steps derivatizations, tedious sample preparation etc. [11,12].

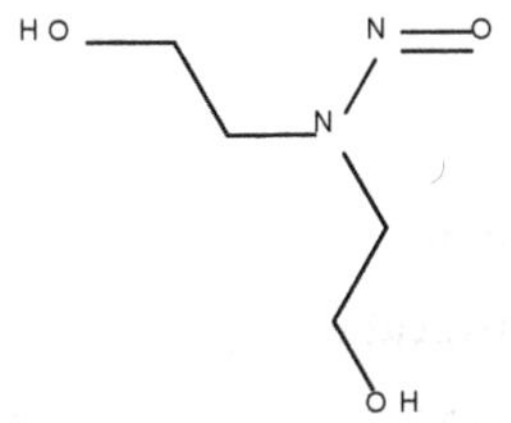

Figure 1: N-nitrosodiethanolamine.

The aim of the work was to develop a simple liquid chromatography mass spectrometry method for direct estimation of NDELA from ethanolamines matrix at <10 ppb level. The required sensitivity for the method was <1 ppb level for neat NDELA standard prepared in water. The required sensitivity was achieved by using multiple reaction monitoring (MRM) mode in LC-MS. The ion suppression due to amine matrix in electrospray positive ionization mode (ESI +Ve) was overcame by using cation exchange columns, switch over valve and followed by washing with acidic aqueous solutions. This method is simple to perform and provides an accurate and precise quantitative results for the measurement of NDELA in ethanolamines.

Materials and Methods

Chemicals and reagents

N-nitrosodiethanolamine was purchased from Sigma-Aldrich, monoethanolamine, diethanolamine and triethaolamine were received

***Corresponding author:** George Kuriakose, ABIC Plastic Application Development Center, King Saud University Campus, Riyadh, KSA 12373, Saudi Arabia, E-mail: kuriakosegk@SABIC.com

from in house synthesized SABIC technology center Riyadh. The purity of standard NDELA was greater than or equal to 90%. Water, acetonitrile, methanol and formic acid (LCMS grade) were purchased from Fisher Scientific.

Instrument and Chromatographic conditions

The method development for quantification of NDELA was performed by using an Agilent 1290 Infinity LC system and Agilent 6460 Triple Quadrupole LC/MS system with Agilent Jet Stream source. The Agilent Mass Hunter Workstation was used to capture and analyze the data. LC/MS parameters used are depicted in Table 1. During the method optimization, LC System with reverse phase chromatographic conditions with C18 column were also used and later changed to Dionex IonPac CS 18 cation exchange column. The mass spectrometry analysis was performed in electrospray positive ionization mode with spray voltage. Nitrogen was used as the sheath, auxiliary and ion sweep gas. The system was operated in selected multiple reaction monitoring (MRM) mode with argon as the collision gas with collision energy of 5V.

Preparation of calibration standards

Stock solution of NDELA (1000 µg/mL) was prepared in water and stored in a refrigerator at 5°C. The purity factor of 90% was applied to calculate the final concentration of the stock solution. From the stock solution of NDELA, the working standards (0.8 to 8 ppb.) were prepared by further dilution in water. 100 µL of this calibration standards were injected to LCMS to generate calibration curve.

Sample preparation

Ethanolamines (MEA, DEA and TEA) was prepared by dissolving in water. 2 g of samples were accurately weighed into 10 mL of standards flask and made up to the mark with water. The samples were vortexed mixed for 1 min and filtered through 0.45 µm syringe filter before injection to LCMS.

Method Validation

The method was validated as per standard LCMS testing guidelines.

LC conditions	
Column	Dionex IonPac CS18 cation exchange column
Column temperature	40°C
Mobile phase A:	0.02% Formic acid in water
Mobile phase B:	20 mmol Methane sulfonic acid in water
Flow-rate	1 mL/min
Gradient Time (min)	% A %B
0	100
5	100
7	100
31	100
32	100
45	100
Injection volumes	100 µL
Agilent 6460 Triple Quadrupole source conditions	
Gas heater	350°C
Gas flow	7 L/min
Nebulizer pressure	45 psi
Sheath gas heater	350°C
Sheath gas flow	11 L/min
Vcap	4,000 V
Nozzle voltage	1000 V
Delta EMV	300 V

Table 1: LC/MS conditions.

The quantification method is fully validated for, linearity, accuracy, precision, recovery and matrix effects.

Linearity

The linearity of different concentration levels for this quantification method was validated by using calibration curves as described in the results and discussion section. The calibration curves were generated by plotting the peak area of NDELA to the concentration in ppb. Linear equation was used to calculate the concentration of NDELA from calibration curve. The limit of quantification (LOQ) was defined as the lowest concentration of analyte where the signal-to-noise ratio (S/N) of NDELA at the LOQ exceeds the minimum requirement.

Accuracy and precision

Accuracy of the measurement was assessed by performing a recovery experiments by spiking studies. 2 g of samples (MEA, DEA and TEA) were accurately weighed into the 10 mL of standards flask, 2 ppb level of NDELA was also spiked into the ethanolamine and finally made up to the mark with water. The samples were vortexed mixed for 1 min and filtered through 0.45 µm syringe filter before injection to LCMS. Three level spiking studies were also conducted for different matrixes. Precision study was conducted using 2 ppb level spiked sample by injecting repeatedly >15 times and the data was evaluated. The mean and standard deviations were determined over the validation period and the precision was calculated.

Matrix Interference and ion suppression

To evaluate the matrix effects (MEA, DEA and TEA) a post column infusion experiment was performed using direct infusion pump. The out let of the syringe from infusion pump was directly inserted into the mass spectrometer. A standard NDELA solution and spiked ethanolamines were separately infused into the eluent stream at a flow rate of 50 µL/min. The signals of the corresponding MRM transitions of the analyte were recorded. The ethanolamine that interfered with the ionization of NDELA analyte would lead to a depression of the MRM signal which would represent matrix effects.

Results and Discussions

Chromatography conditions and MRM analysis

The liquid chromatographic parameters were optimized to maximize sensitivity of the analyte. Initially, the LC MS analysis was carried out with Eclipse XDB C-18, 5 µm, 4.6 × 150 mm column to get the sensitivity of the method and mass transitions to optimize the MS parameters. Various mobile phase compositions, flow rates, MS conditions and profiles were evaluated. The required sensitivity of the method (0.87 ppb of neat NDELA in water) was achieved by separation performed at a flow rate of 1 mL/min with elution of 0.02% HCOOH in water. The optimized mass spectrometer parameters were mentioned in Table 1. The TIC chromatogram for the blank (water) and lowest standard is depicted in Figure 2. There was no interference from blank and signal to noise ratio was >10. The MRM transition was m/z 135.0 to 74.1 and confirmed by m/z 104.0. Further the MRM conditions were optimized (Table 1) to get complete transition of precursor ion (m/z 135.0) to product ions (Figures 3 and 4). This achieved the signal to noise ration far >10, with this conditions the linearity and quantification of NDELA was performed.

Linearity and Quantification

A linear relationship was found between NDELA concentration and peak area throughout the range of 0.87 ppb to 8 ppb. The coefficient

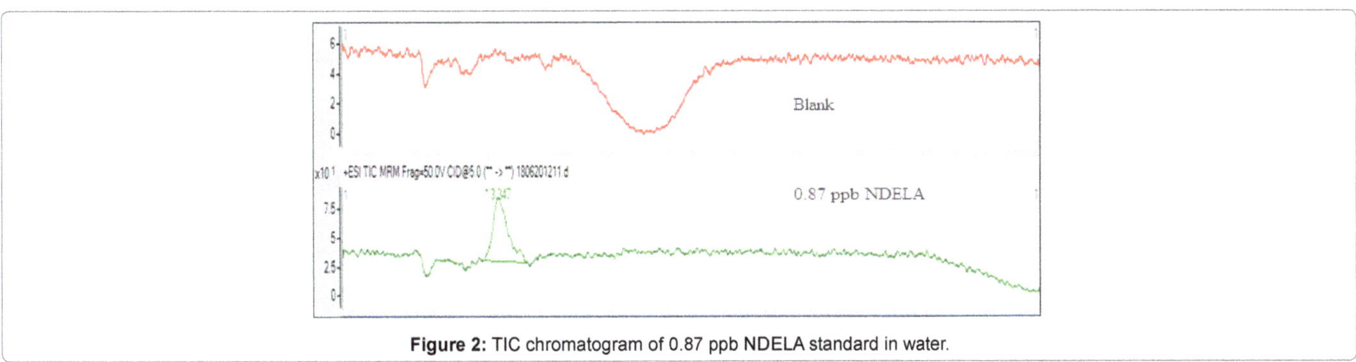

Figure 2: TIC chromatogram of 0.87 ppb NDELA standard in water.

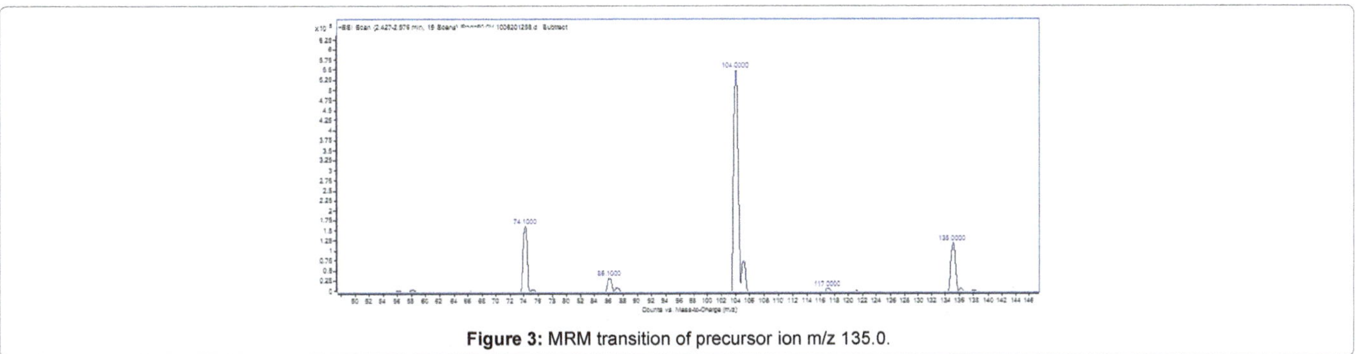

Figure 3: MRM transition of precursor ion m/z 135.0.

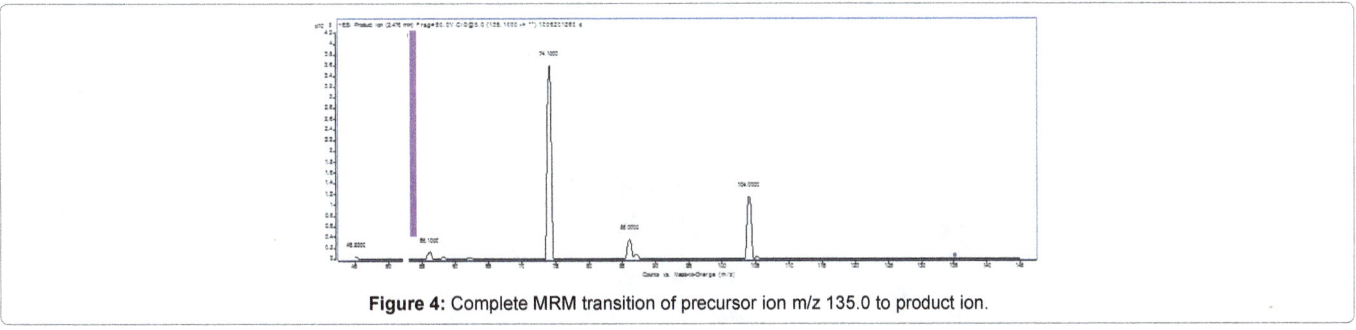

Figure 4: Complete MRM transition of precursor ion m/z 135.0 to product ion.

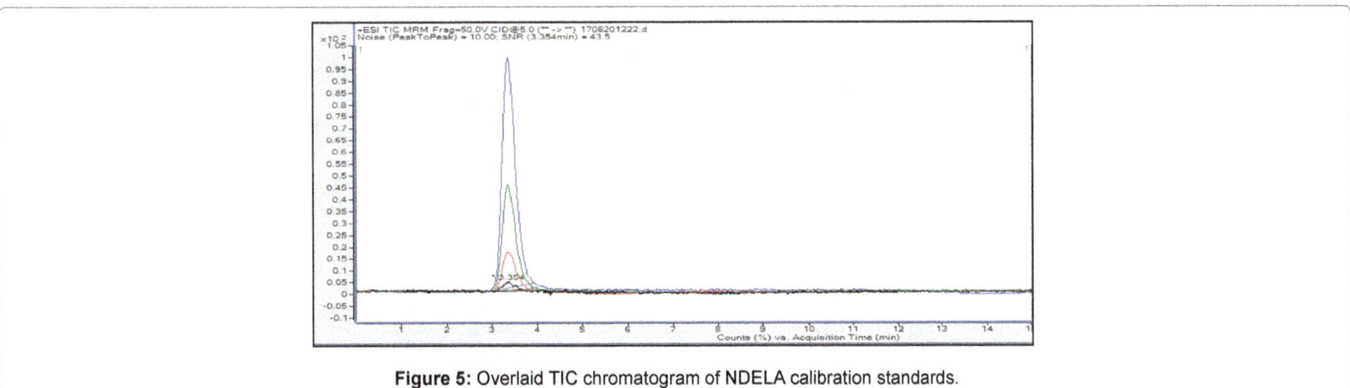

Figure 5: Overlaid TIC chromatogram of NDELA calibration standards.

of correlations R^2 as determined by four point calibration curve was greater than 0.995. The signal to noise (S/N) of NDELA at the lowest limit of quantification (LLOQ) was greater than 100. The overlaid TIC chromatogram of standards and calibration curve generated are depicted in Figures 5 and 6. The sample analysis was performed as per the section materials and methods. The prepared TEA and DEA were injected multiple times to detect the NDELA but the MRM analysis couldn't show any signal towards selected ions (135.0, 104 and 74.1 (Figure 7). In order to confirm the analysis, spiking studies were carried out in the TEA and DEA matrix at 2 ppb level. The percentage recovery for the analysis of TEA is found 50-60 and DEA was <50. In contrast to expectation, the percentage recovery was very low and this leads to detailed investigation of interaction of the matrixes, method and machine parameters, etc.

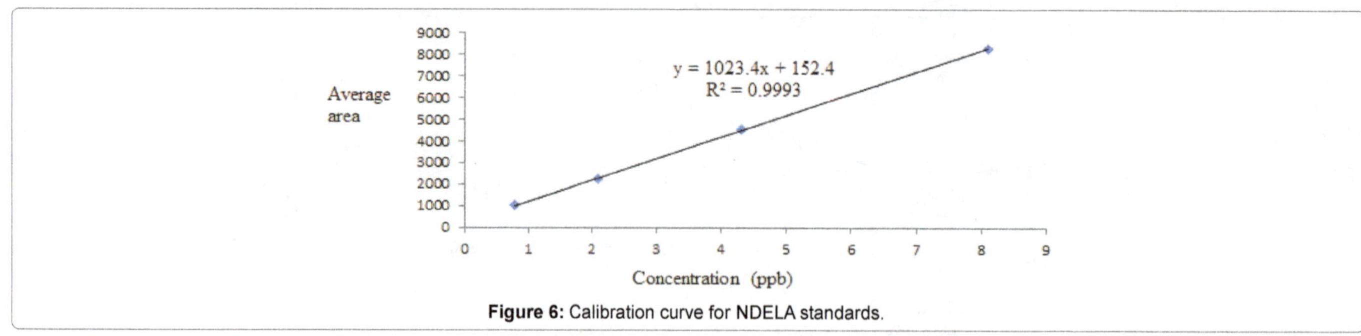

Figure 6: Calibration curve for NDELA standards.

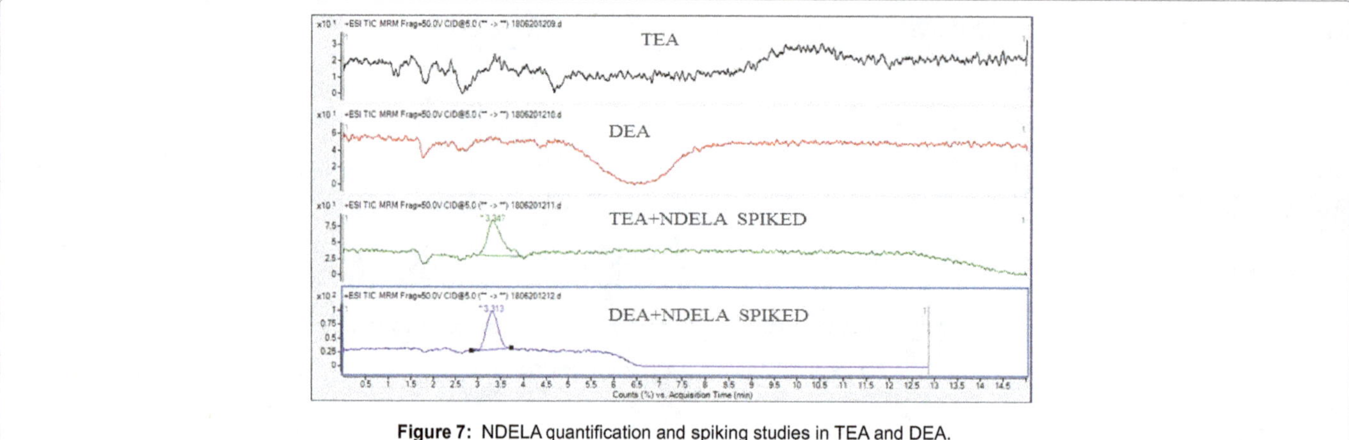

Figure 7: NDELA quantification and spiking studies in TEA and DEA.

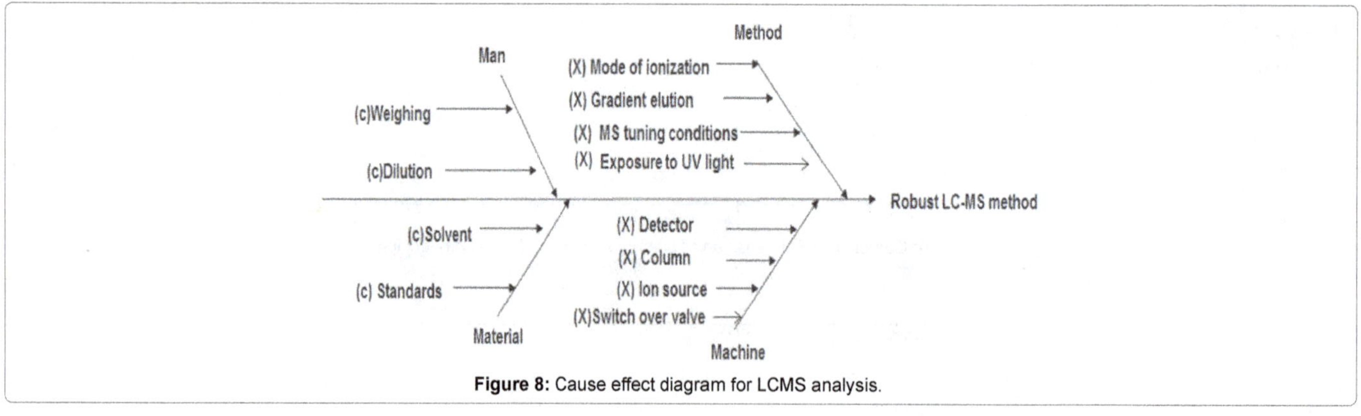

Figure 8: Cause effect diagram for LCMS analysis.

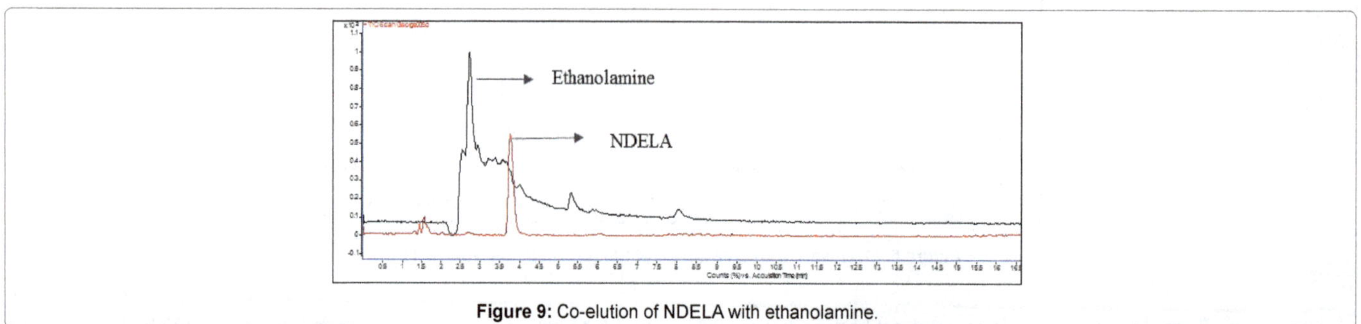

Figure 9: Co-elution of NDELA with ethanolamine.

Matrix interference and ion suppression

A cause effect diagram (Figure 8) was used to evaluate the possible matrix interference and ion suppression during the quantification of NDELA. We have ruled out interference from common endogenous compounds that can potentially cause ion suppression. The main cause for the low recovery due to ion suppression was found to be the analytical column and elution mode used in HPLC and MS detector respectively.

From the low level percentage recovery of NDELA, it was clear that conventional reverse phase HPLC column was leading to co-elution of

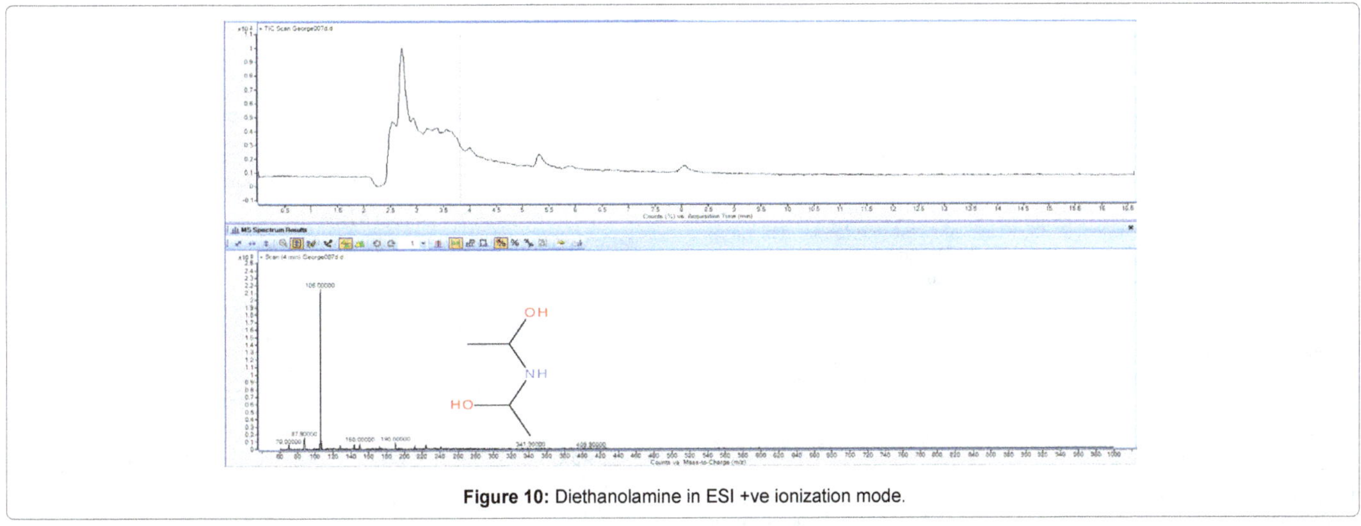

Figure 10: Diethanolamine in ESI +ve ionization mode.

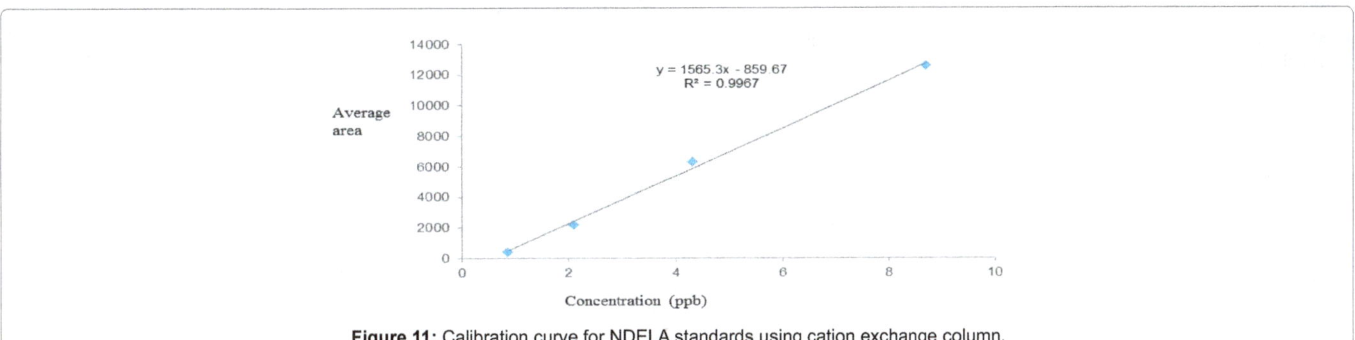

Figure 11: Calibration curve for NDELA standards using cation exchange column.

Analysis result		Spiking studies	
Sample	**NDELA(ppb)**	**Sample**	**%Recovery**
TEA	<10	TEA (10 ppb NDELA)	97.5
DEA	<10	DEA (10 ppb NDELA)	105.7
MEA	<10	MEA (10 ppb NDELA)	105.9

Table 2: Quantification of NDELA and recovery results using cation exchange column.

NDELA with ethanolamine and cause ion suppression (Figure 9). In the TIC chromatogram of Figure 9, it was clear that NDELA is partially coeluted with the huge peak of ethanolamine and the peak retention time from 2.5 minutes to more than 5 minutes were enriched by m/z of 106 which is corresponding to diethanolamine (Figure 10). The co-elution of NDELA with very high concentration of amines can definitely cause ion suppression in ESI positive mode since the ethanolamines are very much susceptible to ionization at positive mode in presence of formic acid. The easy ionization of abundant ethanolamine can reduce the probability of trace level NDELA to get ionized which in turn cause the low level detection and recovery of NDELA in amine matrix. By the usage of cation exchange ion chromatography column enhanced the chromatographic resolution of the NDELA from ethanolamine matrix which in turn reduced the ion suppression. Alternately, the performance of ionization of NDELA was increased by restricting the introduction of amine ions in the MS source by the usage of switch over valves in the MS detector. Also, the column was washed at adequate time using methane sulfonic acid to regenerate the column if any exchange happened. The prepared calibration standards and freshly prepared samples were analyzed with these modifications in the method. The results were validated by spiking of NDELA in matrix to get 10 ppb level of quantification. The calibration curve and analysis results are depicted in Figure 11 and Table 2. All the samples, the level of NDELA was found to be less than 10 ppb.

Accuracy and precision

The accuracy of the assay was evaluated by the recovery experiments, resulted recovery of NDELA ranging between 97.5 to 105.9% (Table 2). The precision of the method was evaluated by repeated injection of NDELA spiked in TEA sample for many days. Since the NDELA was <10 ppb in the amines, NDELA spiked in TEA were used as the sample for calculating the precision of the method. The percentage recovery remains constant (97 to 106%) or all the 10 to 15 data points of repeated analysis.

Conclusion

This article describes the development and validation of a simple direct method for determination of N-nitrosodiethanolamine in ethanolamine by LC-MS. Other methods have used multistep extraction or derivatization steps which are subject to elaborate procedures and additional instrumentation for analysis. This method excludes any prior sample preparation steps and provide direct analysis of NDELA which yields the required sensitivity up to 10 ppb level in ethanolamine matrix. The method is fully validated for specificity of the ions, sensitivity, accuracy linearity and recovery. The method is accurate, with recovery ranging between 97 to 107%. The matrix interferences and ion suppression were minimized by using cation exchange ion chromatographic columns in HPLC and switch over valves in MS detector. The LC-MS/MS assay was extensively used in our technology centers to determine NDELA levels in various ethanolamine batches.

Acknowledgements

The author greatly acknowledges Dr. Khalid Al-Assaf (the former director of SABIC analytical technology department) and Dr. Mohammed Al-Hazmi for their constant encouragement and support to facilitate to develop this method.

References

1. Douglass ML, Kabacoff BL, Anderson GA, Cheng MC (1978) The chemistry of nitrosamine formation, inhibition and destruction. J Soc Cosmet Chem 2: 581-606.
2. Nawrocki J, Andrzejewski P (2011) Nitrosamines and Water. Journal of Hazardous Materials 189: 1-18.
3. Mlongo SH, Mamba BB, Krause RW (2009) Nitrosamines: A Review on Their Prevalence as Emerging Pollutants and Potential Remediation Options. Water SA 35: 735.
4. March J (1968) Advanced organic Chemistry reactions, mechanisms and structure, Mac Graw Hill, New York, pp: 484-486.
5. Schothorst RC, Somers HHJ (2005) Determination of N-Nitrosodiethanolamine in Cosmetic Products by LC-MS-MS. Analytical and Bioanalytical Chemistry 381: 681-685.
6. Rywotycki R (2003) Meat Nitrosamine Contamination Level Depending on Animal Breeding Factors. Meat Science 65: 669-676.
7. Wang X, Gao Y, Xu X, Zhao J, Song G, et al. (2011) Derivatization Method for Determination of Nitrosamines by GC-MS. Chromatographia 73: 321-327.
8. Al-kaseem M, Assaf Z, Karabeet F (2013) Rapid and Simple Extraction Method for Volatile N-Nitrosamines in Meat Products. Pharmacology & Pharmacy 4: 611-618.
9. Incavo JA, Schafer MA (2006) Simplified Method for the Determination of N-Nitrosamines in Rubber Vulcanizates. Analytica Chimica Acta 557: 256-261.
10. Ventanas S, Ruiz J (2006) On-Site Analysis of Volatile Nitrosamines in Food Model Systems by Solid-Phase Micro-Extraction Coupled to a Direct Extraction Device. Talanta 70: 1017-1023.
11. Komarova NV, Velikanov AA (2001) Determination of Volatile N-Nitrosamines in Food by High-Performance Liquid Chromatography with Fluorescence Detection. J Analytical Chem 56: 359-363.
12. Telling GM, Bryce TA, Althorpe JJ (1971) Use of Vacuum Distillation and Gas Chromatography-Mass Spectrometry for Determining Low Levels of Volatile Nitrosamines in Meat Products. J Agricultural and Food Chem 19: 937-940.

Loess Soil Nanoparticles as A Novel Adsorbent for Adsorption of Green Malachite Dye

Heydartaemeh MR[1*], Aslani S[2] and Doulati Ardejani F[2]

[1]*Faculty of Mining, Petroleum and Geophysics, Shahrood University of Technology, Shahrood, Iran*
[2]*College of Mining Engineering, University of Tehran, Tehran, Iran*

Abstract

In this research, Loess Soil Nanoparticles (LSN), as a novel adsorbent nanocomposite was used for removal of green malachite (GM) dye from Aqueous Solutions by in a batch and fixed bed column. Firstly, LSN adsorption properties were investigated. The study investigates the effect of process parameters such as pH, adsorbent dosage, contact time and GM dye initial concentration. Next, GM dye was quantitatively evaluated by using the Langmuir, Freundlich isotherm and pseudo first and second order model. The adsorption data follow the adsorption equilibrium was described well by the Freundlich isotherm model. The results of SEM and AFM analysis show that particle size is less than 100 nm. Also, the BET analysis shows that the surface areas for LSN are 70 m^2/g respectively. The results show that adsorption capacities and removal percentage of GM dye on LSN from wastewaters is about 80%. Consequently LSN is a superior adsorbent for wastewaters purification.

Keywords: Loess soil nanoparticles; Green malachite dye; Adsorption; Waste water treatment

Introduction

Various colorants (dyes and pigments) are being applied in many industries for different coating applications. It is the inevitable reason for existence these materials in industrial wastewater. Colored wastewaters, especially organic ones, are wastes of different industries, such as paper, textile, leather, food, polymers, minerals and plastics [1]. These cause that treatment of water and wastewater, contaminated with colorants, are one of the main concerns of researchers in recent decades. In a real wastewater, there is a complex of different materials, such as colorants, polyacrylates, phosphonates, anti-coagulation factors, and so on. Most of these compounds are poisoning and it is necessary for ecological balances that these dangerous contaminants are being removed from treated wastewater completely. Therefore, the governments and different UN organizations have recently established many rules to prevent and standardize these materials in environment [2,3]. Different physicochemical decolorization processes have been developed to remove contaminants from industrial wastewater in recent years, such as reduction and precipitation [4], coagulation and flotation [5] membrane technologies and electrolysis [6,7], biological treatments [8], advanced oxidation processes [9], chemical and electrochemical techniques [10,11] and adsorption procedures [12-15]. Among all the treatments proposed, adsorption using sorbents is one of the most popular methods. It is now recognized as an effective, efficient and economic method for water decontamination applications and for separation to analytical purpose [16].

But most of these methods are expensive and certain economical foundation is necessary. Among physicochemical processes, adsorption technology has found many applications in water and wastewater treatment, as one of the most efficient and effective technologies [2,17]. Therefore, natural adsorbents have been used to reduce costs and the environmental side effects, such as diatomite [2], red mud [18], chitosan [19], orange skin [20], soy meal hull [21], almond skin [22], sawdust [23], zeolite [24] and clay [25]. These adsorbents have natural base and they are environmental friendly. It is possible to regenerate most of them or be applied in different products.

The nanometer material is a new functional material, which has attracted much attention due to its special properties. Most of atoms on the surface of the nanoparticles are unsaturated and can easily bind with other atoms. Nanoparticles have high adsorption capacity. Besides, the operation is simple, and the adsorption process rapid. So, there is a growing interest in the application of nanoparticles as adsorbents [26].

Nowadays, with helping new methods such as magnetization the efficiency of natural adsorbents to adsorption of pollutants from aqueous solution slightly decreased [27,28] rather natural adsorbent, but the adsorbent material can be easily separated from a mixture of particles using a simple magnet. So, in this research, removal of GM dye has been investigated from simulated textile wastewaters by LSN. Loess Soil is a natural and cheap adsorbent. So, Loess Soil Nano scale size of this adsorbent was prepared by mill (EQ-PC-12 model).

So, the objective of the present study is focused on the development of LSN for removal of GM dye. The dye selected in this study was malachite green because of their environmental significance. The effects of pH, adsorbent dosage, contact time and GM dye initial concentration were investigated. Since optimization of parameters, the characterization of isothermal adsorption and adsorption kinetics and also the SEM, AFM and BET analysis were studied in order to provide a new method and theoretical evidences for wastewater treatment.

Experimental Studies (Materials and Equipment)

Loess Soil Nanoparticles used in this investigation was obtained from NanoMineTech Company (a source in Iran). It can be prepared in any material construction store. Green Malachite (GM) dye was supplied from Ciba Company. Its molecular structure has been shown in Figure 1.

***Corresponding author:** Heydartaemeh MR, Faculty of Mining, Petroleum and Geophysics, Shahrood University of Technology, President and CEO at NanoMineTech Co, Shahrood, Iran
E-mail: m.heydartaemeh@gmail.com, m.heydartaeme@nanominetech.com

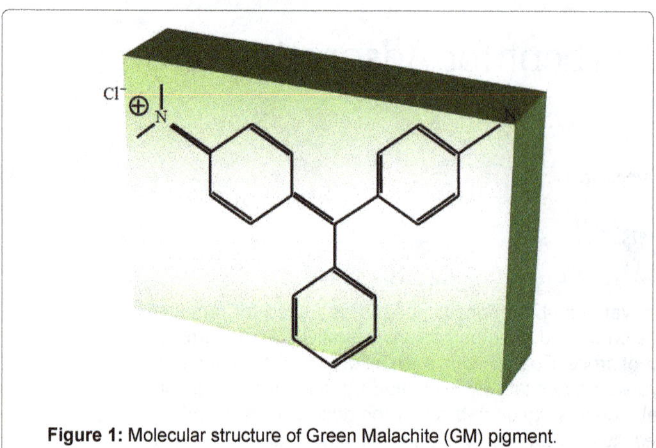

Figure 1: Molecular structure of Green Malachite (GM) pigment.

A laboratory balance (Sartorius-d=0.1 mg, max 120 g model) was used to weight samples. Some simple laboratory heater-stirrer systems were used to mix samples; UV/Visible spectrometer (One beam), high temperature oven 1100°C for drying (Cecil-CE2021-2000 series) was applied to measure the change on concentration of GM dye. In addition, various sieves with different meshes were used to categorize the adsorbent. A Centrifuge (Hettich EBA20, maximum whirl=6000 rpm) was applied to sediment and remove colloidal particles, and pH meter (Metrohm 713) was used to measure and adjusting the pH of simulated wastewaters. A mill (EQ-PC1-12) that crush particle by physical method, and a spray dryer (BUCHI B-191) for drying slurry particles, SEM (Scan Electron Microscope, LEO 1455VP), AFM (Atomic Force Microscope, Model SZMU-L5) and BET analysis was used to increase the knowledge about Loess Soil Nanoparticles microscopic structure and its real nature. Other chemicals, such as Sodium hydroxide and Chloridric acid were supplied from Merck Company, especially to adjust the pH of wastewater.

Adsorption procedure

The adsorption measurements were conducted by mixing various amounts of Loess Soil Nanoparticles (0.05 g) in stirrer containing a dye solution (6, 12, 20 ppm) of GM dye at temperature 25°C and pH=6 for 120 min to attain equilibrium condition. The mixing rate was high enough (>3000 rpm) to minimize the external mass transfer resistance. The changes in adsorption were determined at certain times (0, 10, 30, 60, 90, 120 min) during the removal process. After conducting adsorption experiment, the solution was separated from LSN by centrifugation at 4000 rpm for 5 min using a Hettich EBA20 centrifuge. The percentage adsorption of dye from aqueous solution was computed as follows

$$Adsorption\ (\%) = \frac{C_{int} - C_{fin}}{C_{int}} \times 100 \qquad (1)$$

Where, C_{int} and C_{fin} are the initial and final dye concentrations, respectively. The dye concentrations in aqueous solution were determined using CECIL 2021 spectrophotometer corresponding to the maximum wavelength (λ_{max}=619 nm) of malachite green.

Results and Discussion

Effect of pH

The pH of the solution is an important factor that controls the adsorption process. The effect of pH on the green malachite adsorption by LSN was investigated in the pH of 4, 6, and 8. Figure 2 shows the malachite green adsorption percentage as a function of pH and contact time. As it is shown in Figure 2, dye adsorption fell with an increase in pH parameter. The maximum removal take place at pH=6.

Effect of loess soil nanoparticles dosage and contact time

The effect of the LSN dosage on the removal of GM dye which was based on the contact time was studied by changing at the adsorbent dosage in the range of 0.01, 0.03, 0.05, 0.07 g. The results are presented in Figure 3.

Based on Figure 4, removal trend of GM dye from 20 ppm simulated wastewater in pH=6 by 0.01 g, 0.03 g, 0.05 g and 0.07 g per 25 ml of solution was done. The dosage 20 ppm and 0.05 g LSN have the most effective adsorption value. Additionally, the adsorption yield increased with contact time and attained a maximum value at 120 min. Applying more adsorbent than the optimized dosage (0.05 g) has diminished the capacity of adsorption process. This phenomenon is due to a fluid mechanical problem. By raising the amount of adsorbent, coagulation of the particles would decrease the surface area of the adsorbent and would lessen the capacity of the LSN.

X- ray diffraction

The chemical constituent of the LSN as adsorbent for GM dye was calculated using X-Ray Diffraction (XRD) method. X-Ray diffraction (XRD) study using a Philips X-Ray Diffractometer Xunique 1140 was operated to characterize the minerals existing in LSN phase is the major phase at the Loess Soil Nanoparticles samples (Figure 5). It is obvious that the Quarts phase is the major phase.

Effect of GM dye initial concentration

The effect of GM Dye initial concentration on the adsorption process was investigated by changing the dye initial concentration at

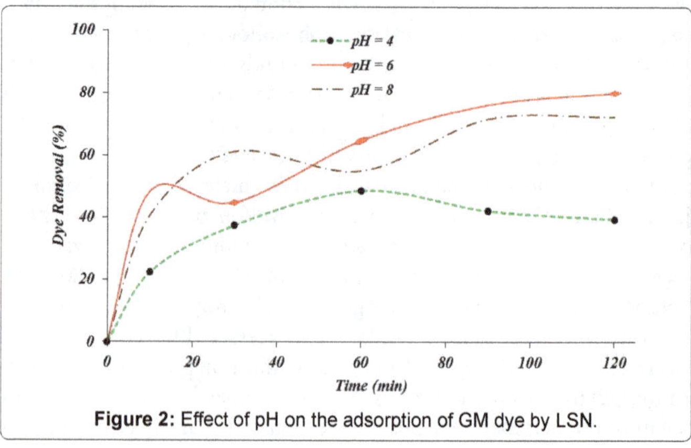

Figure 2: Effect of pH on the adsorption of GM dye by LSN.

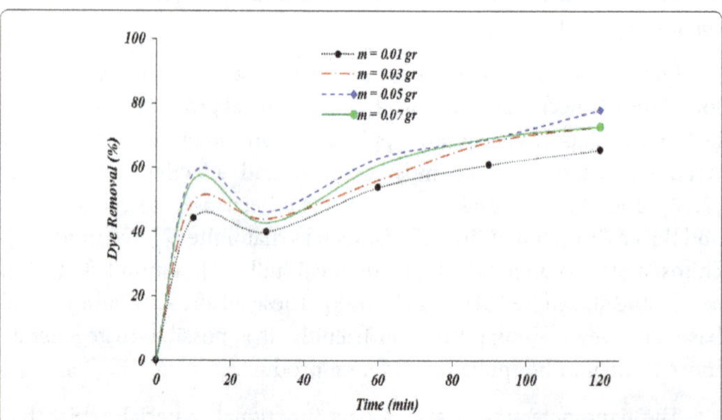

Figure 3: Effect of Loess Soil Nanoparticles dosage on the adsorption of GM dye.

the range of 6, 12 and 20 ppm under optimized conditions (pH=6, contact time of 120 min, adsorbent dosage of 20 ppm and temperature of environment). The GM Dye adsorption efficiency was decreased by increasing the initial dye concentration. When GM Dye concentration increased from 6 to 20 ppm the adsorption percentage increased (Figure 4).

XRF analysis

Chemical composition of Loess Soil Nanoparticles sample was characterized by XRF, (Philips Diffractometer Xunique п) in the Central Laboratory of Amirkabir University of Technology, and the result are presented in Table1. As can be seen in this table, based on XRF analysis, LSN sample is composed mainly of SiO_2 and Al_2O_3.

Minerals under the microscope

Loess is an Aeolian sediment formed by the accumulation of wind-blown silt, typically in the 1–150 micrometer size range, twenty percent or less clay and the balance equal parts sand and silt that are loosely cemented by calcium carbonate. It is usually homogeneous and highly porous and is traversed by vertical capillaries that permit the sediment to fracture and form vertical bluffs. The Loess Consists of Quarts, Tourmaline, Pyrite and Calcite (Figure 6).

SEM analysis of loess soil nanoparticles

The images of the LSN surfaces were obtained using scanning electron microscope (SEM) instrument, model LEO 1455VP, illustrating raw Loess (Figure 7a) and Loess Soil Nanoparticles before (Figure 7b) and after GM dye adsorption (Figure 7c). This characteristic causes that LSN to be a proper adsorbent. Moreover, comparison of the Figures 7b and 7c show that the Nano surfaces have been treated and prepared to adsorb pollutants from the wastewater and the structure of the surface of LSN completely changed before and after process. The extra parts and porosity have been removed.

AFM and BET analysis of loess soil nanoparticles

Atomic force microscopy (AFM) is a powerful tool allowing a variety of surfaces to be imaged and characterized at the atomic level. According to Figure 8, LSN as a Nano adsorbent have a high porosity and this porosity as an effective factor have an important role at the GM dye adsorption. The results of BET analysis surface areas for LSN was 70 m^2/g.

Adsorption isotherms

Adsorption isotherms describe how adsorbents interact with adsorbents. Adsorption isotherms demonstrate the relationships between equilibrium concentrations of adsorbate in the solid phase,

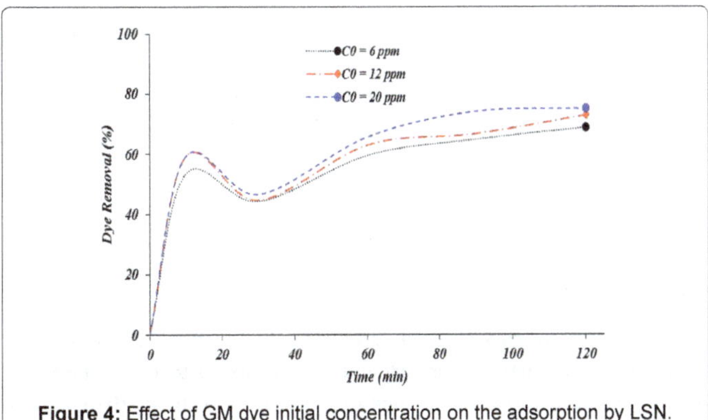

Figure 4: Effect of GM dye initial concentration on the adsorption by LSN.

q and in the liquid phase, C at constant temperature [22,29-31]. Adsorption isotherms are described in many mathematical forms. They are often obtained in the laboratory using batch test in which the equilibrium data are attempted by various isotherm models, such as Langmuir, Freundlich [32-34] isotherms. The Langmuir isotherm model suggests that the uptake of adsorbate occurs on the homogeneous surface by monolayer sorption without interaction between adsorbed molecules. The model assumes that the energies of adsorption on the surface are uniform and no migration of adsorbate happens on it. The linear form of Langmuir isotherm equation is represented by the following equation [35].

$$\frac{C_e}{q_e} = \frac{1}{QK_L} + \frac{C_e}{Q} \tag{2}$$

Where Ce is the equilibrium concentration of adsorbate (mg/L), qe is the amount of adsorbed at equilibrium (mg/g), Q (mg/g) and KL (L/mg) are Langmuir constants related to the adsorption capacity and energy, respectively. When qe is plotted against Ce, a straight line is obtained with slope of 1 Q adsorption of GM dye follows the Langmuir parameters.

The Freundlich equation has been widely used and is applicable to isothermal adsorption. This model is a special case for heterogeneous surface energies in the Langmuir equation. The energy term varies as a function of surface coverage, q_e in this model. q_e Strictly depends on the variations in heat of adsorption [36]. The Freundlich equation has the following general form [22,35].

$$log\ q_e = log\ K_F + (1/\ n\ )log\ C_e \tag{3}$$

where, qe the amount of is adsorbed per until weight (mg/g adsorbent), Ce is the equilibrium concentration of adsorbate (mg/L), KF and n are the Freundlich constants.

The empirical parameters for Loess Soil Nanoparticles are given in Table 2. The fitting of experimental data in each isotherm model were examined by calculation of the correlation factor (R^2). It was found that the Freundlich model better describes the adsorption than the Langmuir models according to correlation factor (R^2=1.0). As can be seen from this table, the highest R^2 values of Freundlich model shows that it is the most suitable equation to describe the adsorption equilibrium. According to this model, the Freundlich constants Kf and n were calculated 7.0345 and 2.987 for LSN. These are relatively uncommon but are often observed at low concentration ranges for compounds containing a polar functional group. Consequently, the adsorption of GM dye by LSN follows the Freundlich isotherm model.

Adsorption kinetics

It is essential to predict the rate at which dye is removed from aqueous solutions in order to design an appropriate treatment system based on adsorption process. Pseudo-first and pseudo-second order models have been applied to describe the adsorption kinetics of GM dye by LSN. The pseudo-first-order Kinetics model can be represented by the following Lagergren's expression.

$$ln(q_e - q_t) = lnq_e - K_{1,ad}(t) \tag{4}$$

where, q_e and q_t are the amounts of dye adsorbed (mg/g) at equilibrium and at time t (min), respectively, and $K_{1,\ ad}$ is the pseudo-first-order rate constant (1/min). The rate of pseudo-second-order model depends on the amount of dye adsorbed on the surface of the adsorbent and its quantity at equilibrium condition [37]. Pseudo-first-order model can be given as follows [38].

$$\frac{1}{q_t} = \frac{1}{K_{2,ad}q_e^2 t} + \frac{1}{q_e} \tag{5}$$

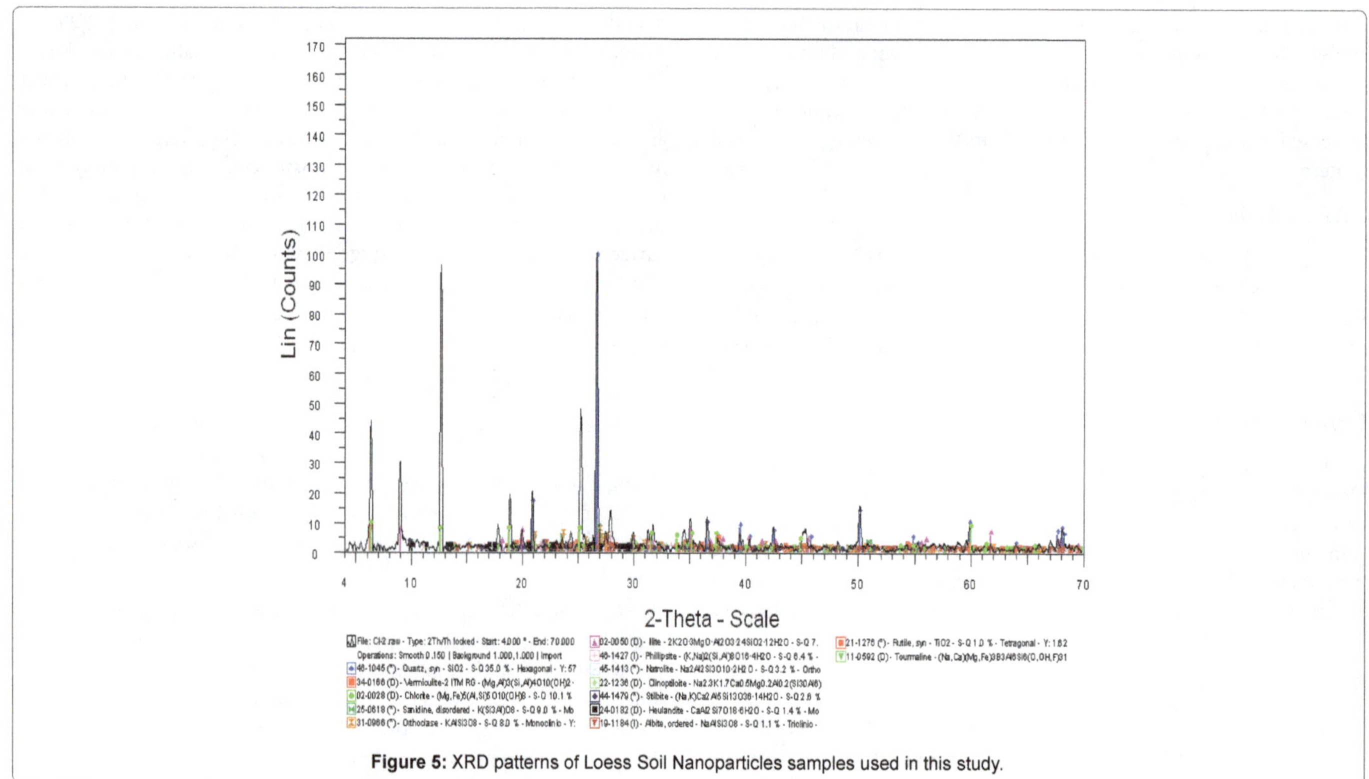

Figure 5: XRD patterns of Loess Soil Nanoparticles samples used in this study.

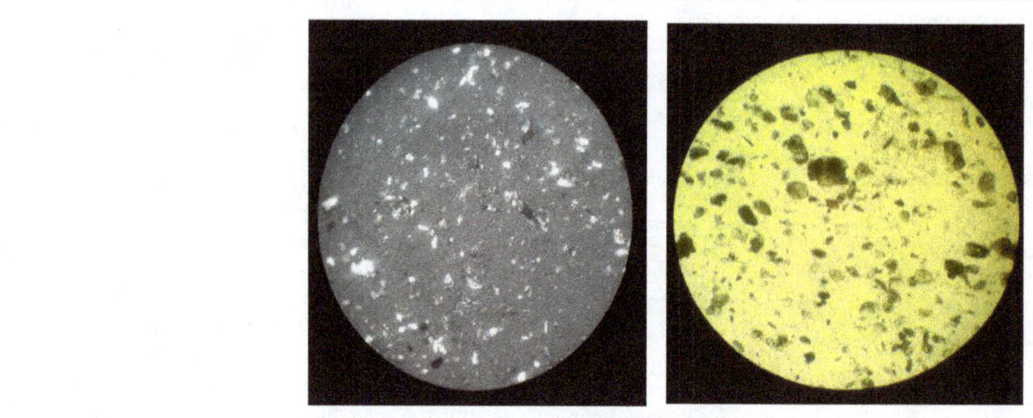

Figure 6: The Loses Soil Nanoparticles Minerals under Microscope.

	SiO_2	$Al2O_3$	$Fe2O_3$	CaO	Na_2O	K_2O	MgO	TiO_2	MnO
Sample (%)	57.32	18.91	10.42	0.96	1.08	2.38	2.70	1.103	0.147
	P_2O_5	S	L.O.I	Cl	Ba	Sr	Cu	Zn	Pb
	0.203	0.002	4.45	53	320	177	35	158	24
Sample (ppm)	Ni	Cr	V	Ce	La	W	Zr	Y	Rb
	102	118	172	114	51	1	180	48	91
	Co	As	U	Th	Mo	Ga	Nb		
	5	4	2	5	3	15	11		

Table 1: Results of XRF analysis, density and color of LSN samples.

Where $K_{2,ad}$ is the rate constant of pseudo-second-order model (g/mg min). The kinetics parameters of the pseudo- first-order and pseudo-second-order models of GM dye at different pH and temperatures are given in Tables 2 and 3.

Conclusions

The results of this research show that the surface morphology of Loess Soil Nanoparticles has very important role on adsorption process in low bulk mass transfer velocity (rotational speed of stirrer). As a

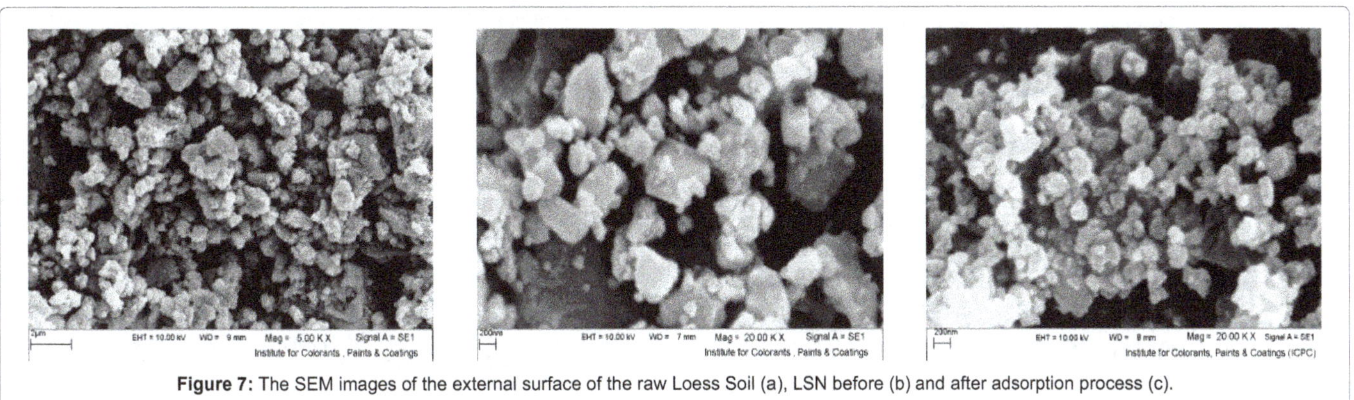

Figure 7: The SEM images of the external surface of the raw Loess Soil (a), LSN before (b) and after adsorption process (c).

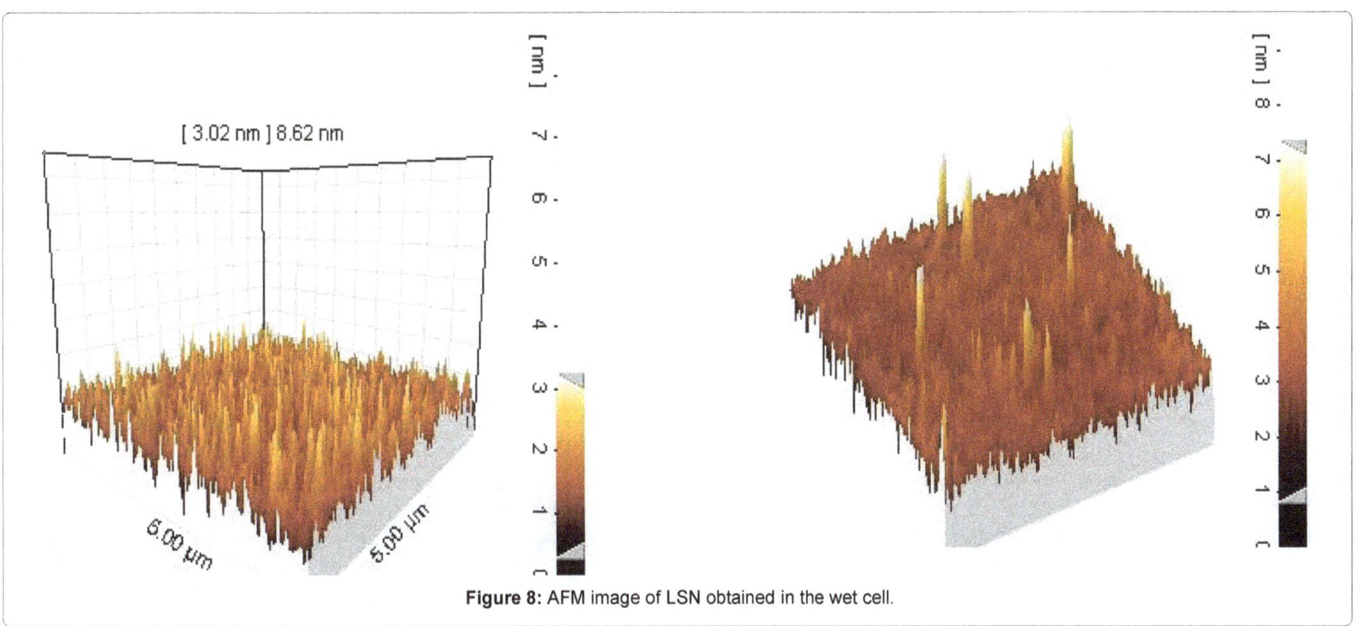

Figure 8: AFM image of LSN obtained in the wet cell.

	Langmuir isotherm				Freundlich isotherm	
Adsorbent type	Q	K_L	$R2^2$	K_F	n	$R1^2$
Loess Soil Nanoparticles	70	0.018	0.921	7.0345	2.987	0.989

Table 2: The isotherm coefficients for adsorption of GM dye on LSN.

	Pseudo-first-order		Pseudo-second-order	
pH	$K_{1,ad}$	R^2	$K_{2,ad}$	R^2
4	0.0038	0.5437	0.0006	0.4583
6	0.0081	0.702	0.002	0.892
8	0.0091	0.6426	0.0025	0.691

Table 3: Kinetics constants for GM obtained at pH=4, 6 and 8.

Novel low-price adsorbent, LSN was applied to remove GM dye from a simulated wastewater. Adsorption process of GM dye by LSN depends on different parameters, such as particles size, pH, adsorbent dosage and temperature. The maximum percentage of GM dye removal was about 80% which was obtained under normal temperature (25°C) and at pH=6. The results of fixed bed column show that 6, 8 and 10 mL/min flow rates the break through curve were saturated in 150, 300 and 450 min, respectively. Based on results the Freundlich model better describes the adsorption than the Langmuir models. Also, the adsorption kinetics of GM dye by LSN adsorbent can be well described by the pseudo-second-order reaction model at various pH values.

Acknowledgements

The authors are thankful to Nano Mine Tech Co. and College of Mining Engineering, University of Tehran, Tehran, Iran for supporting this research.

References

1. Erdem E, Colgeçen G, Donat R (2005) The removal of textile dyes by diatomite earth. J Colloid Interface Sci 282: 314-319.
2. Saberi M, Kor M, Badii KH, Yousefi LN (2009) Applying the Taguchi experimental design method to optimize the dyes removal from simulated wastewater by diatomite. English Proceeding of International Conference on Water Resources. ICWR, pp: 566-573.
3. AL Ghouti MA, Khraisheh MAM, Alien SJ, Ahmad MN (2003) The removal of dyes from textile wastewater: a study of the physical characteristics and adsorption mechanisms of diatomaceous earth. J of Environ Manage 69: 229-238.
4. Esalah OJ, Weber ME, Vera JH (2000) Removal of lead, cadmium and zinc from aqueous solutions by precipitation with sodium di-(n-octyl) phosphinate. Canad J Chem Eng 78: 948-954.
5. Zouboulis AI, Matis KA, Lanara BG, Loos-Neskovic C (1997) Removal of cadmium from dilute solutions by hydroxyapatite, II. Floatation studies. Sep Sci Tech 32: 1755-1767.
6. Canet L, Ilpide M, Seta P (2002) Efficient facilitated transport of lead, cadmium, zinc and silver across a flat sheet-supported liquid membrane mediated by lasalocid A. Sep Sci Tech 37: 1851-1860.
7. Ning RY (2002) Arsenic removal by reverse osmosis. Desalination 143: 237-241.
8. Mcmullan G, Meehan C, Conneely A, Kirby N, Robinson T, et al. (2001) Microbial decolourisation and degradation of textile dyes. Appl Microbiol Biotech 56: 81–87.
9. Lee JM, Kim MS, Hwang B, Bae W, Kim BW (2003) Photo-degradation of acid red 114 dissolved using a photo-Fenton process with TiO2. Dye and Pig 56: 59-67.
10. Von GU (2003) Part 1 Oxidation kinetics and product formation. Water Res 37: 1443-1467.
11. Chen X, Chen G, Yue PL (2002) Novel electrode system for electro flotation of wastewater. Environ Sci Technol 36: 778-783.
12. Ravindran V, Stevens MR, Badriyha BN, Pirbazari M (1999) Modeling the sorption of toxic metals on chelant- impregnated adsorbent. AIChE J 45: 1135-1146.
13. Toles CA, Marshall WE (2002) Copper ion removal by almond shell carbons and commercial carbons: batch and column studies. Sep Sci Tech 37: 2369-2383.
14. Hu Z, Lei L, Li Y, Ni Y (2003) Chromium adsorption on high-performance activated carbons from aqueous solution. Sep Sci Tech 31: 13-18.
15. Varma J, Deshpande SV, Kennedy JF (2004) Metal complexation by chitosan and its derivatives: a review. Carbohydrate Polymers 55: 77-93.
16. Niu Ch, Wu W, Zhu W, Li SH, Wang J (2007) Adsorption of heavy metal ions from aqueous solution by crosslinked carboxymethyl konjac glucomannan. J Hazard Mater 141: 209-214.
17. Ramakrishna KR, Viraraghavan T (1997) Dye removal using low cost adsorbents. Water Sci Tech 36: 189-196.
18. Badii KH, Doulati AF, Norouzi SH (2009) Activation of alumina industrial waste for environmental purposes. Proceedings of Iran International Aluminum Conference, Tehran IR Iran IIAC, pp: 592-595.
19. Chen AH, Liu SC, Chen CY, Chen CY (2008) Comparative adsorption of Cu(II), Zn(II), and Pb(II) ions in aqueous solution on the crosslinked chitosan with epichlorohydrin. J Hazard Mater 154: 184-191.
20. Badii KH, Shafaie TSZ, Tehrani Bagha AR, Yousefi Limaee N (2004) Optimization of condition of decolorization form wastewater containing direct dyes with using orange pills by Taguchi's experimental design method. Proceedings of 1st Seminar on Environment and Color, Tehran, Iran, ICPC, pp: 13-26.
21. Badii KH, Doulati AF, Yousefi LN (2008) A numerical finite element model for the removal of direct dyes from aqueous solution by soy meal hull: Optimization and sensitivity analysis. First Conference and Workshop on Mathematical Chemistry, Tarbiat Modares University, Tehran, Iran, pp: 57-64.
22. Doulati AF, Badii KH, Yousefi LN, Shafaei SZ, Mirhabibi AR (2008) Adsorption of Direct Red 80 dye from aqueous solution on to almond shell: Effect of pH, initial concentration and shell type. J Hazard Mater 151: 730–737.
23. Bulut Y, Tez Z (2007) Removal of heavy metals from aqueous solution by sawdust adsorption. J Environ Sci 19: 160-166.
24. Wang Ch, Li Ji, Wang Li, Sun Xi, Huang Ji (2009) Adsorption of Dye from Wastewater by Zeolites Synthesized from Fly Ash: Kinetic and Equilibrium Studies. Chinese J Chem Eng 17: 513-521.
25. Gürses A, Doğar Ç, Yalçin M, Açikyildiz M, Bayrak R, et al. (2006) The Adsorption Kinetics of The Cationic Dye Methylene Blue onto Clay. J Hazard Mater 131: 217-228.
26. Maria LCS, Santos ALC, Oliveira PC, Valle ASS, Barud HS, et al. (2010) Preparation and antibacterial activity of silver nanoparticles impregnated in bacterial cellulose. Polimeros 20: 72-77.
27. Hashemian S (2010) MnFe2O4/bentonite nano composite as a novel magnetic material for adsorption of acid red138. Afr J Biotech 9: 8667-8671.
28. Martins A, Mata TM, Gallios GP, Václavíková M, Stefusova K (2010) Modeling and Simulation of Heavy Metals Removal from Drinking Water by Magnetic Zeolite. Water Treatment Technologies for the Removal of High-Toxicity Pollutants, project No: APVT-51-017104, 61-84.
29. Markovska L, Meshko V, Noveski V (2001) Adsorption of basic dyes in a fixed bed column. Korean J Chem Eng 18: 190–195.
30. McCabe W L, Smith J C, Harriott P (1985) Unit Operations of Chemical Engineering (eds.) McGraw-Hill.
31. Muhamad H, Doan H, Lohi A (2010) Batch and continuous fixed-bed column biosorption of Cd2+ and Cu2+. Chem Eng J 158: 369–377.
32. Nitzsche O, Vereecken H (2002) Modeling sorption and exchange processes in column experiments and large scale field studies. Mine Water Environ 21: 15-23.
33. Chakraborty S, De S, Dasgupta S, Basu JK (2005) Adsorption study for the removal of a basic dye: experimental and modeling. Chemosphere 58: 1079-1086.
34. Shawabkeh R, Al-Harahsheh A, Al-Otoom A (2004) Copper and zinc sorption by treated oil shale ash. Sep Sci Technol 40: 251-257.
35. Doyurum S, Celik A (2006) Pb2+ and Cd2+ removal from aqueous solution by olive cake. J Hazard Mater 138: 22-28.
36. Adamson AW (1967) Physical Chemistry of Surfaces (eds.) Inter science Publishers Inc, New York.
37. Gucek A, Sener S, Bilgen S, Mazmanci ML (2005) Adsorption and kinetic studies of cationic and anionic dyes on pyrophyllite from aqueous solutions. J Colloid Interface Sci 286: 53-60.
38. Ho YS, Mckay G (1998) Kinetic models for the sorption of dye from aqueous solution by wood. Transac IChemE 76: 183-191.

Determination of Impurities in Bioproduced Succinic Acid

Rousová J[1], Ondrušová K[1], Karlová P[1,2] and Kubátová A[1*]

[1]*University of North Dakota, Department of Chemistry, 151 Cornell Street Stop 9024, Grand Forks, ND 58202, USA*

Abstract

At present, significant research resources are directed towards development of renewable products for replacing petrochemicals such as succinic acid. The critical component of this research is the identification of impurities which have a detrimental impact on further processing of succinic acid. We have adapted derivatization with gas chromatography - mass spectrometry to identify and quantify more than 120 impurities in several succinic acid samples. This study focused on petroleum based succinic acid as well as bio-based samples that use a modified E. coli strain for fermentation. To enable an accurate quantification of both the target product and common impurities, we evaluated the acetonitrile extraction efficiency as an alternative to direct derivatization, and then compared several derivatization agents for trimethylsilylation. A prior acetonitrile extraction was shown to be essential to detect impurities in trace concentrations. N,O-bis(trimethylsilyl)trifluoroacetamide (BSTFA) was most efficient for derivatization of saccharides and low molecular weight monocarboxylic acids. However, the presence of pyridine was necessary for derivatization of saccharides and polyalcohols with BSTFA, whereas low molecular weight acids had to be quantified without pyridine.

Fourteen representative bioproduced succinic acid samples differing in production stage, and cultivation method were characterized. The screening of initial process (1st stage of synthesis) samples showed monocarboxylic acids as most abundant and suggested occurrence of saccharides. Thus we have developed method allowing for quantification of carboxylic acids and saccharides with limits of detection between 0.02-0.3 ng. In initial process bacterial samples and also petrochemical sample, formic, acetic, lactic, oxalic, benzoic, citric and malic acids as well as glycerol, butanediol, and glucose were found in a range of 0.02-1160 µg/g. In final processed samples, formic and acetic acid, and glucose were found in concentration lower than 0.001% demonstrating effectiveness of process as well as applicability of the method as quality control of the process.

Keywords: GC-MS; acids; Saccharides; BSTFA; MSTFA; Succinic acid; Bio-based succinic acid

Abbreviations: GC-MS: Gas Chromatography-Mass Spectrometry; BSTFA: N,O-bis(trimethylsilyl)trifluoroacetamide; MSTFA: N-methyl-N-(trimethylsilyl) trifluoroacetamide; HPLC: High Performance Liquid Chromatography; HMDS: hexamethyldisilazane; TMCS: Trimethylchlorosilane; BSA: N,O-Bis(trimethylsilyl)acetamide; ACN: acetonitrile, I.S.: Internal Standard; TIC: Total Ion Chromatogram; LOD: Limit Of Detection; LOQ: Limit Of Quantification; AMDIS: Automated Mass Spectral Deconvolution and Identification System

Introduction

At present, significant research resources are directed towards development of renewable products for replacing petrochemicals [1-3]. Among them, succinic acid, the precursor of a wide range of polyesters, has a market of 270,000 tons per year [2]. Consequently, bio-based succinate is receiving increasing attention, and with rising oil prices it has become a worthy competitor of petrochemical-based succinate [1,2]. The challenge of being cost competitive with petrochemical-based alternatives is being able to obtain high rates of production with little or no by-products, to efficiently use substrates, and to simplify the purification process [1]. The expected by-product of bioproduced succinic acid is acetic acid; however, other impurities, such as organic acids, amino acids, saccharides and polyalcohols might be present in trace amounts [1].

Chromatography is the preferred method of analysis because it adequately addresses the simultaneous identification and quantification of targeted compounds (i.e., carboxylic acids, saccharides, and polyalcohols) [4]. However, not all chromatographic protocols are suitable for the given task. For example, the high performance liquid chromatography (HPLC) of short-chain carboxylic acids (e.g., acetic or formic) is usually performed in the presence of a strong acid, such as diluted sulfuric acid [4], which is not compatible with mass spectrometry thus preventing the identification of numerous species potentially present in samples. The determination of acetic acid is crucial, because it is considered as the main impurity [1]. The alternative to HPLC is gas chromatography and mass spectrometry (GC-MS). Although the separation using this method generally targets volatile, non-polar species, the use of derivatization for polar low molecular weight species (i.e., the expected impurities) enables detection with a good resolution and sensitivity [4].

Numerous studies addressing acids, saccharides and polyalcohols were performed using GC-MS with trimethylsilylation [5-19] (Supplemental Table S.1 for their overview). Most of these studies characterize food products, focusing on relevant species occurring in fairly high concentration [4-9,16-18]. To our knowledge, no short-chain (i.e., highly volatile) monocarboxylic acids were reported. The shortest-chain acid reported was oxalic acid [11,12,16], which has two carboxylic groups available for derivatization and thus is less volatile than the derivatives of C_1 and C_2 monocarboxylic acids eluting using a non-polar stationary phase after application of a derivatization agent. Similarly, we did not find any study simultaneously addressing

***Corresponding author:** Kubátová A, University of North Dakota, Department of Chemistry, 151 Cornell St. Stop 9024, Grand Forks, ND 58202, USA
E-mail: alena.kubatova@UND.edu

both saccharides and acids. Finally, to our best knowledge, no study has yet addressed the most practical case characteristic for industrial production of pure chemicals when the trace amounts of impurities, such as acids, sugars and polyalcohols, were analyzed in the presence of a high concentration of one major mixture component, e.g., succinic acid.

Several options are available as for selecting the derivatization agents for GC-MS analysis of both acids and saccharides. The most common approach is derivatization with hydroxylamine in pyridine in combination with hexamethyldisilazane (HMDS) with trifluoroacetic acid [5-6,8-10], where hydroxylamine reacts with the saccharide carbonyl group while HMDS functionalizes the moiety containing a reactive hydrogen atom, i.e., carboxyl, hydroxyl and phenyl groups. However, the use of two derivatization agents may lead to uncertainties as the optimal conditions for two different derivatizations may not match. Also, HMDS is not the most efficient derivatization agent, leaving less reactive sources of active hydrogen, e.g., amino groups, unaltered [19]. For a more efficient derivatization of active hydrogen groups, including amino groups, either N-methyl-N-(trimethylsilyl) trifluoroacetamide [13,16] (MSTFA) or N,O-bis(trimethylsilyl) trifluoroacetamide [11,12,14,15] (BSTFA) is typically employed. The derivatization with BSTFA is often catalyzed with trimethylchlorosilane [11] (TMCS) or, in specific cases, trimethylsilylimidazole [7] (TMSI). Because trimethylsilylation is water-sensitive, the most common pretreatment of samples is either evaporation [5,-9,15,16] or lyophilization [13]. However, the short-chain monocarboxylic acids are volatile and thus may be lost together with the solvent, which might lead to underestimation of their content.

Thus, in order to provide a comprehensive characterization of impurities in bioproduced succinic acid samples, we developed a method for simultaneous saccharide and carboxylic acid determination using a GC-MS analysis and ensuring efficient derivatization. The efficiency of prior acetonitrile extraction compared to direct derivatization, and effectiveness of several derivatization agents/conditions for trimethylsilylation was evaluated. Finally, the effectiveness of the manufacturing processing and purification were assessed based on the concentrations of target species found in the samples.

Materials

Studied samples

Fifteen samples of succinic acid were used (labeled A–P; the complete list including detailed sample descriptions is provided in Supplement Table S.2). Samples C–P were produced on a large scale with *E.* coli bacteria using adapted protocol [20]. Briefly, the fermentation took place for 36 hours at 35°C using glucose based media enriched with ammonia as nitrogen source. The purification was accomplished via anion and cation exchange followed by electrodialysis to remove ammonium. Crystallization was used to further improve quality (samples G and L). Samples M–O were produced using a corn steep liquor, which is a by-product of corn wet milling. An analytical standard of succinic acid (99% purity; Sigma-Aldrich, St. Louis, MO, USA) and sample A were used as references, where sample A was petroleum based succinic acid.

Chemicals

Acetonitrile (ACN), methanol (both LCMS Optima grade), and dichloromethane (DCM, GC quality) were purchased from Fisher Scientific (Waltham, MA, USA). Water was purified using a Direct-Q3 water purification system with incorporated dual wavelength UV lamp (Millipore, Billerica, MA, USA) for low total carbon content (the manufacturers claimed impurity is less than 5 ng/g). Derivatization agents N,O-bis(trimethylsilyl)trifluoroacetamide (BSTFA, 99%) with 1% of trimethylchlorosilane (TMCS), BSTFA with 10% of TMCS, N-methyl-N-trimethylsilyl-trifluoroacetamide (MSTFA), were obtained from Sigma-Aldrich. Pyridine (99%) was obtained from Alfa Aesar (Ward Hill, MA, USA). The compounds quantified are listed in Table 2 along with their suppliers.

Sample preparation

Direct BSTFA derivatization: Samples (1.0 mg) were directly mixed with 50 μL BSTFA and derivatized overnight at 60°C. The amount of BSTFA was calculated to be in a 20-fold molar excess, considering the amounts of succinic acid in the samples. Samples were diluted to 200 μL using DCM together with 5.0 μL of an internal standard (*o*-terphenyl) to control the volume changes, and analyzed in vials with 400 μL inserts.

Extraction: Bioproduced succinic acid samples (1.00 ± 0.05 g) were sonicated overnight with 1 mL of acetonitrile. After sonication, the samples were filtered through some purified glass wool inserted into a Pasteur pipette.

BSTFA derivatization: Filtered ACN extracts (100 μL aliquot) were mixed with 50 μL BSTFA (99% + 1% TMCS), then derivatized for 1 h at 60°C. Alternatively, samples were derivatized for 18 h at 70°C in order to achieve a complete derivatization of saccharides and polyalcohols.

BSTFA derivatization with ACN: Acid and saccharides standards (100 μL) were dried and subsequently mixed with 50 μL BSTFA and 100 μL ACN and derivatized for 18 h at 70°C.

BSTFA derivatization with pyridine: Filtered ACN extracts (100 μL aliquot) were mixed with 60 μL BSTFA (99% + 1% TMCS) and 60 μL of pyridine and derivatized for 18 h at 70°C.

MSTFA derivatization: Acid and saccharides standards (100 μL) were mixed with 50 μL MSTFA and derivatized for 18 h at 70°C.

Calibration: Stock solutions of individual compounds were prepared and combined into two mixtures, i.e., acids (the final concentration ~0.5 mg/mL per analyte) and saccharides (the final concentration ~0.2 mg/mL per analyte). The calibration range was between 0.001-50 μg/mL, where the highest calibration point corresponded to ~30 μmoles of carboxylic or hydroxy groups. The list of compounds with their retention times, target and confirmation ions used for data processing is provided in Table 1.

Prior to the analysis an internal standard, *o*-terphenyl (10 μL, ~1 mg/mL), was added to all samples, and the solution was diluted to 1.0 mL using DCM unless stated otherwise.

Instrumentation

GC analyses were performed using a 5890 GC with 5972 MS equipped with an autosampler (6890 series, Agilent Technologies, Santa Clara, CA, USA). Injections were performed in the splitless mode for 0.50 min at 250°C and the injection volume was 1 μL. The separation was performed using a 52-m long DB-5MS capillary column, with 0.25 mm internal diameter (I.D.) and 0.25 μL film thickness (J&W Scientific, Folsom, CA, USA). A constant carrier gas (helium) at a flow rate of 1.0 mL/min was maintained during the analysis. The temperature program used was adapted from our previous work [21,22], and started at 35°C held for 5 min, followed by a gradient of 15°C/min to 300°C and

	Supplier	t_r[a] [min]	r_{12}[b]	MW ion	Target ion	Confirmation ions	LOD [ng]
formic acid	Fluka[c]	2.8	0.1	118	103	73, 45	0.2
acetic acid	Fisher[d]	3.9	0.2	132	117	75, 45	0.3
lactic acid	Sigma-Aldrich[e]	12.3	0.5	230	191	147, 117	0.2
oxalic acid	Sigma-Aldrich	13.6	0.6	230	190	219, 147	0.2
3-hydroxybutyric acid	Sigma-Aldrich	13.8	0.6	244	191	233, 117	0.04
butanediol	Sigma-Aldrich	13.9	0.6	234	177	147, 116	0.02
benzoic acid	Sigma-Aldrich	15.3	0.7	192	179	135, 105	0.4
glycerol	Fisher	15.7	0.7	308	205	218, 117	0.2
proline	Sigma-Aldrich	15.9	0.7	259	142	216, 73	0.1
malic acid	Sigma-Aldrich	18.5	0.8	344	233	245, 147	0.04
phthalic acid	Sigma-Aldrich	20.8	0.9	310	295	147,73	0.1
xylitol	Supelco[f]	21.0	0.9	502	307	319, 217	0.04
arabitol	Supelco	21.1	0.9	502	307	319, 217	0.02
ribitol	Supelco	21.2	0.9	502	319	307, 217	0.06
citric acid	Sigma-Aldrich	22.2	1.0	480	273	465, 73	0.03
glucose	Supelco	23.1	1.0	530	204	191, 147	0.02
sucrose	Supelco	29.8	1.3	902	361	217, 73	19

[a] Retention time
[b] Relative retention time (retention time/IS retention time)
[c] Fluka – St. Louis, MO, USA
[d] Fisher – Waltham, MA, USA
[e] Sigma-Aldrich – St. Louis, MO, USA
[f] Supelco – St. Louis, MO, USA

Table 1: List of acids, saccharides and polyalcohols studied, their suppliers, the GC–MS retention times, target and confirmation ions (used for quantification) of their trimethylsilyl derivatives used for data processing, and limits of detection (LODs).

held for 1 min. The MS data in total ion chromatograms (TIC) were acquired in the mass range of *m/z* of 35–1000 at a scan rate 2.66 scan/s using the EI of 70 eV. The MS was turned off to eliminate signal from the derivatization agents and their by-products in periods determined by observing the increase of pressure in MS. Namely, for BSTFA with pyridine, the MS was off for the first 2.5 min, 2.90-3.60 min, 4.40-7.00 min, 8.00-8.70 min; for MSTFA, the MS was off for the first 4 min.

Data processing

GC-MS data were processed using ChemStation (version E.02.02.1431) and AMDIS software (Automated Mass Spectral Deconvolution and Identification System, version 2.71) [23]. Compounds' identification was based on confirmation with the corresponding analytical standard, or as isomers of standards with similar mass spectra and/or using NIST 05 Mass Spectra library.

AMDIS software was used for the deconvolution of MS ion spectra and tentative identification of impurities for which the analytical standards are not available. The tentative identification was based primarily on the reversed match of >80% and compared to the weighted match requiring at least 80% for both matching methods. Peaks found in the pure succinic acid standard and in the BSTFA blank were not considered. Based on TIC, the AMDIS program provided a percent response, which allowed for semi-quantification of impurities (Table 3) and their comparison between samples, by normalizing to the response of the internal standard.

The limits of detection and quantification (LODs and LOQs) were determined using the target ions *m/z*, which were selected based on the highest signal-to-noise ratio (ions listed in Table 2). The instrumental LODs were calculated from calibration curves (within one order of magnitude of LOD) using the formula $LOD=3.3*s_y/k$, where k is a slope of the calibration curve and s_y is the standard error of the predicted y-value for each x-value; s_y was obtained by a least square linear regression. In order to report the low amounts of impurities we have used for quantification, lower limits of quantification were defined as $LOQ=5*s_y/k$.

The repeatability of the quantification method was evaluated using a representative sample of bioproduced succinic acid (C), which was chosen on the basis of preliminary testing. The sample was prepared in triplicate and analyzed in the following ways: 1) the same sample was analyzed three times in a row to assess the intraday GC repeatability; 2) the same sample was analyzed throughout the sequence on two consecutive days, to evaluate the interday GC repeatability; and 3) the extraction triplicate was analyzed to assess the extraction repeatability.

Results and Discussion

Extraction v/s direct analysis

The selection of a sample preparation method strongly affects the impurities detected. Thus we first compared the extraction using ACN followed by derivatization with BSTFA with direct BSTFA derivatization (no extraction). Figure 1 shows that the ACN extraction was essential for characterization of impurities. A range of peaks representing impurities was observed in the majority of ACN extracted and BSTFA derivatized samples (Figure 1b and Table 3). We expected enhanced derivatization when eliminating the extraction step and using BSTFA in molar excess; however no additional impurities were found when the direct analysis was applied (Figure 1a). The higher responses observed after extraction could be explained by a higher solubility of impurities in acetonitrile than in the derivatization agent alone, combined with a lower solubility of succinic acid in ACN.

Initial identification of impurities

The initial method of analysis was adapted from our previous work [21] allowing for quantification of a wide range of mono- and di-carboxylic acids. Over 120 peaks were observed in the initial process bacterial succinic acid samples upon derivatization with BSTFA. Table 3 shows the normalized data for the most abundant species (the detailed

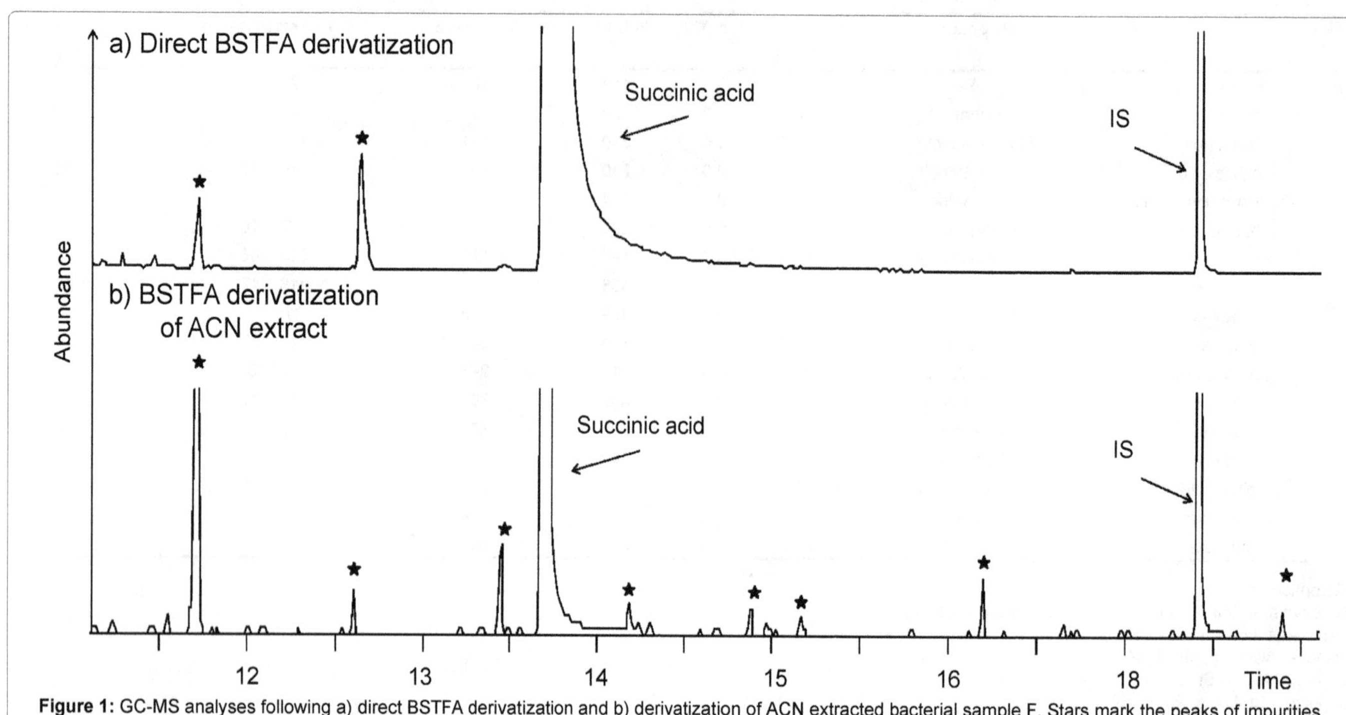

Figure 1: GC-MS analyses following a) direct BSTFA derivatization and b) derivatization of ACN extracted bacterial sample F. Stars mark the peaks of impurities observed in bacterial samples. Chromatograms are scaled to the internal standard height.

r_{12}[a]	Identified compounds	A (petroleum)	F (bacteria)	K (bacteria)	Confirmed[b]
0.319	formic acid	0.01	0.03	0.03	*
0.406	acetic acid	0.02	0.12	0.17	*
0.570	methyl-propanoic acid		0.03		
0.604	alanine			0.03	
0.608	dimethylsulfone			0.01	*
0.631	ethanediol	0.04			*
0.663	butanediol		0.02	0.03	*
0.672	lactic acid		0.74	0.30	*
0.694	alanine			0.01	
0.715	methyl butanol	0.01			
0.720	3-hydroxybutyric acid			0.02	*
0.722	oxypentanoic acid			0.02	
0.724	hydroxymethylbutyric acid		0.05		
0.736	pentenoic acid			0.03	
0.747	L-valine (bisTMS)			0.08	*
0.759	ethyl succinate			0.04	*
0.770	glycerol			0.04	*
0.773	phosphoric acid		0.10		
0.792	methyl succinic acid	0.03			
0.798	pyrimidine			0.02	
0.815	malic acid		0.03	0.08	
0.821	pentanedioic acid			0.02	*
0.854	malic acid	5.40	0.03	0.02	
0.860	hexanedioic acid		0.01		*
0.930	phthalic acid	0.03	0.05		
0.967	citric acid			0.07	*
0.992	heptanol derivative			0.04	
1.000	***o*-terphenyl (IS)**	**1.00**	**1.00**	**1.00**	**IS**
1.012	glucose			0.02	

[a] Relative retention time (retention time/IS retention time)

[b] Confirmed using the analysis of standard.

Table 2: Contaminants and their percent responses, with respect to an internal standard, observed upon BSTFA derivatization of an ACN extract of petroleum produced succinic acid and initial process bio-based succinic acid samples

list is in Supplemental Table S.3). The common impurities of higher abundance in the bacterial samples were formic, acetic, lactic and malic acids, butanediol and L-valine (Figure 2). Using this screening method, we also observed incompletely derivatized saccharides. Other compounds found in a lower abundance were oxalic, benzoic, phthalic, hexadecanoic, and octadecanoic acids (Table 3). These acids might be from the sample preparation contamination; however their abundance in controls (experiment performed without analytes) seemed to be lower.

The screening results showed primarily acids, saccharides and polyalcohols, which are essential for production control on large scale [1,3], and thus, the further quantification efforts targeted these species.

Development of quantification method for analysis of acids and saccharides as the most abundant impurities

Based on our previous work [23] and reported data, several trimethylsilylation methods were compared to determine the most efficient approach for a simultaneous derivatization of saccharides and acids. These methods included the derivatization with MSTFA in the presence of ACN, and BSTFA (1% TMCS) with/without ACN or pyridine. The application of these derivatization agents to saccharides resulted in only an incomplete derivatization in MSTFA with or without ACN and in BSTFA without either pyridine or ACN (Figures 3a and b). Xue et al. [24] reported multiple peaks for glucose derivatized with MSTFA, however, the problem was not addressed. By contrast, BSTFA in the presence of either ACN or pyridine resulted in a complete derivatization of saccharides and polyalcohols (Figures 3c and d). Nevertheless further tests of derivatization evaluation of BSTFA with ACN and pyridine resulted in higher peaks o glucose in presence of pyridine (Figure 4). The comparison of extracted ion chromatograms of acetic acid (ion 117, $[M-15]^+$) demonstrates that the MSTFA (Figure 5a) and BSTFA derivatization with ACN (Figure 5c) resulted in higher peaks compared to the derivatization using BSTFA with pyridine. Perhaps pyridine had a negative effect on the transfer of volatile analytes from the GC injection port to the column due to its relatively high boiling point and tendency to bind acids due to the formation of pyridinium salts. Therefore, the derivatization using BSTFA with ACN seemed to be optimal for acids, while BSTFA with pyridine was more effective for saccharides (Figures 3-5). We also tested the separation of succinic acid and its isomer, methylmalonic acid. Those compounds were completely separated as shown in Supplemental Figure S.1.

Limits of detection and repeatability

Table 2 lists the obtained instrumental LODs, which were in a range of 0.03-0.6 ng for acids and 0.03-0.2 ng for saccharides and polyalcohols. The values obtained for acids are comparable to those reported in our previous study [23], while we achieved ten-fold lower values for sugars than in the study of Adams et al. [10], where HMDS was used as derivatization agent, possibly due to a more effective derivatization or greater calibration range. LOD's in other studies [11,13,15] were not comparable because they have been reported in different units, e.g. Pietrogrande and Bacco [11] reported as air volume concentrations.

The repeatability of the developed quantification method on representative sample C is demonstrated in (Table 4). The GC intra- and interday repeatability as well as sample preparation were similar, with relative standard deviation (RSD) <10%, with exception of glycerol, where intraday reproducibility was 12%

Characterization of succinic acid samples

The developed quantification method was applied to bioproduced succinic acid samples, as an application for monitoring the product quality. The targeted compounds were the most abundant acids, as well as saccharides, and polyalcohols, i.e., formic, acetic, lactic, oxalic, 3-hydroxybutyric, benzoic, malic, phthalic and citric acids, butanediol, glycerol, xylitol, arabitol, glucose, and sucrose (Table 2).

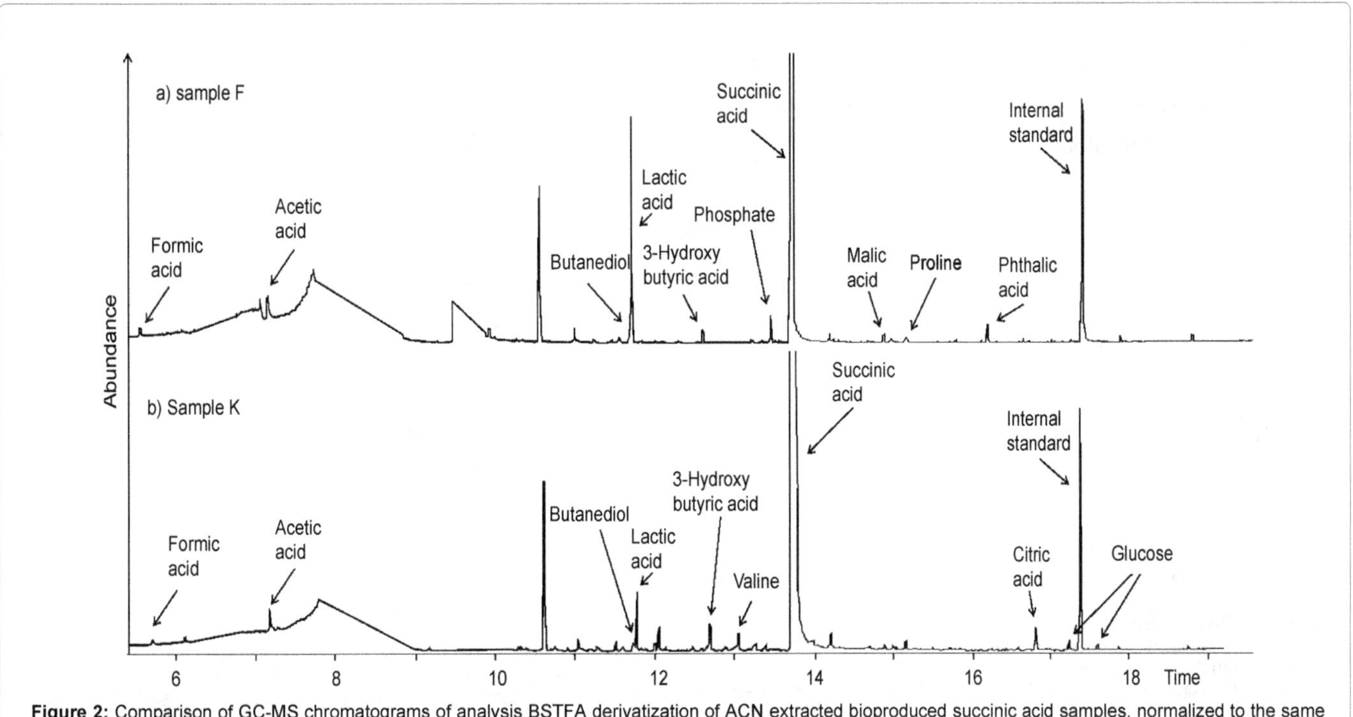

Figure 2: Comparison of GC-MS chromatograms of analysis BSTFA derivatization of ACN extracted bioproduced succinic acid samples, normalized to the same percent response of internal standard. Samples F(a) and K(b) were initial process samples.

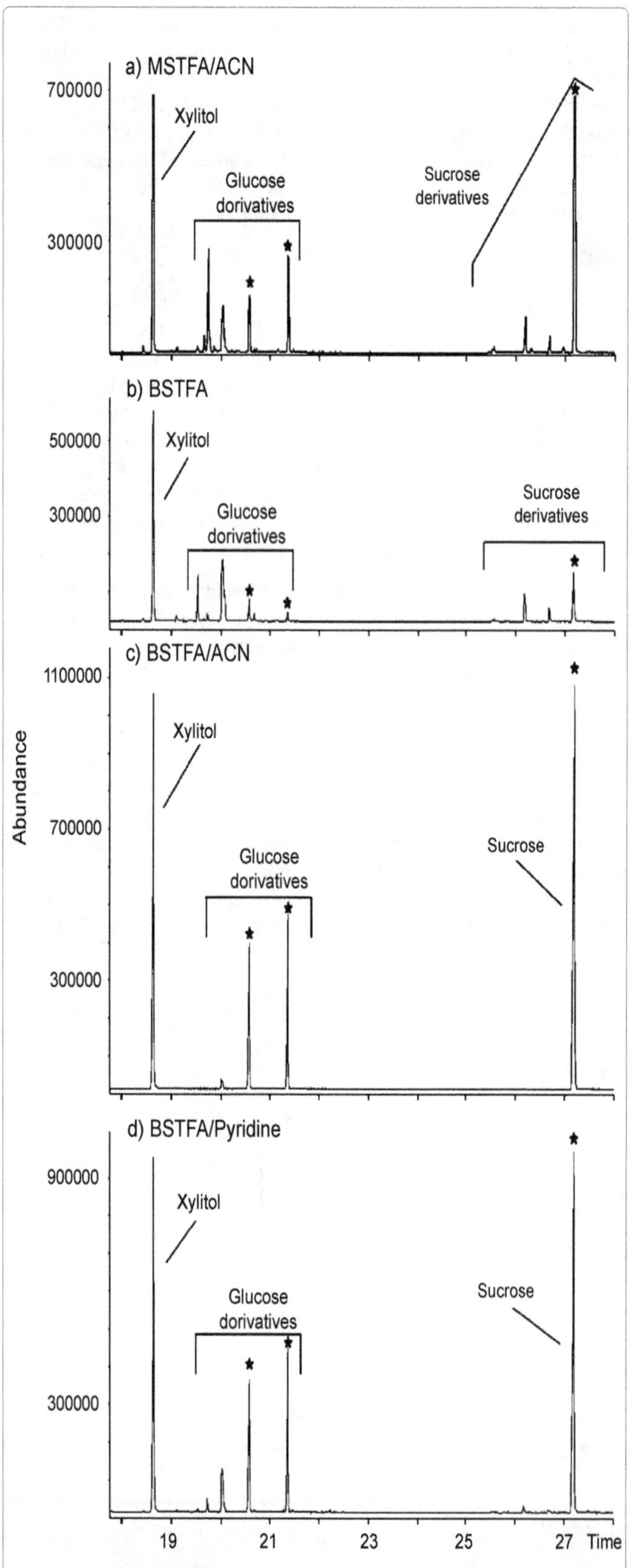

Figure 3: GC-MS extracted ion chromatograms (m/z = 217) of a mixture of standard saccharides and polyalcohols upon derivatization (18 h at 70°C) with a) MSFTA with ACN, b) BSTFA (1%TMCS), c) BSTFA 1% TMCS with ACN, d) BSTFA (1% TMCS) with pyridine. The stars mark peaks of the completely derivatized sucrose and glucose.

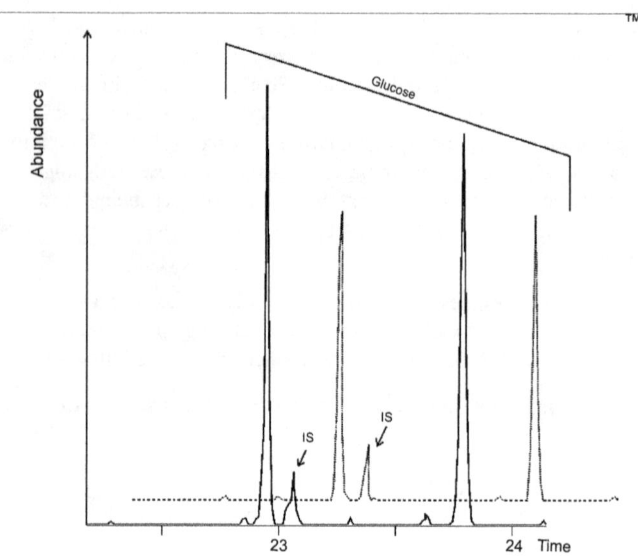

Figure 4: GC-MS extracted ion 204 chromatograms of bio-produced succinic acid (sample F) upon derivatization with various derivatization agents for 18 hours at 70°C. The derivatization with pyridine (solid line) provided a higher response than that with ACN (dashed line). IS denotes internal standard. The IS co-elutes with other derivatized hexose, which is believed to be an impurity in the glucose standard.

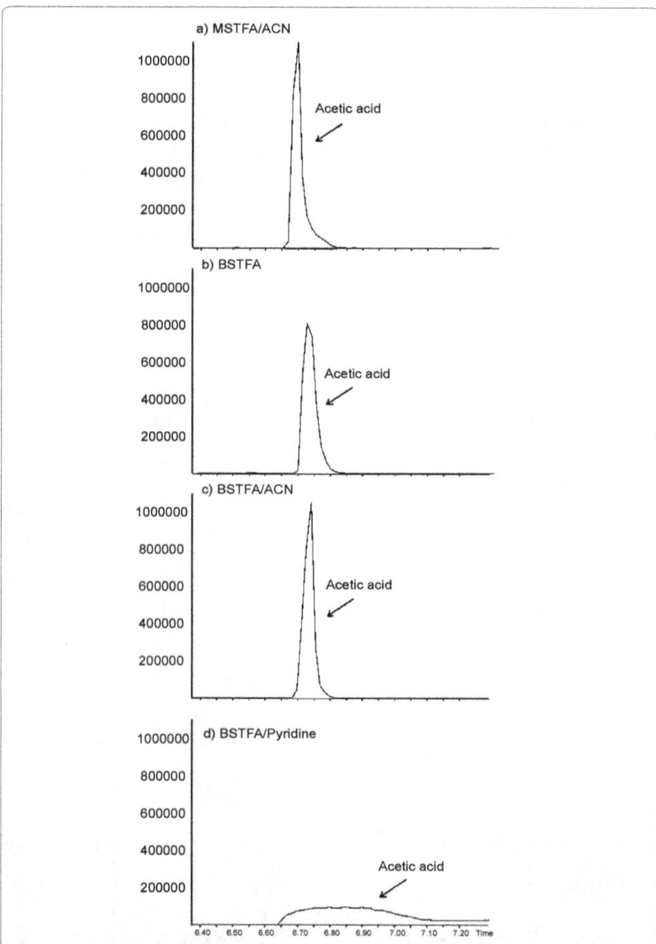

Figure 5: GC-MS extracted ion chromatograms (m/z = 117) of an acetic acid standard upon derivatization (18 h at 70°C) with a) MSFTA with ACN, b) BSTFA (1%TMCS), c) BSTFA 1% TMCS with ACN, d) BSTFA 1% TMCS with pyridine.

Due to the low concentrations of some of these compounds in the samples, quantification is reported only for a narrower range of these compounds featuring the concentrations above the corresponding LODs (Table 4).

Abundance of acids, saccharides, and polyalcohols: Quantification confirmed occurrence of all tested acids and glucose (Tables 3 and 4). The polyalcohols in samples were found as well but xylitol, arabitol and ribitol were below their LOD.

As mentioned above, acids were the prevailing impurities in the bacterial samples. Acetic acid is a common contaminant of biologically produced succinic acid [1], and for its unpleasant smell was an undesirable impurity. It has been abundant in samples F and K (13 μg/g and 20 μg/g, respectively), but its concentration decreased in purified sample G (3 μg/g). Formic acid, which has also undesirable odor, had been determined in all samples between 1 μg/g in samples K, L, and M (Table 4) and 16 μg/g in samples A (petroleum-based sample). Similarly to acetic acid, formic acid concentration decreased after purification from 5 μg/g (sample F) to 1 μg/g (sample G). Malic acid, also used in industry for polymer production [3], was the major impurity in sample A (1.2 mg/g) and lactic acid was found in samples F and K (0.2 mg/g and 27 μg/g, respectively).

Polyalcohols found in the samples were glycerol, butanediol (Table 4). Glycerol was found in samples F and L (0.5 and 0.3 μg/g, respectively). Butanediol was also found in sample F (5 μg/g) and sample K (4 μg/g). Ethanediol was observed in petroleum based sample but it was not quantified in other samples. Sugar polyalcohols were not detected, with exception of arabitol, which was detected in sample N, but it was below its limit of quantification. Glucose was only representative of saccharides with concentration up to 8 μg/g in sample K (Table 4).

The effect of production media on the purity of succinic acid was evaluated for samples K–O comparing the product produced by bacteria in a defined medium (sample K) and in corn steep liquor (samples M, N, O). Corn steep liquor is less expensive as it is a by-product of corn wet milling and so it is preferred in industry; however, the product obtained using this complex organic mixture was expected to contain more impurities. In contrast to this expectation, samples M, N, and O and other initial process samples contained similar impurities (formic acid, acetic acid and glucose), suggesting that the production medium had a lower impact on generation of the observed impurities than the production microorganism. Only oxalic acid was observed in a 4-fold higher abundance in sample M with corn steep, compared to sample K produced using a defined medium.

Final bacterial process samples: The effectiveness of the product purification was evaluated by comparison of samples F and K (initial process), and G and L (final product) where G was purified F. While most of the targeted compounds were detected in initial process samples, only formic and acetic acids were quantified in purified sample G, showing a decrease from 0.13 μg/g to 0.06 μg/g for formic acid and from 0.3 μg/g to 0.1 μg/g for acetic acid. Sample L showed also some glycerol present. Lactic and malic acids were both detected in initial process samples, but were not found in refined samples (Table 4). Thus the developed method was demonstrated to be suitable for the quality control of the process as well as demonstrated purity of the final products.

Conclusions

We have developed a protocol for characterization and quality control of bioproduced succinic acid. A prior ACN extraction was found to be essential to detect impurities. The optimization of derivatization was critical for low molecular weight polar acids as well as saccharides; a procedure using BSTFA with pyridine as a catalyst was determined to be suitable for both polyalcohols and saccharides whereas the BSTFA with ACN treatment was found to be the suitable for quantification

Analyte	GC intraday			GC interday			Extraction		
lactic acid	6.3	±	0.5	6.2	±	0.4	6.0	±	0.1
benzoic acid	0.63	±	0.02	0.67	±	0.06	0.61	±	0.03
glycerol	0.12	±	0.01	0.11	±	0.01	0.12	±	0.01
glucose	0.08	±	0.01	0.08	±	0.01	0.071	±	0.004

Table 3: GC intra, interday, and extraction method repeatability for a bioprocessed sample of succinic acid (sample C) reported as a mean value (in μg/g) ± one standard deviation (n=3).

Analyte	A (petroleum)			F (initial process)			K (initial process)			G (final process)			L (final process)		
formic acid	15	±	5	5	±	2	1.1	±	0.03	1	±	0.06	1.5	±	0.8
acetic acid		Below LOQ		13	±	3	20	±	6	3.2	±	0.6	3.9	±	0.8
oxalic acid	8	±	5		Below LOQ[a]			Below LOQ			ND[b]			ND	
lactic acid		ND		186	±	19	27	±	4		ND			ND	
3-hydroxybutyric acid		Below LOQ			Below LOQ		1.1	±	0.1		ND			ND	
butanediol		Below LOQ		5.1	±	0.3	3.6	±	0.1		ND			ND	
benzoic acid	2.00	±	0.03		Below LOQ			ND			ND			ND	
glycerol		ND		0.49	±	0.06		ND			ND		0.23	±	0.03
malic acid	1159	±	24	10	±	2		Below LOQ			Below LOQ			Below LOQ	
phthalic acid	7	±	2		ND			ND			ND			ND	
citric acid		ND			Below LOQ		8	±	1		ND			ND	
glucose		Below LOQ		3.1	±	0.2	0.07	±	0.01	0.02	±	0.001		Below LOQ	

[a] Below LOQ – below quantification limit
[b] ND – not detected

Table 4: Concentrations of acids and saccharides in bioprocessed succinic acid samples reported as a mean value (in μg/g) ± one standard deviation (n=3).

of low molecular weight carboxylic acids. The presence of short chain monocarboxylic acids, i.e. formic and acetic acid, has an effect on odor of final product, which is undesirable in the industrial process. Presence of saccharides might lead to caramelization or Maillard reactions, resulting in coloring the final product. We achieved LODs as low as 0.02 ng for saccharides and 0.03 ng for acids, which makes the quantification method advantageous for detection of trace-level impurities even in the presence of one major compound at a high concentration, e.g., succinic acid. The final process samples showed removal or decrease of all quantified compounds.

Acknowledgement

We appreciate editorial feedback on the manuscript from Dr. Kozliak at the Chemistry Department, University of North Dakota, Grand Forks, USA. We thank to Andrey Lobashov for his help with obtaining the microscope pictures.

References

1. McKinlay JB, Vieille C, Zeikus JG (2007) Prospects for a bio-based succinate industry. Appl Microbiol Biotechnol 76: 727-740.

2. Willke T, Vorlop KD (2004) Industrial bioconversion of renewable resources as an alternative to conventional chemistry. Appl Microbiol Biotechnol 66: 131-142.

3. Sauer M, Porro D, Mattanovich D, Branduardi P (2008) Microbial production of organic acids: expanding the markets. Trends Biotechnol 26: 100-108.

4. Molnár-Perl I (2000) Role of chromatography in the analysis of sugars, carboxylic acids and amino acids in food. J Chrom A 891: 1-32.

5. Katona ZF, Sass P, Molnár-Perl I (1999) Simultaneous determination of sugars, sugar alcohols, acids and amino acids in apricots by gas chromatography-mass spectrometry. J Chrom A 847: 91-102.

6. Füzfai Z, Katona ZF, Kovács E, Molnár-Perl I (2004) Simultaneous identification and quantification of the sugar, sugar alcohol, and carboxylic acid contents of sour cherry, apple, and ber fruits, as their trimethylsilyl derivatives, by gas chromatography-mass spectrometry. J Agr Food Chem 52: 7444-7452.

7. Ačanski MM, Vujić DN (2014) Comparing sugar components of cereal and pseudocereal flour by GC–MS analysis. Food Chem 145: 743-748.

8. Boldizsár I, Füzfai Z, Molnár-Perl I (2013) Characterization of the endogenous enzymatic hydrolyses of Petroselinum crispum glycosides: Determined by chromatography upon their sugar and flavonoid products. J Chrom A 1293: 100-106.

9. Horváth K, Molnár-Perl I (1998) Simultaneous GC-MS quantitation of o-phosphoric, aliphatic and aromatic carboxylic acids, proline, hydroxymethylfurfurol and sugars as their TMS derivatives: In honeys. Chromatographia 48: 120-126.

10. Adams MA, Chen Z, Landman P, Colmer TD (1999) Simultaneous determination by capillary gas chromatography of organic acids, sugars, and sugar alcohols in plant tissue extracts as their trimethylsilyl derivatives, Anal Biochem 266: 77-84.

11. Pietrogrande MC, Bacco D (2011) GC-MS analysis of water-soluble organics in atmospheric aerosol: Response surface methodology for optimizing silyl-derivatization for simultaneous analysis of carboxylic acids and sugars. Anal Chim Acta 689: 257-264.

12. Schummer C, Delhomme O, Appenzeller BM, Wennig R, Millet M (2009) Comparison of MTBSTFA and BSTFA in derivatization reactions of polar compounds prior to GC/MS analysis. Talanta 77: 1473-1482.

13. Koek MM, Muilwijk B, van der Werf MJ, Hankemeier T (2006) Microbial Metabolomics with Gas Chromatography/Mass Spectrometry. Anal Chem 78: 1272-1281.

14. Rojas-Escudero E, Alarcón-Jiménez AL, Elizalde-Galván P, Rojo-Callejas F (2004) Optimization of carbohydrate silylation for gas chromatography. J Chrom A 1027: 117-120.

15. Medeiros PM, Simoneit BRT (2007) Analysis of sugars in environmental samples by gas chromatography–mass spectrometry. J Chrom A 1141: 271-278.

16. Cerdán-Calero M, Sendra JM, Sentandreu E (2012) Gas chromatography coupled to mass spectrometry analysis of volatiles, sugars, organic acids and aminoacids in Valencia Late orange juice and reliability of the Automated Mass Spectral Deconvolution and Identification System for their automatic identification and quantification. J Chrom A 1241: 84-95.

17. Silva FO, Ferraz V (2004) Microwave-assisted preparation of sugars and organic acids for simultaneous determination in citric fruits by gas chromatography. Food Chem 88: 609-612.

18. Rodríguez-Sánchez S, Hernández-Hernández O, Ruiz-Matute AI, Sanz, ML (2011) A derivatization procedure for the simultaneous analysis of iminosugars and other low molecular weight carbohydrates by GC-MS in mulberry (Morus sp.). Food Chem 126: 353-359.

19. Harvey DJ (2011) Derivatization of carbohydrates for analysis by chromatography; electrophoresis and mass spectrometry. J Chrom B 879: 1196-1225.

20. Glassner D, Elankovan P, Beacom D, Berglund K (1995) Purification process for succinic acid produced by fermentation. Appl Biochem Biotechnol 51-52: 73-82.

21. Šťávová J, Beránek J, Nelson EP, Diep BA, Kubátová A (2011) Limits of detection for the determination of mono- and dicarboxylic acids using gas and liquid chromatographic methods coupled with mass spectrometry, J Chromatogr B Analyt Technol Biomed Life Sci 879: 1429-1438.

22. Stein SE (1999) An integrated method for spectrum extraction and compound identification from gas chromatography/mass spectrometry data. J Am Soc Mass Spectrom 10: 770-781.

23. http://chemdata.nist.gov/mass-spc/amdis/

24. Xue R, Zhang S, Deng C, Dong L, Liu T, et al. (2008) Simultaneous determination of blood glucose and isoleucine levels in rats after chronic alcohol exposure by microwave-assisted derivatization and isotope dilution gas chromatography/mass spectrometry. Rapid Commun Mass Spectrom 22: 245-252.

A Novel Capillary Zone Electrophoresis Method for Simultaneous Separation and Determination of Nalbuphine Hydrochloride and its Related Antagonist Compounds

Alarfaj NA and El-Tohamy MF*

Department of Chemistry, College of Science, King Saud University, PO Box 22452, Riyadh 11495, Saudi Arabia

Abstract

In the present study, we introduced a novel, reliable, sensitive and highly precise capillary zone electrophoresis (CZE) method for simultaneous separation and determination of nalbuphine hydrochloride (NLB) and its related antagonist compounds naloxone hydrochloride (NLX) and naltrexone hydrochloride (NLT) in bulk drug, pharmaceuticals and human biological fluids. The separation conditions using CZE were optimized and the separation process was carried out using fused silica capillary of total and effective lengths 57 and 50 cm, respectively. The applied running buffer was acetate buffer 20 mmol L^{-1} of pH=3.8 at a potential 15 kV. The detection of samples was carried out using Diode array (DAD) at 230 nm with hydrodynamic injection 6 s, under 70 mbar and capillary cartridge 25°C. The applied internal solution (IS) was phenylethylamine (PEA). The developed method displayed an excellent separation of the investigated drugs with linear concentration ranges of 5-200, 20-240 and 10-280 µg mL^{-1} for NLB, NLX and NLT, respectively. Low detection limits were recorded as 0.3, 5.0 and 6.5 µg mL^{-1} while, the quantification limits were 5, 20 and 10 µg mL^{-1} for NLB, NLX and NLT, respectively. Good correlation coefficients were evaluated at 0.9997, 0.9995 and 0.9996, for the previously selected drugs, respectively. The % RSD was in an acceptable limit indicating high precision. Excellent separation and detection with acceptable results were achieved with respect to migration time, peak area and resolution. The obtained results were statistically evaluated and then compared with those obtained from other published methods. The electrophoretic method was validated in compliance with ICH guidelines.

Keywords: Capillary zone electrophoresis; Nalbuphine hydrochloride; Naloxone hydrochloride; Naltrexone hydrochloride; Pharmaceutical formulations; Biological fluids

Introduction

The synthetic opioid agonist-antagonist of the phenanthrene series nalbuphine hydrochloride (NLB), is chemically known as 17-(cyclobutylmethyl)-4,5α-epoxymorphinan-3,6α,14-triol hydrochloride (Figure 1a). It is a member of chemically related drugs to the widely used opioid antagonist, naloxone hydrochloride (NLX) and naltrexone hydrochloride (NLT) [1]. The introduced literature survey showed a number of analytical methods concerning the determination of NLB in different matrices including pharmaceutical formulations and biological samples. Among these methods are the chromatographic separation methods, namely high performance liquid chromatography [2,3], thin layer chromatography [4], gas chromatography-mass spectrometry [5] and liquid chromatography coupled with mass spectrometry [6]. Also, other analytical methods such as spectrophotometry [7] and potentiometry [8] have been published. NLX is a narcotic antagonist drug used for treatment of overdoses of narcotic medications [9]. It is a member of isoquinoline compounds (Figure 1b). Many analytical methods have been reported for the detection of NLX mainly by high performance liquid chromatography [10,11], high performance liquid chromatography coupled with mass spectrometry [6,12-15], chemiluminescence [16] and potentiometry [17].

NLT, (Figure 1c) is an opioid receptor antagonist which has a long action to block the subjective effect of alcohol dependence and opioid dependence [18]. The chemical name of naltrexone is 17-(cyclopropylmethyl)-4,5-epoxy-3,14-dihydroxymorphinan-6-one,(5α)-hydrochloride [19]. Many analytical methods have been reported for determination of naltrexone hydrochloride including RP-high performance liquid chromatography [20,21], liquid chromatography-mass spectrometry [22], spectrophotometry [23], spectrofluorimetry [24] and potentiometry [25].

In recent years, much attention has been undertaken for using CZE in wide fields of chemical analysis [26,27]. Although, the chromatographic separation methods such as LC-MS has a high sensitivity and selectivity for drug analysis, the proposed CZE method offered many advantages, including cost benefit technique, improved separation speed and less solvent consuming rather than other chromatographic separation techniques.

Regarding to the studied drugs NLB, NLX and NLT no spectroscopic analysis methods have been reported yet for simultaneous separation and determination of the selected drugs. Therefore, the aim of the present study is to develop a novel, rapid, sensitive and reliable method for simultaneous separation and estimation of NLB and its related antagonist compounds NLX and NLT. The described method was applied in the detection of the investigated drugs and further extended to be validated according to ICH guidelines [28].

***Corresponding author:** Maha El-Tohamy, Department of Chemistry, College of Science, King Saud University, PO Box 22452, Riyadh 11495, Saudi Arabia
E-mail: star2000star@gmail.com

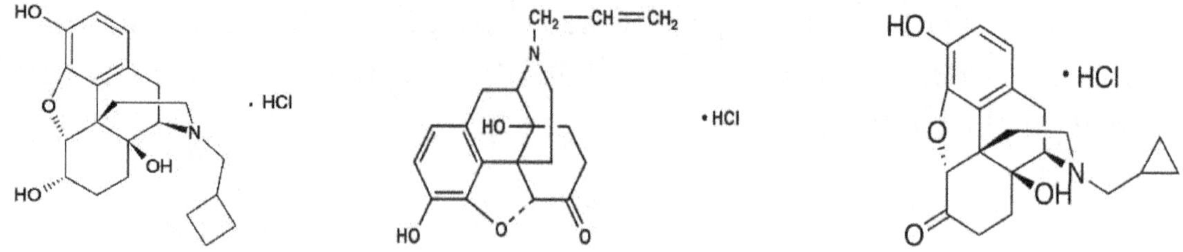

Figure 1: Chemical structures of nalbuphine hydrochloride, naloxone hydrochloride and naltrexone hydrochloride.

Experimental

Instrumentation and software

The electrophoretic separation was done using (PrinCE 770-Technology) instrument, which comprises a fused silica capillary with a total and effective lengths of 57 and 50 cm, respectively and 75 µm i.d. It was connected with diode-array detector (DAD) for peak detection. The CZE PrinCE 770-Technology system was equipped with Autosampler and a thermostated column cartridge was connected to adjust the temperature. Also, it has a high voltage built in power supply. Automated, controlled PC for electrophoretic separation was applied using a WinPrinCE-770, DA × 3D software for data acquisition. HANNA 211-pH meter connected with Ag/AgCl reference electrode was used for adjusting the electrolyte pH. The HPLC-DAD detection was carried out using Agilent 1200 series (Agilent Technologies, Santa Clara, California, USA), coprises of pump, vacuum degasser and diode array detector. The column used Zorbax SB-C18 (4.6 × 250 mm, 5.0 mm particle size). The system was PC controlled and the computer was loaded with Agilent ChemStation Software.

Materials and reagents

Analytical grade of all materials and reagents were used and all solvents were of HPLC spectroscopic grade. The electrophoretic separation was performed using acetate buffer pH=3.8 (freshly prepared using 0.1 mol L^{-1} acetic acid and 0.1 mol L^{-1} sodium acetate). Phosphate buffer using mono and dibasic hydrogen phosphate solutions with pH 5.8 - 8.0, glacial acetic acid ≥ 99.0%, phenylethylamine hydrochloride (PEA) ≥ 99.0%, orthophorphoric acid and ammonium acetate 99.0% were purchased from (Sigma-Aldrich, Hamburg, Germany). Also, acetonitrile, methanol, ethanol and isopropanol were supplied from (BDH, Philadelphia, USA). Zinc sulfate ≥ 99.0%, sodium hydroxide ≥ 99.0%, sodium acetate, boric acid and sodium dihydrogen phosphate ≥ 90.0 % were purchased from (WinLab, East Midlands, UK). Pure grade of nalbuphine hydrochloride was kindly provided from (Amoun Pharmaceutical Co., Cairo, Egypt). While, naloxone hydrochloride and naltrexone hydrochloride were supplied by (Bristol Myer Squibb Co., Giza, Egypt). Nalufin® ampoules 20 mg mL^{-1} of NLB, Vivitrol vials 380 mg/vial of NLT and Narcan® ampoules, 400 µg mL^{-1} of NLX were purchased from local drug stores. Deionized water was used throughout the experiments. The urine samples were provided from healthy volunteers. Informed consent was obtained from all of them prior to the start of the study. The study was approved by the Medical Ethics Committee in the College of Medicine, King Saud University. Commercial sources were supplied the human serum samples (Multi-Serum Normal, Randox Laboratories, Crumlin, Antrim, UK).

Preparation of analytical samples

Standard solutions: Freshly, 300 µg mL^{-1} stock solutions of NLB, NLX and NLT were prepared by dissolving 30 mg of each pure drug in 100 mL deionized water. Phenylethylamine hydrochloride 100 µg mL^{-1} was applied as (IS) and prepared by dissolving 10 mg of PEA in 100 mL deionized water. Deionized water was used for serial dilution of daily working solutions.

Preparation of authentic mixtures: The electrophoretic analysis was performed using a working solution 100 µg mL^{-1} of each drug in the presence of 1.0 mL of 100 µg mL^{-1} IS. Aliquots of NLB, NLX and NLT standard solutions in the final concentration ratios of 1:1:1, 1:2:2, 1:4:4, 1:6:6, 1:8:8 and 1:10:10 (w/w) respectively were mixed and subjected to analysis. The regression equation was applied to calculate the percentage recoveries of each drug.

Preparation of Nalufin, Narcan and Vivitrol injection solutions: The standard Nalufin injection solution was prepared by transferring the content of two ampoules Nalufin ampoules into a 100-mL volumetric flask and diluted with deionized water to obtain a solution containing 400 µg mL^{-1} NLB. In case of Narcan ampoules the content of ten ampoules was diluted with deionized water in 20-mL volumetric flask to obtain 200 µg mL^{-1} NLX. A stock solution of the NLT was prepared by transferring the content of one vial of Vivitrol injection into a 100-mL volumetric flask. Then, it was diluted with deionized water to obtain a solution containing 3800 µg mL^{-1} of NLT. Working solutions were prepared in the ranges of 50-200, 100-240 and 50-280 for NLB, NLX and NLT, respectively. The electrophoretic detection of the investigated drugs was carried out in the presence of 1.0 mL of 100 µg mL^{-1} of IS.

Preparation of biological samples

The proposed CZE method was successfully applied for determination of NLB, NLX and NLT in human serum and urine. Spiking technique method was used for detection of the investigated drugs in biological fluids. 1.0 mL of human serum was spiked with different aliquots of each drug. Serum deprotination was performed by adding 1.0 mL of acetonitrile, 0.1 mL of NaOH (0.1 mol L^{-1}) followed by 1.0 mL of $ZnSO_4$.7 H_2O (5.0% w/v). The prepared solution was centrifuged at 3500 rpm for 30 min. Then the clear layer was filtered using 0.5 Milli-pore a membrane filter. The human urine samples were collected from healthy volunteers, 5.0 mL of urine was spiked with accurately measured aliquots of the investigated drugs separately. Then the solutions were diluted with deionized water no further treatment was required. Working solutions were obtained by serial dilutions with the same solvent.

Electrophoretic conditions

In the present study, a new developed CZE method for simultaneous separation and determination of NLB, NLX and NLT was applied. CZE separation of the selected drugs was performed under optimum conditions using 20 mmol L^{-1} acetate buffer pH=3.8. The capillary should be conditioned before carrying the separation process using 0.1 mol L^{-1} sodium hydroxide for 2 min followed by deionized water for 2 min and then equilibrated with running electrolyte for 5 min. Hydrodynamic injection of the samples was carried out under applied voltage 15 kV, capillary cartridge temperature of 25°C and applied pressure of 70 mbar for 6 s. To ensure the separation reproducibility throughout the experiment, the capillary was replenished by 0.1 mol L^{-1} sodium hydroxide for 5 min, deionized water for 5 min and running buffer electrolyte for 10 min.

Calibration curve

The calibration curves of the investigated drugs were plotted using different concentration ranges of 5-200, 20-240 and 10-280 μg mL^{-1} for NLB, NLX and NLT, respectively. The obtained data was recorded in the presence of 1.0 mL of 100 μg mL^{-1} of PEA as IS. The sample injection was triplicated for each concentration. The peak area ratio of each concentration with respect to the IS vis. corresponding standard concentration was plotted to obtain the calibration graphs. Then, the corresponding regression equations were derived.

Results and Discussion

The developed CZE method was employed for simultaneous separation and determination of a mixture of the investigated drugs NLB, NLX and NLT of (100:40:40) μg mL^{-1}, respectively, in the presence of 100 μg mL^{-1} PEA as IS. Figure 2 showed the typical electropherogram for the laboratory mixture of the selected drugs. It was found that under optimum conditions the described method exhibited excellent separation of NLB, NLX and NLT at retention times 5.19, 4.21 and 4.55 min, respectively. The proposed method was encouraged to determine the selected drugs with high sensitivity and accuracy in dosage forms and biological fluids.

Optimization of CZE conditions

The key strategy to optimize the CZE separation, the degree of ionization of the investigated drugs and their electrophoretic mobility is the type, pH and the concentration of the running buffer solution used. The selected drugs (NLB, NLX and NLT) have pka 8.71, 7.94 and 8.13, respectively. By using an acidic buffer these compounds can be positively changed and separated using CZE technique.

Selection of running buffer solution

Owing to the selection of the suitable running buffer electrolyte considered as one of the most important parameters in electrophoretic separation, the concentration range of 5-50 mmol L^{-1} of each phosphate, acetate and borate buffer solution was investigated. Under constant instrumental conditions (applied voltage, applied pressure, injection time, temperature and wavelength, etc.), each selected buffer was tested. The recorded results indicated that the most reasonable resolution, signal intensity and migration time was achieved by using acetate buffer solution. Therefore, it was selected for further studies.

Effect of pH

The separation in CZE is very sensitive to pH changes rather than in HPLC. Therefore, small change can greatly affect the separation. Also, one of the typical or the most common buffer used in CZE separation is acetate buffer pH 3.8. The pH value of the running buffer was investigated to ensure excellent separation of the selected drugs. The mobility (u_{eff}) curve of NLB, NLX and NLT and IS was plotted. As shown in Figure 3, the investigated drugs were separated using acetate buffer of pH value 3.8. At pH less than 3.0 no possible separation was obtained, this may be attributed to the interaction of the drugs with internal capillary wall. Therefore, the pH interval 3-7 was tested in the preliminary studies.

Effect of running buffer concentration

The relation of the buffer concentration and the separation process in CZE is very important and the mechanism of action based on the stacking phenomenon which explained by keeping the conductivity of the sample less than the conductivity of the buffer. Also, there are other factors which affect the separation and they are related to the buffer concentration such as EOF. The increasing of buffer concentration will increase the separation process but also it will cause a decrease in the EOF through the capillary. The produced current as well as the electroosmotic flow (EOF) in the capillary was greatly influenced by the concentration of the running buffer solution which applied during the electrophoretic separation. Therefore, to investigate the effect of the acetate buffer concentration on the electrophoretic separation of the selected drugs, 5-50 mmol L^{-1} of acetate buffer solutions were tested. Figure 4, demonstrated that a high separation performance was obtained under constant conditions (pH 3.8, 70 mbar, 15 kV, 25°C) by using 20 mmol L^{-1} acetate buffer.

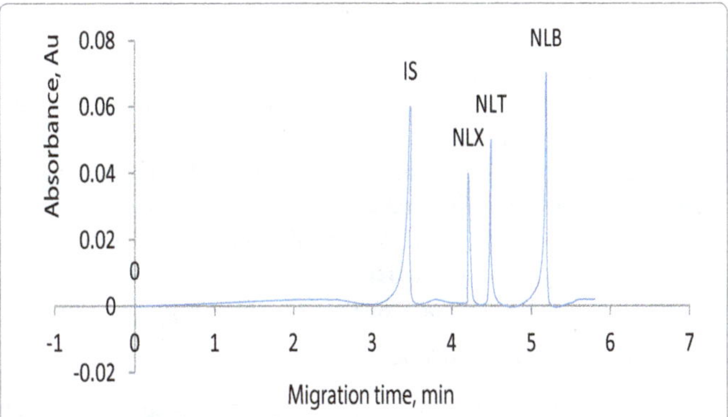

Figure 2: Typical electropherogram of a mixture of NLB (100 μg mL^{-1}), NLX (40 μg mL^{-1}), NLT (40 μg mL^{-1}) and 100 μg mL^{-1} IS.

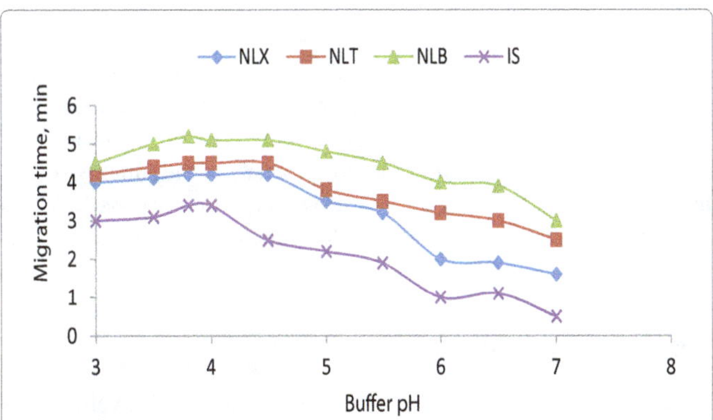

Figure 3: Effect of buffer pH on the migration time: optimum conditions 20 mmolL^{-1} acetate buffer pH=3-7, injection time 6 s, 25°C, 230 nm, 15kV and 50 μg mL^{-1} of each of the tested drugs and 100 μg mL^{-1} of IS

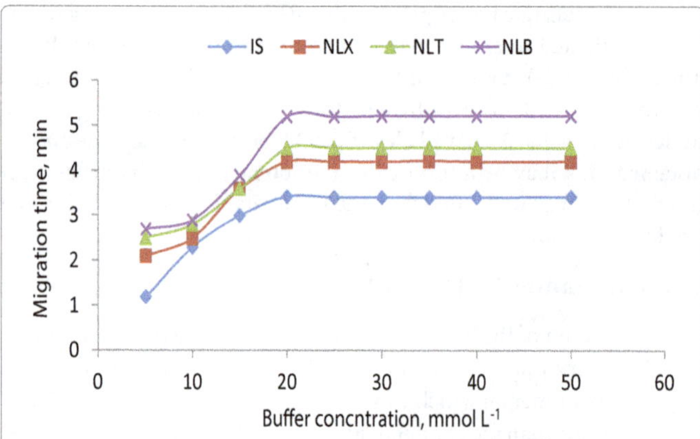

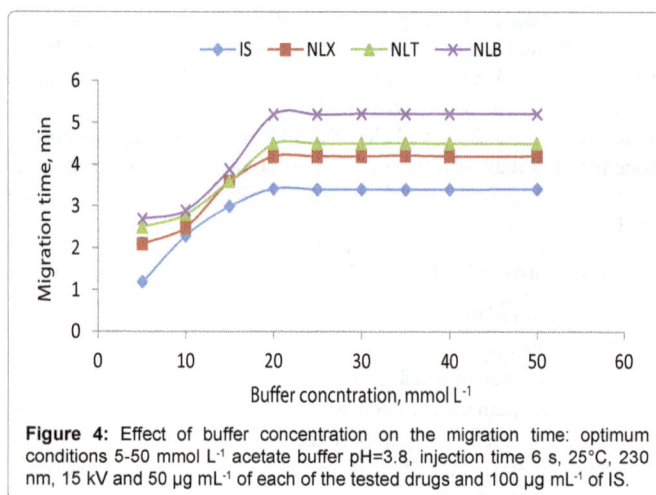

Figure 4: Effect of buffer concentration on the migration time: optimum conditions 5-50 mmol L^{-1} acetate buffer pH=3.8, injection time 6 s, 25°C, 230 nm, 15 kV and 50 µg mL^{-1} of each of the tested drugs and 100 µg mL^{-1} of IS.

Effect of additives and organic modifiers

The effect of some additives to the system, electrolyte was tested by adding 5-25 mol L^{-1} sodium dodecyl sulfate (SDS) and beta-cyclodextrin (β-CD). The obtained results revealed that no significant improvement in the separation of the investigated drugs was recorded by adding (β-CD). On the other hand, it was found that the addition of SDS in the level above the critical micelle concentration promotes the aggregation of the surfactant molecules, hence causes, interaction of hydrophobic molecules leading to change in the mobility of the analytes [29]. Therefore, the experiment was carried out in the absence of SDS or (β-CD). Moreover, one of the most critical parameters which should be investigated is the addition of organic modifiers such as methanol (MeOH), ethanol (EtOH), isopropanol (IPA) and acetonitrile (ACN) due to their effect on the electroosmotic mobility, zeta potential and dielectric constant of the CZE. To our knowledge, the analytes move across the capillary under the effect of electroosmotic and electrophoretic forces. The velocity of solute was calculated using the algebraic sum of electrophoretic velocity (V_{ef}) and electroosmotic velocity (V_{eo}) according to the following equation:

$$V_{nett}=V_{ef}+V_{eo}=[D\zeta_{ef}/4\pi\eta+D\zeta_{eo}/4\pi\eta]\ E$$

Where, η is the medium viscosity, ζ is the Zeta potential and E is the strength of the electric field. Moreover, the increase of proportion of organic modifiers added causes a significant decrease in zeta potential and dielectric constant [30].

To investigate the effect of adding organic modifiers to the buffer electrolyte, different % (10-70 v/v) of each organic modifier (MeOH, EtOH, IPA and ACN) was added. As shown in Figure 5, it was found that increasing the proportion of the organic modifier caused a considerable increase in the migration time and the viscosity of the running buffer, but no significant improvement in drugs separation was recorded by adding the organic modifiers. Therefore, the electrophoretic analysis of the investigated drugs was performed without adding organic modifiers and this considered as an Eco friendly method.

Effect of applied voltage

Under optimum conditions the effect of the applied voltage was tested by performing several runs with gradual increase of the applied voltage from 10-35 kV. For our knowledge, direct relationship was obtained between the efficiency of the resolution (Rs) of analysis and the applied voltage [31]. Therefore, the resolution efficiency was increased by increasing the applied voltage in the range of 15-25 kV. While, it was noticed that excessive Joule heat was generated with further increase of the applied voltage more than 30 kV, which gave a significant decrease in the Rs efficiency of the capillary. As indicated in Figure 6, 15 kV was selected to be suitable for further detection.

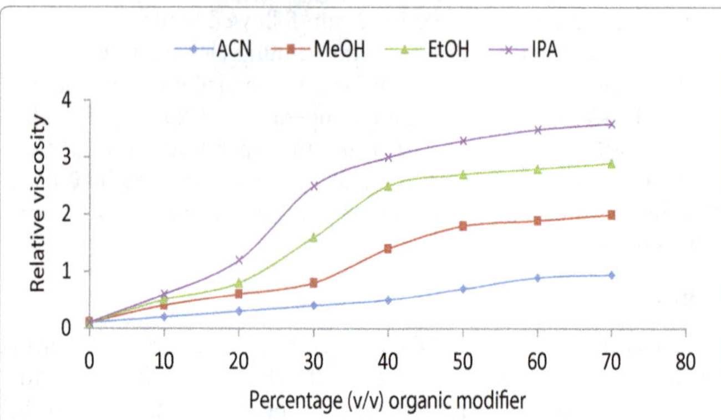

Figure 5: Effect of percentage (v/v) organic modifiers on the viscosity of the running buffer solution.

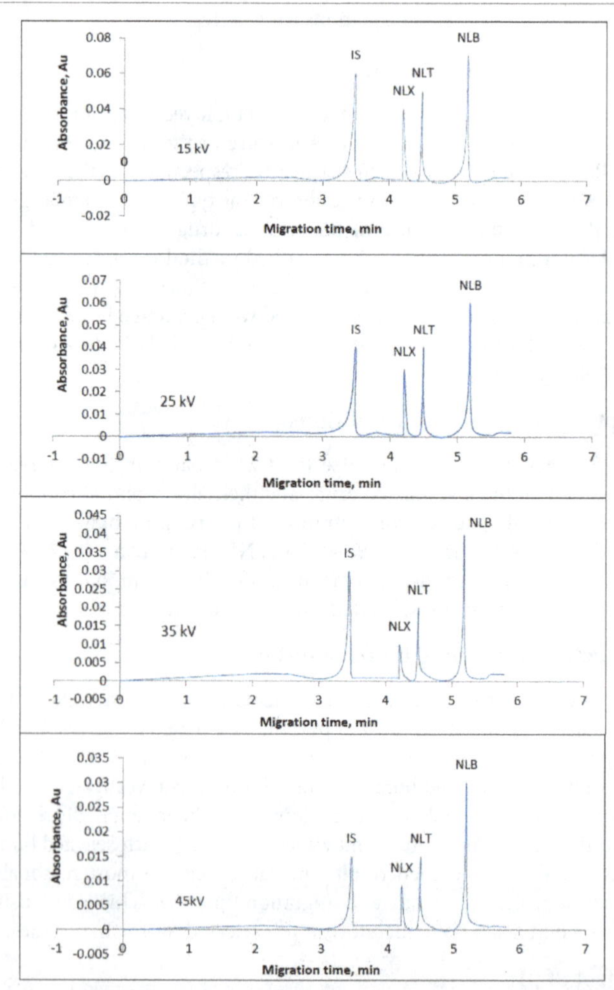

Figure 6: Effect of different voltages on separation of NLB (100 µg mL^{-1}); NLX (40 µg mL^{-1}), NLT (40 µg mL^{-1}) and 100 µg mL^{-1} IS; running buffer; 20 mmol L^{-1} acetate buffer; injection 70 mbar for 6 s; separation voltage (15-45 kV); capillary temperature 25°C and DAD detection at 230 nm.

Effect of capillary cartridge temperature

Due to the influence of the temperature of the capillary on the EOF and electrophoretic mobility, the capillary cartridge temperature should be controlled and optimized. The temperature of the capillary cartridge was investigated in the range of 25-35°C. It was obtained that excellent separation with good resolution and short migration time was recorded at 25°C.

Selection of injection time

In electrophoretic analysis, the peak width and peak height were affected by injection time. So, the samples of the investigated drugs were hydrodynamically injected under 70 mbar and injection time varied from 2-10 s. It was found that after 6 s peaks deformation was seen. Therefore, 6 s was selected for further studies.

Selection of internal standard

The role of IS in electrophoretic separation is to set off the injection errors, improves the performance of the electrophoretic separation and detection. Also, it lowers the migration time of the separation process. 9.78 is the pKa value of phenylethylamine and its molecular weight is less than the investigated drugs. It will be expected that under acidic conditions the IS positively charged and eluted before the investigated drugs.

Selection of the detection wavelength

The suitable wavelength for electrophoretic separation and detection of the investigated drugs should be optimized. The electrophoretic separation was performed at 190-400 nm. It was found that excellent separation and best signal to noise ratio were achieved at 230 nm.

Method Validation

According to ICH guidelines [28], the developed CZE method was validated with respect to system stability, linear concentration range, accuracy, precision, specificity, limits of detection and quantification and robustness.

Linearity

In order to establish the linear relationship of the developed CZE method, the peak area ratio of the studied drug/IS as a function of drug concentration was plotted. The developed method was established linear relationships at concentration ranges of 5-200, 20-240 and 10-280 μg mL^{-1} for NLB, NLX and NLT, respectively. Good correlation coefficients (r) were found to be 0.9997, 0.9994 and 0.9996 with regression equations Y_{NLB}=0.0024x+0.4718, Y_{NLX}=0.0018x+0.3183 and Y_{NLT}=0.0011x+0.3898 for the three mentioned drugs, respectively (Figure 7).

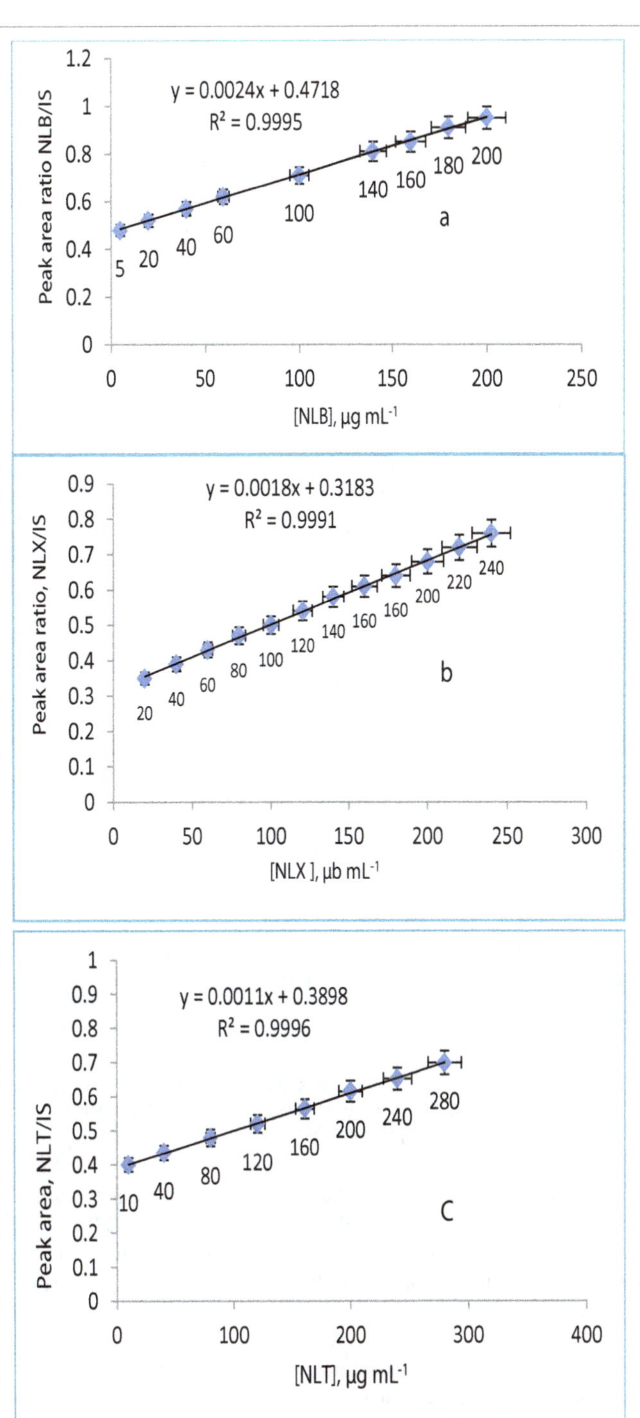

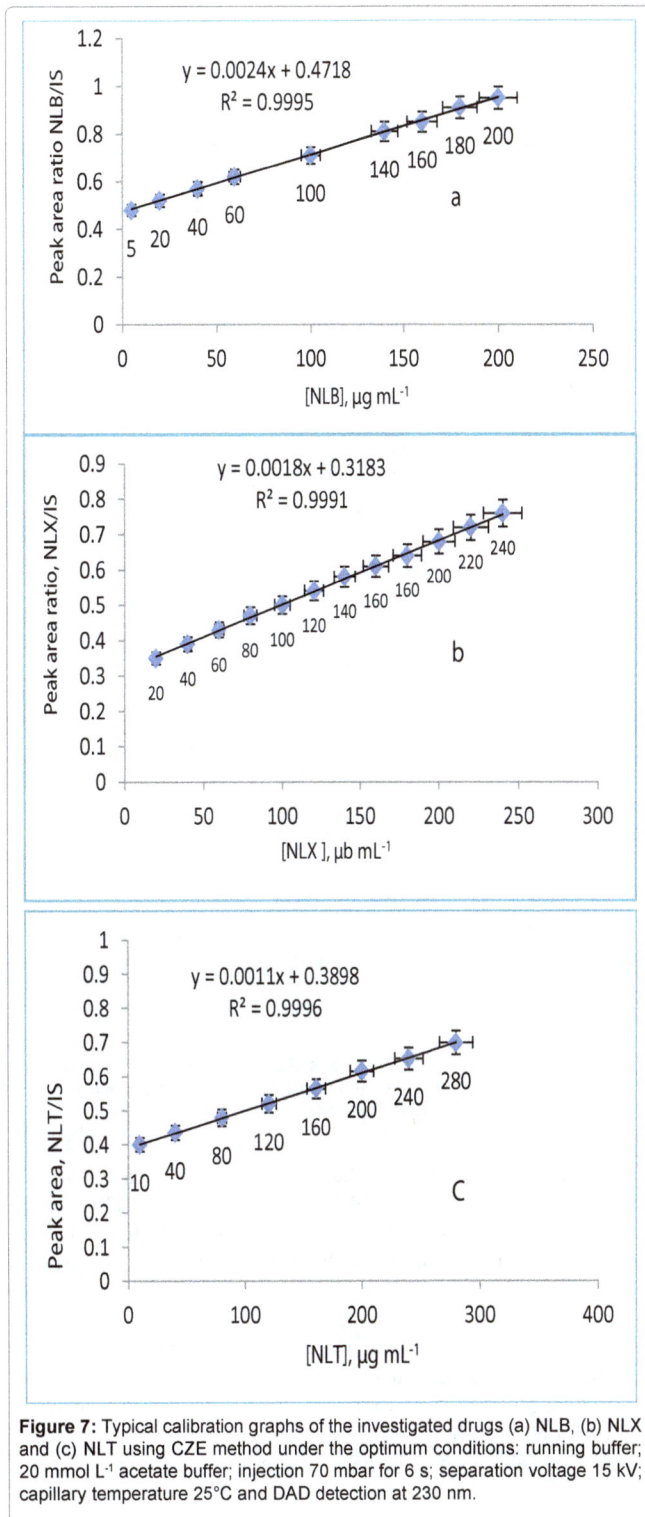

Figure 7: Typical calibration graphs of the investigated drugs (a) NLB, (b) NLX and (c) NLT using CZE method under the optimum conditions: running buffer; 20 mmol L^{-1} acetate buffer; injection 70 mbar for 6 s; separation voltage 15 kV; capillary temperature 25°C and DAD detection at 230 nm.

Limit of detection (LOD) and limit of quantification (LOQ)

The guidelines ICH Q2 (R1) were used for calculation of LOD and LOQ according to the following equations: LOD=3.3 S_a/b and LOQ=10 S_a/b. Where, S_a is the standard deviation of the intercept and b is the slope. Under optimum electrophoretic conditions it was found that the investigated drugs were detected with LOD of 0.3, 5.0 and 6.5 μg mL^{-1} and quantification limits of 5, 20 and 10 μg mL^{-1} for NLB, NLX and NLT, respectively (Table 1). Also, Figure 8, showed the electropherogram of blank sample, LOQ and LOD solutions.

Accuracy and precision

The accuracy of the developed CZE method for separation and determination of NLB, NLX and NLT was evaluated as mean percentage recoveries (n=6) using standard solutions of the investigated drugs. The obtained results were statistically treated and compared with those obtained from other reported methods, high-performance liquid chromatography determination of nalbuphine hydrochloride using

Parameter	NLB	NLX	NLT
Linearity range (μg mL^{-1})	5-200	20-240	10-280
Regression equation	y=0.0024x+0.4718	y= 0.0018x+0.3183	y=0.0011x+0.3898
Correlation coefficient (r)	0.9997	0.9995	0.9996
Standard deviation of residuals, $S_{y/x}$	0.0041	0.0042	0.0062
Standard deviation of intercept, S_a	0.0002	0.0025	0.0015
Standard deviation of slope, (S_b)	0.0002	0.0001	0.0001
Limit of detection (LOD)	0.3	4.6	6.5
Limit of quantification (LOQ)	5	20	10
%RSD	0.2	0.2	0.7
%Error*	0.0001	0.0001	0.0064

*% Error was calculated by SD/$\sqrt{n}$

Table 1: Critical performance data of the determination of the investigated drugs using CZE method.

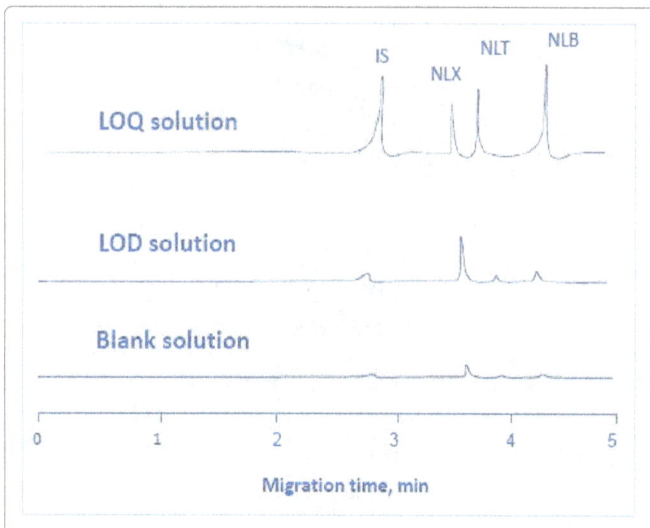

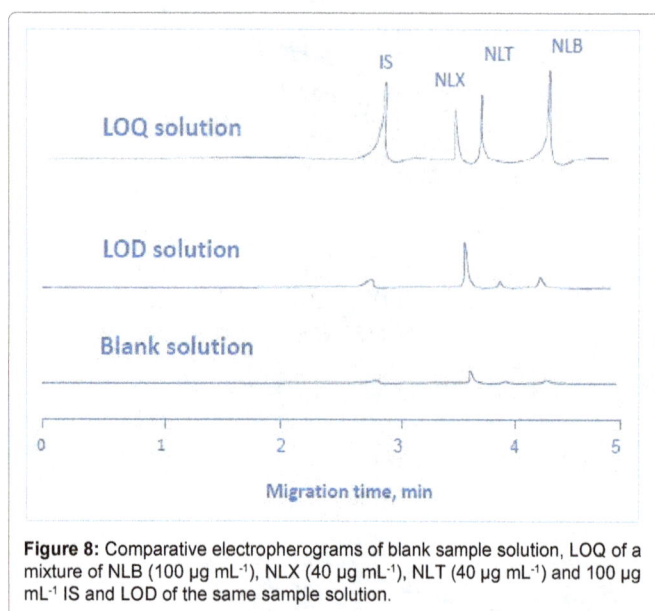

Figure 8: Comparative electropherograms of blank sample solution, LOQ of a mixture of NLB (100 μg mL^{-1}), NLX (40 μg mL^{-1}), NLT (40 μg mL^{-1}) and 100 μg mL^{-1} IS and LOD of the same sample solution.

C18 column and mobile phase sodium acetate buffer pH 5.5: acetonitrile (40:60 v/v), flow rate 1.0 mL/min and UV-detection at 210 nm [2], high performance liquid chromatography determination of NLX using C18 column, mobile phase 10 mmol L^{-1} potassium phosphate buffer pH 6.0 with orthophorphoric acid: acetonitrile (17:83 v/v), flow rate 1 mL/ min and UV detection at 210 nm [10] high performance liquid chromatography determination of NLT which based on the using of C18 column, mobile phase ammonium acetate buffer pH 5.8 : acetonitrile (60:40 v/v), flow rate 1 mL/min and UV detection at 220 nm [21] for NLB, NLX and NLT, respectively). The obtained data were presented in Table 2 which indicated that no significant difference was recorded.

To prove the precision of the developed CZE method the intermediate precision was applied using inter-day and intra-day assay and the % RSD was evaluated. As listed in Table 3, the obtained data provided a good precision of the developed CZE method.

Robustness of the developed method

To evaluate the robustness of the described CZE method, minor deliberated changes in method parameters were introduced. These parameters include the change in the running buffer solution pH (3.8 ± 0.2), the concentration of the running buffer (20 ± 5 mmol L^{-1}), the temperature of capillary cartridge 25 ± 2°C, injection time 6 ± 1 s and applied voltage 15 ± 2 kV with only one parameter at a time was changed. As reported in Table 4, the peak area ratio and the migration time were not significantly affected by these changes.

Specificity

To discriminate the investigated drugs from all other interfering species, standard drugs and some common spiking interfering species such as citric acid, sodium citrate dehydrate, sodium chloride, methylparaben and carboxymethylcellulose sodium salt were used. The peak purity against the pure standard drugs was recorded using a diode array detector and prinCE-770 DA × 3D software. The obtained results demonstrated that no separation peaks were detected at the retention time of each investigated drug and the IS at the lower limit of quantification. Therefore, the developed CZE method for determination of NLB, NLX and NLT was selective and specified for separation and detection of the previously mentioned drugs.

Analytical Applications

Quantification of nalbuphine, naloxone and naltrexone hydrochloride in authentic mixture

The developed electrophoretic method was employed for simultaneous separation and estimation of the tested drugs in their laboratory authentic mixture. Triplicate injection was applied for each sample. The percentage recoveries of each concentration were calculated and found to be 99.00 ± 0.8, 99.65 ± 0.4 and 99.62 ± 0.2 for NLB, NLX and NLT, respectively. The obtained results as reported in Table 5 were statistically treated using student's t-test and variance F-test [32], and then compared with data obtained from other reported methods [2, 10 and 21 for the three drugs, respectively]. Regarding to the accuracy and precision good agreements and no significant difference was recorded. Furthermore, the three drugs were determined in pharmaceutical formulations, human serum and urine.

The evaluated data were summarized in Tables 6 and 7. Firstly, in pharmaceutical preparations, percentage recoveries were calculated and it was found to be 99.43 ± 0.5, 99.36 ± 0.6 and 99.73 ± 0.3. While, in biological fluids the recorded results in human serum were 99.21 ± 0.6, 99.45 ± 0.6 and 99.24 ± 0.7 and in human urine were 99.33 ± 0.5, 99.03 ± 0.6 and 99.50 ± 0.4 for NLB, NLX and NLT, respectively.

Conclusion

The present study introduced a new electrophoretic method for simultaneous separation and determination of opioid agonist nalbuphine hydrochloride and its related antagonists naloxone hydrochloride and naltrexone hydrochloride. The developed method was employed for determination of the selected drugs in their pharmaceutical preparations and biological fluids. Under optimum conditions the proposed method gave excellent separations and detection of the investigated drugs and the obtained results were acceptable with respect to migration time, resolution, and peak area. The described method was also employed for determination of the drugs of interest in human serum and urine, excellent results were obtained. The method was very sensitive, simple, less time consumed and highly precise. Method validation was made to ensure the suitability of the developed method for detection of nalbuphine hydrochloride, naloxone hydrochloride and naltrexone hydrochloride using CZE.

Drug	Proposed CZE method			Reference method			
	Taken µg mL^{-1}	Found µg mL^{-1}	% Recovery	Taken µg mL^{-1}	Found µg mL^{-1}	% Recovery	
NLB	5	4.99	99.80	2	1.99	99.50	Ref. [2]
	50	49.85	99.70	4	3.98	99.50	
	100	98.88	98.88	6	5.87	97.83	
	140	139.47	99.62	8	8.00	100.00	
	160	159.53	99.70	10	9.99	99.90	
	200	199.74	99.87	15	14.88	99.20	
Mean ± SD	99.60 ± 0.4			99.32 ± 0.7			
n	6			6			
Variance	0.16			0.49			
**%SE	0.16			0.29			
t-test	0.845(2.228)*						
F-test	3.06(5.05)*						
NLX	20	19.77	98.85	10	10.00	100.00	Ref [10]
	60	59.28	98.80	20	19.99	99.95	
	100	98.65	98.65	40	39.85	99.62	
	140	139.47	99.99	60	59.78	99.63	
	160	158.98	99.36	80	79.63	99.53	
	180	179.85	99.92	100	98.85	98.85	
Mean ± SD	99.26 ± 0.6			99.59 ± 0.4			
n	6			6			
Variance	0.36			0.16			
**%SE	0.24			0.16			
t-test	1.144(2.228)*						
F-test	2.25(5.05)*						
NLT	10	10.00	100.00	12	11.95	99.58	Ref [21]
	50	49.58	99.16	16	15.99	99.93	
	80	79.89	99.86	20	19.68	98.40	
	100	99.85	99.85	24	23.98	99.92	
	150	149.57	99.71	28	28.00	100.00	
	200	199.74	99.87	32	31.67	98.97	
Mean ± SD	99.74 ± 0.3			99.47 ± 0.6			
n	6			6			
Variance	0.09			0.36			
**%SE	0.12			0.24			
t-test	1.006(2.228)*						
F-test	4.00(5.05)*						

*Figures in parentheses are the tabulated values of t-and F- testes at 95% confidence limit [32]; **%SE= SD / $\sqrt{n}$

Table 2: Analytical results of the determination of NLB, NLX and NLT in bulk powder using CZE method and reference methods.

Parameter	NLB			NLX			NLT		
	Taken µg mL^{-1}			Taken µg mL^{-1}			Taken µg mL^{-1}		
Inter-day	5	50	150	80	160	200	50	200	240
% Found	99.5	99.68	99.26	99.75	99.23	100	99.82	99.96	99.84
	100	99.47	99.86	100	99.74	99.48	99.26	99.24	99.12
	99.87	99.12	99.98	99.52	99.24	99.86	99.42	99.87	99.63
Mean	99.79	99.43	99.7	99.76	99.4	99.78	99.5	99.69	99.53
± SD	0.26	0.28	0.39	0.24	0.29	0.27	0.29	0.39	0.37
%RSD	0.26	0.28	0.39	0.24	0.29	0.27	0.29	0.39	0.37
%SE	0.15	0.16	0.23	0.14	0.18	0.16	0.17	0.23	0.21
Intra-day	5	50	150	80	160	200	50	200	240
% Found	100	99.96	100	99.25	99.64	99.26	100	99.27	99.32
	99.98	99.98	100	99.69	99.87	99.78	99.65	99.65	99.89
	99.97	99.99	99.58	99.97	99.27	100	99.87	99.97	99.47
Mean	99.98	99.98	99.86	99.64	99.59	99.68	99.84	99.63	99.56
± SD	0.02	0.02	0.24	0.36	0.3	0.38	0.18	0.35	0.29
%RSD	0.02	0.02	0.24	0.36	0.3	0.38	0.18	0.35	0.29
%SE	0.01	0.01	0.14	0.21	0.17	0.22	0.1	0.2	0.17

Table 3: Analytical data of inter-day and intra-day precisions for the determination of NLB, NLX and NLT using CZE method.

Parameter	Migration time, min			Peak area ratios		
	NLB	NLX	NLT	NLB	NLX	NLT
Standard	5.19	4.21	4.55	0.71	0.57	4.35
Acetate buffer pH						
3.6	5.24	4.14	4.62	0.72	0.57	4.35
4.0	5.12	4.22	4.46	0.71	0.57	4.35
Acetate buffer concentration, mmol L^{-1}						
15	5.15	4.12	4.35	0.70	0.55	4.34
25	5.30	4.23	4.62	0.71	0.57	4.35
Injection time, s						
5	5.02	4.20	4.56	0.69	0.56	4.33
7	5.17	4.16	4.48	0.70	0.57	4.35
Applied voltage, kV						
13	5.34	4.50	4.68	0.71	0.62	4.37
17	5.13	4.30	4.66	0.74	0.57	4.36
Capillary cartridge temperature, °C						
23	5.36	4.32	4.69	0.71	0.63	4.37
27	5.20	4.24	4.63	0.69	0.64	4.38

Table 4: Robustness data using 100 µg mL^{-1} NLB, 40 µg mL^{-1} of NLX and 40 µg mL^{-1} in the presence of 100 µg mL^{-1} IS.

Ratio NLB:NLX:NLT % w/w	Taken µg mL^{-1}	NLB		Reference method [2] Taken 2-12 µg mL^{-1}		NLX		Reference method [10] Taken 10-100 µg mL^{-1}		NLT		Reference method [21] Taken 12-32 µg mL^{-1}	
		Found µg mL^{-1}	% Recovery	Found µg mL^{-1}	% Recovery	Found µg mL^{-1}	% Recovery	Found µg mL^{-1}	% Recovery	Found µg mL^{-1}	% Recovery	Found µg mL^{-1}	% Recovery
1:1:1	20:20:20	19.98	99.9	1.97	98.5	19.85	99.3	9.99	99.9	19.87	99.4	11.98	99.8
1:2:2	20:40:40	19.87	99.4	3.99	99.8	39.57	98.9	19.98	99.9	39.74	99.4	15.86	99.1
1:4:4	20:80:80	19.52	97.6	5.87	97.8	79.99	99.9	39.86	99.7	79.63	99.5	20.00	100.0
1:6:6	20:120:120	19.69	98.5	7.95	99.4	120.00	100.0	59.87	99.8	119.82	99.9	23.87	99.5
1:8:8	20:160:160	19.84	99.2	9.86	98.6	159.96	99.9	79.86	99.8	159.99	99.9	27.93	99.8
1:10:10	20:200:200	19.88	99.4	11.96	99.7	199.89	99.9	98.95	98.9	199.25	99.6	32.00	100.0
Mean ± SD		99.00 ± 0.8		98.97 ± 0.7		99.65 ± 0.4		99.67 ± 0.3		99.62 ± 0.2		99.70 ± 0.3	
Variance		0.64		0.49		0.16		0.09		0.40		0.09	
**%SE		0.33		0.29		0.16		0.12		0.08		0.12	
t-test		0.068(2.228)*				0.100(2.228)*				0.555(2.228)*			
F-test		1.31(5.05)*				1.78(5.05)*				4.44(5.05)*			

* Figures in parentheses are the tabulated values of t-and F- testes at 95% confidence limit [32]. ** %SE= SD/ $\sqrt{n}$.

Table 5: Analytical results of the determination of NLB, NLX and NLT in authentic mixture using CZE method and reference methods.

Drug	Proposed CZE method			Reference method			
	Taken µg mL^{-1}	Found µg mL^{-1}	% Recovery	Taken µg mL^{-1}	Found µg mL^{-1}	% Recovery	
NLB®Nalufin 50- 200 µg mL^{-1}	50	49.15	98.30	2	2.00	100.0	
	100	99.82	99.82	4	3.96	99.0	
	140	139.36	99.54	6	5.87	97.8	
	160	159.12	99.37	8	8.00	100.0	
	180	179.75	99.86	10	9.96	99.6	
	200	199.32	99.66	15	14.99	99.9	
Mean ± SD		99.43 ± 0.5			99.38 ± 0.8		Ref. [2]
n		6			6		
Variance		0.25			0.64		
%SE		0.20			0.33		
t-test		0.129 (2.228)*					
F-test		2.56 (5.05)*					
NLX® Narcan 100-240 µg mL^{-1}	100	98.35	98.35	10	9.99	99.9	
	140	139.12	99.37	20	19.89	99.5	
	160	158.53	99.08	40	39.95	99.9	
	180	179.76	99.87	60	59.85	99.8	
	200	199.45	99.73	80	79.99	99.9	
	240	239.47	99.78	100	98.58	98.6	
Mean ± SD		99.36 ± 0.6			99.60 ± 0.5		Ref [10]
n		6			6		
Variance		0.36			0.25		
%SE		0.24			0.20		
t-test		0.768(2.228)*					
F-test		1.44 (5.05)*					

NLT® Vivitrol 50-280 μg mL^{-1}	50	49.95	99.90	12	11.95	99.6	
	100	99.14	99.14	16	15.89	99.3	
	150	149.69	99.79	20	20.00	100.0	
	200	200.00	100.00	24	23.78	99.0	
	240	239.36	99.73	28	27.94	99.8	
	280	279.47	99.81	32	31.95	99.8	
Mean ± SD		99.73 ± 0.3			99.58 ± 0.4		Ref [21]
n		6			6		
Variance		0.09			0.16		
%SE		0.12			0.16		
t-test		0.750(2.228)*					
F-test		1.78(5.05)*					

* Figures in parentheses are the tabulated values of t-and F- testes at 95% confidence limit [32] ** %SE= SD/$\sqrt{n}$

Table 6: Analytical results of the determination of NLB, NLX and NLT in dosage forms using CZE method and reference methods.

Samples	Ratio of NLB:NLX:NLT	Taken μg mL^{-1}	NLB		NLX		NLT	
			Found μg mL^{-1}	% Recovery	Found μg mL^{-1}	% Recovery	Found μg mL^{-1}	% Recovery
Serum	1:1:1	20:20:20	19.85	99.25	19.69	98.46	19.60	98.00
	1:2:2	20:40:40	39.79	99.47	39.82	99.54	39.84	99.59
	1:4:4	20:80:80	78.56	98.20	79.75	99.69	79.49	99.36
	1:5:5	20:100:100	99.50	99.50	99.90	99.90	99.70	99.70
	1:6:6	20:120:120	119.57	99.64	119.62	99.68	119.48	99.57
Mean ± SD			99.21 ± 0.6		99.45 ± 0.6		99.24 ± 0.7	
n			5		5		5	
Variance			0.36		0.36		0.49	
**%SE			0.27		0.27		0.31	
Urine	1:1:1	20:20:20	19.69	98.48	19.60	98.00	20.00	100.00
	1:2:2	20:40:40	39.94	99.87	39.65	99.12	39.74	99.35
	1:4:4	20:80:80	79.50	99.38	79.32	99.15	79.50	99.38
	1:5:5	20:100:100	99.33	99.33	99.50	99.50	99.79	99.79
	1:6:6	20:120:120	119.58	99.65	119.28	99.40	118.79	98.99
Mean ± SD			99.33 ± 0.5		99.03 ± 0.6		99.50 ± 0.4	
n			5		5		5	
Variance			0.25		0.36		0.16	
**%SE			0.22		0.27		0.18	

**%SE= SD/$\sqrt{n}$

Table 7: Analytical results of the determination of NLB, NLX and NLT in human serum and urine using CZE method.

Acknowledgments

This project was supported by King Saud University, Deanship of Scientific Research College of Science Research Center.

References

1. Goodman Gilman's (2007) The Pharmacological basis of therapeutics. 10th edn, McGraw-Hill: London, p: 569.

2. Attia KA, Nassar MW, El-Olemy A (2014) Stability- indicating HPLC for determination of nalbuphine hydrochloride. Int J Res Pharm Biosci 1: 15-22.

3. Groenendaal D, Blom-Roosemalen MCM, Margret CM, Danhof M (2005) High-performance liquid chromatography of nalbuphine, butorphanol and morphine in blood and brain microdialysate samples: Application to pharmacokinetic/pharmacodynamics studies in rats. J Chromatogr B 822: 230-237.

4. Fouad MM, Abdel Razeq SA, Elsayed ZA, Hussin LA (2013) Stability indicating methods for determination of nalbuphine-hydrochloride. Brit J Pharm Res 3: 259-272.

5. Kim JY, In MK, Paeny KJ, Chung BC (2004) Simultaneous determination of nalbuphine and opiates in human hair by gas chromatography-mass spectrometry. Chromatogr 59: 219-226.

6. Cai LJ, Zhang J, Wang XM, Zhu RH, Yang J, et al. (2011) Validation LC-MS/MS assay for the quantitative determination of nalbuphine in human plasma and its application to a pharmacokinetic study. Biomed Chromatogr 25: 1308-1314.

7. Attia KA, Nassar MW, El-Olemy A (2014) Stability-indicating spectrophotometric determination of nalbuphine hydrochloride using first derivative of ratio spectra and ratio difference methods. Eur J Biomd Pharm Sci 1: 1-11.

8. El-Tohamy MF, El-maamly M, Shalaby AA, Aboul-Enein H (2007) Development of nalbuphine-selective membrane electrode and its applications in pharmaceutical analysis. Anal Lett 40: 1569-1578.

9. Kendrick WD, Woods AM, Daly MY, Birch RF, DiFazio C (1996) Naloxone versus nalbuphine infusion for prophylaxis of epidural morphine-induced pruritus. Anesth Analg 82: 641-647.

10. Tzatzarakis MN, Vakonaki E, Kovatsi L, Belivanis S, Mantsi M, et al. (2015) Determination of buprenorphine, norbuprenorphine and naloxone in fingernail clippings and urine of patients under opioid substitution therapy. J Anal Tox 39: 313-320.

11. Tawakkol MS, Mohamed ME, Hassan MMA (1983) Determination of naloxone hydrochloride in dosage form by high-performance liquid chromatography. J Liq Chromatogr 6: 1491-1497.

12. Zhurkovich IK, Rudenko AO, Chelovechkova VV, Merkusheva IA, Lugokina NV, et al. (2015) Determination of buprenorphrine and naloxone in patient blood plasma using HPLC-MS. Pharm Chem J 48: 690-695.

13. Al-Tannak NF (2014) An LC-MS method for evaluating photostability of naloxone hydrochloride in I.V. infusion. J Chem Chem Eng 8: 524-529.

14. Jiang H, Wang Y, Shet MS, Zhang Y, Zenke D, et al. (2011) Development and validation of a sensitive LC/MS/MS method for the simultaneous determination

of naloxone and its metabolites in mouse plasma. J Chromatogr B Analyt Technol Biomed Life Sci 879: 2663-2668.

15. Fang WB, Chang Y, McCance-Katz EF, Moody DE (2009) Determination of naloxone and nornaloxone (noroxymorphone) by high-performance liquid chromatography-electrospray ionization- tandem mass spectrometry. J Anal Toxicol 33: 409-417.

16. Alarfaj NA, El-Tohamy MF (2015) A high throughput gold nanoparticles chemiluminescence detection of opioid receptor antagonist naloxone hydrochloride. Chem Cent J 9:6.

17. Alarfaj NA, El-Tohamy MF (2015) Comparative electrochemical studies of modified 2-hydroxypropyl beta cyclodextrin modified carbon nanotubes sensors for determination of naloxone hydrochloride. Sensor Lett 13: 199-208.

18. The Merck Index (2001) An encyclopedia of chemicals, drugs, and biological. 13th Edition, Merck research laboratories, 1141.

19. Rockville MD (2004) The United State Pharmacopoeia, 26th Revision, US Pharmacopoeial Convention Inc., p: 279.

20. Srikalyani V, Tejaswi M, Srividya P, Nalluri N, Buchi N (2013) Simultaneous analysis of naltrexone hydrochloride and bupropion hydrochloride in bulk and dosage forms by RP-HPLC-PDA method. J Chem Pharm Res 5: 429-435.

21. Sarsambi PS, Faheem A, Sonawane A, Gowrisankar D (2010) Reverse Phase-HPLC method for the analysis of Naltrexone hydrochloride in bulk drug and its pharmaceutical formulations. Der Pharm Let 2: 294-299.

22. Clavijo C, Bendrick-Peart J, Zhang YL, Johnson G, Gasparic A (2008) An automated, highly sensitive LC-MS/MS assay for the quantification of the opiate antagonist naltrexone and its major metabolite 6 beta-naltrexol in dog and human plasma. J Chromatogr B 874:33-41.

23. El-didamony MA, Khater HM, Ali II (2013) New sensitive bromatometric assay methods for the determination of four analgesic drugs in pharmaceutical formulations and biological fluids. J Pharm Edu Res 4: 54-63.

24. El-didamon AM, Hassan WS (2012) Spectrophotometric and fluorimetric methods for determination of naltrexone in urine, serum and tablets by oxidation with cerium (IV). J Chil Chem Soc 57: 1404-1408.

25. Ganjali MR, Alipour A, Riahi S, Norouzi P (2009) Design and Construction of a Naltrexone Selective Sensor Based on Computational Study for Application in Pharmaceutical Analysis. Int J Electrochem Sci 4: 1153-1166.

26. Kasawar GB, Farooqui MN (2010) Validated capillary electrophoresis method for the simultaneous determination of formic acid and acetic acid in cephalosporin drug substances using indirect UV detection. Arch Appl Sci Res 2: 106-111.

27. Salim M, El-Enany N, Belal F, Walash M, Patonay G (2012) Simultaneous determination of sitagliptin and metformin in pharmaceutical preparations by capillary zone electrophoresis and its application to human plasma analysis. Anal Chem Insights 7: 31-46.

28. ICH (1996) Technical requirements for registration of pharmaceuticals for human use, complementary guidelines on methodology, Washington, DC, 13.

29. Schmitt-Kopplin P, Burhenne J, Freitag D, Spiteller M, Kettrup A (1999) Development of capillary electrophoresis methods for the analysis of Fluroquinolones and application to the study of influence of humic substances on their photo degradation in aqueous phase. J Chromatogr A 837: 253-256.

30. Janini GM, Chan KC, Barnes JA, Muschik GM, Issaq HJ (1993) Effect of organic solvents on solute migration and separation in capillary zone electrophoresis. Chromatogr 35: 479-502.

31. MacLaughlin M, Nolan A, Lindahl L, Palmieri H, Anderson W, et al. (1992) Pharmaceutical drug separations by HPLC, practical guidelines. J Liq Chromatogr 15: 961-1021.

32. Miller JC, Miller JN (1993) Statistics for Analytical Chemistry. 3rd edition, Ellis Horwood-Prentice Hall, Chichester.

Micellar High Performance Liquid Chromatographic Method for Simultaneous Determination of Clonazepam and Paroxetine HCl in Pharmaceutical Preparations Using Monolithic Column

Fawzia Ibrahim, Nahed El-Enany, Shereen Shalan and Rasha Elsharawy*

Department of Analytical Chemistry, Faculty of Pharmacy, University of Mansoura, Mansoura, Egypt

Abstract

A new and accurate micellar high performance liquid chromatographic method coupled with ultraviolet detection was developed for simultaneous determination of Clonazepam (CLZ) and Paroxetine (PRX) using a monolithic C-18 column, and mobile phase consisting of 0.175M Sodium dodecyl sulphate, 12% n-propanol prepared in 0.02M phosphoric acid at pH 6.0. The analysis was performed at a flow rate of 1 mL/min with ultraviolet- detection at 300 nm. The method was linear over the concentration range (1.0-20 μg) and (4.0-250 μg) with limits of detection of 0.277, 2.675 μg/mL and limits of quantification of 0.838, 8.106 μg/mL for CLZ and PRX respectively. The average % recovery was found to be 100.08 ± 1.31 and 100.22 ± 1.15 for CLZ and PRX respectively. Method validation according to ICH Guidelines recommendation was evaluated. Statistical analysis of the results obtained by the proposed method was compared successfully with those obtained using the reference one. There was no significance difference between the two methods regarding accuracy and precision respectively.

Keywords: Micellar liquid chromatography; Ultraviolet detection; Clonazepam; Paroxetine; Co-formulated tablets; Method validation

Introduction

Clonazepam (5-(2-Chlorophenyl)-7-nitro-2,3-dihydro-1,4-benzodiazepin-2-one) (Figure 1) is a benzodiazepine drug having anxiolytic, anticonvulsant, muscle relaxant, sedative, and hypnotic properties [1]. Clonazepam is used to eliminate seizure activity, anxiety, mania, panic disorders and schizophrenia as it calms brain and nerves [2]. For these reasons, Clonazepam has been identified as a promising drug. Some of the commonly side effects are restless, changing mood, hyperactive and aggressive [3]. The British Pharmacopoeia [4] recommends non-aqueous titration with perchloric acid for the determination of clonazepam. For dosage forms, high-pressure liquid chromatography (HPLC) is recommended by US Pharmacopoeia [5]. Several methods have been reported for determination of this compound including spectrophotometric methods [6-8], potentiometric methods [9,10], voltammetry [11], palarography [12], HPLC [12-14] and gas chromatography coupled with mass spectrometry (GC-MS) [13,15].

Paroxetine ((3S,4R)-3-[(2H-1,3-benzodioxol-5-yloxy)methyl]-4-(4-fluorophenyl) piperidine) (Figure 1) is an antidepressant drug of the selective serotonin reuptake inhibitor (SSRI) type. Paroxetine is used to treat major depression, obsessive-compulsive disorder, panic disorder, anxiety, posttraumatic, generalized anxiety disorder and vasomotor symptoms (e.g., hot flashes and night sweats) associated with menopause [16,17] in adult outpatients. Paroxetine is primarily used to treat major depression, obsessive-compulsive disorder (OCD), post-traumatic stress disorder (PTSD), panic disorder, generalized anxiety disorder (GAD) [18] social phobia/social anxiety disorder [19] premenstrual dysphoric disorder (PMDD) [20] and menopausal hot flashes.

Paroxetine was the first antidepressant formally approved in the United States for the treatment of panic attacks [21].

The previously published methods that concerned with quantitative determination of PRX in tablets include voltammetry [22,23], densitometry [24,25], high-performance liquid chromatography [26-31], gas chromatography [32-34], capillary electrophoresis [35] and spectrofluorimetry [36].

Paroxetine co administered with clonazepam demonstrated significant improvement by endpoint. Combined treatment with paroxetine and clonazepam resulted in more rapid response than with the SSRI alone [37].

Up till now there was not any micellar HPLC method for simultaneous determination of Paroxetine and Clonazepam in the tablet dosage forms. There is one method for the simultaneous determination of clonazepam and paroxetine by RP-HPLC [38]. Our method is more sensitive and we use a smaller amount of organic solvent comparing to the other method (using 60% acetonitrile). For these reasons we are encouraged to perform the present work which determines CLZ and PAX simultaneously using monolithic column.

Experimental

Apparatus

HPLC using a Chromatograph separation was carried out using a (Merck Hitachi model L-7100) equipped with a Rheodyne injector valve with a 20 μL loop, and an ultraviolet detector (Merck Hitachi L-7400), operated at 300 nm. The chromatograms were recorded on a Shimadzu C-R6A integrator. Mobile phase was filtered using membrane filters (Millipore, Ireland) and degassed using Merck solvent L-7612 degasser. pH-meter used is Consort P-901. Ultrasonic bath, model SS 101 H 230, USA.

***Corresponding author:** Rasha Elsharawy, Department of Analytical Chemistry, Faculty of Pharmacy, University of Mansoura, 35516, Mansoura, Egypt
E-mail: rasha.elsharawy@yahoo.com

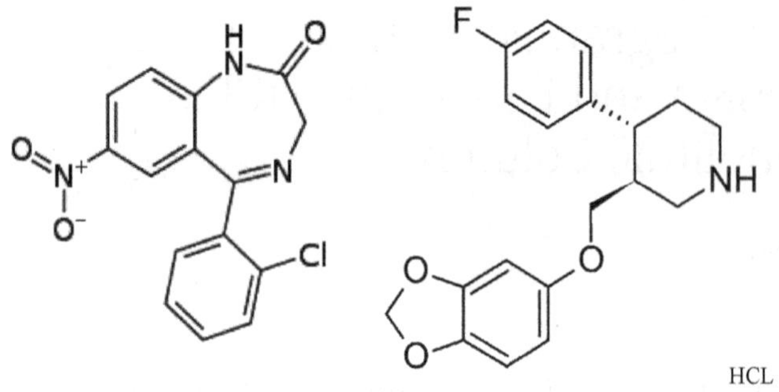

(a) (b)

(a) Clonazepam (CLZ) (b) Paroxetine HCL (PRX)

Figure 1: Structure formula of the studied drugs.

Materials and reagents

- Clonazepam CLZ: was kindly provided from Egyptian International Pharmaceutical Industry Company (EIPICO), with a purity of 99.3% as determined by the comparison method.
- Paroxetine HCl: was kindly provided by Pharaonia pharmaceuticals company (Alexandria, Egypt), with a purity of 99.8% as determined by the comparison method.
- Sodium dodecyl sulphate (SDS) 90% and orthophosphoric acid 85% were obtained from Riedel-deHäen (Germany).
- Methanol and n-propanol (HPLC grade) were obtained from Sigma-Aldrich (Germany).
- Amotril® tablets product of (Amoun Pharmaceutical Co. El Obour city, Cairo, Egypt), labeled to contain 0.5 mg clonazepam. (Batch no. ≠131777).
- Seroxat® CR tablets product of (GlaxoSmithKline Inc, 7333 Mississauga Road North, Canada) labeled to contain 12.5 mg Paroxetine HCl. (batch no. ≠A103035).
- Laboratory prepared co-formulated tablets (0.25 mg clonazepam, 12.5 mg Paroxetine Hcl, 20 mg talc powder, 15 mg starch, 15 mg lactose and 10 mg magnesium stearate per tablet).
- All the pharmaceuticals used were obtained from Egyptian market.

Chromatographic conditions: Chromolith® speed ROD C-18 (50 mm × 4.6 mm i.d., 2 μm particle sizes), Merck, Germany. Mobile phase: a solution consists of a mixture of 12% n-propanol and 0.175 MSDS and the pH was adjusted to 6.0 using orthophosphoric acid. The mobile phase was filtered through Millipore membrane filter. Flow rate: 1 mL/min. Ultraviolet detection: 300 nm.

Standard solutions: Stock solutions were prepared by dissolving either 10.0 mg of CLZ or 10.0 mg PRX in 100.0 mL of methanol to give solution containing 100 μg/ml using ultrasonic bath for good solubility. Working standard solutions were prepared by appropriate dilution of the stock solutions with methanol. All solutions were stored in the refrigerator at 2°C and found to be stable for at least 7 days.

Procedures

Different volumes of the drug working standard solutions were accurately conveyed into a series of 10.0 mL volumetric flasks to obtain the final concentrations in the range of 1.0-20 μg/mL for CLZ and 4.0-250 μg/mL for PRX. Then the volumes of solutions were completed to the mark with the mobile phase and pH was adjusted at 6.0 and mixed well.

Volumes of 20.0 μL were injected (triplicate) and eluted with the mobile phase under the optimum chromatographic conditions. The peak areas against the final concentration of the drugs in μg/mL were drawn. And the corresponding regression equations were obtained.

Analysis of CLZ/PRX laboratory prepared mixtures by the proposed method: Aliquots of CLZ and PRX standard solutions at a pharmaceutical ratio of 1: 50 [39]. Were transferred into a series of 10.0 mL volumetric flasks. The solutions were diluted to the volume with the mobile phase and mixed well. Procedure described under "Construction of the Calibration Graphs" was then applied. The mean percentage recoveries were calculated by referring to the calibration graphs, or using the corresponding regression equations.

Analysis of the two drugs in their single tablets by the proposed method: The content of ten tablets (Amotril®, Seroxat® CR) were accurately weighed, finely powdered, and thoroughly mixed. Accurately weighed amounts of the powdered tablets equivalent to 0.5 mg of CLZ or 12.5 mg of PRX were transferred into a 100 ml volumetric flask and extracted with 80 mL of methanol. The contents of the flask were sonicated for 30 min, completed to the volume with the same solvent and filtered. Filtration utilizing syringe filter was performed to get clear solutions. Aliquots containing suitable concentrations of the studied drugs were analyzed as described under "*construction of the calibration graphs*". The nominal content of each drug was calculated either from a previously plotted calibration graph or using the corresponding regression equation.

Analysis of the two drugs in their laboratory prepared co-formulated tablets by the proposed method: Co-formulated tablets were prepared, according to their pharmaceutical ratio (1:50). Weighed quantity of mixed laboratory prepared tablets equivalent to 0.25 mg CLZ and 12.5 mg PRX were transferred into 100 mL volumetric flasks and about 80 mL of methanol were added. The contents of the flask were sonicated for 30 min, completed to the volume with the same solvent and filtered twice using syringe filters to get highly clear filtrate. Volumes containing different concentrations of CLZ and PRX were taken and analyzed as described under construction of the calibration graphs. The content of each drug was calculated either from the already plotted calibration graphs or by using the corresponding regression equations.

Results and Discussion

A micellar HPLC method coupled with ultraviolet detection was developed and fully validated for the simultaneous determination of CLZ and PRX.

The proposed method can separate CLZ and PRX with good resolution, as the retention time between CLZ and PRX less than 5 min. The experimental parameters influencing the chromatograms of the studied drugs were accurately considered and optimized. The optimal parameter gives the highest number of theoretical plates and the best resolution within a reasonable time. Figure 2 shows a typical chromatogram for laboratory prepared mixture of CLZ and PRX under the described chromatographic conditions and the detection was performed at 300 nm. The separation was achieved within short retention time (t_r=1.36 and 4.9 min) for CLZ and PRX, respectively.

Optimization of the chromatographic performance and system suitability

Study of experimental parameters: Different experimental conditions affecting chromatographic behavior and determination of

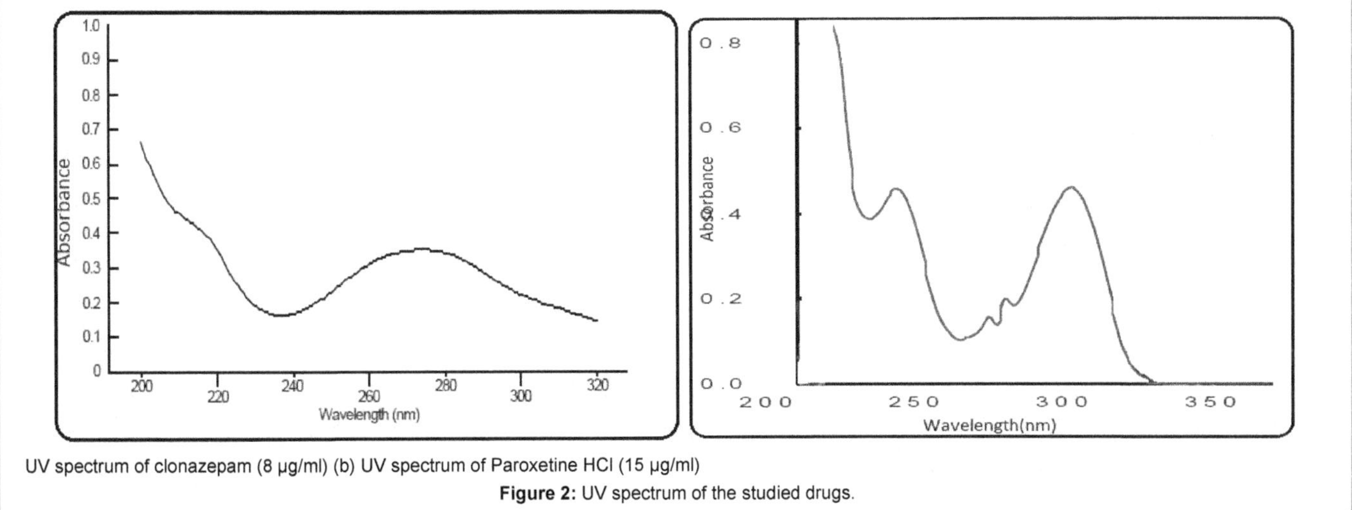

UV spectrum of clonazepam (8 μg/ml) (b) UV spectrum of Paroxetine HCl (15 μg/ml)

Figure 2: UV spectrum of the studied drugs.

the studied drugs including; type of column, concentration of SDS, pH of the mobile phase, concentration of the organic modifier and detection wavelength were carefully studied and optimized.

Choice of column: Two different columns were tried for performance investigations, including: Promosil ODS C18 column (250 × 4.6 mm i.d., 5 μm particle size), Agela Technologies, USA and Chromolith® speed ROD C-18 (50 mm × 4.6 mm i.d., 2 μm particle sizes), Merck, Germany. Experimental studies revealed that, the second column was the appropriate one, giving symmetrical, well defined peaks with good resolution within reasonable time. The first column was not suitable as it showed disturbed overlapped peaks.

Concentration of SDS: The effect of the concentration of SDS on the selectivity and retention time of the drugs was studied using mobile phase containing concentration of 0.1 to 0.2M SDS. It was found that 0.175M of SDS was the optimum conc.

pH of the mobile phase: The influence of pH changing on the retention time of CLZ and PRX was studied over the range of (4.0-6.5), pH 6.0 was the most appropriate pH as it gives symmetrical peaks within reasonable time and high number of theoretical plates as shown in Table 1.

Type of organic modifier: Different organic modifiers were tried including methanol, n-propanol and n-butanol. It was found that n-propanol was the organic modifier of choice as it gives the most symmetrical separated peaks. Methanol was found to give a precipitate in the mobile phase and n- butanol gives non symmetrical peaks.

Concentration of organic modifier: The effect of increasing the % concentration of n-propanol on the chromatographic behavior was studied over the range of (8% -14%). It was found that 12% (v/v) was the most appropriate concentration as it gives the highest number of theoretical plate and good resolution. As shown in Table 1.

Choice of detection wavelengths: The effect of changing wavelength on the chromatographic behavior of both drugs was investigated over the range (290-320 nm). We use the wavelength that gives the maximum peak for paroxetine to enable us to determine both drugs simultaneously. It was found that 300 nm was the most suitable wavelength found for the determination and separation since it gives the symmetrical peaks for both drugs with high number of theoretical plates and good resolution.

Development and validation of the analytical method

The validity of the proposed method was tested regarding linearity, specificity, accuracy, repeatability and precision according to ICH Q2R1 recommendations [40].

Linearity and range: Using the proposed procedure, a linear regression equation was obtained. The regression plot showed that there was a linear relationship established by plotting the peak area against the drug concentration μg/mL. Linear regression analysis of the data gave the following equation:

P=-20657.1+53108C (r=0.9998) for CLZ

P=-19856.2+12613.3C (r=0.9999) for PRX

Where the P is the peak area, C is the concentration of the drug in μg/ mL and r is the correlation coefficient.

Statistical analysis [41] of the data gave a reasonable value of the correlation coefficient (r) of the regression equation, accepted values of the standard deviation of residuals ($S_{y/x}$), standard deviation of intercept (S_a), and standard deviation of slope (S_b), and accepted value of the percentage relative standard deviation and the percentage relative error (Table 2). These data proved the linearity of the calibration curve and low scattering of the points around the calibration curve.

Limit of Quantitation (LOQ) and limit of detection (LOD): The limit of quantitation (LOQ) is the minimal concentration which can be determined based on ICH Q2R1 recommendations (40) under which the calibration plot is non linear.

The limit of detection (LOD) is the minimum analyte concentration which can be detected (41).

$LOQ=10\ S_a/b$

$LOD=3.3\ S_a/b$

Where S_a=standard deviation of the intercept of the calibration curve and b=slope of the calibration curve.

LOQ and LOD values for CLZ and PRX by the suggested method were showed in Table 2. LOQ values are 0.838 and 8.106 μg/mL while LOD values are 0.277 and 2.675 μg/mL for CLZ and PRX, respectively.

Accuracy: The accuracy of the proposed method was proved by comparing the results of the proposed method with those obtained

Parameter		No. of theoretical plates (N)		Mass distribution ratio (Dm)		Resolution (Rs)	Relative retention (α)
		CLZ	PRX	CLZ	PRX		
pH of the mobile phase	4	494	767	0.77	5.62	6.46	7.298
	4.5	1098	807	0.76	5.41	7.08	7.12
	5	1117	745	0.775	5.53	6.9	7.14
	5.5	663	877	0.71	5.29	6.98	7.45
	6	1036	1129	0.71	5.24	8.06	7.39
	6.5	667	862	0.715	5.23	6.89	7.32
Conc. of SDS (M)	0.1	667	862	0.71	5.2	6.89	7.3
	0.125	1044	834	0.71	5.13	7.07	7.22
	0.15	653	950	0.69	5.14	7.11	7.45
	0.175	1058	860	0.727	5.23	7.2	7.19
	0.2	1051	856	0.722	5.21	7.19	7.22
Conc. of n-propanol	8%	1288	659	0.9	6.5	6.8	7.22
	10%	740	844	0.8	5.9	7.15	7.375
	12%	613	1015	0.6	4.9	7.2	8.16
	14%	455	850	0.7	5.17	6.5	7.38
Effect of flow rate(ml/min)	0.8	464	868	0.716	5.26	6.6	7.35
	1.0	1021	904	0.697	5.3	7.5	7.73
	1.2	1029	871	0.7	5.27	7.33	7.52

Where: Number of theoretical plates (N)=$5.54(t_R/W_{h/2})^2$
Mass distribution ratio (Dm)=t_R-t_m/t_m
Relative retention (α)=Dm_2/Dm_1
Resolution (R)=$2\Delta t_R/W_1+W_2$

Table 1: Optimization of the chromatographic conditions for clonazepam and Paroxetine mixture by the proposed method.

Parameter	CLZ	PRX
Linearity range (μg/mL)	1.0-20	4.0-250
Intercept (*a*)	-20657.1	-19856.2
Slope (*b*)	53108	12613.3
Correlation coefficient (*r*)	0.9998	0.9999
S.D. of residuals ($S_{y/x}$)	6962.97	17960
S.D. of intercept (S_a)	4448.77	10224.1
S.D. of slope (S_b)	411.73	76.408
Percentage relative standard deviation, % RSD	1.310	1.143
Limit of detection, LOD (μg/mL)	0.277	2.675
Limit of quantitation, LOQ (μg/mL)	0.838	8.106

Table 2: Analytical performance data for the determination of the studied drugs by the proposed method.

using the comparison chromatographic methods [42,43] for CLZ and PAX respectively. The comparison method for CLZ involved the use of methanol and ammonium phosphate (50:50 v/v) adjusted to pH 8.0 and detected ultravioletly at 254 nm, and for PAX involved the use of dipotassium hydrogen phosphate and Acetonitrile (90: 10 v/v), adjusted to pH 6.5 and detected spectrophotometrly at 295 nm. Statistical analysis of the results obtained by the proposed method and comparison methods using Student's t-test and variance ratio F-test [41] revealed no significant difference between the performance of the two methods regarding the accuracy and precision, respectively (Table 3).

Precision

I. **Intra-day precision:** Intra-day precision was assessed through replicate analysis of three concentrations of the studied drugs on three successive times within the same day. The results are shown in Table 4.

II. **Inter-day precision:** Inter-day precision was carried out through replicate analysis of three concentrations of the studied drugs on three successive days. The results are summarized in Table 4.

Robustness of the method: The robustness of the method was assessed by evaluating the influence of small variation of experimental variables: concentrations of organic modifier (12% ± 1), pH (6.0 ± 0.1), and conc. of SDS (0.175M ± 0.01) on the analytical performance of the method. In these experiments, one experimental parameter was changed while the other parameters were kept unchanged, and the recovery percentage was calculated each time. The minor changes in these experiment parameters did not significantly affect the peak areas; recovery percentage in case of CLZ is 100.08 ± 1.31 and in case of PRX is 100.22 ± 1.15 respectively.

Selectivity: The proposed MLC method was considered selective by detecting any change resulted from common tablet additives such as lactose, starch, magnesium stearate, and talc. The high mean % recovery and high accuracy with low SD indicated that excipients did not affect the results of the proposed method.

Applications

Application of the proposed method to the analysis of CLZ/PRX laboratory prepared mixtures: CLZ and PRX can be determined in laboratory prepared mixtures simultaneously by the suggested method in ratios of 1:50 (Figure 3). Both drugs can be quantitated in the laboratory prepared mixtures concerning the linear regression equations of the calibration plots.

The results shown in Table 5 are in good agreement with those obtained using the official methods [42,43]. Student's t-test and variance ratio F-test statistically analyzed the results [41] and showed no significant difference between the two methods concerning the accuracy and precision, respectively. Good recoveries, 99.86 ± 1.58 and 100.25 ± 1.443 were achieved for CLZ and PRX, respectively.

Pharmaceutical application

Application of the proposed method to the analysis of laboratory prepared co-formulated tablets: The proposed method was successfully applied to the estimation of CLZ and PRX in their laboratory prepared co-formulated tablets. The results listed in Table 6 show a good agreement with those by the official method [42,43]. Student's t-test and variance ratio F-test [41] analyzed the results obtained by the proposed method and showed no significant difference

Compound	Proposed Method			Comparison method (42)
	Amount taken (µg/mL)	Amount found (µg/mL)	% Found	% Found
CLZ	1	1.16	101.59	100.4 99.85 99.36 101.65
	2	2.01	100.40	
	5	4.92	98.39	
	8	7.89	98.59	
	10	9.96	99.62	
	12	12.23	101.88	
	14	14.12	100.83	
	20	19.87	99.33	
Mean			100.08	100.32
± S.D.			1.31	0.99
t-test			0.315 (2.23)*	
F-test			1.77 (8.89)*	
				Comparison method (43)
PRX	4	4.04	100.96	100.2 100.36 99.32 101.27
	10	10.13	101.33	
	50	50.39	100.78	
	75	74.09	98.79	
	100	98.51	98.51	
	150	151.74	101.16	
	200	201.77	100.88	
	250	248.33	99.33	
Mean			100.22	100.61
± S.D.			1.15	0.58
t-test			0.109 (2.23)*	
F-test			2.05 (8,89)*	

Table 3: Determination of the studied drugs in pure form by the proposed and comparison methods.

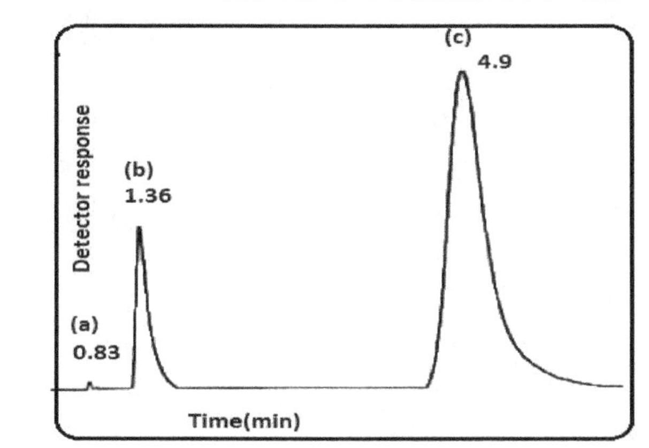

(a) Solvent front (b) Clonazepam (3 µg/ml) (c) Paroxetine HCl (150 µg/ml) (1:50

Figure 3: Typical chromatogram of the laboratory prepared mixture of clonazepam and Paroxetine.

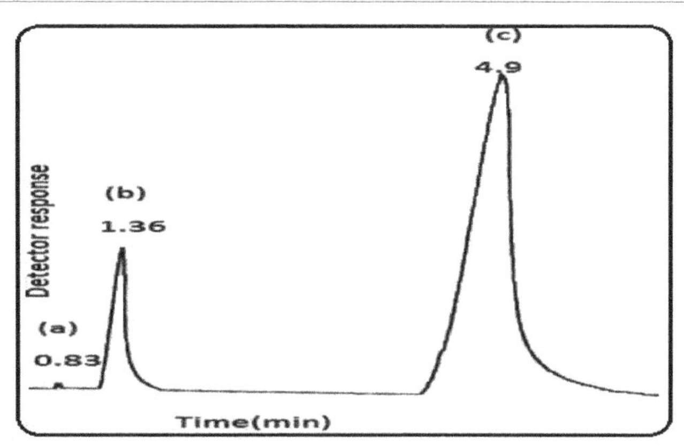

(a) Solvent front (b) Clonazepam (3 µg/ml) (c) Paroxetine HCL (150 µg/ml) (1:50)

Figure 4: Typical chromatogram of the laboratory prepared co-formulated tablet of clonazepam and Paroxetine.

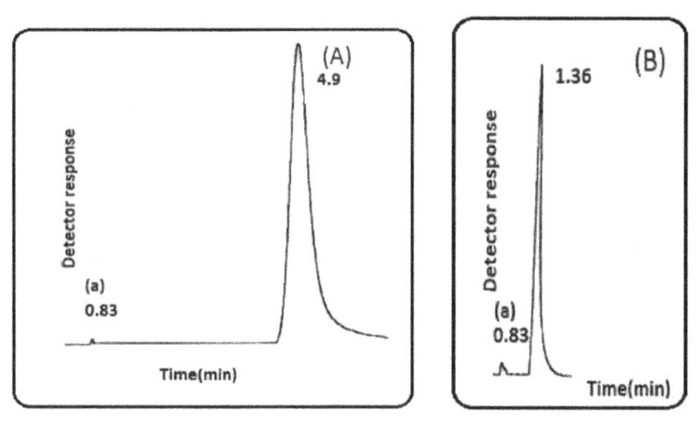

(a) Solvent front (A) Paroxetine HCL (200 µg/ml) in Seroxat® 12.5 mg tablet (B) Clonazepam (10 µg/ml) in Amotril ®0.5 mg tablet.

Figure 5: Typical Chromatograms of the studied drugs in their single pharmaceutical preparation.

between it and the official method [42,43] concerning the accuracy and precision, respectively. Good recoveries, 100.02 ± 0.59 and 100.19 ± 1.33 were achieved for CLZ and PRX respectively, from their prepared tablets in 1:50 ratio, respectively. Figure 4 shows chromatograms indicating good resolved peaks of CLZ and PRX in their laboratory prepared co-formulated tablets.

Analysis of CLZ and PRX tablets: The proposed method was successfully applied to the assay of the studied drugs in single dosage forms. The results of the proposed method were favorably compared with those obtained using the comparison method [42,43]. Mean percent recoveries from Amotril® 0.5 mg CLZ tablets and Seroxat® CR 12.5 mg PRX tablets were 99.96 ± 0.59 and 99.99 ± 1.27, respectively.

The results shown in Table 7 are in good agreement with those obtained with the comparison method [42,43]. Statistical analysis of the results obtained using Student's t-test and variance ratio F-test [41] revealed no significant difference between the performance of the two methods regarding the accuracy and precision, respectively. Figure 5 shows chromatograms indicating good resolved peaks of CLZ and PRX in their dosage forms.

Conclusion

A simple, accurate and validated chromatographic method with ultraviolet detection was proposed for the simultaneous determination

Parameters		CLZ Conc. (μg/mL)			PRX Conc. (μg/mL)		
		2	4	6	50	100	150
Intraday	% Found	99.44	100.74	99.59	101.91	99.80	98.69
		99.17	100.03	100	100.62	98.81	98.41
		100.04	98.97	100.76	99.92	100.36	99.79
	($\bar{x}$)	99.54	99.91	100.12	100.82	99.66	98.96
	± S.D.	± 0.46	± 0.89	± 0.59	± 1.01	± 0.79	± 0.73
	%RSD	0.46	0.89	0.59	1.00	0.79	0.74
Interday	% Found	99.44	100.74	99.59	101.91	99.80	98.41
		99.74	99.32	100.29	101.33	99.39	98.13
		100.66	100.38	101.23	100.98	100.22	99.24
	($\bar{x}$)	99.95	00.15	100.37	101.41	99.80	98.59
	± S.D.	± 0.64	± 0.74	± 0.82	± 0.47	± 0.42	± 0.58
	%RSD	0.64	0.74	0.82	0.46	0.42	0.59

Table 4: Precision data for the determination of CLZ and PRX in pure form by the proposed method.

Laboratory prepared mixtures	Proposed method						Comparison Method (42)	Comparison Method (43)
	Conc. taken (μg/mL)		Conc. found (μg/mL)		% Found		% Found	% Found
	CLZ	PRX	CLZ	PRX	CLZ	PRX	CLZ	PRX
0.25 mg CLZ + 12.5 mg PRX (1 : 50 ratio)	2	50	1.96	50.96	98.21	101.91	98.51	100.33
	4	100	4.08	99.66	101.93	99.66	100.28	100.65
	6	150	5.95	147.82	99.23	98.55	100.55	98.90
	8	200	8.00	201.57	100.07	100.78	100.61	100.15
X;‾					99.86	100.23	99.99	100.01
± SD					1.58	1.45	1.00	0.77
% RSD					1.578	1.443		
% Error					0.788	0.723		
T-test					0.136 (2.4)*	0.27 (2.4)*		
F-test					2.5 (9.27)*	3.56 (9.27)*		

N.B. *The value between parenthesis are the tabulated t and F values at *P* = 0.05

Table 5: results for the analysis of CLZ and PRX in their laboratory prepared mixtures.

Laboratory prepared co-formulated tablets	Proposed method						Comparison Method (42)	Comparison Method (43)
	Conc. taken (μg/mL)		Conc. found (μg/mL)		% Found		% Found	% Found
	CLZ	PRX	CLZ	PRX	CLZ	PRX	CLZ	PRX
0.25 mg CLZ + 12.5 mg PRX /tablet	2 4	50 100	2.01 3.97	50.97 98.75	100.36 99.28	101.93 98.75	101.4 98.85	101.2 99.36
(1 : 50 ratio)	6	150	6.04	149.6	100.61	99.74	100.36	101.32
	8	200	7.99	200.68	99.81	100.34	100.65	100.36
X;‾					100.02	100.19	100.32	100.56
± SD					0.59	1.33	1.07	0.91
% RSD					0.593	1.329		
% Error					0.297	0.666		
T-test					0.49 (2.44)*	0.46 (2.44)*		
F-test					3.26 (9.27)*	2.15 (9.27)*		

N.B.

*The value between parenthesis are the tabulated t and F values at *P*=0.05

Table 6: Results for the analysis of CLZ and PRX in their prepared tablets by the proposed and comparison methods.

Pharmaceutical preparation	Proposed method			Comparison method
	Conc. taken (μg/mL)	Conc. found (μg/mL)	% Found	% Found
Amotril® tablets (0.5 mg CLZ/tablet)	2	1.99	99.44	100.4
	4	4.03	100.74	101.98
	6	5.98	99.59	99.65
	8	8.01	100.08	100.12
AX;¯E			99.96	100.54
± SD			0.59	1.01
% RSD			0.586	
% Error			0.293	
t-test			0.98 (2.44)*	
F-test			2.97 (9.27)*	
Seroxat® CR tablets (12.5 mg PRX/tablet)	50	50.08	100.16	98.96
	80	79.01	98.77	99.43
	100	101.69	101.69	101.39
	120	119.21	99.34	99.98
AX;¯E			99.99	
± SD			1.27	
% RSD			1.269	
% Error			0.635	
T-test			0.06 (2.44)*	
F-test			1.45 (9.27)*	

N.B.
*The value between parenthesis are the tabulated t and F values at P=0.05

Table 7: Determination of CLZ and PRX in their single tablets by the proposed and comparison methods.

of CLZ and PRX in binary mixtures. In addition, it could be applied to the analysis of both drugs in their single and laboratory prepared co-formulated dosage forms without any interference from the common excipients and the results show good agreement with those obtained by the comparison method.

References

1. Cowen PJ, Green AR, Nutt DJ (1981) Ethyl beta-carboline carboxylate lowers seizure threshold and antagonizes flurazepam-induced sedation in rats. Nature 290: 54-55.
2. Browne TR (1976) Clonazepam A review of a new anticonvulsant drug. Arch Neurol 33: 326-332.
3. Tomson T, Svanborg E, Wedlund JE (1986) Nonconvulsive status epilepticus: high incidence of complex partial status. Epilepsia 27: 276-285.
4. The British Pharmacopoeia 98/34/EEC (2005) London: The Stationery Office.
5. United States Pharmacopoeia (2004) USP-27/NF-22. Rockville: Authority of the United States Pharmacopeia Convention.
6. El-Brashy A, Aly FA, Belal F (1993) Determination of 1,4-benzodiazepines in drug dosage forms by difference spectrophotometry. Microchim Acta 110: 55-60.
7. Salem AA, Barsoum BN, Izake EL (2002) Determination of bromazepam and clonazepam in pure and pharmaceutical dosage forms using chloranil as a charge transfer complexing agent. Anal Lett 35: 1631-1648.
8. Salem AA, Barsoum BN, Izake EL (2004) Spectrophotometric and fluorimetric determination of diazepam, bromazepam and clonazepam in pharmaceutical and urine samples. Spectrochim Acta Part A 60: 771-780.
9. Nie L, Liu D, Yao SJ (1990) Potentiometric determination of diazepam with a diazepam ion-selective electrode. Pharm Biomed Anal 8: 379-383.
10. Salem AA, Barsoum BN, Izake EL (2003) Potentiometric determination of diazepam, bromazepam and clonazepam using solid contact ion-selective electrodes. Anal Chim Acta 498: 79-91.
11. Correia dos Santos MM, Famila V, Goncalves ML (2002) Square-wave voltammetric techniques for determination of psychoactive 1,4-benzodiazepine drugs. Anal Bioanal Chem 374: 1074-1081.
12. Wilhelm M, Battista HJ, Obendorf D (2000) Development of Indirect Spctrophotometric Method for the Determination of Clonazepam in pharmaceutical preparation Using Resorcinol. J Chromatogr A 897: 215-225.
13. Cavedal LE, Mendes FD, Domingues CC, Patni AK, Monif T, et al. (2007) Clonazepam quantification in human plasma by high-performance liquid chromatography coupled with electrospray tandem mass spectrometry in a bioequivalence study. J Mass Spectrom 42: 81-88.
14. Gandhi SV, Dhavale ND, Jadhav VZ, Sabnis SS (2008) Spectrophotometric and Reversed-Phase High-Performance Liquid Chromatographic Methods for Simultaneous Determination of Escitalopram Oxalate and Clonazepam in Combined Tablet Dosage Form. JAOAC Int 91: 33-38.
15. Pujadas M, Pichini S, Civit E, Santamarina E, Perez K, et al. (2007) A simple and reliable procedure for the determination of psychoactive drugs in oral fluid by gas chromatography-mass spectrometry. J Pharm Biomed Anal 44: 594-601.
16. Katzman MA (2009) Current considerations in the treatment of generalized anxiety disorder. CNS Drugs 23: 103-120.
17. Food and Drug Administration (2013) FDA News release: FDA approves the first non-hormonal treatment for hot flashes associated with menopause.
18. Baldwin DS, Anderson IM, Nutt DJ, Bandelow B, Bond A, et al. (2005) Evidence-based guidelines for the pharmacological treatment of anxiety disorders: recommendations from the British Association for Psychopharmacology. Journal of Psychopharmacology 19: 567-596.
19. Baldwin D, Bobes J, Stein DJ, Scharwächter I, Faure M (1999) Paroxetine in social phobia/social anxiety disorder. Randomised,double-blind, placebo-controlled study. Paroxetine Study Group. The British Journal of Psychiatry 175: 120-126.

20. Yonkers KA, Gullion C, Williams A, Novak K, Rush AJ (1996) Paroxetine as a treatment for premenstrual dysphoric disorder. Journal of Clinical Psychopharmacology 16: 3-8.

21. Turner FJ (2005) Social Work Diagnosis in Contemporary Practice. Oxford University Press, USA.

22. Nouws HPA, DelerueMatos C, Barros AA, Rodrigues JA (2006) Electroanalytical determination of paroxetine in pharmaceuticals. Journal of Pharmaceutical and Biomedical Analysis 42: 341-346.

23. Erk N, Biryol J (2003) Voltammetric and HPLC techniques for the determination of paroxetine hydrochloride. Pharmazie 10: 699-704.

24. Robert S, Genowefa M, Marcin K (2003) Determination of fluoxetine and paroxetine in pharmaceutical formulations by densitometric and videodensitometric TLC. Journal of Planar Chromatography-Modern TLC 1: 19-22.

25. Venkatachalam A, Chatterjee VS (2007) Stability indicating high performance thin layer chromatography determination of Paroxetine hydrochloride in bulk drug and pharmaceutical formulations. Analytica Chimica Acta 2: 312-317.

26. Zainaghi IA, Lanchote VL, Queiroz RHC (2003) Determination of paroxetine in geriatric depression by high performance liquid chromatography. Pharmacological Research 2: 217-221.

27. Zhu Z, Neirinck L (2002) Highperformance liquid chromatographymass spectrometry method for the determination of paroxetine in human plasma. Journal of Chromatography 2: 295-300.

28. Massaroti P, Cassiano NM, Duarte LF (2005)Validation of a selective method for determinationof Paroxetine in human plasma by LC MS/MS. Journal of Pharmacy and Pharmaceutical Sciences 2: 340-347.

29. Jhee OH, Seo HK, Lee MH (2007) Determination of paroxetine in plasma by liquid chromatography couled to tandem mass spectrometry for pharmacokinetic and bioequivalence studies. Arzneimittelforschung 7: 455-461.

30. British Pharmacopoeia (2003) The Stationary Office, London, UK.

31. United States Pharmacopeial Convention (2008) The United States Pharmacopeia 31. The National Formulary 26, United States Pharmacopeia, Rockville, MD, USA.

32. Eap CB, Bouchoux G, Amey M, Cochard N, Savary L, et al. (1998) Simultaneous determination of human plasma levels of citalopram, paroxrtine , sertraline, and their metabolites by gas chromatography mass spectrometry. Journal of Chromatographic Science 7: 365-371.

33. Hans JL, Werner W, Günter F (2002) Improved sample preparation for the quantitative analysis of paroxetine in human plasma by stable isotope dilution negative ion chemical ionisation gas chromatography mass spectrometry. Journal of Chromatography 2: 353-357.

34. Chien L, Emily SG, Sidney HK, Alan N, Ronald TC, et al. (2000) Determination of paroxetine levels in human plasma using gas chromatography with electron capture detection. Journal of Chromatography 2: 275-279.

35. Labat L, Deveaux M, Dallet P, Dubost JP (2002) Separation of new antidepressants and their metabolites by micellar electrokinetic capillary chromatography. Journal of Chromatography 1: 17-23.

36. Walsh M, Belal F, El-Enany N, Elmansi H (2011) Spectrofluorimetric determination of paroxetine HCl in pharmaceuticals via derivatization with 4-chloro-7- nitrobenzo-2-oxa-1,3-diazole (NBD-Cl). J Fluoresc 21: 105-112.

37. Pollack MH, Simon NM, Worthington JJ, Doyle AL, Peters P, et al. (2003) Combined paroxetine and clonazepam treatment strategies compared to paroxetine monotherapy for panic disorder. J Psychopharmacol 17: 276-282.

38. Geetharam Y, Praveen S (2014) Development and validation of a stability-indicating HPLC method for the simultaneous determination of paroxetine hydrochloride and clonazepam in pharmaceutical dosage forms. International J Pharm 4: 448-457.

39. http://www.medguideindia.com

40. ICH Harmonized Tripartite Guideline (1996) Validation of Analytical Procedure. Text and Methodology, Q2 (R1) Current Step 4 Version, Parent Guidelines on Methodology.

41. Miller JC (2005) Statistics and Chemometrics for Analytical Chemistry. In: Harlow. 5th edn. Pearson Education Limited.

42. British Pharmacopeia (2010) 1: 548.

43. British Pharmacopeia (2010) 2: 1618-1619.

Determination of Radiochemical Purity of Radioactive Microspheres by Paper Chromatography

Ghahramani MR*, Garibov AA and Agayev TN

Institute of Radiation Problems, Azerbaijan national academy of sciences, Baku, Azerbaijan

Abstract

Simple methods to quantify radiochemical impurities existing in glass microspheres were tested. A system of two solvent mobile phases (acetone and 0.9% NaCl) were used with Whatman no. 3 paper strips as the stationary phase. Paper chromatography was used to identify radiochemical impurities in the radioactive microspheres. These impurities were found to be <0.01% for Azer spheres in saline and acetone solution. The results confirmed that glass microspheres prepared using the new procedures are convenient for radiotherapy purposes.

Keywords: Microspheres; Radioactive; Paper chromatography

Introduction

Chromatography is the collective term for a set of laboratory techniques for the separation of mixtures into their components. All forms of chromatography work on the same principle. They all have a stationary phase (a solid or a liquid supported on a solid) and a mobile phase (a liquid or a gas). The mixture is dissolved in a fluid called the mobile phase, which carries it through a structure holding another material called the stationary phase. The mobile phase flows through the stationary phase and carries the components of the mixture with it. The various constituents of the mixture travel at different speeds, causing them to separate. The separation is based on differential partitioning between the mobile and stationary phases. There are different types of chromatography, including column chromatography and paper chromatography [1-10].

Paper chromatography is a solid-liquid form of chromatography where the stationary phase is a specially manufactured porous paper and the mobile phase can be a single solvent or a combination of solvents. Paper chromatography is a method used to separate mixtures into their different parts [1-10]. Paper chromatography has been most commonly used to separate pigments, dyes and inks.

Chemists frequently use paper chromatography to follow the progress of a reaction by monitoring the disappearance of a reactant or the appearance of a product. Autoradiography can be used to investigate the distribution of radioactive materials [11].

Radiochemical purity is an important quality parameter for radiopharmaceuticals [12]. The radiochemical purity (RCP) of compounded radiopharmaceuticals should be monitored before administration to patients [13]. A number of analytical methods can be used to detect and determine the radiochemical impurities in a given radiopharmaceutical. Particularly important are precipitation, paper, thin-layer, and gel chromatography, paper and gel electrophoresis, ion exchange, solvent extraction, high performance liquid chromatography and distillation [8].

Thin-layer and paper chromatography is mostly used in a hospital environment. In paper and thin-layer chromatography, a volume equal to that described in the monograph is deposited on the starting-line. After development, the support is dried and the positions of the radioactive areas are detected by autoradiography or by measurement of the radioactivity over the length of the chromatogram using suitable collimated counters or by cutting strips and counting each portion. The positions of the spots or areas permit chemical identification by comparison with the solutions of the same chemical substance (non-radioactive) using a suitable detection method.

In the present study, paper chromatography was used to separate Y^{3+} ions into different structures. Depending on their molecular structures and interactions with the paper and mobile phase, they adhered to the paper more or less than the other compounds to allow quick and efficient separation. Autoradiography was used to investigate the distribution of radioactive Y^{3+} ions. This study developed a type of paper chromatography to determine the radiochemical purity of radioactive microspheres.

Materials and Method

Microsphere preparation

The present study developed a type of paper chromatography for determination and comparison of the radiochemical purity of radioactive yttrium-90 microspheres produced by sol gel method and different composition 1) TEOS: H_2O: HCl (Azer spheres) [14,15], 2) TEOS: CH_3COOH:H_2O (Azar spheres) [16], 3) TEOS: H_2O: H_3PO_4 (Az spheres) [17].

Neutron activation

For this procedure, 200 mg of each microsphere powder was poured into a quartz ampoule (6 cm in height, 0.7 cm in diameter) and the ampoule was sealed using an oxygen flame. The sample was irradiated inside a sealed aluminum container (7 cm in height, 3 in cm diameter) for 10 h in the research reactor. The medium neutron flux was about 3.0 $\times 10^{13}$ n/(s.cm^2).

After cooling for 3 d, the sealed aluminum container was cut and the quartz ampoule was crushed. These microspheres were subsequently added to a 10 ml vial with 5 ml of saline to form an elution. These

***Corresponding author:** Ghahramani MR, Institute of Radiation Problems, Azerbaijan national academy of sciences, Baku, Azerbaijan
Email: ghahramani.mr@gmail.com

radioactive microspheres were then analyzed to determine their radiochemical purity.

Radiochemical purity

Paper chromatography was used to determine the radiochemical impurities in the radioactive microspheres. The radiochemical purity of a radiopharmaceutical is defined as the fraction of total activity in the desired chemical form in the sample [18-20]. These impurities arise from incomplete labeling, breakdown of the labeled products over time caused by instability, and introduction of extraneous labeled ingredients during synthesis. Impurities can cause altered *in vivo* bio distribution after administration which can result in an unnecessary radiation dose to the patient. For these reasons, the US Pharmacopeia and US Food and Drug Administration have set limits on impurities in different radiopharmaceuticals that must not be exceeded in clinical operations.

Radiochemical impurities were checked using the two-solvent system: (a) saline solution (NaCl 0.9%) as the mobile phase on Whatman no. 3 paper; (b) acetone solution on Whatman no. 3 paper. The Whatman no. 3 chromatography paper sheets were cut into 1 cm × 8 cm strips and placed into empty 10 ml glass pharmaceutical vials. Approximately 1 ml of the appropriate solvent was placed into the vial and two droplets of radioactive solution were dropped onto it, forming a spot about 1 cm in diameter. After drying the droplets, the strips were placed in the appropriate solvent and the solvent was allowed to migrate until it reached the top of the strip.

This procedure generally requires 30 min. After the strip had dried, it was cut into eight 1 cm pieces and the radioactivity of each segment was measured in an alpha and beta counter (LB123 UMO Berthold) for 1 min. The data was expressed in counts per minute (cpm). The chromatography spectra are shown in Figures 1 and 2 and indicate low levels of impurities for these microspheres. The impurities were calculated by expressing the percentage of activity corresponding to the total activity on the plate (Tables 1 and 2).

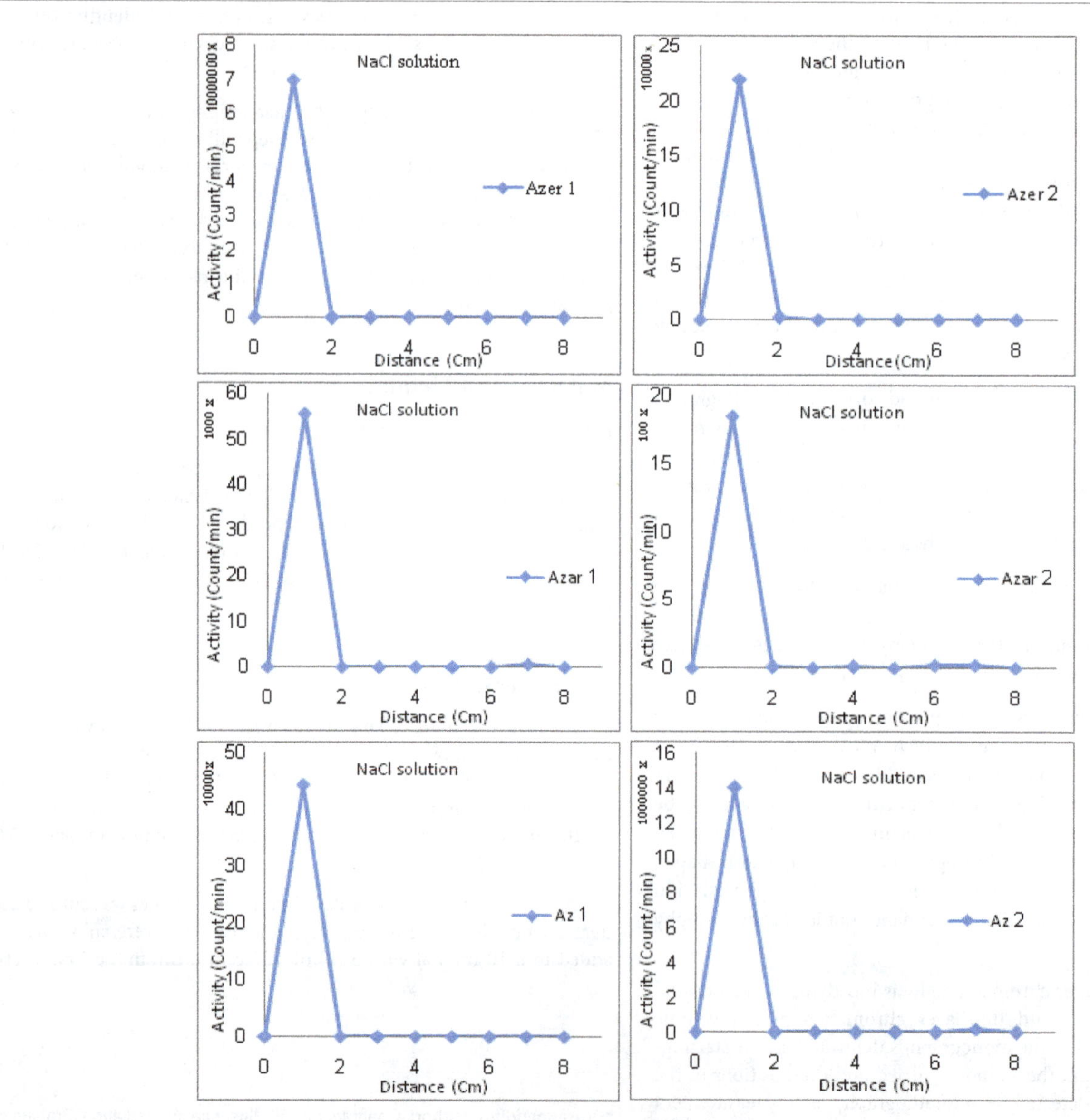

Figure 1: Chromatography analysis of different microspheres in saline solution.

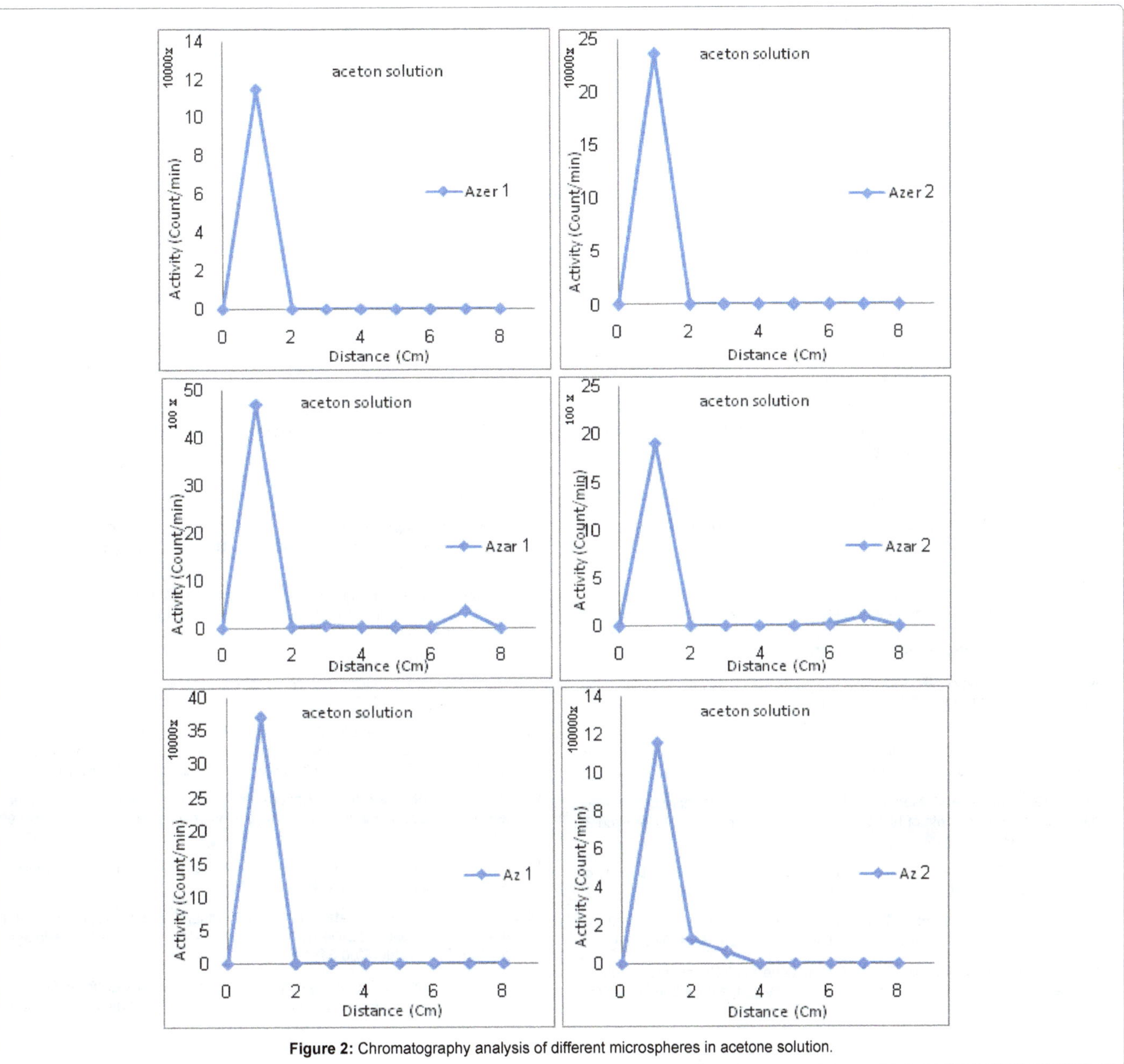

Figure 2: Chromatography analysis of different microspheres in acetone solution.

	Net activity of 1/3 strip in origin **(Count/min)**	***All activity of strip*** **(Count/min)**	***Radiochemical Purity***
Azer-spher 1	115005	115008.7	0.9999
Azer-sphere 2	236030	236035	0.9999
Azar-sphere 1	4712	5121	0.9201
Azar-sphere 2	1919	2040	0.9407
Az-sphere 1	370015	370022.4	0.9999
Az-sphere 2	1290000	1339992	0.9627

Table 1: Radiochemical purity of different microspheres in saline solution.

For protection from ingestion, inhalation and irradiation of the radioactive product, these processes were performed in a box made of Plexiglas.

Results and Discussion

The chromatography spectra shown in Figures 1 and 2 indicate low levels of impurities for these microspheres. The impurities were calculated by expressing the percentage of activity corresponding to the total activity on the plate.

The chromatography spectrums of Azer sphere 1, Azar sphere 1 and Az sphere 1 and those for free Al Azer sphere 2, Azar sphere 2 and Az sphere 2 are shown in Figures 1 and 2. The radiochemical purity was found to be more than 99% for the Azer spheres in saline and acetone solution. And it was less than 95% for the Azar spheres in

	Net activity of 1/3 strip in origin (Count/min)	*All activity of strip* (Count/min)	*Radiochemical Purity*
Azer-sphere 1	70000700	70000707.7	0.9999
Azer-sphere 2	222400	222446.6	0.9998
Azar-sphere 1	55605	56316	0.9874
Azar-sphere 2	1853	1898	0.9763
Az-sphere 1	445019	445092	0.9998
Az-sphere 2	14000035	14213742	0.9849

Table 2: Radiochemical purity of different microspheres in acetone solution.

saline solution. The radiochemical purity of Az sphere 1 was more than 99% in the saline and acetone solutions.

Conclusion

The results indicate that the low level of chemical impurities in the Azer and Az spheres make them suitable for medical applications. Paper chromatography with acetone and 0.9% NaCl was shown to be a simple and quick method for routine quality control of radiopharmaceuticals. It confirmed that glass microspheres prepared by the new procedure are a radiopharmaceutical of high efficiency and radiochemical purity with a satisfactory number of particles of the required size. These qualities promise good results for applications in radiotherapy where the radiochemical purity for radiopharmaceutical products must be higher than 95% [9].

References

1. Kumar S, Jyotirmayee K, M. Sarangi (2013) Thin Layer Chromatography: A Tool of Biotechnology for Isolation of Bioactive Compounds from Medicinal Plants. Int J Pharm Sci Rev Res 18: 126-132.
2. Ivanova NV, Trofimova NN, Es'kova LA, Babkin VA (2012) The Study of the Reaction of Pectin-Ag(0) Nanocomposites Formation. Int J Carbohydrate Chem 1-9.
3. Eldahshan OA (2010) Isolation and Structure Elucidation of Phenolic Compounds of Carob Leaves Grown in Egypt. Curr Res J Biol Sci 3: 52-55
4. Emam SS, El-Moaty Abd, Mohamed HI, Abd El S, (2010) Primary metabolites and flavonoid constituents of Isatis microcarpa J. Gay ex Boiss J Nat Prod 3: 12-26.
5. Himesh S, Ak S, Sarvesh S (2012) Quantification of Ascorbic Acid in Leaves of *Annona squamosa*. Int J Pharmacy and Pharmaceut Sci 4: 144-147.
6. Kaur P, Sharma P, Ahmed F, Tembhurkar V (2011) Optimization of Subtilin Production by *Bacillus Subitilis*. Indo Global J Pharmaceut Sci 1: 362-368.
7. Djoki D, Jankovi D, Maksin T (2002) Radiochemical Purity and Particles Number Determinations of Modified ^{99m}tc-Macroaggregated Albumin. J Serb Chem Soc 67: 573-579.
8. Saha GB (2010) Fundamentals of Nuclear Pharmacy 6th (Edn) Springer Science+Business Media, LLC, 233 Spring Street, New York, USA.
9. Zolle I (2007) Technetium-99m Pharmaceuticals Preparation and Quality Control in Nuclear Medicine. Department of Medicinal/Pharmaceutical Chemistry University of Vienna AlthanstraBe, Vienna, Austria.
10. Loveless VS (2009) Quality Control of Compounded Radiopharmaceuticals. University of New Mexico Health Sciences Center Pharmacy Continuing Education 1-34.
11. Laboratory course manual (2012) Autoradiography. Radioactive Isotopes and Ionizing Radiation Departmant of Biology, University of Copenhagen.
12. Decristoforo C, Siller R, Chen F, Riccabona G (2000) Radiochemical purity of routinely prepared 99Tcm radiopharmaceuticals: a retrospective study. Nucl Med Commun 21: 349-54.
13. American Pharmaceutical Association. Nuclear Pharmacy Guidelines for the Compounding of Radiopharmaceuticals. Nuclear Pharmacy Compounding Practice Guidelines 5.2: 1-29
14. Ghahramani MR, Garibov AA, Agayev TN (2014) Production and Quality Control of Radioactive Yttrium Microspheres for Medical Application. Appl Radiat Isot 85: 87-91.
15. Ghahramani MR, Garibov AA, Agayev TN, Mohammadi MA (2014) A Novel Way to Produce Yttrium Glass Microspheres for Medical Applications. Glass Phys and Chem 40: 283-287.
16. Ghahramani MR, Garibov AA, Agayev TN (2014) Production of yttrium aluminum silicate microspheres by gelation of an aqueous solution containing yttrium and aluminum ions in silicone oil. Int J Radiat Res 12: 189-197.
17. Ghahramani MR, Garibov AA, Agayev TN (2014) Preliminary results on a new method for producing yttrium phosphorous microspheres. Appl Radiat Isot 92: 46-51
18. Robbins PJ (1984) Chromatography of Technetium-99m Radiopharmaceuticals – A Practical Guide. Soc Nucl Med, New York, USA.
19. Furumoto S, Shinbo R, Iwata R, Ishikawa Y, Yanai K, et al (2013) In Vitro and In Vivo Characterization of 2-Deoxy-2-^{18}F Fluoro-D-Mannose as a Tumor-Imaging Agent for PET. J Nucl Med 54: 1-8
20. International Atomic Energy Agency (2006) Quality assurance for radioactivity measurement in nuclear medicine - Vienna, Technical reports series, no. 454

14

A Simple and Sensitive Method for Quantitative Measurement of Methylmalonic Acid by Turbulent Flow Chromatography and Tandem Mass Spectrometry

Tecleab AG[1], Schofield RC[2], Ramanathan LV[2] and Dean C Carlow[2]*

[1]Department of Pathology and Laboratory Medicine, Staten Island University Hospital, Staten Island, New York 10305, USA
[2]Department of Laboratory Medicine, Memorial Sloan Kettering Cancer Center, New York, NY 10065, USA

Abstract

A simple and sensitive method for the detection of methylmalonic acid in serum without derivatization has been developed. This method implements protein precipitation using methanol followed by additional sample clean up by turbulent flow liquid chromatography (TFLC). The sample was directly injected into the turbulent flow liquid chromatography tandem mass spectrometry system (TFLC-MS/MS) for online extraction followed by HPLC separation. The eluent was transferred to the mass spectrometer and ionized by heated electrospray negative ionization (HESI) and the analyte was quantified using a six-point calibration curve. The validated analytical measurement range (AMR) is 30-1,000 nMol/L. Dilutions of 10 and 200-fold were validated giving a clinical reportable range (CRR) of 30-200,000 nMol/L. The between-day and within-day imprecision values at concentrations spanning the AMR were less than 15%. This method was compared to an established LC-MS/MS method at a CLIA certified national reference laboratory and shows an excellent correlation with our TFLC-MS/MS method.

Keywords: Methylmalonic acid; Method validation; Mass spectrometry; Turbulent flow liquid chromatography; Protein precipitation

Abbreviations: AMR: Analytical measurement range; CLIA: Clinical laboratory improvement amendments; CRR: Clinical reportable range; HESI: Heated electrospray ionization; LLOQ: Lower limit of quantitation; MMA: Methyl malonic acid; MMA-D3: Deuterated methyl malonic acid; SA: Succinic acid; SRM: Selected reaction monitoring; TFLC: Turbulent flow liquid chromatography.

Introduction

Methylmalonic acid (MMA) levels in serum are commonly used as a marker for cobalamin (vitamin B12) deficiency or to diagnose patients with inherited defects in methylmalonyl CoA mutase enzyme activity. Lack of vitamin B12 causes an increase in the concentration of MMA, and its measurement in serum plays an important role in diagnosing B12 deficiency, which can lead to megaloblastic anemia and irreversible neurological disorders [1,2]. A mild elevation of MMA (>400 nMol/L) is an early indicator of vitamin B12 deficiency while a large elevation (>40,000 nMol/L) is indicative of methylmalonic acidemia, which is an inborn metabolic disorder [3].

Quantitative methods for the measurement of MMA have previously been described. These earlier methods used either protein precipitation [4], solid phase extraction [5-7], ultra-filtration [8-11], or chemical derivatization [12-14] which was followed by MS/MS or GC/MS. One of the major difficulties in measuring MMA is the presence of similar organic acids such as succinic acid (SA) which is present in much greater concentrations. SA is chromatographically difficult to separate from MMA and the mass spectra are very similar. These previously described methods require extensive sample preparation and lengthy extractions protocols and large sample volumes. The advantage of turbulent flow chromatography is that it allows for the injection of sample extract with minimal preparation. The turbulent flow column is packed with large porous particles which allow the retention of small molecules while large proteinaceous molecules are discarded into the waste thereby allowing the removal of large molecules, such as proteins, from the sample [15]. A previously described TFLC method for the quantitation of MMA has been published however it requires large specimen volume and a laborious multi wash SPE extraction prior to analysis [7]. In the present study we utilize turbulent flow chromatography to obviate the need for chemical derivatization and simplified the sample preparation procedure. The small sample was prepared by precipitating proteins using methanol followed by centrifugation and the supernatant was directly injected into the turbulent flow column. This method is rapid, simple to perform and provides an accurate and precise quantitative method for the measurement of MMA in serum.

Materials and Methods

Chemicals and reagents

Methylmalonic acid (MMA) was obtained from Sigma-Aldrich (Saint-Louis, MO, USA). Methylmalonic acid (1 mg/mL in acetonitrile) and deuterated methylmalonic acid (MMA-D_3) at a concentration of 1 mg/ml in acetonitrile were obtained from Cerilliant (Round Rock, TX, USA). All compounds had a purity greater than or equal to 99%. Acetone, formic acid, methanol, acetonitrile, 2-propanol, and water (LCMS grade) were from Fisher Scientific (Fair Lawn, NJ, USA). Commercially available human serum that had been chemically treated to remove endogenous small molecules (DC Mass Spect Gold®, stripped human serum) was obtained from Golden West Biologicals, Inc, (Temecula, CA, USA). Ammonium acetate was from Fisher Scientific (Fair Lawn, NJ, USA). Succinic acid (SA) and deuterated SA were from Sigma-Aldrich (Saint-Louis, MO, USA).

***Corresponding author:** Dean C Carlow, Department of Laboratory Medicine, Memorial Sloan Kettering Cancer Center, New York, NY 10065, USA
E-mail: carlowd@mskcc.org

Instrument and analytical conditions

A Thermo Scientific Aria TLX-2 turbulent flow chromatography system (Franklin, MA, USA) comprised of a CTC analytics PAL auto sampler, a low-pressure mixing quaternary pump (loading pump), a high-pressure mixing binary pump (eluting pump) and a three-valve switching device unit (VIM) containing six-port valves were operated in accordance with manufacturer recommendations. An in-depth description of the system was previously published [16]. The triple quadrupole mass spectrometer was a Thermo Scientific TSQ Quantum Ultra (San Jose, CA, USA) and it implemented a heated electrospray ionization probe that was maintained at 380°C. The analysis was performed in a negative ionization mode with a spray voltage of 4500 V. Nitrogen was used as the sheath, auxiliary and ion sweep gas at 60, 15, and 2 arbitrary units, respectively. The system was operated in selected reaction monitoring (SRM) mode with argon as the collision gas at a pressure of 1.5 mTorr. The ion transfer tube was maintained at 235°C. The entire system was controlled using Aria 1.6.2 software. The Turbo Flow column used was a Cyclone Max® (0.5 mm × 50 mm) from Thermo Scientific (San Jose, CA, USA) and the analytical column was an Allure Organic Acid® 5μm (150 mm × 3.0 mm) from Restek (Bellefonte, PA, USA). The mobile phase consisted of loading pump A (water), loading pump B (water containing 1% formic acid), and loading pump C (acetonitrile, 2-propanol and acetone in 1:1:1 ratio). Eluting pump A contained 15.2 mM ammonium acetate in water containing 0.06% formic acid and eluting pump B contained methanol.

Preparations of stock solutions, calibration standards and quality control samples

The internal standard (MMA-D_3) was prepared by diluting the stock solution (100 μg/mL) to 50 ng/mL in methanol and stored at -20°C. Two stock solutions of MMA were prepared gravimetrically in methanol (1 mg/mL) and were used to prepare calibrators and quality control materials. MMA obtained from Sigma-Aldrich was used for calibrators while MMA from Cerilliant was used for quality controls, recovery and imprecision studies.

Calibrators and quality controls were prepared by spiking MMA into the stripped human serum. Prior to the addition of MMA the stripped human serum was tested to ensure that the concentration of endogenous MMA was below the lower limit of quantitation (LLOQ) of the assay. Briefly, the stripped human serum was spiked to obtain a concentration of 1000 nMol/L of MMA and then further diluted to obtain a six point calibration curve (500, 250, 125, 62.5, and 50 nMol/L). The three levels of controls were prepared by spiking MMA into the stripped human serum to yield concentrations of 800, 400, 100, and 30 nMol/L. Calibrators and controls were aliquotted and stored at -80°C.

Sample preparation

Samples were prepared by protein precipitation using methanol. Calibrators, controls or patient specimen (100 μL) were aliquotted into 1.5 mL micro centrifuge tubes followed by the addition of 200 μL of methanol containing MMA-D_3 (50 ng/mL). Subsequently, samples were vortexed mixed for five seconds and incubated on ice for five minutes. The samples were then centrifuged at 13,000 RPM for ten minutes. The supernatant (150 μL) was transferred into auto sampler vials containing inserts and 60 μL was injected into the system for analysis by TFLC-MS/MS.

Method validation

The method was validated per U.S. Food and Drug Administration (FDA) [17], Clinical Laboratory Standards Institute (CLSI) [18] and Clinical Laboratory Improvement Amendments (CLIA) guidelines. The assay is fully validated for imprecision, accuracy, linearity, recovery, and carryover, specificity and matrix effects.

Specificity

The assay was validated for its ability to selectively detect MMA over SA, a compound which is found in patient samples at elevated concentrations and is difficult to separate chromatographically and is known to interfere with MMA in several assays. The specificity was tested by spiking both MMA and SA into stripped human serum and extracting the samples following the protocol described above. The chromatograms of both analytes were identified based on the SRM responses and retention times (Figure 1).

Linearity

The linearity of this method was evaluated by using calibration curves as described in section 2.3. The calibration curves were generated by plotting the peak area ratios of MMA to the internal standard, MMA-D_3. Weighted linear regression models with weights inversely proportional to the X values were used. The LLOQ was defined as the lowest concentration of analyte where the coefficient of variation (CV) was below 20%, per FDA guidelines [17]. In addition, the signal-to-noise ratio (S/N) of analyte at the LLOQs exceeded the minimum requirement [18].

Accuracy and precision

Accuracy of the assay was assessed in two ways. The first way was by performing a recovery experiment, and additionally by comparing the results of 63 residual patient samples that were analyzed at a CLIA certified national reference laboratory using LC-MS/MS. Stripped human serum was spiked with MMA at three different concentrations spanning the analytical measurement range (AMR). Each level was then extracted in triplicate and the accuracy was calculated by determining the percent recovery. The correlation study was performed by extracting each patient sample as described above and comparing the results to the result obtained from the national reference laboratory and plotting the data using linear regression analysis.

A within-day imprecision study was conducted using stripped human serum spiked with MMA at concentrations of 30, 100, 400 and 800 nMol/L. For each level, ten separate extractions were prepared and analyzed in one day and the data was evaluated. Between-day imprecision was evaluated by extracting these same samples over 20 days in triplicate. The mean and standard deviation were determined over the validation period and the imprecision was calculated. The CV for the between-day imprecision should not be greater than 15% [17].

Matrix interference and ion suppression

To evaluate matrix effects a post column infusion experiment was performed using an experimental set-up as described previously [16]. A tee was inserted between the outlet of the analytical column and a syringe was inserted into the tee to allow for the infusion of analyte into the mass spectrometer. A standard mixture solution of MMA and MMA-D_3 were infused into the eluent stream at a flow rate of 10 μL/min. The signals of the corresponding MRM transitions of the analytes were recorded. After obtaining a steady baseline, a blank serum sample was extracted and injected and processed by the TFLC-MS/MS system. Any eluting compound that interfered with the ionization of target analytes would lead to an elevation or depression of the baseline which would represent matrix effects.

Several commonly occurring endogenous compounds which have

the potential to interfere with assays were evaluated. Interferences from lipemic, hemolyzed, and icteric samples were tested by spiking MMA into patient samples containing high levels of triglyceride (926 mg/dL), hemoglobin (4.6 g/dL), and bilirubin (34.9 mg/dL). MMA was measured before and after the addition of MMA and the effect of the indices was assessed by subtracting the basal value from concentration determined after the addition of MMA.

Quantitation of sample carryover

Sample carryover can be an issue with on-line sample preparation systems. To rule out carryover a blank stripped serum sample was injected after the highest calibrator during the validation period to confirm it did not exceed 20% of the LLOQ [18].

Results and Discussion

Chromatographic conditions

The turbulent flow chromatography parameters were adapted from Yuan et al. [7] and optimized to maximize sensitivity for all analytes. It was determined that the combination of a Cyclone-MAX (50 mm × 0.5 mm) and an Allure Organic Acids column (3 mm × 150 mm) produced adequate retention and separation of the compounds. Various mobile phase compositions, flow rates and profiles were evaluated. The desired sensitivity was achieved using water and water containing 0.1% formic acid in the quaternary pump for the loading and eluting solvents, respectively. While 15.2 mM ammonium acetate containing 0.06% formic acid in water and methanol were used in the binary pump for the loading and eluting solvents, respectively. The analytes were loaded on the turbulent flow column in 100% mobile phase A and transferred to the HPLC column with 100% mobile phase B using a 200 µL transfer loop. The loading and eluting mobile phase composition for the HPLC column are depicted in Table 1. The optimal injection volume was 60 µL and the analytical column was maintained at 70°C. The integration parameters for all analytes were similar with a baseline window of 50, area noise factor of 5, peak noise factor of 10 and an integration window of 15 seconds (Table 1).

MS/MS detection

The TFLC-MS/MS analysis was performed as described in section 2.2. The optimization of the SRM parameters was executed by direct infusion of the standards using negative electrospray ionization. The transitions monitored for MMA and MMA-D_3 in SRM mode were 117>73 and 120>76, respectively. Collision induced dissociation (CID) mass spectra were recorded. The optimal collision energies (CE) and tube lens values were 10 and 85 V, respectively. The skimmer offset was determined in MS mode to be optimal at a value of 10 V. The scan time was 0.05 (s) and scan width was 0.05 (m/z). The data was processed using LCquan software version 2.6.

Method validation

As previously stated the method was validated per U.S. Food and Drug Administration [17], Clinical Laboratory Standards Institute [18] and CLIA guidelines. The assay is fully validated for imprecision, accuracy, linearity, recovery, and carryover, specificity and matrix effects.

Specificity

A bioanalytical method should be selective for a specific analyte and not be affected by interfering or co-eluting components in a biological matrix. A TFLC-MS/MS SRM chromatogram for MMA, MMA-D_3, and SA in human serum is shown in Figure 1. The retention times for MMA, MMA-D_3, and SA were 0.76, 0.76, and 1.0 minutes, respectively. It can be observed in the chromatogram that separation of MMA from the isobaric interference from SA was achieved (Figure 1).

Linearity

A linear relationship was found between analyte concentrations

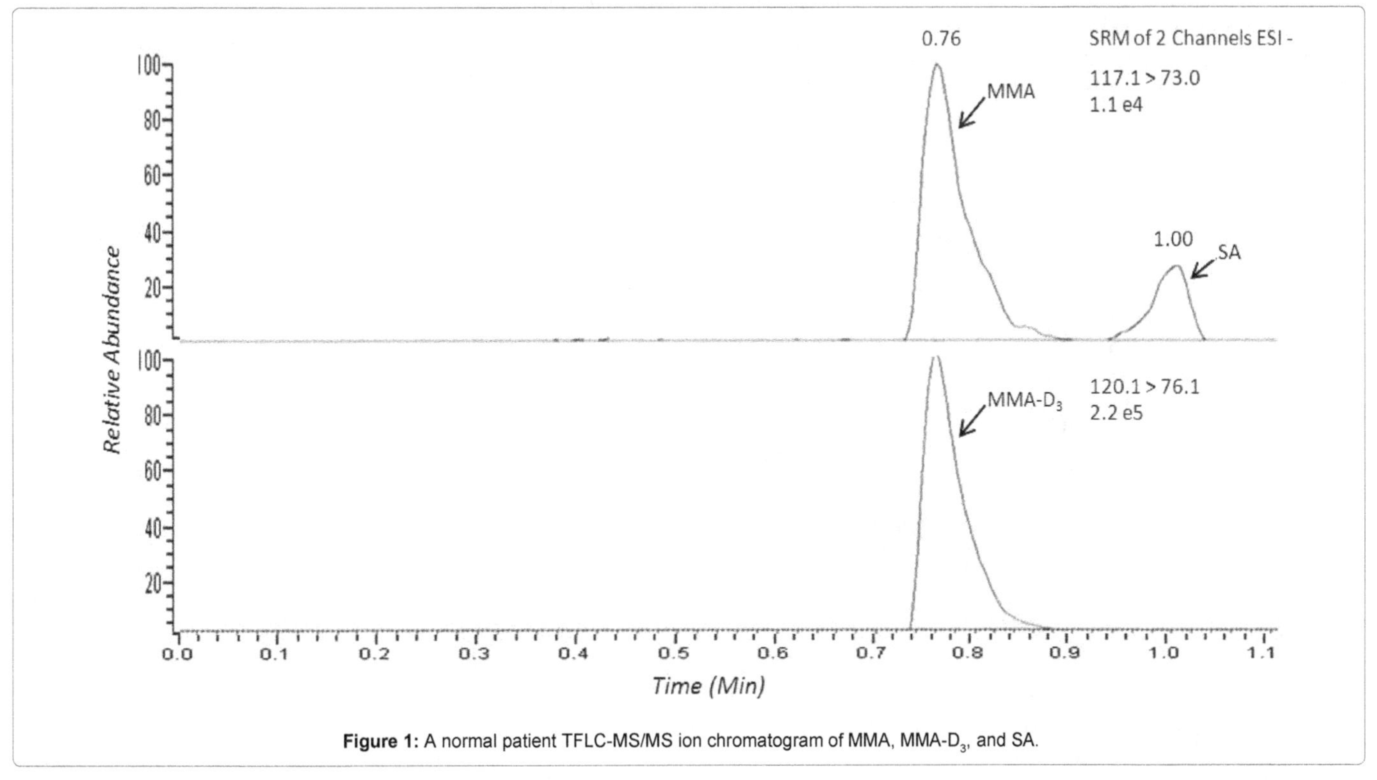

Figure 1: A normal patient TFLC-MS/MS ion chromatogram of MMA, MMA-D_3, and SA.

and peak area ratios throughout the AMR. The validated AMR of this method is 30-1,000 nMol/L. The coefficient of correlations (r) as determined by a six-point calibration curve was greater than >0.995. The LLOQ for this assay was determined to be 30 nMol/L. The signal-to-noise (S/N) of MMA at the LLOQ was 118; more than 5 times the S/N requirement with a between-day CV of <20% [17].

Accuracy and precision

To assess the accuracy and precision of this method, QC samples at three different concentrations spanning the AMR were analyzed; 100, 400, and 800 nMol/L. The concentration of MMA in each sample was determined by comparing MMA concentration to internal standard response based on the calibration curve. The imprecision was calculated as the %CV for both the within-day and between-day batches. The within-day and between-day imprecision for the assay was less than 15% for the three levels of controls while the %CV at the LLOQ was less than 20% (Table 2).

The accuracy of the assay, as evaluated by the recovery experiments, resulted in a recovery of MMA ranging between 95.8 to 110.9% (Table 3).

We developed and validated a dilution protocol to allow the measurement of patient specimens with values greater than 1000 nMol/L. Specimen containing MMA at 100,000 and 5,000 nMol/L were diluted 200 and 10-fold, respectively. The data showed <7% deviation from the expected values.

Correlation to alternative methods

We further analyzed the accuracy of the assay by comparing the results from 63 patients to a LC-MS/MS method from a CLIA certified national reference laboratory. The slope of the linear regression curve was within 2% and exhibited an excellent correlation coefficient as shown in Figure 2.

Matrix interference and ion suppression

Evaluation of possible matrix interference and ion suppression was conducted using a tee-infusion experiment which did not detect ion suppression or enhancement at the point of elution. We have also ruled out interferences from common endogenous compounds that can potentially cause interferences. No interferences were observed from triglycerides (926 mg/dL), bilirubin (34.9 mg/dL), or hemoglobin (4.6 g/dL).

Quantitation of sample carryover

Sample carryover was evaluated by running a serum blank after the highest calibrator on all calibration curves during the validation period. The average carryover was determined to be less than 20% of the calculated response of the LLOQ. As mentioned previously the LLOQ of this assay is 30 nMol/L with a S/N greater than five times the minimal requirement.

Conclusion

This article describes the development and validation of a simplified TFLC-MS/MS method for the quantification of MMA in human serum. TFLC for analyte extraction allows for reduced sample preparation and sample clean-up. Other methods have used solid phase extraction, ultra-filtration or derivatizations which are subject to elaborate procedures and additional instrumentation. This method allows for a simple protein precipitation which yields the required sensitivity for clinical applications. The assay is fully validated for specificity, sensitivity, accuracy, precision, linearity, and recovery. The method is accurate, with recoveries ranging between 96 and 111% at concentrations spanning the AMR and shows excellent agreement with an alternate LC-MS/MS assay form a CLIA certified national reference laboratory. The MMA assay is linear between 30-1000 nMol/L, and has excellent performance characteristics. There are no matrix interferences observed or interferences from other co-eluting compounds such as SA. Unlike other LC-MS/MS assays that require extensive sample preparation techniques we have development an application Utilizing turbulent flow chromatography that implements a simple protein precipitation to achieve the clinically required performance characteristics.

Acknowledgements

This work was supported by the Department of Laboratory Medicine of Memorial Sloan Kettering Cancer Center and through the NIH/NCI Cancer Center Support Grant P30 CA008748.

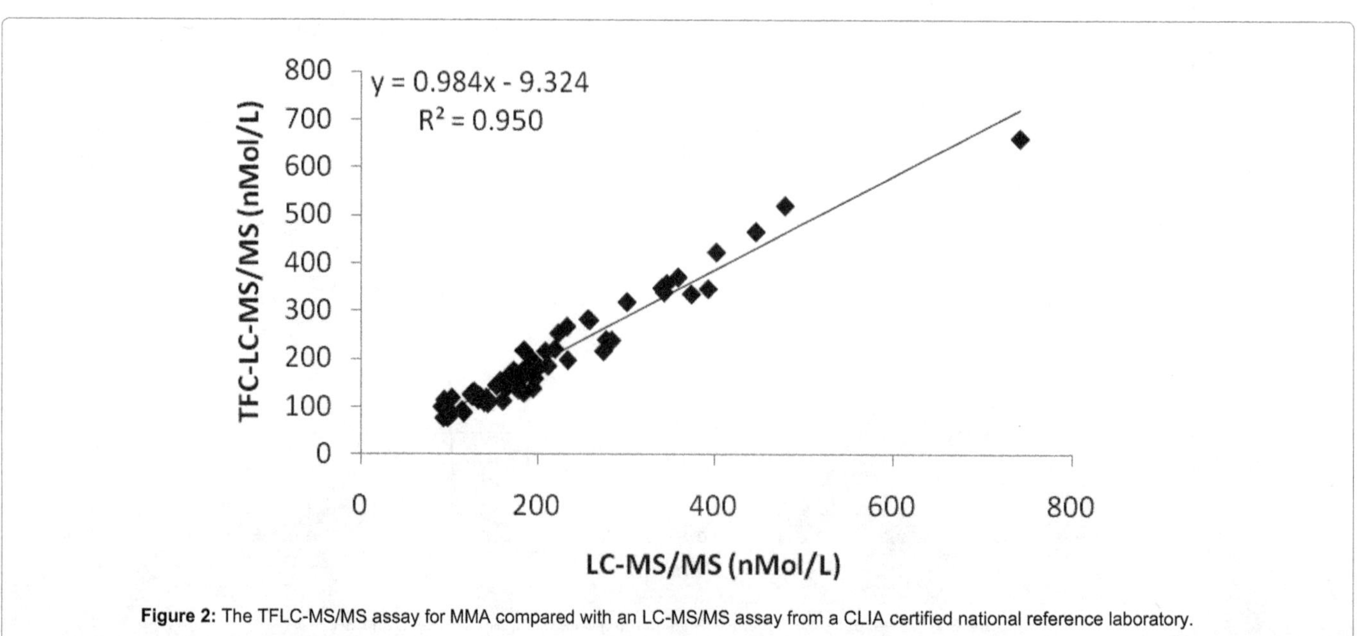

Figure 2: The TFLC-MS/MS assay for MMA compared with an LC-MS/MS assay from a CLIA certified national reference laboratory.

Step	Time (s)	Flow (mL/min)	%A	%B	%C	Flow (mL/Min)	Grad.	%A	%B
1	30	3.00	100	-	-	0.8	Step	92	8
2[1]	30	0.90	100	-	-	0.2	Step	92	8
3[2]	90	2.00	-	50	50	0.8	Step	92	8
4	20	2.00	-	50	50	0.8	Step	50	50
5	40	3.50	-	100	-	0.8	Step	50	50
6	20	3.50	-	100	-	0.8	Step	92	8
7	120	2.00	100	-	-	0.8	Ramp	92	8
8	10	3.00	100	-	-	0.8	Ramp	92	8

Table 1: Gradient used on the loading and eluting pumps for the TFLC-LS/MS analysis of MMA.

Sample	Nominal value (nMol/L)	Mean (nMol/L) Within-Day	Within-day CV (%) n=10	Mean (nMol/L) Between-Day	Between-day CV (%) n=20
LLOQ	30	29.8	19.4	32.8	19.4
Level 1	100	89.3	11.6	107.8	8.8
Level 2	400	415.2	2.8	414.5	5.0
Level 3	800	825.1	4.6	796.3	5.8

[1]During this step valve "A" and valve "B" are in line and the analyte is transferred from the turbulent flow column to the analytical column.
[2]Data Acquisition window: 1.5-3.0 minutes.

Table 2: Within and between-day precision of MMA in human serum.

Sample	Nominal value (nMol/L)	Measured value (nMol/L)	%Recovery (measured/nominal)
Low	100.0	95.8 ± 13.3	95.8 ± 13.9
Medium	500.0	531.7 ± 11.6	106.3 ± 2.2
High	800.0	887.2 ± 78.4	110.9 ± 8.8

Table 3: Method accuracy.

References

1. Cox EV, White AM (1962) Methylmalonic acid excretion: an index of vitamin B12 deficiency. Lancet 2: 853-856.
2. Fenton W, Rosenblatt D (2001) Disorders of propionate and methylmalonate metabolism. The Metabolic and Molecular Bases of Inherited Disease 2: 2165-2193.
3. Kushnir M, Nelson G, Frank E, Rockwood A (2016) High-throughput analysis of methylmalonic acid in serum, plasma, and urine by LC-MS/MS. Method for analyzing isomers without chromatographic separation. Clinical Applications of Mass Spectrometry in Drug Analysis: Methods and Protocols. Methods in Molecular Biology 1378: 159-173.
4. Lakso HA, Appelblad P, Schneede J (2008) Quantification of methylmalonic acid in human plasma with hydrophilic interaction liquid chromatography separation and mass spectrometric detection. Clinical Chemistry 54: 2028-2035.
5. Schmedes A, Brandslund I (2006) Analysis of methylmalonic acid in plasma by liquid chromatography-tandem mass spectrometry. Clinical Chemistry 52: 754-757.
6. Magera MJ, Helgeson JK, Matern D, Rinaldo P (2000) Methylmalonic acid measured in plasma and urine by stable-isotope dilution and electrospray tandem mass spectrometry. Clinical Chemistry 46: 1804-1810.
7. Yuan C, Gabler J, El-Khoury JM, Spatholt R, Wang S (2012) Highly senstive and selective measurement of underivatized methylmalonic acid in serum and plasma by liquid chromatography-tandem mass spectrometry. Analytical and Bioanalytical Chemistry 404: 133-140.
8. Blom HJ, van Rooij A, Hogeveen M (2007) A simple high-throughput method for the determination of plasma methylmalonic acid by liquid chromatography-tandem mass spectrometry. Clinical Chemistry and Laboratory Medicine 45: 645-650.
9. Fasching C, Singh J (2010) Quantitation of methylmalonic acid in plasma using liquid chromatography-tandem mass spectrometry. Methods in Molecular Biology 603: 371-378.
10. Hempen C, Wanschers H, Veer G (2008) A fast liquid chromatographic tandem masss spectrometric method for the simultaneous determination of total homocysteine and methylmalonic acid. Analytical and Bioanalytical Chemistry 391: 263-270.
11. Fu X, Xu Y, Chan P, Pattengale P (2013) Simple, fast, and simultaneous detection of plasma total homocysteine, methylmalonic acid, methionine, and 2-methylcitric acid using liquid chromatography and mass spectrometry (LC/MS/MS). JIMD Reports 10: 69-78.
12. Kushnir MM, Komaromy-Hiller G, Shushan B, Urry FM, Roberts WL (2002) Analysis of dicarboxylic acids by tandem mass spectrometry. High-throughput quantitative measurement of methylmalonic acid in serum, plasma, and urine. Clinical Chemistry 47: 1993-2002.
13. Pedersen TL, Keyes WR, Shahab-Ferdows S, Allen LH, Newman JW (2011) Methylmalonic acid quantification in low serum volumes by UPLC-MS/MS. Journal of chromatography B Analytical Technologies in the Biomedical and Life Sciences 879: 1502-1506.
14. Mineva E, Zhang M, Rabinowitz D, Phinney K, Pfeiffer C (2015) An LC-MS/MS method for serum methylmalonic acid suitable for monitoring vitamin B12 status in population surveys. Analytical and Bioanalytical Chemistry 407: 2955-2964.
15. Couchman L (2012) Turbulent flow chromatography in bioanalysis: a review. Biomedical Chromatography 26: 892-905.
16. Schofield RC, Ramanathan LV, Murata K, Grace M, Fleisher M, et al. (2015) Development and validation of a turbulent flow chromatography and tandem mass spectrometry method for the quantitation of methotrexate and its metabolites 7-hydroxy methotrexate and DAMPA in serum. Journal of chromatography B Analytical Technologies in the Biomedical and Life Sciences 1002: 169-175.
17. FDA Guidance for Industry Bioanalytical Method Validation (2001) US department of health and human service, Food and Drug Administration. Center for drug evaluation and research (CDER).
18. CLSI (2013) Liquid Chromatography-Mass Spectrometry Methods; Approved Guideline. CLSI document C62-A. Wayne, PA: Clinical and Laboratory Standards Institute.

15

Estimation of Octanol-Water Partition Coefficient Using Cationic Gemini Surfactants by Micellar Electrokinetic Chromatography

Mustafa GUZEL[1], Cevdet AKBAY[2*], Yatzka HOYOS[2], David H. AHLSTROM[2]

[1]*Istanbul Medipol University, International School of Medicine, Department of Medical Pharmacology (Chair), Regenerative and Restrorative Medicine Research Center (REMER) (Molecular Discovery and Development Group), Kavacık Campus, Kavacık/Beykoz-ISTANBUL 34810*

[2] *Department of Chemistry and Physics, Fayetteville State University, Fayetteville, NC 28301, USA*

Abstract

Micellar electrokinetic chromatography (MEKC) provides a simple and rapid approach for determining n-octanol-water partition coefficients (log P_{ow}). A set of non-hydrogen bonding (NHB), hydrogen bond accepting (HBA) and hydrogen bond donating (HBD) benzene derivatives with known log P_{ow} values was used as sample solutes. Two novel cationic gemini surfactants with different head groups were used as pseudostationary phases. Sodium dodecyl sulfate (SDS) was also used for comparison. Two approaches were applied for the determination of log P_{ow} values: calibration curve and phase ratio. In calibration curve approach, the MEKC retention factors (log k) of six alkyl phenyl ketones were plotted against their literature log P_{ow} values for constructing the calibration curve. Log P_{ow} values of benzene derivatives were then determined from the slope and the y-intercept of the linear calibration line. In the phase ratio approach, total surfactant concentration, critical micelle concentration, partial specific molar volume and experimental log k values were utilized for estimation of the log P_{ow} values. Both approaches provided comparable results for HBA solutes; however, the calibration curve approach and phase ratio approach were found to be more successful for NHB and HBD solutes, respectively. In general, gemini surfactants provided better estimated log P_{ow} values for NHB and HBA solutes while SDS gave better values for HBD solutes.

Keywords: Gemini surfactants; Micellar electrokinetic chromatography; n-Octanol-water partition coefficient; log P_{ow}; Partial specific volume; Phase ratio

Abbreviations: CMC: Critical Micelle Concentration; HBA: Hydrogen Bond Accepting; HBD: Hydrogen Bond Donating; NHB: Non-Hydrogen Bonding; log P_{ow}: n-octanol-water Partition Coefficients; MEKC: Micellar Electrokinetic Chromatography; PSV: Partial Specific Volume

Introduction

Lipophilicity, which correlates with the bioactivity of chemicals, is an important molecular descriptor and its determination constitutes an important element in pharmaceutical characterization of drug candidates since drugs must pass across various biological membranes to reach its site of action. Its determination is of great importance in a variety of fields such as in micellar catalyst, [1] in estimation of the toxic effect of substances in animals and plants, [2] in prediction of chemical adsorption in soil, [3] and in method development and optimization in micellar electrokinetic chromatography (MEKC) [4]. Introduced by Hansch and Fujita for biological activity of chemicals, the logarithm of the partition coefficient between n-octanol and water (log P_{ow}) has been widely used as a general measure of lipophilicity [5-7]. Shake-flask methods is well-known method for determination of log P_{ow} values of chemicals [8,9]. However, this method is cumbersome, time-consuming, needs skilled operator, and requires relatively large amount of pure compounds. After equilibrium between n-octanol and water, the relative concentration of the sample in each layer needs to be determined using spectroscopic or chromatographic techniques. In addition, n-octanol and water system does not mimic the biological model because biomembranes consisting of relatively rigid phospholipids are different from n-octanol-water system in terms of physicochemical property and size.

Due to the drawbacks of the direct measurement of log P_{ow}, alternative methods such as high performance liquid chromatography (HPLC) [10-12] and theoretical calculation methods [13,14] have been introduced. High correlations are observed between the logarithms of retention factors in reversed-phase HPLC (RP-HPLC) and log P_{ow} values [10]. However, due to the excessive retention in RP-HPLC at a purely aqueous mobile phase, the direct measurement of log P_{ow} values is not achievable for many compounds.

As an alternative to RP-HPLC, electrokinetic chromatography (EKC) with micelles [15,16] and microemulsion [17,18] has been introduced as a simple and inexpensive analytical tool for log P_{ow} determination. Owing to its high efficiency and resolving power, requirement for small sample and buffer size, ease and speed of separation, MEKC has been a technique of choice for separation of a variety of charged and neutral compounds since its introduction by Terabe et al. [19] In MEKC, solutes are separated based on their differential partitioning between the aqueous phase and the micellar phase (i.e., pseudostationary phase). One of the major advantages of MEKC over other separation techniques is the feasibility of manipulating the selectivity by simply rinsing the capillary with the solution of a new micellar phase with diverse physicochemical properties [16].

Correlation between micelle-water partition coefficient, log P_{mw}, and log P_{ow} has been known since 1975 [20]. Good correlations between log P_{mw}, and log P_{ow} for phthalate esters have been shown in

***Corresponding author:** Cevdet Akbay, Department of Chemistry and Physics, Fayetteville State University, Fayetteville, NC 28301, USA
E-mail: cakbay@uncfsu.edu

the literature using MEKC [21]. Because the retention factor (log k) in MEKC is directly related to the log P_{mw}, a linear relationship between log k and log P_{ow} for aromatic solutes in anionic surfactant systems has also been confirmed [22]. In addition to the anionic surfactants, several other surfactant systems such as cationic, anionic-nonionic mixed micelles, [23,24] bile salts, [25] and micro emulsions [26,27] have also been utilized for log P_{ow} determination.

Due to their unique properties, gemini surfactants have been introduced as alternative pseudo stationary phases in MEKC [28-30]. Gemini surfactants are made up of two hydrophobic carbon chains and two polar head groups covalently linked to each other through a spacer. As compared to their single-chain analogues with the same chain length and head group, geminis generally exhibit superior properties [31,32]. They possess remarkably lower critical micelle concentration (CMC), low Krafft point and C_{20} values (surfactant concentration that reduces the surface tension of the solvent by 20 mNm^{-1}). They have better wetting, solubilizing and foaming properties, closer packing of the hydrophobic groups, and stronger interaction with the oppositely charged surfactants. Head group could be anionic, cationic, zwitterionic and nonionic; spacer can be polar or nonpolar, flexible or rigid, short or long. The nature and length of spacer can have a significant effect on the physicochemical properties and morphology of the gemini aggregates [32].

In present study, two cationic gemini surfactants, 1,1'-didodecyl-1,1'-but-2-yne-1,4-diyl-bis-pyrrolidinium dibromide (G1) and N,N'-didodecyl-N,N,N',N'-tetramethyl-N,N'-but-2-ynediyl-di-ammonium dibromide (G2) were used as pseudostationary phases in MEKC for P_{ow} determination. Sodium dodecyl sulfate (SDS), a commonly used anionic conventional pseudostationary phase with identical hydrocarbon chain length, was also used for comparison. Both gemini surfactants contain 2-butyne spacer and the same hydrocarbon chain lengths (C12), however, G1 has pyrrolidinium while G2 has dimethyl ammonium head group (chemical structures of surfactant systems are provided in Figure 1). To the best of our knowledge, no other study has yet been reported in the literature using cationic gemini surfactants for P_{ow} determination.

Experimental

Chemicals

All benzene derivatives, alkyl phenyl ketone (APK) homologues, disodium hydrogenphosphate, sodium dihydrogenphosphate, and sodium hydroxide were obtained from Alfa Aesar (Ward Hill, MA, USA). Deionized water was obtained from a water purification system from Millipore (Milford, MA, USA). SDS was purchased from EMD Chemicals (Gibbstown, NJ, USA). The gemini surfactants G1 and G2 were donated by Professor Fredric M. Menger's Research Laboratory at Emory University (Atlanta, GA, USA). All chemicals were used as received without any further purification.

Characterization of surfactants

Surface tension measurement was used for CMC determination of the gemini surfactants and SDS. This method is based on the change in surface tension as a factor of surfactant concentration. The surface tension of surfactant solutions with given concentrations were measured at ambient temperature by a KSV Sigma 703D digital tensiometer (Monroe, CT, USA) using a DuNoüy ring. Surface tension values were plotted against surfactant concentration and the CMC value was taken as the breakpoint of the curve. The details of the experiment are explained elsewhere, [29] thus, are not reported here. Partial specific volume, PSV, is defined as the increase in volume upon dissolving 1.0 g of a dry material in a large volume of a solvent at constant temperature and pressure. Since the measurement of such small volume change is nearly impossible, an approach based on density measurement of surfactant solutions was used for determination of PSV. Five solutions with varied surfactant content were prepared in deionized water and their densities were measured at 25°C using a high-precision digital DMA 4500 density meter (Anton Paar, Ashland, VA, USA). The PSV values were obtained from the y-intercept of a graph of reciprocal of density against weight fraction of solvent. Experimental details and calculations on PSV are discussed elsewhere [29] and thus are not repeated here.

Capillary electrophoretic separations

Instrumentation

An Agilent CE system (Agilent Technologies, Palo Alto, CA, USA) equipped with a diode array detector was used for MEKC separations. The system control and data handling were done using 3D-CE ChemStation software. The MEKC separations were performed in fused-silica capillaries (Polymicro Technologies, Tucson, AZ, USA) with dimensions of 66.0 cm total length (57.5 cm effective length) × 50 µm ID (360 µm OD). Capillaries used in this study were cut from the same capillary bundle and were reactivated thoroughly after each surfactant system using deionized water (10 min) and 1.0M NaOH (ca. 20 min) to eliminate possible cross contamination. Each new capillary was activated with 1M NaOH (30 min at 40°C) and deionized water (10 min at 25°C) before use. For a typical MEKC run, the capillary was rinsed for 3 min with triply deionized water and for 3 min 0.1M NaOH followed by 3 min rinse with separation buffer between injections. Each day, the capillary was reactivated by rinsing with 1M NaOH (10 min) and triply deionized water (5 min). Unless otherwise noted, the applied voltage was -30 kV for cationic geminis and +30 kV for anionic SDS. The injection size was 50 mbar for 1 s. Peaks were identified by comparison of their individual UV-spectrum obtained from diode array detector or in case of confusion the individual solute was spiked into the mixture.

Preparation of separation buffers and solute solutions

The background electrolyte (BGE) was prepared by dissolving appropriate amounts of anhydrous NaH_2PO_4 and anhydrous Na_2HPO_4 in deionized water to obtain 100 mM solution of each. A stock solution of 10 mM phosphate buffer with pH of 7.0 was prepared from the mixture of 42.3 mL NaH_2PO_4 and 57.7 mL Na_2HPO_4. When necessary, dilute HCl or NaOH was used for adjustment of pH. Run buffers were prepared by addition of various amount of surfactant to the BGE. The final concentration of the two geminis in run buffers was 6.0 mM each and that of SDS was 40.0 mM. All run buffers were filtered through a 0.45 µm syringe filter (Nalgene, Rochester, NY, USA) followed by degassing using ultrasonication for about one min before used in MEKC experiments. All stock solutions of the test solutes were prepared in methanol with a concentration of ca. 20 mg/mL each and diluted with 50:50 methanol:deionized water before injection. The final solute concentration ranged from ca. 0.2 to 0.5 mg/mL.

Calculations

The retention factor values, k, of neutral solutes were calculated by use of the following equation [33]:

$$k = \frac{t_R - t_{eof}}{t_{eof}\left[1-\left(\frac{t_R}{t_{psp}}\right)\right]} \quad (1)$$

Where t_R, t_{eof} and t_{psp} are the migration times of solute, EOF, and the pseudostationary phase, respectively. Methanol and undecanophenone were used as t_{eof} and t_{psp} markers, respectively. Solute partition coefficient between bulk aqueous and micellar phase, P_{mw}, is directly related to k and the phase ratio, β (Equation 2) [4].

$$P_{mw} = \frac{k}{\beta} \tag{2}$$

The β is defined as the ratio of the volume of micellar phase (V_{mc}) over that of aqueous phase (V_{aq}) and is related to the total concentration, C_{surf}, the partial specific molar volume, $\overline{V}$, and the CMC of the surfactant (Equation 3) [33].

$$\beta = \frac{V_{psp}}{V_{aq}} = \frac{\overline{V}(C_{surf} - CMC)}{1 - \overline{V}(C_{surf} - CMC)} \tag{3}$$

The relationship between log k and log P_{ow} may be expressed using the following functional form [34,35]:

$$\log P_{ow} = a \log k + b \tag{4}$$

Where a and b are constants that represent the slope and intercept of a linear calibration line.

Results and discussion

Characterization of surfactant systems

The physicochemical properties of the surfactants are listed in Table 1. As compared with SDS, a conventional surfactant with the same carbon chain length, geminis have lower CMC and phase ratio but higher PSV values. The PSV values (in mL·g^{-1}) are converted to partial specific molar volumes, PMV, (in L·mol^{-1}) using molar masses of the surfactants (PMV=PSV × MM × 0.001 L/mL).

Estimation of octanol-water partition coefficients using alkyl phenyl ketones calibration curves

Like most other separation techniques, electrophoretic technique requires very small amount of compound (which does not have to be pure), is fast compared to traditional methods used for P_{ow} determination, and are relatively easy to automate. As presented in Equation (4), this method is indirect. In other words, it is based on the construction of a correlation between a retention property characteristic of the solute (e.g., log k) and the separation system for a training set of solutes with known log P_{ow} values. Further measurements of log k in the separation system can be used to estimate log P_{ow} values for other compounds of interest.

Six APKs, i.e., acetophenone, propiophenone, butyrophenone, valerophenone, hexanophenone and heptanophenone, with known log P_{ow} values were selected as training solutes to construct the calibration curve needed for estimation of log P_{ow} values of 29 sample benzene derivatives. The sample benzene derivatives used in this study are characterized as non-hydrogen bond donors (NHBs; 10 solutes), hydrogen bond acceptors (HBAs; 9 solutes), and hydrogen bond donors (HBDs; 10 solutes). The NHB solutes include alkyl- and halo-substituted benzenes and polycyclic aromatic hydrocarbons (e.g., naphthalene) and do not hold any hydrogen bonding functional groups. However, due to the aromatic ring(s), they are considered to be weak hydrogen bond acceptors. The HBAs possess only hydrogen bond accepting functional groups on the aromatic ring, whereas, the HBDs have both hydrogen bond donating and hydrogen bond accepting functional groups. Based on their pKa values, all test solutes are believed to be neutral under experimental conditions.

The six APK training solutes were analyzed and their log k values were determined under the given MEKC conditions. The log k values were then plotted against their literature log P_{ow} values for construction a linear calibration graph (Figure 2). High correlations between log k and log P_{ow} values for ketones were obtained with correlation coefficients (R^2) greater than 0.99 in all surfactant systems. Linear regression analysis yielded the following equations:

$$\text{G1 system:} \quad \log P_{ow} = 1.308 \log k + 2.587,\ R^2 = 0.993 \tag{5}$$

$$\text{G2 system:} \quad \log P_{ow} = 1.419 \log k + 2.521,\ R^2 = 0.995 \tag{6}$$

$$\text{SDS system:} \quad \log P_{ow} = 1.360 \log k + 1.953,\ R^2 = 0.992 \tag{7}$$

Based on the correlation coefficients obtained, the gemini surfactants provided relatively better linear equations than SDS. The log k values for 29 test solutes were determined under the same MEKC conditions. The estimated log P_{ow} values, or more precisely, the micelle-water partition coefficient, log P_{mw}, for benzene derivatives were then using the experimental log k values in Equations 5-7. The estimated log P_{mw} and the differences between estimated log P_{mw} and log P_{ow} (Δ) values are listed in Table 2. The best estimated log P_{mw} values were obtained for HBA solutes (including APKs) in all three surfactant systems (Figure 3), as indicated by their smaller Δ values. The absolute mean Δ values are 0.08, 0.06, and 0.18 log units for G1, G2 and SDS surfactant systems, respectively. In log P_{ow} estimation studies, it is important to remember that the structures of the training set and samples must be similar [11]. Since the training ketones show hydrogen bond accepting characteristics, superior log P_{mw} estimates for HBA samples are not surprising. Furthermore, scientifically sound log P_{mw} values were determined for NHB solutes, which are weak hydrogen bond acceptors, due to the benzene ring(s) in their structures. The

Physicochemical property	Pseudostationary phase		
	G1	G2	SDS
Chemical formula	$C_{36}H_{70}N_2Br_2$	$C_{32}H_{66}N_2Br_2$	$C_{12}H_{25}NaO_4S$
Molar mass (g mol^{-1})	690.76	638.69	288.38
CMC[a*] in pure water (mM)	0.82	0.71	8.0
CMC[a*] in 10 mM phosphate buffer (pH 7.0) (mM)	0.21	0.11	3.0
Partial specific volume[b*] (mL·g^{-1})	0.91	1.05	0.85
Partial specific molar volume[c] (L·mol^{-1})	0.63	0.58	0.25
Phase ratio[d]	0.0037	0.0034	0.0092

*Values from reference [29].

[a]Critical micelle concentrations were determined in deionized water or 10 mM phosphate buffer (pH 7.0) by surface tensiometer at ambient temperature.

[b]Partial specific volume was determined in deionized water by density meter at 25°C.

[c]Partial specific molar volume was calculated using PSV and molar mass of the surfactant (i.e., PMV=PSV × MM × 0.001 L/mL).

[d]Phase ratio was determined from Equation 3.

Table 1: Physicochemical properties of investigated surfactants.

average Δ values for NHB solutes are 0.19, 0.27, and 0.67 log units in G1, G2 and SDS surfactant systems, respectively. Conversely, the poorest estimates were obtained for HBD solutes, especially in gemini surfactant systems, as can be seen from relatively higher average Δ values. This signifies that when structurally unrelated compounds are correlated via Equation (4), incorrect estimates of log P_{ow} values are obtained. This is because separation mechanisms that influence log k are not usually same as those influence log P_{ow} [11]. These results demonstrate G1 and G2 gemini surfactant systems are suitable for the high throughput estimation of log P_{ow} of weakly basic (i.e., HBA) and, to some extent, nonpolar (i.e., NHB) compounds, but not acidic compounds; while SDS system is suitable for log P_{ow} estimation of both HBA and HBD solutes.

Estimated log P_{mw} values versus literature log P_{ow} values plots show apparent differences between the two values (Figure 4). The divergence is due to the fact that the nature of the interactions between solute-octanol and solute-surfactant systems is different. As seen in Figure 4, three distinct congeneric lines can be observed for sample solutes using the three pseudostationary phases. This shows that the factors that influence retention in G1, G2 and SDS surfactant systems are notably different from those that influence octanol-water partitioning. Figure 4 also suggests that nonpolar NHB solutes have stronger interaction with octanol whereas polar HBD solutes tend to interact more with G1 and G2 surfactants. In addition, octanol and geminis are found to have similar affinities for polar HBA solutes while SDS has relatively higher affinity for the same solutes. It is also important to note that the slope of the regression line for HBA solutes is close to unity in all three surfactant systems, suggesting that these surfactants and octanol-water systems possess very similar partitioning mechanisms for these solutes. Also, unlike G1 and G2 geminis, SDS and octanol-water systems have similar partitioning mechanism for HBD solutes.

To better understand the origins of the congeneric behavior for surfactant systems studied here, it is helpful to compare the linear solvation energies (LSER) results discussed in our previous report [29]. It has been shown in the literature that polarizability and volume of solute enhance the log P_{ow}, while dipolarity and hydrogen bond accepting ability of solute inhibit it. The solute hydrogen bond donating ability, however, has been found to have no significant influence on log P_{ow} values [16,36,37]. The geminis, SDS and octanol-water systems show large positive values, indicating that the cohesive energy density and dispersion interaction term has a great amount of influence in MEKC retention. The magnitude of coefficients v suggests that hydrophobic solutes prefer to interact more with octanol and SDS systems. It is suggested by Abraham et al. that the relative values of coefficients are more descriptive than their absolute values [37]. Similar ratios indicate similar interactions between solutes and pseudostationary phase (or octanol phase). Since hydrophobic interaction has the major influence on partitioning, the e, s, a, and b coefficients have been normalized against v coefficient for G1, G2 and SDS systems in this study (Table 3). For comparison, normalized values for octanol-water system and the differences between the ratios were also included. The results in Table 3 show that the b/v ratios for G1 and G2 as well as e/v for G2 are practically similar to that for octanol-water system and the remaining ratios are somehow different. The major difference is in a/v ratios of G1 (0.33) and G2 (0.35) as well as in b/v ratio of SDS (0.38) seem to be the sources of the observed congeneric behavior and different lines in Figure 4.

Estimation of octanol-water partition coefficients using phase ratios

To minimize the errors due to the application of the training ketones, a new approach was applied to the estimation of log P_{ow} values. As mentioned earlier, P_{mw} is directly related to k and the phase ratio (Equation 2). Retention factor, k, can be easily obtained in any chromatographic technique; however, the phase ratio cannot be measured accurately in conventional (e.g., liquid) chromatography due to the fact that the phase ratio varies from column to column and even with time for a given column [4]. Thus, the use of Equation 2 for estimation of log P_{ac} in conventional chromatography is not feasible. However, since the pseudostationary phase remains constant under given experimental conditions and the physicochemical properties of the micellar phase do not depend on the capillary system and separation column, unlike conventional chromatographic methods, MEKC can be used for log P_{ac} estimations using the phase ratio. Since k is directly related to solute partition between the bulk aqueous buffer solution and the micellar phase, Equation 2 is applicable in MEKC. As stated in Equation 3, the phase ratio is related to total surfactant concentration, critical micelle concentration, and partial specific molar volume of the surfactant. Since it is a characteristic of the micellar phase, the phase ratio remains constant at a given MEKC conditions. Unlike HPLC, it does not vary from capillary to capillary or with time. Before using Equation 2 for log P_{ow} estimations, the phase ratio was determined first by using the total surfactant concentrations (6.0×10^{-3} mol·L^{-1} geminis and 4.0×10^{-2} mol·L^{-1} SDS), CMC values of surfactants under experimental conditions (2.1×10^{-4} mol·L^{-1} G1, 1.1×10^{-4} mol·L^{-1} G2 and 3.0×10^{-3} mol·L^{-1} SDS) and partial specific molar volume values listed in Table 1.

The estimated log P_{mw} obtained from the phase ratio equation and the differences between estimated log P_{mw} and log P_{ow} (Δ) values are listed in Table 4. As seen in Figure 5, the estimated log P_{mw} values are very comparable with those obtained from alkyl phenyl ketones calibration curves (Figure 3). The absolute mean Δ values for HBA solutes (0.10 log units for both G1 and G2 systems and 0.17 log units for SDS) show the feasibility of the phase ratio approach for log P_{mc} estimations. As compared with calibration curve approach, slightly poorer estimates were obtained for NHB (Δ values are 0.35, 0.38, and 0.70 log units for G1, G2 and SDS, respectively) and APK solutes. It is worth mentioning that significant improvements were observed for HBD solutes. Similar to the calibration curves, estimated log P_{mw} values obtained from phase ratio calculations are plotted against the

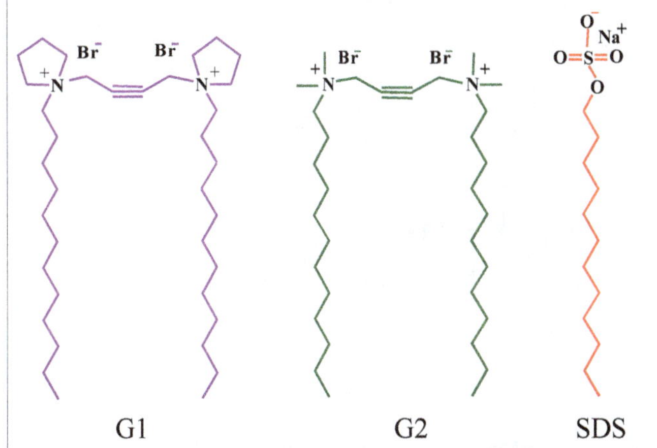

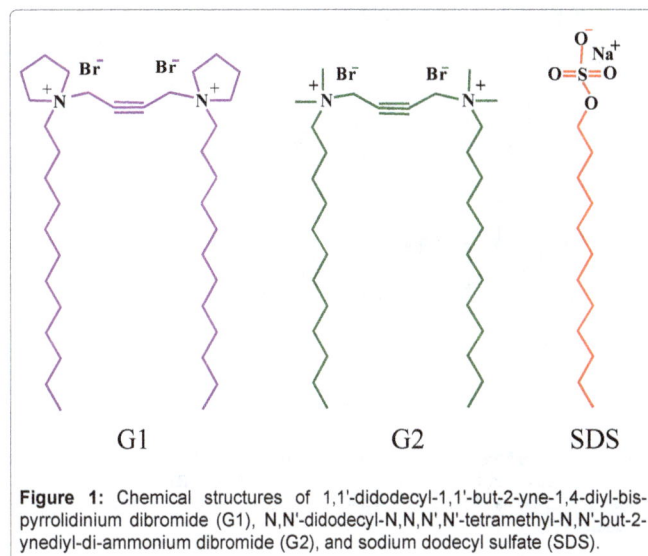

Figure 1: Chemical structures of 1,1'-didodecyl-1,1'-but-2-yne-1,4-diyl-bis-pyrrolidinium dibromide (G1), N,N'-didodecyl-N,N,N',N'-tetramethyl-N,N'-but-2-ynediyl-di-ammonium dibromide (G2), and sodium dodecyl sulfate (SDS).

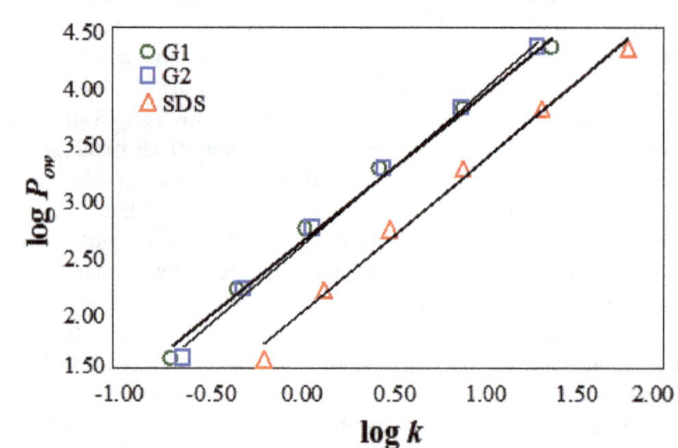

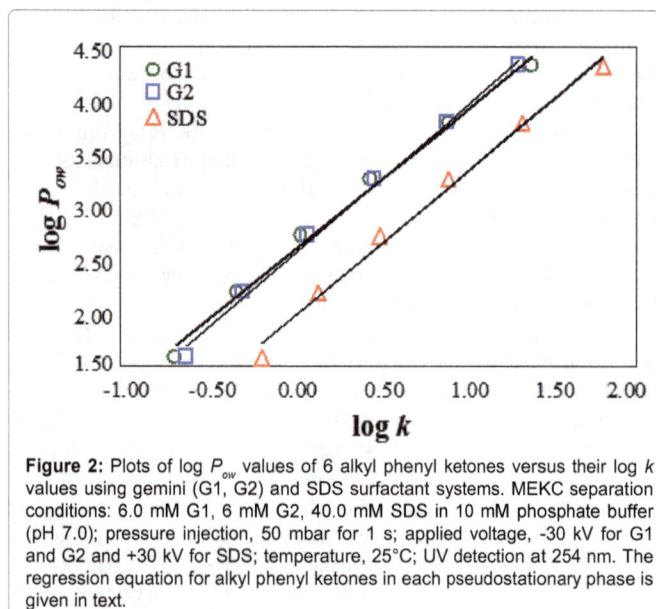

Figure 2: Plots of log P_{ow} values of 6 alkyl phenyl ketones versus their log k values using gemini (G1, G2) and SDS surfactant systems. MEKC separation conditions: 6.0 mM G1, 6 mM G2, 40.0 mM SDS in 10 mM phosphate buffer (pH 7.0); pressure injection, 50 mbar for 1 s; applied voltage, -30 kV for G1 and G2 and +30 kV for SDS; temperature, 25°C; UV detection at 254 nm. The regression equation for alkyl phenyl ketones in each pseudostationary phase is given in text.

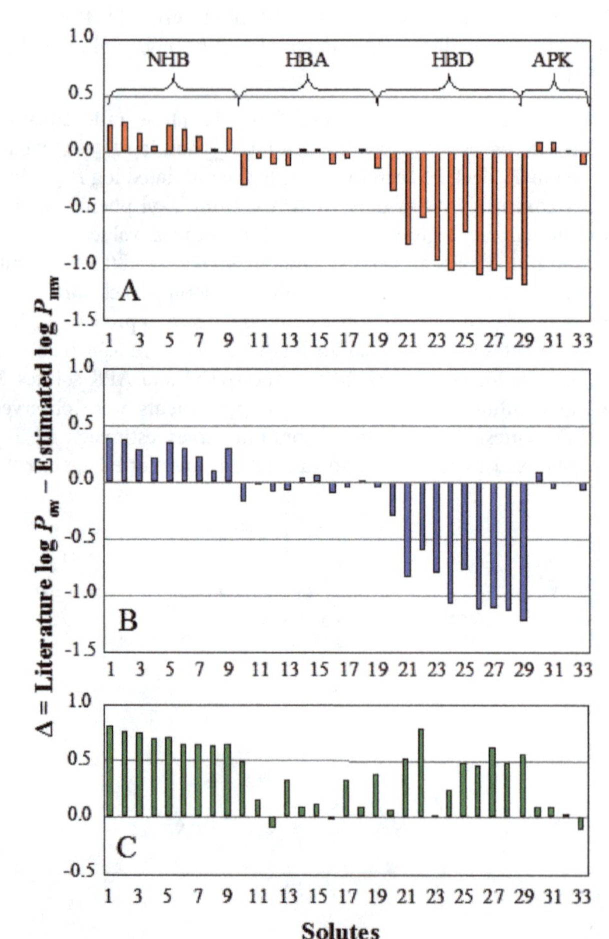

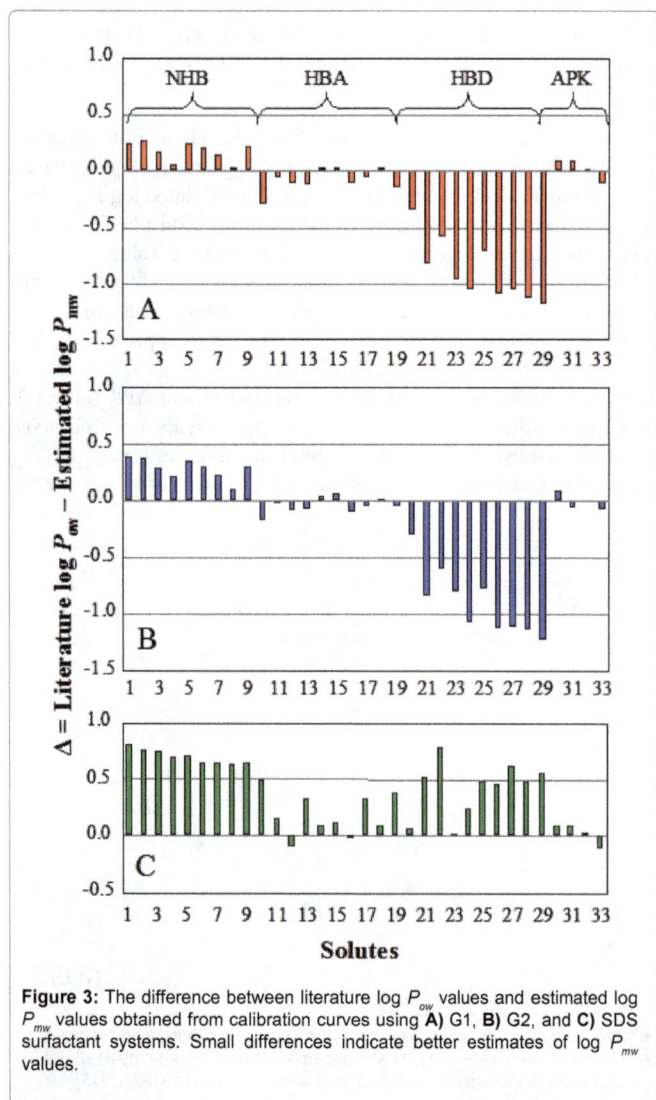

Figure 3: The difference between literature log P_{ow} values and estimated log P_{mw} values obtained from calibration curves using **A)** G1, **B)** G2, and **C)** SDS surfactant systems. Small differences indicate better estimates of log P_{mw} values.

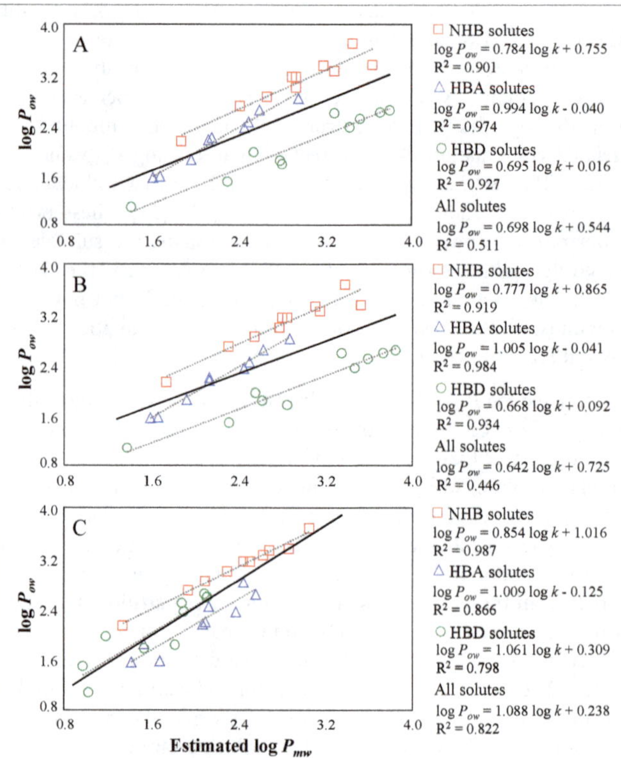

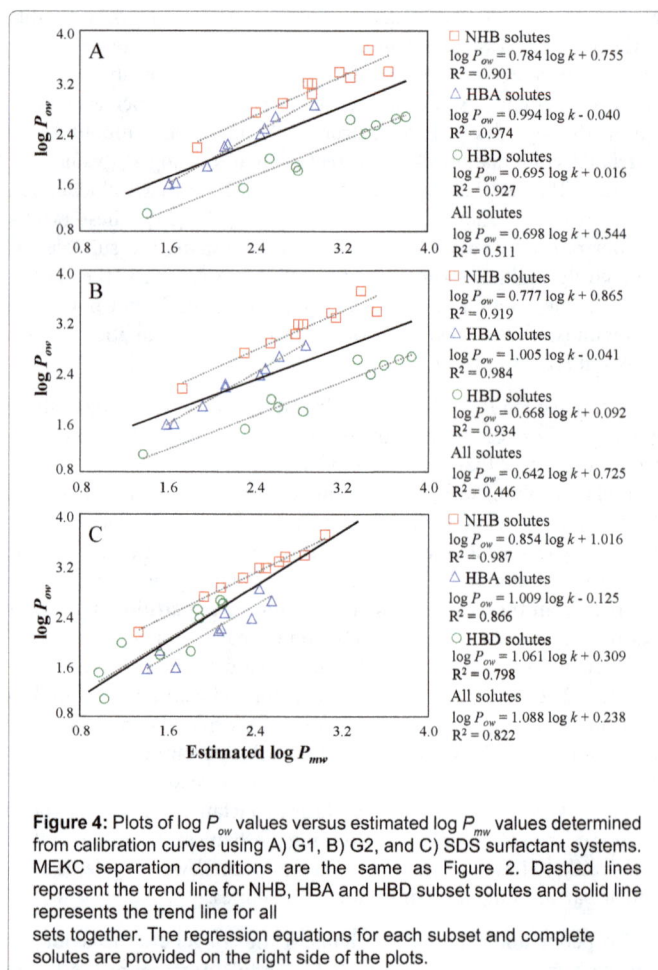

Figure 4: Plots of log P_{ow} values versus estimated log P_{mw} values determined from calibration curves using A) G1, B) G2, and C) SDS surfactant systems. MEKC separation conditions are the same as Figure 2. Dashed lines represent the trend line for NHB, HBA and HBD subset solutes and solid line represents the trend line for all sets together. The regression equations for each subset and complete solutes are provided on the right side of the plots.

literature log P_{ow} values (Figure 6). Based on the correlation coefficient values, the estimated log P_{mw} results were found to be slightly better than the previous values. These preliminary estimates can be further improved by careful determination of phase ratios and partial specific molar volumes.

Conclusion

Two cationic gemini surfactants, 1,1'-didodecyl-1,1'-but-2-yne-1,4-diyl-bis-pyrrolidinium dibromide (G1) and N,N'-didodecyl-N,N,N',N'-tetramethyl-N,N'-but-2-ynediyl-di-ammonium dibromide (G2) were used as pseudostationary phases in MEKC for P_{mw} determinations. Sodium dodecyl sulfate (SDS) was also used for comparison. Both gemini surfactants contain 2-butyne spacer with twelve hydrocarbon chain length. However, G1 has pyrrolidinium but G2 has dimethyl ammonium head groups. MEKC was successfully applied as a simple and rapid approach for determining log P_{ow} using two novel gemini surfactants and SDS. Two approaches were applied for determination of log P_{ow} values: calibration curve and phase ratio. In calibration curve approach, the log k values of six alkyl phenyl ketones were plotted against their literature log P_{ow} values for constructing linear calibration curve. Log P_{ow} values of 29 sample benzene derivatives were then determined from the slope and the y-intercept of the calibration line. In the phase ratio approach, total surfactant concentration, critical micelle concentration, partial specific molar volume and experimental log k values were utilized for estimation of the log P_{ow} values. Both approaches were found to provide very comparable results. In general,

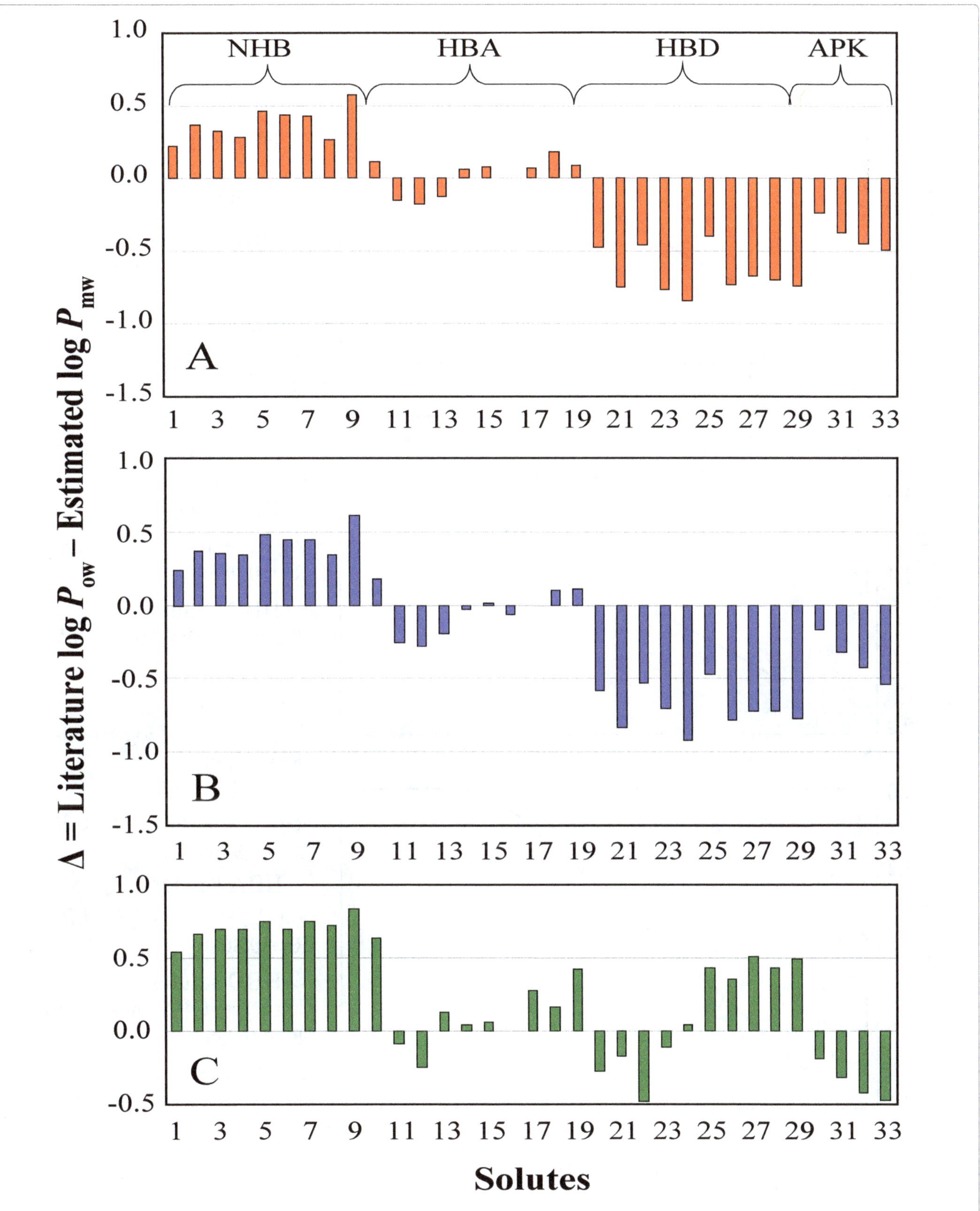

Figure 5: The difference between literature log P_{ow} values and estimated log P_{mw} values obtained from phase ratio approach using **A)** G1, **B)** G2, and **C)** SDS surfactant systems. Small differences indicate better estimated log P_{mw} values.

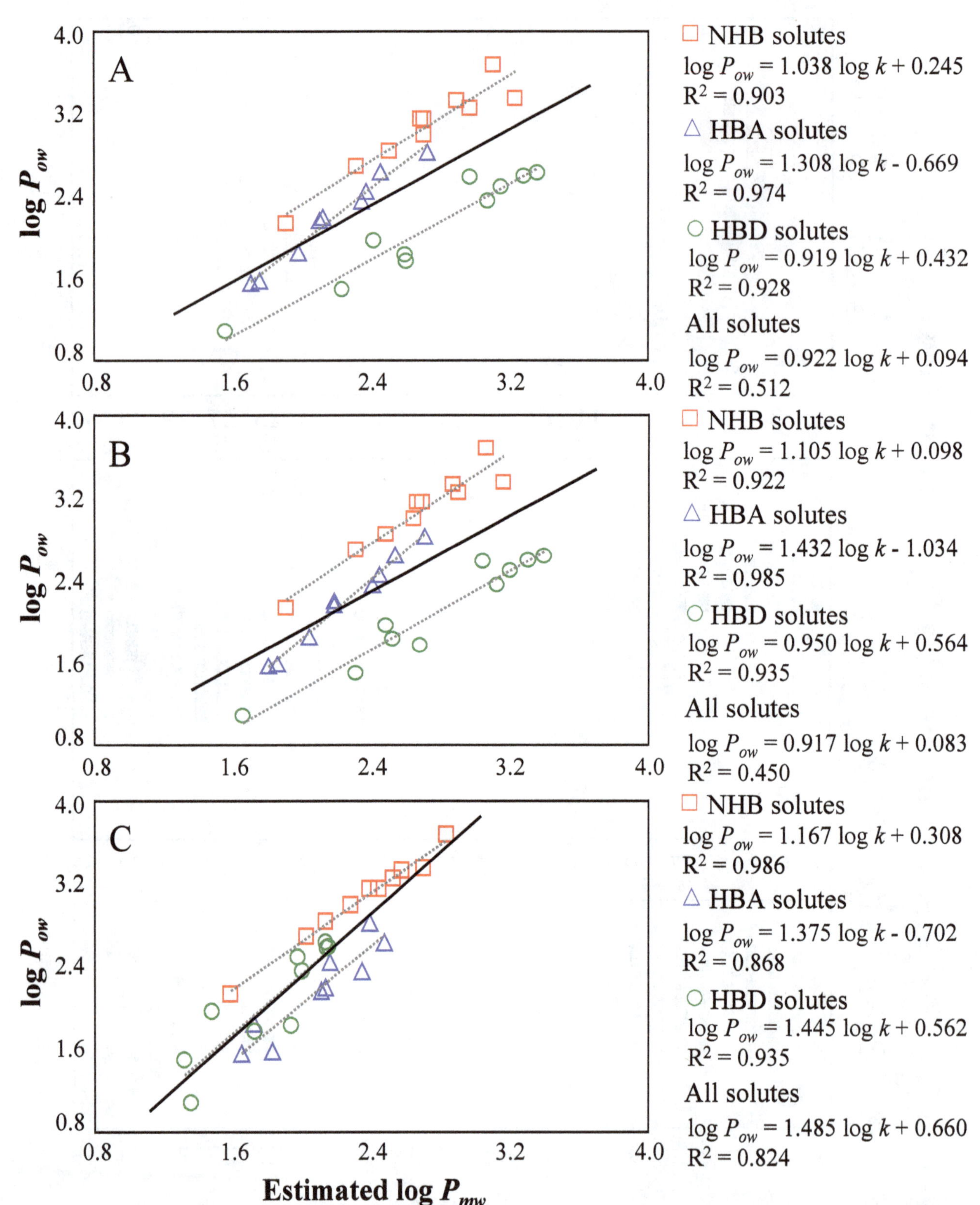

Figure 6: Plots of log P_{ow} values versus estimated log P_{mw} values determined from phase ratio approach using **A)** G1, **B)** G2, and **C)** SDS surfactant systems. MEKC separation conditions are the same as Figure 2. Dashed lines represent the trendline for NHB, HBA and HBD subset solutes and solid line represents the trendline for all sets together. The regression equations for each subset and complete solutes are provided on the right side of the plots.

			G1		G2		SDS	
No	Analytes	log P_{ow} [a]	log P_{mw}	Δ[b]	log P_{mw}	Δ	log P_{mw}	Δ
	NHB solutes							
1	Benzene	2.13	1.88	0.25	1.74	0.39	1.34	0.79
2	Toluene	2.69	2.42	0.27	2.31	0.38	1.94	0.75
3	Chlorobenzene	2.84	2.67	0.17	2.55	0.29	2.10	0.74
4	Bromobenzene	2.99	2.94	0.05	2.78	0.21	2.30	0.69
5	Ethylbenzene	3.15	2.90	0.25	2.81	0.34	2.45	0.70
6	p-Xylene	3.15	2.94	0.21	2.85	0.30	2.51	0.64
7	4-Chlorotoluene	3.33	3.19	0.14	3.11	0.22	2.69	0.64
8	Iodobenzene	3.25	3.29	0.04	3.15	0.10	2.63	0.62
9	Propylbenzene	3.68	3.46	0.22	3.38	0.30	3.05	0.63
10	Naphthalene	3.35	3.64	-0.29	3.53	-0.18	2.87	0.48
	Absolute mean Δ[c]			0.19		0.27		0.67
	HBA solutes							
11	Benzonitrile	1.56	1.62	-0.06	1.59	-0.03	1.42	0.14
12	Acetophenone	1.58	1.68	-0.10	1.66	-0.08	1.68	-0.10
13	Nitrobenzene	1.85	1.97	-0.12	1.92	-0.07	1.53	0.32
14	Methyl benzoate	2.16	2.13	0.03	2.13	0.03	2.07	0.09
15	Propiophenone	2.20	2.16	0.04	2.13	0.07	2.10	0.10
16	4-Chloroacetophenone	2.35	2.46	-0.11	2.45	-0.10	2.38	-0.03
17	4-Nitrotoluene	2.45	2.50	-0.05	2.50	-0.05	2.13	0.32
18	Ethyl benzoate	2.64	2.60	0.04	2.63	0.01	2.56	0.08
19	4-Chloroanisole	2.82	2.96	-0.14	2.87	-0.05	2.45	0.37
	Absolute mean Δ[c]			0.08		0.06		0.18
	HBD solutes							
20	Benzyl alcohol	1.08	1.42	-0.34	1.38	-0.30	1.03	0.05
21	Phenol	1.49	2.31	-0.82	2.32	-0.83	0.98	0.51
22	3-Methylphenol	1.96	2.55	-0.59	2.56	-0.60	1.19	0.77
23	4-Chloroaniline	1.83	2.79	-0.96	2.62	-0.79	1.82	0.01
24	4-Flourophenol	1.77	2.81	-1.04	2.85	-1.08	1.54	0.23
25	4-Ethylphenol	2.58	3.29	-0.71	3.35	-0.77	2.11	0.47
26	4-Chlorophenol	2.35	3.43	-1.08	3.47	-1.12	1.90	0.45
27	3-Chlorophenol	2.49	3.53	-1.04	3.59	-1.10	1.88	0.61
28	4-Bromophenol	2.59	3.71	-1.12	3.73	-1.14	2.12	0.47
29	3-Bromophenol	2.63	3.80	-1.17	3.85	-1.22	2.09	0.54
	Absolute mean Δ[c]			0.89		0.89		0.41
	Alkyl phenyl ketones							
30	Butyrophenone	2.73	2.64	0.09	2.65	0.08	2.64	0.09
31	Valerophenone	3.26	3.17	0.09	3.20	-0.06	3.18	0.08
32	Hexanophenone	3.79	3.77	0.02	3.79	0.00	3.77	0.02
33	Heptanophenone	4.32	4.42	-0.10	4.39	-0.07	4.42	-0.10
	Absolute mean Δ[c]			0.07		0.06		0.08
	Overall absolute mean Δ[d]			0.31		0.32		0.33

[a]Literature values. From reference [16].
[b]Δ=log P_{ow}(Lit)-log P_{mw}(MEKC)
[c]Absolute mean of differences for each subset solutes.
[d]Absolute mean of differences for all solutes.

Table 2: Estimated log P_{mw} values obtained from calibration curves using six alkyl phenyl ketones as training solutes.

Surfactant system	System constant ratios				
	v	*e/v*	*s/v*	*a/v*	*b/v*
1,1'-didodecyl-1,1'-but-2-yne-1,4-diyl-bis-pyrrolidinium dibromide (G1)	3.10	0.19	-0.07	0.35	-0.97
N,N'-didodecyl-N,N,N',N'-tetramethyl-N,N'-but-2-ynediyl-di-ammonium dibromide (G2)	3.05	0.13	-0.03	0.37	-0.92
Sodium dodecyl sulfate (SDS)	3.16	0.04	0.00	-0.06	-0.56
Octanol–water partition coefficient (log P_{ow})	4.07	0.11	-0.22	0.02	-0.94
Differences between ratios					
G1 and log P_{ow}	0.97	0.08	0.15	0.33	0.03
G2 and log P_{ow}	1.02	0.02	0.19	0.35	0.02
SDS and log P_{ow}	0.91	0.15	0.22	0.08	0.38

*LSER coefficients obtained from geminis and SDS are taken from reference [29]. Coefficients for octanol-water system were calculated using MS Excel (n=29).

Table 3: LSER coefficient ratios for surfactant systems and octanol-water system*.

			G1		G2		SDS	
No	Analytes	log P_{ow} [a]	log P_{mw}	Δ[b]	log P_{mw}	Δ	log P_{mw}	Δ
NHB solutes								
1	Benzene	2.13	1.91	0.22	1.91	0.22	1.59	0.54
2	Toluene	2.69	2.32	0.37	2.32	0.37	2.03	0.66
3	Chlorobenzene	2.84	2.51	0.33	2.49	0.35	2.14	0.70
4	Bromobenzene	2.99	2.71	0.28	2.65	0.34	2.29	0.70
5	Ethylbenzene	3.15	2.69	0.46	2.67	0.48	2.40	0.75
6	p-Xylene	3.15	2.71	0.44	2.70	0.45	2.45	0.70
7	4-Chlorotoluene	3.33	2.90	0.43	2.88	0.45	2.58	0.75
8	Iodobenzene	3.25	2.98	0.27	2.91	0.34	2.53	0.72
9	Propylbenzene	3.68	3.11	0.57	3.07	0.61	2.84	0.84
10	Naphthalene	3.35	3.24	0.11	3.17	0.18	2.71	0.64
Absolute mean Δ[c]				0.35		0.38		0.70
HBA solutes								
11	Benzonitrile	1.56	1.71	-0.15	1.81	-0.25	1.65	-0.09
12	Acetophenone	1.58	1.76	-0.18	1.86	-0.28	1.83	-0.25
13	Nitrobenzene	1.85	1.98	-0.13	2.04	-0.19	1.72	0.13
14	Methyl benzoate	2.16	2.10	0.06	2.19	-0.03	2.12	0.04
15	Propiophenone	2.20	2.12	0.08	2.19	0.01	2.14	0.06
16	4-Chloroacetophenone	2.35	2.35	0.00	2.41	-0.06	2.35	0.00
17	4-Nitrotoluene	2.45	2.38	0.07	2.45	0.00	2.17	0.28
18	Ethyl benzoate	2.64	2.46	0.18	2.54	0.10	2.48	0.16
19	4-Chloroanisole	2.82	2.73	0.09	2.71	0.11	2.40	0.42
Absolute mean Δ[c]				0.10		0.10		0.17
HBD solutes								
20	Benzyl alcohol	1.08	1.56	-0.48	1.66	-0.58	1.36	-0.28
21	Phenol	1.49	2.24	-0.75	2.32	-0.83	1.32	-0.17
22	3-Methylphenol	1.96	2.42	-0.46	2.49	-0.53	1.48	-0.48
23	4-Chloroaniline	1.83	2.60	-0.77	2.53	-0.70	1.94	-0.11
24	4-Flourophenol	1.77	2.61	-0.84	2.69	-0.92	1.73	0.04
25	4-Ethylphenol	2.58	2.98	-0.40	3.05	-0.47	2.15	0.43
26	4-Chlorophenol	2.35	3.08	-0.73	3.13	-0.78	2.00	0.35
27	3-Chlorophenol	2.49	3.16	-0.67	3.21	-0.72	1.98	0.51
28	4-Bromophenol	2.59	3.29	-0.70	3.31	-0.72	2.16	0.43
29	3-Bromophenol	2.63	3.37	-0.74	3.40	-0.77	2.14	0.49
Absolute mean Δ[c]				0.65		0.70		0.33
Alkyl phenyl ketones								
30	Butyrophenone	2.73	2.49	-0.24	2.56	-0.17	2.54	-0.19
31	Valerophenone	3.26	2.89	-0.37	2.94	-0.32	2.94	-0.32
32	Hexanophenone	3.79	3.34	-0.45	3.36	-0.43	3.37	-0.42
33	Heptanophenone	4.32	3.83	-0.49	3.78	-0.54	3.85	-0.47
Absolute mean Δ[c]				0.30		0.30		0.28
Overall absolute mean Δ[d]				0.35		0.37		0.37

[a]Literature values. From reference [16].
[b]Δ=log P_{ow}(Lit)- log P_{mw}(MEKC)
[c]Absolute mean of differences for each subset solutes.
[d]Absolute mean of differences for all solutes.

Table 4: Estimated log P_{mw} values obtained from phase ratios.

gemini surfactants provided better estimated log P_{ow} values for NHB and HBA solutes while SDS gave better values for HBD solutes.

Acknowledgments

This work was supported by a Support of Competitive Research (SCORE) Program grant from the National Institute of General Medical Sciences, one of the National Institutes of Health (Grant No. S06GM078246). The authors acknowledge the financial support of the Fayetteville State University Research Initiative for Scientific Enhancement (FSU-RISE) Program. The authors are thankful to Dr. Menger's Research Team at Emory University (Atlanta, Georgia) for donating the cationic gemini surfactants.

References

1. Cordes EH, Gitler C (1973) Prog Bioorg Chem 2.
2. Di Marzio W, Saenz ME (2004) Quantitative structure-activity relationship for aromatic hydrocarbons on freshwater fish. Ecotoxicol Environ Saf 59: 256-262.
3. Kah M, Brown CD (2007) Prediction of the adsorption of ionizable pesticides in soils. J Agric Food Chem 55: 2312-2322.
4. Kelly KA, Burns ST, Khaledi MG (2001) Prediction of retention in micellar electrokinetic chromatography from solute structure. 1. Sodium dodecyl sulfate micelles. Anal Chem 73: 6057-6062.
5. Leo A, Hansch C, Elkins D (1971) Partition coefficients and their uses. Chem Rev 71: 525-616.
6. Hansch C, Fujita T (1964) p-σ-π Analysis. A Method for the Correlation of Biological Activity and Chemical Structure. J Am Chem Soc 86: 1616-1626.
7. Hansch C, Maloney PP, Fujita T, Muir RM (1962) Correlation of Biological

Activity of Phenoxyacetic Acids with Hammett Substituent Constants and Partition Coefficients. Nature 194: 178-180.

8. Kah M, Brown CD (2008) Log D: lipophilicity for ionisable compounds. Chemosphere 72: 1401-1408.

9. Organization for Economic Cooperation and Development (OECD) Guidelines for the Testing of Chemicals, 1995. Test No. 107: n-Octanol/Water Partition Coefficient: Shake Flask Method.

10. Dorsey JG, Khaledi MG (1993) Hydrophobicity estimations by reversed-phase liquid chromatography. Implications for biological partitioning processes. J Chromatogr 656: 485-499.

11. Poole SK, Poole CF (2003) Separation methods for estimating octanol-water partition coefficients. J Chromatogr B Analyt Technol Biomed Life Sci 797: 3-19.

12. Organization for Economic Cooperation and Development (OECD) Guidelines for the Testing of Chemicals, 2004. Test No. 117: n-Octanol/Water Partition Coefficient: HPLC Method.

13. Kong XQ, Shea D, Baynes RE, Riviere JE, Xia XR (2007) Regression method of the hydrophobicity ruler approach for determining octanol/water partition coefficients of very hydrophobic compounds. Chemosphere 66: 1086-1093.

14. Fatemi MH, Karimian F (2007) Prediction of micelle-water partition coefficient from the theoretical derived molecular descriptors. J Colloid Interface Sci 314: 665-672.

15. Herbert BJ, Dorsey JG (1995) n-Octanol-water partition coefficient estimation by micellar electrokinetic capillary chromatography. Anal Chem 67: 744-749.

16. Trone MD, Leonard MS, Khaledi MG (2000) Congeneric behavior in estimations of octanol-water partition coefficients by micellar electrokinetic chromatography. Anal Chem 72: 1228-1235.

17. Poole SK, Durham D, Kibbey C (2000) Rapid method for estimating the octanol--water partition coefficient (log P ow) by microemulsion electrokinetic chromatography. J Chromatogr B Biomed Sci Appl 745: 117-126.

18. Klotz WL, Schure MR, Foley JP (2001) Determination of octanol-water partition coefficients of pesticides by microemulsion electrokinetic chromatography. J Chromatogr A 930: 145-154.

19. Terabe S, Otsuka K, Ichikawa K, Tsuchiya A, Ando T (1984) Electrokinetic separations with micellar solutions and open-tubular capillaries. Anal Chem 56: 111-113.

20. Collett JH, Koo L (1975) Interaction of substituted benzoic acids with polysorbate 20 micelles. J Pharm Sci 64: 1253-1255.

21. Takeda S, Wakida S, Yamane M, Kawahara A, Higashi K (1993) Migration behavior of phthalate esters in micellar electrokinetic chromatography with or without added methanol. Anal Chem 65: 2489-2492.

22. Chen N, Zhang Y, Terabe S, Nakagawa TJ (1994) Effect of physico-chemical properties and molecular structure on the micelle—water partition coefficient in micellar electrokinetic chromatography. Chromatogr 678: 327-332.

23. Ishihama Y, Oda Y, Uchikawa K, Asakawa N (1994) Correlation of octanol-water partition coefficients with capacity factors measured by micellar electrokinetic chromatography. Chem Pharm Bull (Tokyo) 42: 1525-1527.

24. Yang S, Bumgarner JG, Kruk LF, Khaledi MG (1996) Quantitative structure-activity relationships studies with micellar electrokinetic chromatography. Influence of surfactant type and mixed micelles on estimation of hydrophobicity and bioavailability. J Chromatogr A 721: 323-335.

25. Ibrahim WAW, Hermawan D, Hasan MN, Aboul-Enein HY, Sanagi MM (2008) Rapid Estimation of Octanol–Water Partition Coefficient for Triazole Fungicides by MEKC with Sodium Deoxycholate as Surfactant. Chromatographia 68: 415-419.

26. Chen K, Locke DC, Maldacker T, Lin JL, Aawasiripong S, et al. (1998) Separation of Ergot Alkaloids By Micellar Electrokinetic Capillary Chromatography Using Cationic Gemini Surfactants. J Chromatogr A 822: 281-290.

27. Akbay C, Hoyos Y, Hooper E, Arslan H, Rizvi SA (2010) Cationic gemini surfactants as pseudostationary phases in micellar electrokinetic chromatography. Part I: effect of head group. J Chromatogr A 1217: 5279-5287.

28. Ahmed HH, Ahlstrom DM, Arslan H, Guzel M, Akbay C (2012) Study of chemical selectivity of molecular binary mixed micelles of sodium 10-undecenyl sulfate and sodium N-undecenyl leucinate using linear solvation energy relationships model. J Chromatogr A 1236: 207-214.

29. Biesen VG, Bottaro CS (2008) Linear solvation energy relationships of anionic dimeric surfactants in micellar electrokinetic chromatography III. Effect of fluorination. J Chromatogr A 1202: 90-95.

30. Rosen MJ, Tracy DJ (1998) Gemini surfactants. J Surfact Deterg 1: 547-554.

31. Menger FE, Keiper JS (2000) Gemini Surfactants. Angew Chem Int Ed 39: 19061920.

32. Terabe S, Otsuka K, Ando T (1985) Electrokinetic chromatography with micellar solution and open-tubular capillary. Anal Chem 57: 834-841.

33. Collander R (1950) The Distribution of Organic Compounds Between iso-Butanol and Water Acta Chem Scand 4: 1085-1098.

34. Collander R (1951) The Partition of Organic Compounds Between Higher Alcohols and Water. Acta Chem Scand 5: 774-780.

35. Abraham MH, Chadha HS, Whiting GS, Mitchell RC (1994) Hydrogen bonding. 32. An analysis of water-octanol and water-alkane partitioning and the delta log P parameter of seiler. J Pharm Sci 83: 1085-1100.

36. Abraham MH, Chadha HS, Dixon JP, Leo AJ (1994) Hydrogen bonding.39. The partition of solutes between water and various alcohols. J Phys Org Chem 7: 712-716.

37. Abraham MH, Treiner C, Roses M, Rafols C, Ishihama Y (1996) Linear free energy relationship analysis of microemulsion electrokinetic chromatographic determination of lipophilicity. J Chromatogr A 752: 243-249.

Comprehensive Isotopic and Elemental Analysis of a Multi-Oxide Glass By Multicollector ICP-MS in Isotope Substitution Studies

Mitroshkov AV*, Ryan JV, Thomas ML and Neeway JJ

Environmental Systems Group, Pacific Northwest National Laboratory, USA

Abstract

Multicollector ICP-MS was used to comprehensively analyze different types of isotopically modified glass created to investigate the processes of glass corrosion in the water. Analytical methods were developed for the analyses of synthesized, isotopically modified solid glass and the release of glass constituents upon contact with deionized water. To validate the methods, results from an acid digestion sample of the Analytical Reference Glass showed good agreement when compared to data from multiple prior analyses on the same glass [1]. In this paper, we present the results of the comprehensive analysis of the acid digestion of six types of isotopically modified glass and the release of glass constituents into water corrosion after one year of aqueous corrosion.

Keywords: Isotopically modified glass; Multicollector; Isotopic ratios; Water corrosion

Abbreviations: ARG: Analytical Reference Glass; DSN: Desolvating Nebulizer; HR: High Resolution; ICP: Inductively Coupled Plasma; IS:Internal Standard; LA: Laser Ablation; MC: Multicollector; MS: Mass Spectroscopy; RR: Round Robin.

Introduction

Vitrification is used by many countries to immobilize radioactive wastes due to the technical reliability of the process and long-lasting stability of the material. Nevertheless, glass can corrode in aqueous environments over long time periods. To investigate the mechanisms of glass corrosion in water, experiments were conducted in which coupons of glasses with nominally equivalent chemical compositions but two different isotopic ratios were immersed in 90°C water for extended times. Periodically, small aliquots of solution were sampled and analyzed. After a certain period of time, the coupon solutions were swapped, effectively creating a system in which the chemical composition of the solution was unchanged but in which the isotopic differences enabled tracking of the provenance of ions into and out of the corroding glass systems. To be specific, the diffusion and exchange of ions between solid and liquid phases is measurable if there is a way to distinguish between ions in the solid phase, either from the pristine glass or from alteration layers formed as a result of aqueous corrosion, and those in solution.

Glass is one of the most difficult objects for elemental analysis because of its complexity and the presence of many elements in high concentrations that can produce interferences with the detection of other elements in lower concentrations. In recent years, Laser ablation inductively coupled plasma-mass spectroscopy (LA-ICP-MS) has been considered the method of choice for elemental analyses of solid glass samples [2,3]. The LA-ICP-MS method does not require time- and labor-consuming sample preparation; it can be conducted on relatively low-volume samples with satisfactory levels of accuracy and precision. Although, as we can see from Trejos et al. [2,3] the level of precision for the LA-ICP-MS method (~10%) is noticeably higher than the precision normally achieved in the liquid solution aspiration ICP-MS method (0.1-1.0%). In our isotope substitution study, we had very limited amounts of each glass type and needed to analyze the full content of the small glass samples with a simple sample preparation. Simple sample preparation techniques were developed and the solid glass and solutions in acidic water were analyzed by three different methods: 1) isotopic ratio analysis for seven elements with modified isotopic ratios: Li, B, Si, Ca, Fe, Zn, and Mo; 2) compositional analysis of all elements of interest in the glass; and 3) group analysis of five selected elements of interest, where the combination of elemental and isotopic methods was used. During the method development, Analytical Reference Glass (ARG) was used as a model sample. The composition of ARG had previously been analyzed by seven different laboratories during six rounds of round-robin tests [1]. A wide range of analytical procedures have been used by the participating laboratories, including atomic absorption spectroscopy, inductively coupled plasma-atomic emission spectroscopy, direct current plasma-emission spectroscopy, and ICP-MS techniques. The consensus average relative error for round-robin tests 1 through 6 was 5.4%, with values ranging from 9.4 to 1.1%. The results of our ARG analyses are in good agreement with previously reported results of the round-robin research.

Experimental

Instrumentation

The instrument used in our study was a multicollector (MC) ICP-MS Nu Plasma HR (Nu Instruments, Wrexham, UK) fitted with a desolvating nebulizer (DSN) (Aridus II, CETAC, NB, USA). The MC is equipped with 14 Faraday cups and 3 ion counters. The use of the Aridus II was essential because it accounts for much lower molecular ion interferences relative to the use of wet plasma.

***Corresponding author:** Alex V Mitroshkov, Research Scientist, Environmental Systems Group, Pacific Northwest National Laboratory, USA
E-mail: alex.mitroshkov@pnnl.gov

The sampler cone and the skimmer cone diameters for dry plasma were 0.9 mm and 0.6 mm, respectively. Aridus II was used in all analyses, except the analyses for boron. Typical instrument and acquisition settings have had relatively small variations from one element to another and are listed in Table 1.

Sample preparation

The amounts of glass material available for analysis were very small-a few hundreds of milligrams. Therefore, there was no possibility for replicate sample preparations with fractionations. The glass was ground before digestion. The 20-100 mg samples of glass were digested first with 1 mL of concentrated HF with 0.1 g of mannitol at 50°C for 48 hr. After digestion, the samples were evaporated to 25-75 µL of liquid and further digested with 1 mL of concentrated HNO_3 at 50°C for 9 hr. After the second digestion, 15 mL of deionized water was added to the digests and the solutions were sonicated. This was considered a stock solution, which were diluted 200-10,000 times for various analyses.

Mannitol plays a significant role by retaining the boron in solution, as was mentioned in Ref. [4]. In addition, the results of Ref. [5] indicated that the addition of mannitol and proper control in the evaporation process were effective in preventing the loss of several other elements, including Ti, Ge, Sn and Sb. In Ref. [6], the authors suggest that closed-vessel microwave digestion and H_3PO_4–HNO_3–HF reagents may limit boron volatilization during both digestion and subsequent acid-drying processes. We have also found that the presence of mannitol is essential for preventing the loss of Si. The losses of Si in the absence of mannitol were observed to be in the range of 20-50%.

ARG also was used to evaluate the fusion method as an alternative to the acid digestion method [7]. The analytical results for the fusion method in the experiments done with ARG turned out to be unsatisfactory: the sample dissolutions were significantly contaminated by the impurities in the fusion reagents and some of the elements exhibited poor recoveries.

Standards

For quantitative and isotopic analyses, separate standards with natural ratios for each element at 1000 ppm concentration were used (High Purity Standards, SC, USA). The natural isotopic ratios of the standards were confirmed when the samples with natural isotopic ratios were analyzed after calibration and determination of mass bias factors. The difference in the isotopic ratios of interest between the standards and the samples with presumably natural ratios did not exceed 0.2%, which was sufficient for the work conducted in this experiment.

HR MC-ICP-MS Settings	
Forward power	1050–1300 W
Coolant gas	13–14 L/min
Auxiliary gas	1.2–2.0 L/min
Nebulizer pressure	33–39 psi
Resolution	400
Aridus II Desolvating Nebulizer Settings	
Sweep gas flow	4–6 L/min
Spray chamber	110°C
Desolvator	160°C
Solution introduction rate	150–200 µL/min.

Table 1: Instrument and acquisition settings for the Nu Plasma HR MC-ICP-MS (the optimal setting values change from day to day).

As a control, two glasses with the same nominal composition but natural isotopic ratios were also tested under the same corrosion conditions as the isotopically modified samples. It was confirmed that, despite the growing concentration of all elements in the water solution during the experiments, the isotopic ratios of the elements of interest remained natural for two non-modified glasses. This fact confirms the validity of the method, indicating that the correct isotopic analyses are possible even in such complex mixtures without significant interferences from other elements.

Quantitation

During the method development, the Internal Standard (IS) method for quantitative analyses was attempted through the addition of an IS, containing eight elements: Bi, Ho, In, Lu, Rh, Sc, Tb, and Y (High Purity Standards, SC, USA). In order to cover the low side of the mass scale, Be was also added to this standard. The routine external standard method does not work accurately for ICP-MS analyses because of the instability of the flow rate of the solution and corresponding change of the MS response.

With the IS method, aliquots of IS solution were added to both the analyzed solution and the Calibration Standard so that the concentration of the IS was 20 ppb in both. The closest IS was used for the quantitation of the analytes. For example, Be was used for quantitation of Li, Sc was used for quantitation of Ca, etc. It was found that, although in many applications such an approach can work, in comprehensive glass analyses the implementation of this method is problematic. The reason is that IS contains some elements in relatively low concentrations, which we have to analyze in the glass, and glass samples contain some elements of IS.

Therefore, the use of an External Standard method was required, and for more accurate analyses every sample was bracketed with two Calibration Standard runs, before and after the sample. The bracketing compensates very significantly for the drift of flow rate of the DSN and, accordingly, for the drift of MS response. As a result of this approach, the accuracy of our quantitative elemental analyses was typically in the range of 2-5%.

The isotopic ratios of some of the elements in water changed by 2-3 orders of magnitude over the course of these experiments. Therefore, the requirement for isotopic analysis accuracy was not as strict and the mark was set at 1%. In most cases, our accuracy significantly exceeded the requirements. When the signals from the isotope ions were comparable in value and significantly exceeded the background level, like for Li and Mo, our accuracy was better than 0.1% and precision was in the range of 0.01%.

Zinc and iron were present in the solutions in much lower concentrations than other elements of interest. For Zn and Fe, when the intensities of measured ions were in 1×10^{-3} – 1×10^{-4} V range, the precision and accuracy for isotopic ratios were significantly lower and in the range of 2-3%.

Background

The MS reading for a certain m/z value, when only the blank solution of 2% HNO_3 is aspirated, is considered a background or zero-point baseline. It consists of elements present in the Milli-Q water and nitric acids used in the process of preparing 2% nitric acid as well as some single atom or polyatomic interferences (e.g., ArO for ^{56}Fe). Whatever it is, we use on-peak zeros when we measure the signal in the 2% nitric acid blank solution, save it in the memory, and subtract it from the signal measured during the standard or sample runs. In cases when the background was more or less substantial (in other words, was significantly higher than the off-peak background, as for example, for Li and B), the sample dilution was adjusted such that the signal from the measured element was much higher than the ever-changing background. In this way, the effect of the instability of the background on the results is minimized. For Li, for example, the samples were diluted so that the measured value for ^{7}Li was in the region of 2-5 V. The background peak for ^{7}Li was in the region 2-5 × 10^{-3} V. It is obvious that the observed changes in the background-in the range of 5 × 10^{-5} to 2 × 10^{-4} V-could not have significant effect of the final results of the analyses.

Analytical Methods

Li isotopic analyses

For Li analyses, two Faraday cup detectors on opposite sides of the detector array were used: L4 and H6 (Table 2). Other Faraday cups (H5-H1 and L1-L3) and ion counters (IC0-IC2 detectors) were not used in this work.

The tuning parameters of MS for Li isotopic analyses are rather extreme, because Li is on the lower limit of the mass range of MS. The electronic lenses of MS are involved, which are not normally involved in the tuning for other elements. Once tuning values are determined, it is important to keep once found tuning values saved in the tuning file.

The most important part of Li isotopic and quantitative analyses is the minimization of the background, which consists of Li impurities present in acid and water used in preparation of solutions, some scattered ions, and C^{2+} and N^{2+} [8]. The problem is that when minimizing the interferences the sensitivity of the system decreases. Therefore, a compromise between minimization of the background and optimization of sensitivity must be found.

The DSN, Aridus II, is also used for Li analyses. Normally, for other elements, we use a QuickWash attachment with a 5% HNO_3/0.02% HF washing solution to wash the previous analytical run materials from the system, This washing solution cannot be used for Li analyses, because when we use it the background rises significantly and does not diminish for a few minutes. Instead of an acid solution, pure Milli-Q water was used for quick removal of Li from the spray chamber of the DSN.

One of the main sources of instability in Li and other element isotopic ratio measurements is the instability of the DSN flow rate. This instability causes changes in the mass bias correction factor. Although our method of bracketing the sample with two standards runs partially compensates for this instability, stabilizing the solvent flow rate can significantly increase the accuracy of analyses. Currently, the DSN uses the self-aspiration mode for delivering solvent into the MS.

Boron isotopic analysis

Boron cannot be analyzed with the DSN because of losses in the process of desolvation, so the direct self-aspiration method was used. The same detector layout was used for B as for Li (Table 3).

The main problem with boron analyses is the very slow washout from the spray chamber of the MS. According to Ref. [6] and Ref. [9], the use of 2% ammonia as the rinse solution may effectively eliminate memory effects and accelerate the washout of boron during ICP-MS analysis. We have found that maintaining the spray chamber at room temperature, when Peltier is turned off, can reduce washout time by approximately 50%.

Silicon isotopic analyses

Isotopic analyses of silicon are complicated by the presence of polyatomic interferences such as $^{14}N_2^+$ and $^{12}C^{16}O^+$ for ^{28}Si, $^{14}N_2{}^{1}H^+$ and $^{12}C^{16}O^{1}H^+$ for ^{29}Si, and very strong $^{14}N^{16}O^+$ and some minor interferences like $^{60}Ni^{2+}$ for ^{30}Si [10,11]. It is possible, by using higher resolution and a 0.05 mm source slit, to partially resolve $^{30}Si^+$ from $^{14}N^{16}O^+$. The interferences for $^{28}Si^+$, $^{29}Si^+$, and $^{30}Si^+$ are minimized by adjusting the parameters of the DSN, mostly sweep gas flow rate. Interferences measured during the blank run are subtracted from the measurements obtained during the sample or standard run. As one can see, the proper conditions for the DSN allow for a nearly complete elimination of interferences for ^{29}Si and ^{30}Si. It is worthwhile to mention that we do not use nitrogen as makeup gas for DSN, which is recommended to be added at a small rate to the flow Ar to reduce the noise and increase the response of MS. The absence of nitrogen in this particular case is partially responsible for minimizing the interferences. Previously, the reduction of 90% of polyatomic interferences for Si by the Aridus II DSN compared to wet plasma was reported [10]. In our experiments, the reduction of interferences is estimated to be two orders of magnitude as compared to the wet plasma. As shown in Figure 1, the total coincidence of all peaks was not required, which is different than our typical procedure for most elements. Calcium is a good example of good coincidence of all peaks (Figure 2). In the case of Si, the peaks were adjusted in such a way that only part of the flat top of ^{30}Si coincided with ^{28}Si and ^{29}Si, leaving the part of peaks affected by the interference aside.

The bottom part of Figure 1 shows the analysis of 10 ppb with interferences. This is not the optimal tuning of MS for Si. These interferences can be reduced significantly by adjusting the parameters of MS and DSN. This condition was chosen for demonstration that even in the presence of relatively high interferences one can do isotopic analyses using the left part of the peaks free of interferences.

Ca isotopic analysis

Calcium is one of the elements most difficult to analyze using ICP-MS methods because of significant interferences for practically all Ca isotopes. $^{40}Ca^+$ has a very strong interference from $^{40}Ar^+$, $^{42}Ca^+$ is affected by $^{40}Ar^{1}H_2^+$ and $^{14}N_3^+$, ^{43}Ca is influenced by $^{27}Al^{16}O$ and $^{14}N_3{}^{1}H^+$, and ^{44}Ca is influenced by $^{12}C^{16}O_2$, $^{14}N_2{}^{16}O$ and $^{28}Si^{16}O$ [11,12]. Because of the abundance of Ar in the plasma, the influence of ^{40}Ar on ^{40}Ca is so strong that in normal conditions it is practically impossible to analyze ^{40}Ca. It is possible, though, when the RF Power is low and we work with cold plasma. In this case, the ^{40}Ca signal can be measured at the top of the ^{40}Ar peak. This method has been described in the literature [13] and the results are satisfactory for some applications. The shortcoming of this

H6	H5	H4	H3	H2	H1	Ax	L1	L2	IC0	L3	IC1	IC2	L4
^{7}Li													^{6}Li

Table 2: Detector layout for Li analysis.

H6	H5	H4	H3	H2	H1	Ax	L1	L2	IC0	L3	IC1	IC2	L4
^{11}B													^{10}B

Table 3: Detector layout for B analysis.

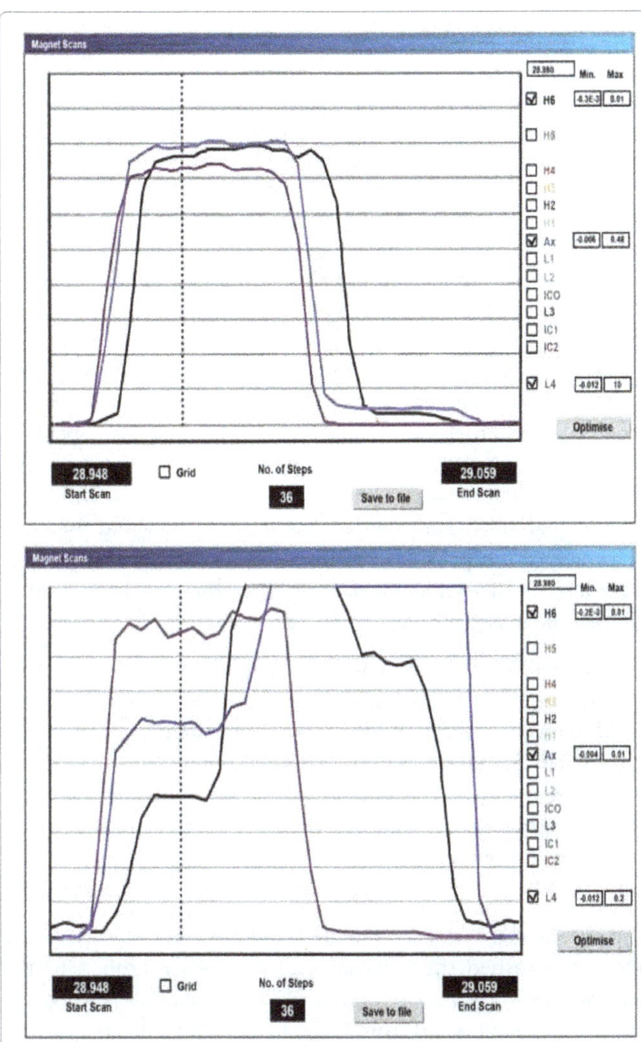

Figure 1: Analysis of Si. Top: 400 ppb of Si natural; L4- 28Si (purple), Ax- 29Si (blue), H6- 30Si (dark). Bottom: 10 ppb of Si with interferences. Intensities for each ion are shown in small boxes against every Faraday cup.

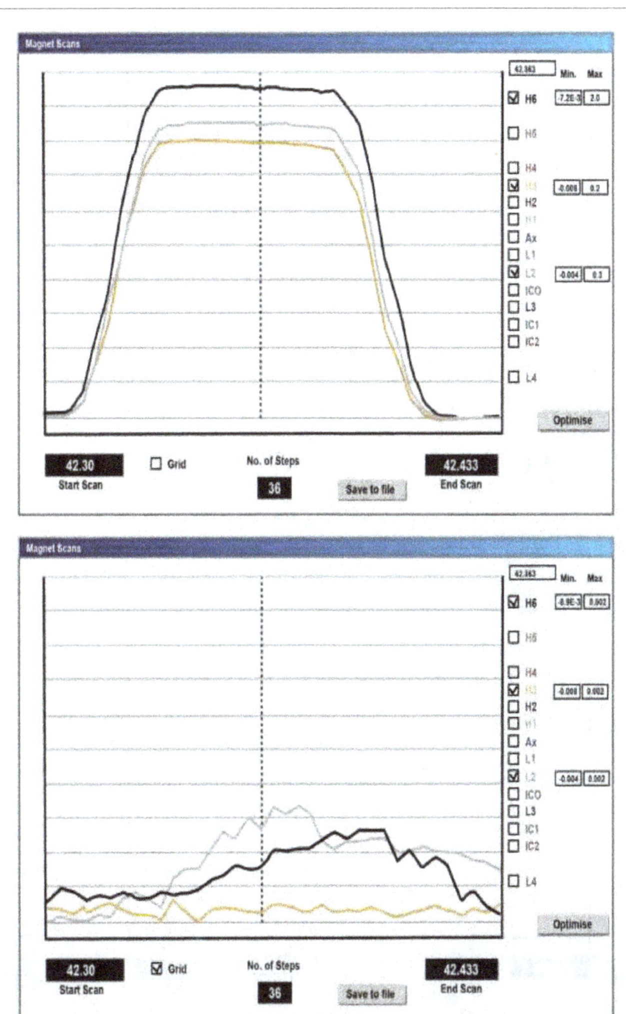

Figure 2: Analysis of Ca. Top: 400 ppb of Ca natural; L2- 42Ca (grey), H3- 43Ca (orange), H6- 44Ca (dark). Bottom: interferences. Intensities for each ion are shown in small boxes against every Faraday cup. Interferences are given on the scale of 0.002 V, which is magnified 100−1000 times compared to the standard.

approach is its relatively low sensitivity and we could not afford it in our experiments. We have used hot plasma and concentrated our efforts on measuring $^{42}Ca/^{44}Ca$ and $^{43}Ca/^{44}Ca$ ratios. In these experiments, the analyzed mixtures contained natural Ca and isotopically modified Ca, the ratios of which were known. In these circumstances, knowledge of $^{42}Ca/^{44}Ca$ and $^{43}Ca/^{44}Ca$ allows for simple calculations of $^{42}Ca/^{40}Ca$, $^{43}Ca/^{40}Ca$, and $^{44}Ca/^{40}Ca$ ratios, which were essential for the experiments. Figure 3 demonstrates the analysis of the 400 ppb Ca standard and shows interferences on different scales, because the interferences are 100-1000 times lower in intensity than in the Ca isotope readings. The interferences are minimized by optimizing the parameters of the plasma and AridusII DSN.

Iron isotopic analyses

The problems associated with isotopic analyses of iron are related to strong interferences from $Ar^{40}O^{16+}$ for $^{56}Fe^{+}$ and $Ar^{40}N^{14+}$ for $^{54}Fe^{+}$ [14,15]. The molecular mass of $Ar^{40}O^{16}$ is so close to ^{56}Fe that the maximum resolving power available is required. For these purposes, the source slit is set up in middle position (0.05 mm slit) and alpha slits are set up in the positions: α1=65 and α2=75. The ratio of interest was $^{57}Fe/^{56}Fe$, but the ratio $^{54}Fe/^{56}Fe$ was also measured. This ratio is strongly affected by the presence of ^{54}Cr in the solution [15] so the data obtained for this ratio were not used in the final analysis.

Zn and Mo isotopic analysis

Compared to Ca, Si, and Fe isotopic analyses, the Zn and Mo isotopic analyses are much more straightforward. The detector layouts for Zn and Mo are shown in Tables 4a and 4b.

The significant ratios for the experiments were $^{68}Zn/^{64}Zn$ for Zn and $^{95}Mo/^{98}Mo$ for Mo, although other ratios, such as

H6	H5	H4	H3	H2	H1	Ax	L1	L2	IC0	L3	IC1	IC2	L4
		^{68}Zn			^{67}Zn			^{66}Zn					^{64}Zn
								^{70}Zn					

Table 4a: Detector layout for Zn analyses. Mass separation: 0.333. Two cycles are implemented with ^{70}Zn measured in the second cycle.

H6	H5	H4	H3	H2	H1	Ax	L1	L2	IC0	L3	IC1	IC2	L4
^{100}Mo		^{98}Mo		^{97}Mo		^{96}Mo		^{95}Mo		^{94}Mo			^{92}Mo

Table 4b: Detector layout for Mo analyses. Mass separation: 0.5. All isotopes are analyzed in one cycle.

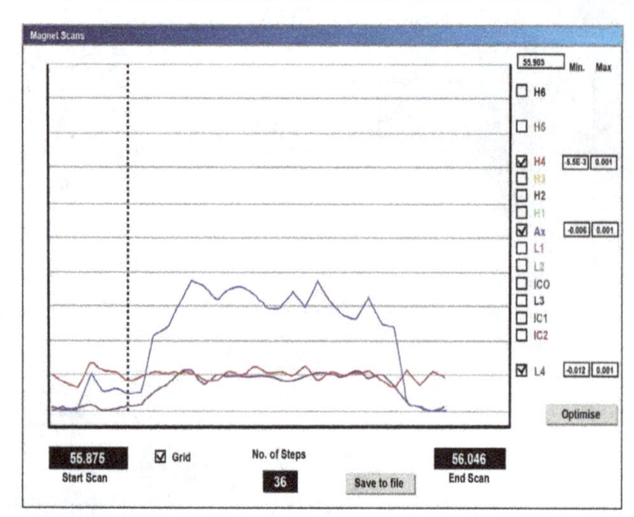

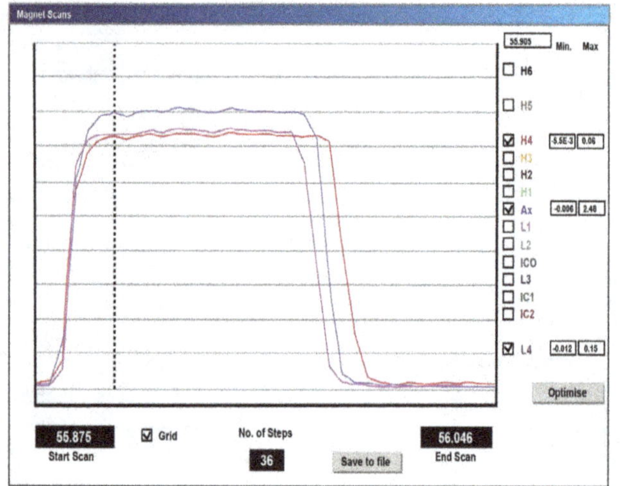

Figure 3: Analysis of Fe. Top: 400 ppb of Fe natural; L4- ^{54}Fe (purple), Ax- ^{56}Fe (blue), H3- ^{57}Fe (orange). Bottom: interferences. Intensities for each ion are shown in small boxes against every Faraday cup. Interferences are given on the scale of 0.001 V, which is magnified 100–1000 times compared to the standard.

$^{66}Zn/^{64}Zn$, $^{67}Zn/^{64}Zn$, and $^{70}Zn/^{64}Zn$ for Zn and $^{92}Mo/^{98}Mo$, $^{94}Mo/^{98}Mo$, $^{96}Mo/^{98}Mo$, $^{97}Mo/^{98}Mo$ and $^{100}Mo/^{98}Mo$ for Mo have been measured.

Group analysis

During the experiment, a group of five elements (La, Ba, Ce, Pr, Nd) had to be analyzed quantitatively with accuracy better than the accuracy of elemental analyses, but they did not require isotopic analyses. For this purpose, we developed a method we term "group analysis." Normally, MC-ICP-MS has two different modes of analyses: isotopic and elemental. As mentioned before, when using the elemental method the peaks are quickly scanned from the bottom to the top, but there is no centering on the middle of the peak. For the above-mentioned elements, we used the isotopic mode with centering on one of the peaks of one isotope of every element with five different cycles and the detector layout shown in Table 5. This allowed for quick elemental-mode-like jumps from one element to another with the advantage of using the centering made possible by using the isotopic method.

It is possible to run these kinds of analyses in two or three cycles [16] by involving detectors other than Ax Faraday cups. However, it was decided to use this slower method to allow for precise centering on every chosen isotope.

H6	H5	H4	H3	H2	H1	Ax	L1	L2	IC0	L3	IC1	IC2	L4
						^{146}Nd							
						^{141}Pr							
						^{140}Ce							
						^{139}La							
						^{137}Ba							

Table 5: The detector layout for group analysis.

Elemental analysis

In the elemental analysis mode, only the central Faraday cup (Ax) is used and the system quickly jumps from one isotope peak of a certain element to the isotope peak of another element. The scan rate is 1000 ms/amu and the dwell time is 2 s. Unlike in isotopic analyses, where the MS has the ability to find precisely the center of the peak, the accurate centering of every peak is unavailable in the elemental analysis. Therefore, it is important to calibrate the mass scale very accurately.

To do so, two different standards are used: 1) the IS, containing eight elements: Bi, Ho, In, Lu, Rh, Sc, Tb, Y (High Purity Standards, SC, USA) and 2) the ICP-MS standard, containing 68 elements (High Purity Standards, SC, USA). Beryllium is also added to the IS to cover the low side of mass scale. The IS provides a 9-point mass range calibration, followed by the ICP-MS standard, which provides a 66-point calibration. The concentration of the ICP-MS standard is 10 ppb, and it is used also as a quantification standard.

After the mass scale is calibrated, and the AridusII and MS are thoroughly washed, the blank sample (2% HNO_3) is run. After the blank sample is run, the background values, obtained during the zero run, are subtracted from the standard and from the sample values. The blank cannot be run before the standard, because the precise mass range calibration is not available before the standard is run.

Results and Discussion

The actual results of the isotope substitution studies will be reported separately in a paper dealing specifically with what those findings reveal about the mechanisms behind long-term glass corrosion. As an example, however, the results for boron are presented here. The measured value of $^{10}B/^{11}B$ as a function of time is presented in Figure 4 for the two non-exchanged solutions (SN1-7 and SE1-7) and their exchanged counterparts (SN8-14 and SE8-14). The initial $^{10}B/^{11}B$ ratios for both the enriched samples (SE1-7 and SE8-14) and the non-enriched samples (SN1-7 and SN8-14) are also presented as solid lines to lead to the eye. As expected, the $^{10}B/^{11}B$ ratio for the non-exchanged samples remains relatively constant throughout the duration of experiment. On the other hand, the $^{10}B/^{11}B$ ratio for the exchanged samples changes dramatically, by roughly two orders of magnitude, following the exchange of the two solutions containing the different boron isotopic signatures. After the solution exchange, the glass coupons continue to corrode and the $^{10}B/^{11}B$ ratio slowly evolves as a greater amount of either enriched or non-enriched boron is leached into

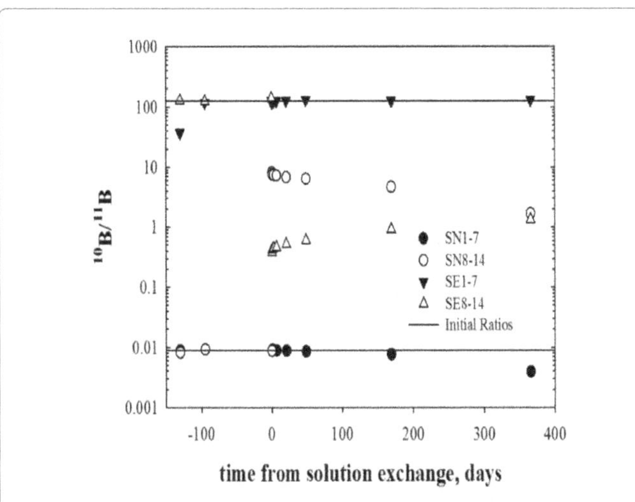

Figure 4: The 10B/11B ratio as a function of time for the non-exchanged samples (SN 1-7 and SE 1-7) and the exchanged samples (SN 8-14 and SE 8-14).

the contacting solution.

Verification of analytical meth

Table 6 shows a comparison of the results of our analysis of ARG glass and the results of the round-robin test. The average difference for the concentrations of the element oxides between the round-robin (RR) results and ours is 0.80%. The results for each element were calculated as a percentage of the mass of each element oxide relative to the total sample mass. In theory, if the results are correct, the sum of all results for all of the elements should be close to 100%, depending on the accuracy of the analytical method. As can be seen from in Table 7, the total sum for the RR tests was 99.48% while the results from this study total 100.92%. It appears that sensitivity of our method was somewhat higher than the RR test and we were able to quantify more elements.

Conclusion

Methods for comprehensive analyses of glass-sourced solutions by MC-ICP-MS have been developed and the validity of the methods was confirmed on different glass samples, including digestions, fusions, and solutions resulting from aqueous glass corrosion. Although some errors in isotopic analyses of are still possible with such complex samples due to mutual interferences of different elements, it was proven for seven selected elements that accurate isotopic analyses for the ratios of interest are

	Concentration Mass% Oxide			
		This Research		
	Round Robin	Isotopic Method	Elemental Method	Difference, %
Ag2O			nd	
Al2O3	4.66	4.20		-9.87%
As2O3			0.0006	
B2O3	8.54	8.34		-2.34%
BaO	0.09		0.078	
BeO			0.005	
Bi2O3			0.0001	
CaO	1.42	1.524		7.32%
CdO				
CeO2			0.0013	
CoO			0.0068	
Cr2O3	0.1		0.1038	3.84%
Cs2O			0.0008	
CuO	0.01		0.0059	
Er2O3				
Eu2O3				
Fe2O3	14.02	15.05		7.35%
Ga2O3			0.011	
Gd2O3				
HfO2			0.0026	
K2O	2.68		2.48	-7.46%
La2O3			0.0006	
Li2O	3.18	3.58		12.58%
MgO	0.87		0.74	
MnO2	2.32		2.24	-3.50%
MoO3			0.0007	
Na2O	11.2	12.14		8.39%
Nb2O5			0.0079	
Nd2O3			0.0004	
NiO	1.04		1.0578	1.72%
P2O5	0.27		0.4359	
PbO			0.0005	

PdO				
Pr2O3			0.00015	
Rb2O			0.0068	
Re2O7				
Rh2O3				
RuO2				
SO3				
Sb2O3			0.00027	
Sc2O3			0.00070	
SeO2				
SiO2	47.75	47.50		-0.52%
Sm2O3				
SnO2			0.0090	
SrO	0.0037		0.0030	
Ta2O5			0.0017	
TeO2			0.0067	
TiO2	1.17		1.2077	3.22%
Tl2O3			0.00007	
U3O8			0.00067	
V2O5			0.0242	
WO3			0.0006	
Y2O3			0.0007	
ZnO	0.02		0.0246	
ZrO2	0.14		0.1256	-10.31%
Sum	99.4837	92.33	8.59	0.80%
		100.92		Sum for both methods

Table 6: Comparison of the results of our analysis of ARG and the results of round-robin test.

Concentration reported as Wt% Oxide	AFC1 natural	AFC1 full	AFC1 enriched	SON68 natural	SON68 full	SON68 enriched
Ag2O	0.037	0.053	0.074	0.035	0.050	0.058
Al2O3	8.16	10.33	11.32	6.36	6.46	5.92
As2O3	0.0014	0.0014	0.0015	0.0005	0.0039	0.0214
B2O3	9.89	11.67	8.43	14.69	17.30	12.46
BaO	0.662	0.716	1.012	0.506	0.551	0.678
BeO						
Bi2O3	0.00002	0.00019				0.00015
CaO	3.200	5.463	6.645	4.581	2.907	5.774
CdO	0.0389	0.0447	0.0504	0.0353	0.0349	0.0339
CeO2	0.124	0.124	1.301	0.049	0.348	1.000
CoO	0.00054	0.00096	0.00060	0.00069	0.00073	0.00111
Cr2O3	0.0012	0.0007	0.0015	0.581	0.611	0.596
Cs2O	1.224	1.248	1.266	1.306	1.241	1.298
CuO				0.0026	0.0041	0.0050
Dy2O3	0.0007	0.0012	0.0039	0.0014	0.0037	0.0002
Er2O3						
Eu2O3	0.016	0.028	0.129			
Fe2O3	0.052	0.059	0.027	3.40	3.36	3.31
Ga2O3	0.0825	0.0759	0.0949	0.0485	0.0914	0.0961
Gd2O3	0.0285	0.0319	0.1370			
HfO2	0.0153	0.0159	0.0191	0.0436	0.0473	0.0463
K2O						
La2O3	0.0460	0.0410	0.4940	0.0190	0.2160	0.7110
Li2O	3.993	4.424	4.382	1.913	2.039	4.006
MgO	0.0208	0.0401	0.0168	0.0211	0.0381	0.0088
MnO2	0.0007	0.0009	0.0007	0.4389	0.4692	0.4206
MoO3	1.891	1.770	1.794	1.964	2.109	2.254
Na2O	7.911	7.862	8.918	12.003	11.276	12.074

Nb2O5						
Nd2O3	0.2077	0.3036	1.9011	0.1888	0.9879	1.8935
NiO				0.4893	0.4731	0.5059
P2O5	0.0703	0.1225	0.1653	0.3425	0.4508	0.3835
PbO	0.0002	0.0003	0.0004	0.0002	0.0005	0.0005
PdO						
Pr2O3	0.0545	0.0677	0.5344	0.0305	0.1863	0.4280
Rb2O	0.1840	0.1726	0.1816	0.0007	0.0010	0.0005
Re2O7						
RhO2	0.0192	0.0211	0.0194			
RuO2	0.0461	0.0444	0.0419			
SO3						
Sb2O3						
Sc2O3	0.0015	0.0014	0.0018	0.0034	0.0044	0.0045
SeO2	0.0075	0.0073	0.0302			
SiO2	57.00	55.46	51.56	43.20	41.77	40.30
Sm2O3	0.0408	0.0698	0.3309			
SnO2	0.0306	0.0289	0.0318	0.0245	0.0244	
SrO	0.2520	0.3230	0.4037	0.2249	0.3266	0.4171
Ta2O5						
TeO2	0.0847	0.0762	0.0867	0.0755	0.0759	0.0571
TiO2	0.0119	0.0119	0.0032	0.0094	0.0129	0.0047
Tb2O3	0.0005	0.0009	0.0034			0.0002
U3O8		0.0003				
V2O5	0.0015	0.0014	0.0011	0.0009	0.0012	0.0007
WO3	0.0051	0.0036	0.0488	0.0059	0.0090	0.0448
Y2O3	0.0217	0.0272	0.2060	0.0160	0.1081	0.1895
ZnO	0.0014	0.0013	0.0065	2.832	3.061	2.628
ZrO2	0.73	0.74	0.99	1.99	2.50	2.58
Sum, %	96.17	101.48	102.66	97.43	99.17	100.21

Table 7: The elemental analyses results obtained for six different isotopically modified glass samples.

achievable. All major and relatively minor components of the glass were identified and their concentrations were measured, and the total summarized content of six different isotopically modified glasses was close to 100%. The capabilities of MC-ICP-MS for comprehensive analyses of glass dissolution and glass corrosion samples with simple sample preparation have been demonstrated.

Acknowledgements

These studies were supported by the U.S. Department of Energy (DOE) through the Office of Nuclear Energy. We would like to thank Carmen Rodriguez of PNNL for help in glass preparation

References

1. Smith GL, Marschman SC (1993) Nuclear Waste Analytical Round Robins 1-6 Summary Report, MRS Proceedings, pp: 333-461.
2. Trejos T, Koons R, Becker S (2013) Cross-validation and evaluation of the performance of methods for the elemental analysis of forensic glass by μ-XRF, ICP-MS, and LA-ICP-MS. Anal Bioanal Chem 405: 5393-5409.
3. Trejos T, Montero S, Almirall JR (2003) Analysis and comparison of glass fragments by laser ablation inductively coupled plasma mass spectrometry (LA-ICP-MS) and ICP-MS. Anal Bioanal Chem 376: 1255-1264.
4. Ishikawa T, Nakamura E (1990) Suppression of boron volatilization from a hydrofluoric acid solution using a boron-mannitol complex. Anal Chem 62: 2612-2616.
5. Takeda K, Watanabe S, Naka H, Okuzaki J, Fujimoto T (1998) Determination of ultra-trace impurities in semiconductor-grade water and chemicals by inductively coupled plasma mass spectrometry following a concentration step by boiling with mannitol, Analytica Chimica Acta 377: 47-52.
6. Dai SF, Song WJ, Zhao L, Li X, Hower CR, et al. (2014) Determination of Boron in Coal Using Closed-Vessel Microwave Digestion and Inductively Coupled Plasma Mass Spectrometry (ICP-MS). Energy & Fuels 28: 4517-4522.
7. ASTM Book of Standards (1967) Part 13, C169-13, American Society for Testing of Materials, Philadelphia.
8. Choi MS, Shin HS, Kil YW (2010) Precise determination of lithium isotopes in seawater using MC-ICP-MS. Microchemical Journal 95: 274-278.
9. Al-Ammar A, Gupta RK, Barnes RM (1999) Elimination of boron memory effect in inductively coupled plasma-mass spectrometry by addition of ammonia. Spectrochim Acta Part B 54: 1077-1084.
10. van den Boorn SHJM, Vroon PZ, van Belle CC, van der Wagt B, Schwieters J, et al. (2006) Determination of silicone isotope ratios in silicate materials by high-resolution MC-ICP-MS using a sodium hydroxide digestion method. J Anal At Spectrom 21: 734-742.
11. May TW, Wiedmeyer RH (1998) A Table of Polyatomic Interferences, in ICP-MS. Atomic Spectroscopy 19: 150-155.
12. Wieser ME, Buhl D, Bouman C, Schwieters J (2004) High precision calcium isotope ratio measurements using a magnetic sector multiple collector inductively coupled plasma mass spectrometer. J Anal At Spectrom 19: 844-851.
13. Wollenweber D, Strassburg S, Wunsch G (1999) Determination of Li, Na, Mg, K, Ca and Fe with ICP-MS using cold plasma condition. Fresenius J Anal Chem 364: 433-437.
14. Weyer S, Schwieters J (2003) High presicion Fe isotope measurements with high mass resolution MC-ICPMS. International Journal of Mass Spectrometry 226: 355-369.
15. John SG, Adkins JF (2010) Analysis of dissolved iron isotopes in seawater. Marine Chemistry 119: 65-76.
16. Baker J, Waight T, Ulebeck D (2002) Rapid and highly reproducible analysis of rare earth elements by multiple collector inductively coupled plasma mass spectrometry. Geochimica et Cosmochimica Acta 66: 3635-3646.

Notes on Synthesis of Perdeutero-5-^{13}C,5,5,5-Trifluoroisoleucine VI[56]

Naugler DG* and Prosser RS

Department of Chemistry, University of Toronto, ON, Canada

Abstract

The $^{13}CF_3$ group is a promising label for heteronuclear (^{19}F,^{13}C) NMR studies of proteins. Desirable locations for this NMR spin label include the branched chain amino acid methyl groups. It is known that replacement of CH_3 by CF_3 at such locations preserves protein structure and function and enhances stability. $^{13}CF_3$ may be introduced at the δ position of isoleucine and incorporated biosynthetically in highly deuterated proteins. This paper reports our work in synthesis and purification of 5,5,5-trifluoroisoleucine, its perdeutero and 5-^{13}C versions and of 2-^{13}C-trifluoroacetate and its utility as a precursor for introduction of the $^{13}CF_3$ group into proteins.

Keywords: ^{13}C; ^{19}F; Heteronuclear; Paramagnetic; NMR; Spin label; Methyl TROSY; Unnatural amino acid; Fluorous; Lipophilic; Protein expression

Introduction

Fluorine NMR and labelling strategies in proteins

Fluorine NMR spectroscopy is a powerful method for the study of both structure and dynamics of proteins and their interactions with other proteins or ligands [1-4]. Because of the ability of the ^{19}F lone-pair electrons to participate in non-bonded interactions with the local environment, ^{19}F chemical shifts are sensitive to changes in van der Waals contacts, electrostatic fields and hydrogen bonding. As such, ^{19}F chemical shifts (or changes in shifts) are often indicative of conformational changes [5-8] binding [9] and protein folding or unfolding events [10-12]. Fluorinated probes are also frequently used to assess solvent exposure, via chemical shift changes or relaxation effects resulting from: 1) substituting H_2O for ^2H_2O [2] paramagnetic additives such as Gd^{3+}: EDTA[2] heteronuclear nuclear Over Hauser effects [13]. In membranous systems, analogous paramagnetic effects are also observed upon addition of nitroxide spin-labels [14] or dissolved oxygen [15,16] facilitating the study of topology and immersion depth via fluorinated probes. Finally, associated ^{19}F spin-spin and spin-lattice relaxation rates are useful for studying conformational dynamics over a wide range of timescales, due to the significant chemical shift dispersion, shift anisotropy, and large heteronuclear dipolar relaxation terms [17-20].

Fluorine labelling of proteins is achieved in many ways. Via biosynthetic means, monofluorinated versions of tyrosine, phenylalanine, and tryptophan may be substituted for their nonfluorinated equivalents, often with little effect on overall expression yields [2,4]. Fluorinated versions of methionine [7,21] proline [22], leucine [23] and isoleucine [24] have also been successfully incorporated into proteins. An alternative approach to ^{19}F labelling of proteins is to make use of a thiol specific fluorinated probe, which frequently consists of a terminal trifluoromethyl group [4]. In this way, a fluorine tag may be placed at virtually any site in the protein, using successive single cysteine mutations of the protein under study. Thus, the majority of ^{19}F labels used in protein NMR fall under the category of isotopic fluoroaromatic or trifluoromethyl species. For very large proteins, spectral overlap may become problematic for biosynthetically labelled proteins. However, a doubly (^{13}C, ^{19}F) labeled amino acid should provide greater resolution in two-dimensional (^{19}F,^{13}C) NMR spectra, with further possibilities of assignment without mutational analysis. Moreover, such two-dimensional NMR schemes should benefit from the relatively large one bond (^{13}C,^{19}F) coupling which is between 265 and 285 Hz for both fluoroaromatics and trifluoromethyl groups. Finally, the possibilities for studying dynamics from a ^{13}C,^{19}F pair encompasses a much greater range, since various zero, single, and double quantum coherences in addition to Zeeman and two-spin longitudinal order, may be separately evolved and studied [25,26]. In this paper, we present a method for the preparation of perdeuterated isoleucine, in which the terminal trifluoromethyl group consists of a ^{13}C-^{19}F pair. The motivation for this work is to develop a useful doubly labelled species for subsequent nD NMR studies of proteins, whose isoleucine residues have been fluorinated.

Advantages of a trifluoromethyl group

The trifluoromethyl group is expected to be a useful probe of molecular structure and dynamics, particularly in the hydrophobic core of proteins, at the interface between protein complexes, and in the membrane or detergent interior in studies of integral membrane proteins. Expressed within proteins, the CF_3 group offers additional

***Corresponding author:** Naugler David G, Department of Chemistry, University of Toronto, UTM, 3359 Mississauga Road, North Mississauga, ON, Canada, L5L 1C6, E-mail: david.naugler@utoronto.ca

benefits of sensitivity and relatively long transverse relaxation times. However, the inherent slow rotational tumbling associated with large proteins or protein complexes, and membrane proteins, results in line broadening and reduced sensitivity. ^{19}F spin labels also suffer extensively from dipolar relaxation with nearby proton spins of the protein [2], which may be largely avoided by extensive deuteration. Furthermore, in situations where ^{13}C,^{19}F two-dimensional NMR schemes are employed, the use of transverse relaxation optimized spectroscopy (TROSY) techniques [27,28], may be considered. The TROSY effect in methyl groups, results from interference between intra-methyl dipolar interactions [29]. As such, the effect is independent of field, to the extent that chemical shift anisotropy does not contribute to relaxation. Since the geometry of the trifluoromethyl group is like that of a CH_3 group, while the gyromagnetic ratio of the ^{19}F nucleus is 0.83 times that of ^{1}H, the methyl TROSY effect would be expected to be preserved in appropriate (^{19}F,^{13}C) two-dimensional schemes. In particular, in the rigid limit, the maximum peak intensities in the (^{1}H,^{13}C) HMQC are predicted to depend on terms which are derived from relaxation via reorientation of either the CF or FF intramolecular bonds i.e., transverse rates proportional to $S^2_{axis}\gamma_F^2\gamma_C^2\hbar^2\tau_C / r_E^6$ and, to a lesser extent, $S^2_{axis}\gamma_F^4\hbar^2\tau_C / r_F^6$ [29]. In this case, S_{axis} and τc represent the local order parameter of the methyl rotor and the global correlation time associated with rotational tumbling of the protein. Finally, γC and γF represent the magnetogyric ratios of ^{13}C and ^{19}F, respectively while r_{CF} and r_{FF} designate the intramolecular bond lengths. In the (^{19}F,^{13}C) HMQC, where we estimate the CF and FF intramolecular bond lengths in a trifluoromethyl group to be 1.33 Å and 2.12 Å respectively, the above transverse rates are predicted to be more than three times smaller than those for the CH_3 groups, in the absence of external dipolar relaxation or relaxation due to chemical shift anisotropy.

The trifluoromethyl probe in isoleucine

Considering these anticipated advantages for 1D and 2D NMR, we have developed a protocol for the synthesis and purification of perdeuterated 5,5,5-trifluoroisoleucine, in which the carbon nucleus of the trifluoromethyl group is ^{13}C enriched. Incorporation, using a cell-free protein expression technique, is reported [30]. We also describe herein a synthesis strategy for 2-^{13}C-trifluoroacetate and purification of the ammonium salt (or hypothetically CF_3CO_2H), to produce perdeutero 5-^{13}C-5,5,5-trifluoroisoleucine. This report is intended to communicate some of the subtleties involved in these efforts.

Isoleucine has some additional features that make it attractive for (^{19}F, ^{13}C) double labelling. At neutral pH, isoleucine is the most hydrophobic of the aliphatic side chain amino acids and it is often highly conserved at positions involved in hydrophobic interactions. Consequently, isoleucine often functions at the hydrophobic binding cleft. For example, isoleucine residues 737 and 898 in the human androgen receptor ligand binding domain mediates interdomain communication with the NH_2-terminal domain, which in turn mediates transcriptional activation [31]. Calmodulin provides another example. ^{1}H NMR studies of amide proton exchange rates of Ca^{2+}-saturated calmodulin and a Ca^{2+}-saturated calmodulin-mastoparan complex showed a reduction in solvent accessibility of Ile27 upon mastoparan binding [32]. Genomic analysis of membrane protein families informs us of amino acid abundances and conserved motifs. The most abundant amino acids in transmembrane regions are leucine, isoleucine, valine, phenylalanine, alanine, glycine, serine, and threonine. Taken together, these amino acids account for 75% of the amino acids in transmembrane regions. In contrast, isoleucine (10% abundance versus 4% conserved), valine (8% versus 4%), methionine (4% versus 1%) and threonine (7% versus 4%) are less prevalent in conserved positions [33]. Large hydrophobic (Phe, Leu, Ile, Val) residues show a clear preference for the protein surfaces facing the lipids for β-barrels, but in α-helical proteins, no such preference is seen, with these residues equally distributed between the interior and the surface of the protein [34].

Synthesis outline

The scheme shown below shows the route used to make 5,5,5-trifluoroisoleucine as a racemic mixture of diastereomers [35]. This scheme is modified for perdeuteration by substitution with cyanoacetic acid-d_3, acetone-d_6, CF_3CO_2D, ammonium-d_4 acetate-d_3 and D_2/Pd. Modification of this scheme for the $^{13}CF_3$ amino acid requires electrochemical trifluoromethylation using $^{13}CF_3CO_2^-$.

Acetone, NH_4Ac / Benzene

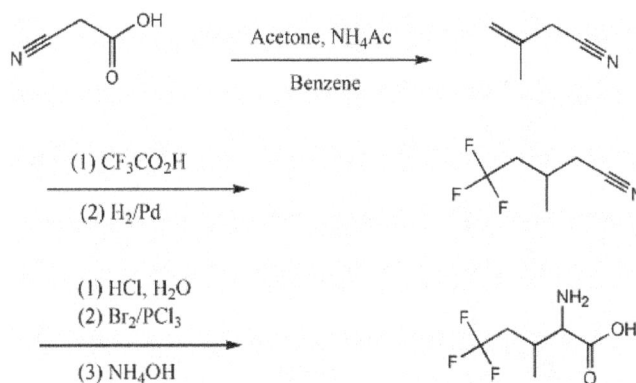

(1) CF_3CO_2H (2) H_2/Pd

(1) HCl, H_2O (2) Br_2/PCl_3 (3) NH_4OH

Experimental

Melting points are uncorrected. ^{1}H, ^{19}F and ^{13}C NMR spectra were recorded on Varian Gemini 200 MHz and Unity Inova 600 MHz spectrometers. ^{2}H NMR spectra were recorded on the Varian Unity Inova 600 MHz spectrometer. UV spectra were recorded using a Biochrom Libra S22 UV/Visible spectrophotometer. IR spectra were recorded using a Nicolet Avatar 360 E.S.P. FT-IR spectrometer. GC/MS was done using a Shimadzu GCMS-QP5050 system. LC/MS was performed using a system comprised of a Waters/Alliance 2690 liquid chromatography module and a Waters Micromass ZQ ESI mass spectrometer. Hydrogenation was performed with a Parr high-pressure hydrogenation apparatus. Electrochemical trifluoromethylation was performed with a Sargent Slomin analyzer with (one spinning) platinum mesh electrodes.

Synthesis of methallylcyanide and methallylcyanide-d_7

Synthesis of methallylcyanide and methallylcyanide-d_7 relies on 'active hydrogen' chemistry, i.e., exchangeable protons/deuterons. Wang et al. cite a synthesis reported by Marson et al. [36]. Cyanoacetic acid (e.g., 50.0 g, 0.59 mol), acetone (34.2 g, 0.59 mol), and ammonium acetate (4.0 g, 0.05 mol) (or their perdeutero counterparts) in dry benzene (100 mL) were refluxed with a Dean-Stark trap. Reflux was performed until one equivalent of water (or D_2O) was collected in the trap. Due to the cost of perdeuteroammonium acetate, a minimal amount of this catalyst was used. The quantity was observed not to be critical to yield. One gram of ammonium-d_4 acetate-d_3 will suffice for the synthesis of methallylcyanide-d_7. The Dean-Stark unit was replaced with a distillation head and the fraction collected between 110°C and

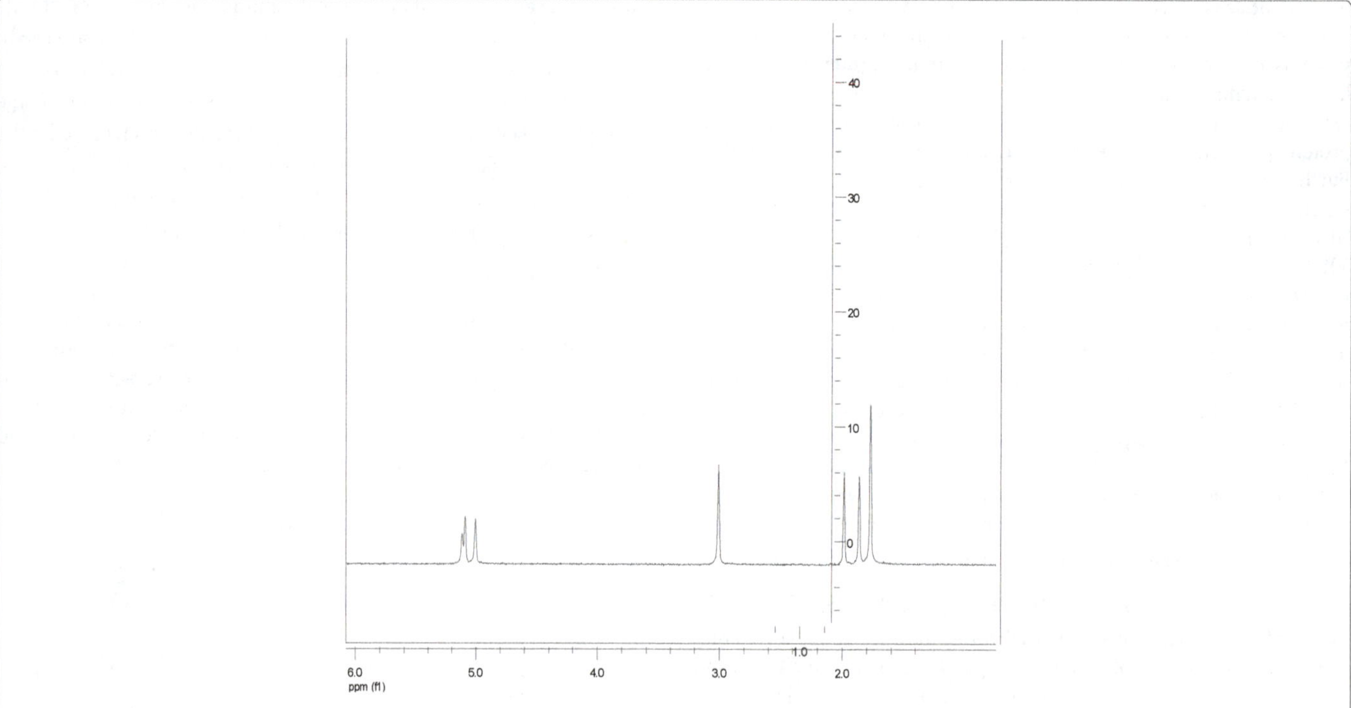

Figure 1: Deuterium NMR spectrum of methyallylcyanide-d_7, an equilibrium mixture of 3-methyl-2-butenenitrile-d_7 and 3-methyl-3-butenenitrile-d_7. The predicted boiling points of these two components are within 2°C, thus these isomers cannot be separated by fractional distillation.

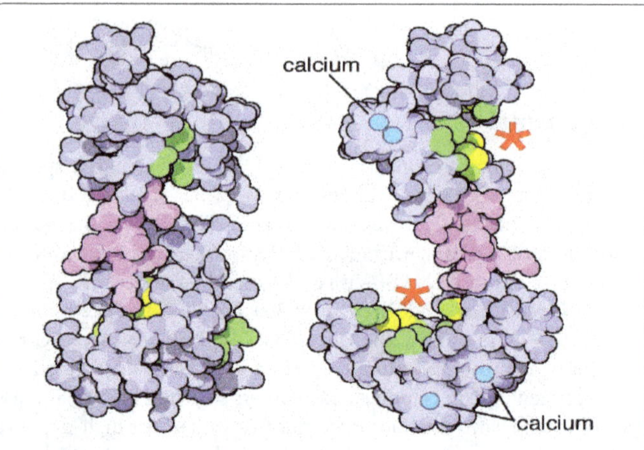

Figure 2: These images show conformational changes in calmodulin. On the left is calmodulin without calcium and on the right, is calmodulin with calcium. Sites that bind target proteins are indicated by red stars.

115°C.

Compton et al. reports the existence of both 3-methyl-2-butenenitrile and 3-methyl-3-butenenitrile in a sample of methallylcyanide. This is consistent with an equilibrium arising from 'active hydrogen' chemistry as evidenced by the ^{2}H NMR spectrum in Figure 1, below. The predicted boiling points for these isomers are within 2°C, so they cannot be separated by fractional distillation. Accordingly, this product is properly called methallylcyanide [37] rather than 3-methyl-but-3-enenitrile [27]. Appreciation of the active hydrogen nature of this product is required to make methallylcyanide-d_7.

Synthesis of methallylcyanide-d_7 requires the preparation of cyanoacetic acid-d_3, first and then its condensation with acetone-d_6, catalyzed by ammonium-d_4 acetate-d_3. 50 grams of cyanoacetic acid was dissolved with warming in a minimum quantity of D_2O, then left overnight to affect exchange. Excess H_2O/D_2O was subsequently removed with a rotary-evaporator, and the wet crystals further dried overnight in a vacuum desiccator. Because not all H_2O/D_2O can be removed by this procedure, subsequent steps require less D_2O to affect dissolution. The proton/deuteron exchange procedure was executed three times and the product submitted for mass spectral analysis. Three more H/D exchange steps were performed. The resulting cyanoacetic acid-d_3 was condensed with a molar excess of actetone-d_6 in the presence of ammonium-d_4 acetate-d_3 and the water of condensation was analyzed by ^{1}H NMR. The trace aqueous solubility of benzene is known. Comparison of integrated peak intensities showed that the water of condensation contained 0.6% H_2O and 99.4% D_2O. Given that all H/D sites are exchangeable, it is concluded that methallylcyanide-d_7 obtained after decarboxylation and distillation of the condensation product was 99.4% isotopically pure. This isotopic purity is consistent with that of the reagents used. See the ^{2}H NMR spectrum in Figures 1 and 2.

Electrochemical trifluoromethylation of methallylcyanide-d_7

Muller chose to perform electrochemical trifluoromethylation of methallylcyanide in aqueous 90% methanol between platinum electrodes whereas we chose a later system of CH_3CN and H_2O (8:1) [35]. An electrochemical solvent system of acetonitrile and water necessitates further use of acetonitrile and D_2O for methallylcyanide-d_7.

Because of hydrogen exchange chemistry of methallylcyanide-d_7, special care needs to be applied to the adaptation of this reaction to a perdeuterated substrate. H_2O needs to be replaced by D_2O. Both MeOD and CH_3CN methyls ought to be resistant to H/D exchange under the conditions of the reaction but CH_3CN is cheaper than MeOD and so is a better choice. Muller chose to use a 73% molar excess of CF_3CO_2H relative to methallylcyanide and to use a slight coulombic excess [37]. Dmowski chose to use a 20 times molar excess of TFA relative to substrate and 1.5 Faradays per mole of TFA. Because both methallylcyanide-d_7 and 2-^{13}C-trifluoracetate are very expensive these should be used in 1:1 molar ratio and with 50% excess current. Both authors achieved mild basic condition with 10% Na.

We used a Sargent Slomin S-29460 electrolytic analyzer, for radical trifluoromethylation. Both rotating and stationary electrodes were platinum. The rotating platinum electrode was chosen to be the anode. At the anode, reactions $CF_3CO_2^- \rightarrow CF_3CO_2 + e^- \rightarrow CF_3 + CO_2\uparrow$ and at the cathode $2H^+ + 2e^- \rightarrow H_2\uparrow$, both yield gases. The spinning electrode helps stirring and shakes off gas bubbles. Upon entering the limiting current region increasing the applied voltage is counterproductive. Gas production observation aids adjustment.

Muller's workup [37] was followed, as described below. For methallylcyanide-d_7, H_2O should be replaced by D_2O, H_2 by D_2 and MeOH by MeOD or some other exchange resistant solvent such as CH_3CN. An iron (steel) cathode would minimize loss of valuable substrate by electrolytic hydrogenation.

The mixture was poured into 700 mL of water, the dense oil collected, and the aqueous layer extracted with two 40 mL portions of dichloromethane. The combined organic layers from three identical runs were distilled to remove the solvent and then steam distilled. The non-aqueous layer was a mixture of 3-methyl-5,5,5-trifluoropentanonitrile, several isomeric 3-methyl-5,5,5-trifluoropentenonitriles, methallyl cyanide, and unidentified by-products. It was diluted with methanol and hydrogenated at low pressure over 5% Pd/C. Distillation afforded 66 g of nearly pure 3-methyl-5,5,5-trifluoropentanonitrile, b.p. 166-171°C.

For the perdeutero route, once the various isomeric perdeutero-3-methyl-5,5,5-trifluoropentenonitriles are reduced by D_2, exchange is no longer possible. Subsequent steps closely follow the protocol outlined by Muller as does this exposition. Quantities were adjusted proportionately.

The distillation product, 3-methyl-5,5,5-trifluoropentanonitrile was stirred with sufficient concentrated aqueous hydrochloric acid to bring most of the organic material into solution, for several days, then diluted with water and refluxed overnight. The separated organic layer was isolated, dried over Na_2SO_4, and distilled at 6 torr, giving 58.3 g of 3-methyl-5,5,5-trifluoropentanoic acid, b.p. 76-81°C.

58.3 g (0.343 mol) of 3-methyl-5,5,5-trifluoropentanoic acid, 19.1 mL of bromine and 0.6 mL of phosphorus trichloride were refluxed with a trap to absorb gaseous hydrogen bromide until the colour of bromine had disappeared. On distilling at 6 torr, about 6 g of 3-methyl-5,5,5-trifluoropentanoic acid were recovered. 60 g of distillate boiling at 100-106°C consisted mainly of nearly equal amounts of the two diastereomers of 2-bromo-3-methyl-5,5,5-trifluoropentanoic acid. This mixture was used for the preparation of 2-amino-3-methyl-5,5,5-trifluoropentanoic acid without further purification. 64.4 g (0.259 mol) of this nearly pure 2-bromo-3-methyl-5,5,5-trifluoropentanoic acid and 225 mL of concentrated aqueous ammonia were cautiously mixed and stored in a closed flask at 44-48°C for 5 days. The stopper was removed and the mixture gently heated with a water bath to drive off excess ammonia and reduce the volume to about 90 mL.

We observed the formation of polymeric materials and so we could not crystallize the product from our reaction mixture. Accordingly, we developed several purification methods discussed in the next section.

Purification of a reaction mixture containing 2-amino-3-methyl-5,5,5-trifluoropentanoic acid

Two factors are likely to alter the physical properties of 5,5,5-trifluoroisoleucine relative to native isoleucine. The inductive effect of the CF_3 group will make the amino acid and amino groups more acidic and the greater hydrophobicity of the CF_3 group will enhance the hydrolytic stability of its polymers [38,39].

In our hands, reaction of a mixture of diastereomers of 2-bromo-3-methyl-5,5,5-trifluoropentanoic acid with aqueous ammonia produced a product mixture that contained a significant quantity of polymeric material. Some of this material was readily soluble in diethyl ether and $CDCl_3$. That fraction that was soluble in organic solvent could be hydrolysed by dissolution in TFA followed by gradual addition of water. We found that the reaction mixture could be stabilized by formation of the TFA salt.

We first chose a method of chemical purification that was appropriate for the partial fluorous character of the amino acid [40-42]. The use of a C_8F_{17}BOC derivative [43] proved to be problematic because the only rational method for its hydrolysis was the use of TFA and we already had made the TFA salt. Exploitation of the C_8F_{17}Cbz derivative [43] proved to be more fruitful. A product mixture was dissolved in THF and a minimal quantity of water and reacted with a small molar excess of N-[4-(1H,1H,2H,2H-perfluorodecyl) benzyloxycarbonyloxy] succinimide. The resulting C_8F_{17} Cbz derivative was purified by solid phase extraction on fluorous silica [44]. This derivative was subjected to atmospheric pressure hydrogenolysis over 5%Pd/C and the subsequent product mixture subjected to fluorous SPE. Lyophilization yielded a soluble product that was white. Incorporation of this product in calmodulin [30] (M. Kainosho: private communication) using a cell free protein expression system proved that the product mixture contained ≤ 25% 5,5,5-trifluoro-L-isoleucine and proved the efficacy of the chemical purification.

During lyophilization, much of the product was lost due to sublimation. Sublimation has recently been revisited as a means for purification of amino acids [43]. We found that our chemically purified product mixture could be sublimed at 150°C and 6 mm Hg, but that fractional sublimation would require better vacuum and temperature control.

Finally, we exploited ion exchange chromatography using cellulose phosphate to separate monomeric and polymeric fractions. Elution with distilled H_2O yielded a microcrystalline fraction while elution with dilute aqueous ammonia yielded a waxy fraction identifiable as polymeric material. Refinement of ion exchange column purification should be developed using ion exchange TLC on cellulose phosphate paper.

We explored the use of analytical HPLC-MS first using an acetonitrile gradient in 0.1% aqueous TFA on a C_{18} column. 2-amino-3-methyl-5,5,5-trifluoropentanoic acid has a molecular weight of 185. We identified two separated peaks corresponding to [M+] and [M+H$^+$] on an ESI-MS instrument. We conclude that these are the diastereomers and that differential inductive effects due to CF_3 cause differing acid/base properties of the diastereomers. These peaks were followed at longer time by peaks due to polymers. We wished to explore the possibility of preparative scale chromatography where TFA would be counterproductive. An isocratic method was developed using 5% acetonitrile in 0.2% aqueous formic acid on an analytical C_{18} column.

Ammonium 2-^{13}C-trifluoroacetate

To embed the $^{13}CF_3$ group into a synthetic scheme for

5,5,5-trifluoroisoleucine following the synthetic scheme above, a route to 2-^{13}C-trifluoroacetate is required. Trifluoro acetic acid has been made via electrochemical fluorination [44]. Since this electrolysis would entail use of anhydrous hydrogen fluoride within a customized Teflon reaction vessel, a refrigeration unit, a high current power supply and a process control system, we chose to explore an alternate route involving halogen exchange with 2-^{13}C-tribromoacetic acid. We devised a new route to 2-^{13}C-tribromoacetic acid starting with 2-^{13}C-ethanol, discussed below.

Halogen exchange reaction design considerations

Preliminary experiments were performed following the halogen exchange reaction originated by Ref. [45]. Mass spectral analysis of an early natural abundance test reaction showed that trifluoro acetic acid was formed by reaction of $AgBF_4$ with CBr_3CO_2H in DCM. However, ^{13}C NMR multiple analysis of a test reaction with $^{13}CBr_3CO_2H$ showed a conversion to $^{13}CF_3CO_2H$ of only 18% after stirring at for 10 days at room temperature. A longer reaction time in glassware was found to be counterproductive because of failure of containment. $AgBF_4$ releases highly aggressive BF_3 during the reaction. The reaction between BF_3 and silica gel [46,47], is useful at the purification stage, however, reaction with ground glass joints may give rise to leakage.

A procedure for making 2-^{13}C-trifluoroacetic acid

The following procedure was designed to avoid the necessity of handling hydrogen fluoride, either as a solvent, reagent, or product. Synopsis: 2-^{13}C-ethanol is converted to the tribromoacetaldehyde, 2-^{13}C-bromal (hydrate) using a molar excess of bromine, Br_2. One equivalent of water converts the product mixture to bromal hydrate. Reaction with excess nitric acid at a temperature less than 50°C converted bromal hydrate to 2-^{13}C-bromoacetic acid. This product is isolated and converted to 2-^{13}C-trifluoroacetic acid using $AgBF_4$ under pressure in dichloromethane. The 2-^{13}C-trifluoroacetate is extracted into ammonia. Impure ammonium 2-^{13}C-trifluoroacetate can be enhanced in purity by sublimation at 85°C and with a vacuum less than 10 microns Hg. The yields were very low.

Conversion of 2-^{13}C-ethanol to 2-^{13}C-bromal (hydrate)

The reaction proceeds according to:

2-^{13}C-ethanol+4Br_2 → 2-^{13}C-bromal+5HBr

In a closed system, the above is an equilibrium reaction. To drive the reaction to completion, product HBr gas must be permitted to escape. This loss of mass results in a considerable reduction in the volume of the reaction mixture. We have tried sulphur and I_2 as catalysts. TFA is probably a better catalyst for this reaction. The oxidation potential of Br_2 is not sufficient to carry oxidation beyond the aldehyde. The aldehyde is required for tribromination, because each bromination step proceeds via the enol. To avoid loss of volatiles, 2-^{13}C-ethanol and excess Br_2 were combined at liquid nitrogen temperature and warmed very slowly to reflux temperature. When the reaction has been driven to completion, ^{13}C NMR shows the presence of only 2-^{13}C-bromal (~40 ppm) and its hydrate (~12 ppm). Prior to the next step, it may be desirable to isolate 2-^{13}C-bromal via distillation, and its hydrate by crystallization but it is not essential. One equivalent of H_2O is added to convert all to hydrate.

Conversion of 2-^{13}C-bromal hydrate to 2-^{13}C-tribromoacetic acid

$$^{13}CBr_3CH(OH)_2+HNO_3 \rightarrow {}^{13}CBr_3CO_2H+½N_2O_4+H_2O$$

The more stable hydrate of tribromoacetaldehyde is the species oxidized. Completion of this reaction can be determined by ^{13}C NMR (~34 ppm). At completion of reaction, excess nitric acid is removed under vacuum. Purity can be determined by melting point, 128°C.

Conversion of 2-^{13}C-tribromoacetic acid to 2-^{13}C-trifluoroacetic acid

$$13CBr_3CO_2H+3AgBF_4+3(C_2H_5)2O \rightarrow 13CF_3CO_2H+3AgBr+3[BF_3\cdot(C_2H_5)_2O]$$

This reaction is conducted in dichloromethane. This three-step reaction is exceedingly slow. For practical synthesis, it is necessary to conduct this reaction in a sealed pressure reaction vessel at above 75°C. The completion of this reaction can be determined by ^{13}C NMR (quartet at 116.6 ppm is dominant) or ^{19}F NMR (doublet at -76.55 ppm is dominant). Under standard conditions BF_3 is a gas, whereas $BF_3\cdot(C_2H_5)_2O$ is liquid. A strong Lewis acid, BF_3 forms a complex with diethyl ether. This reduces the pressure. At the completion of the reaction, the DCM reaction mixture is eluted through silica gel. SiO_2 reacts with BF_3. 2-^{13}C-trifluoroacetate is extracted into aqueous ammonia and excess removed under vacuum. Ammonium 2-^{13}C-trifluoroacetate was converted into a fine powder by lyophilization, to aid in sublimation. Ammonium 2-^{13}C-trifluoroacetate was sublimed to improve purity by sublimation at 85°C and less than 10 microns Hg vacuum. Because initial purity was poor, yield was poor. Gram scale quantities of material could be processed in this way.

Manipulation of ^{13}C haloacetates

The halogen exchange reaction between 2-^{13}C-tribromoacetic acid and $AgBF_4$ proceeds in a stepwise fashion and hence yields a mixture of haloacetates. The target compound, 2-^{13}C-trifluoroacetic acid is too volatile and so the haloacetate mixture is best manipulated as a salt. The ammonium salts have volatilities that were exploited for purification by high vacuum sublimation and the progress of purification was monitored by ^{19}F and ^{13}C NMR. Whereas high vacuum sublimation has only one theoretical plate, the greater load capacity compared to chromatography affords it an advantage for the first stages of purification. Repeated stages of high vacuum sublimation yielded a product that gave three spots on silica gel TLC. Neutral alumina TLC gave streaks. One spot showed an R_f=0.78 on silica gel TLC that was identical to the R_f of natural abundance ammonium trifluoroacetate, eluted with 10% aqueous ammonia in methanol. Based on this observation, the ammonium ^{13}C haloacetate mixture was subjected to preparative TLC under the same conditions. ^{19}F and ^{13}C NMR of the product from the preparative TLC target band showed that the mixture consisted of ammonium salts of 2-^{13}C-trifluoroacetate, 2-^{13}C-bromodifluoroacetate and 2-^{13}C-dibromofluoroacetate (Table 1).

It is interesting to note that CF_3I has been enriched to 86% in ^{13}C by selective multiphoton dissociation of $^{12}CF_3I$ at pressures less than 1.0 torr [48]. A more practical method would rely on enrichment in the condensed phase. An isotope effect is often observed on melting

	δ (^{19}F) ppm*	δ (2-^{13}C) ppm	$^1J_{CF}$ Hz
$^{13}CF_3CO_2^-$	-76.68	116.6	291.58
$^{13}CBrF_2CO_2^-$	-60.18	112.76	318.39
$^{13}CBr_2FCO_2^-$	-56.51	89.21	327.75

*Relative to TFA, -76.55 ppm

Table 1: ^{19}F and ^{13}C NMR of the product from the preparative TLC target band showed the mixture consisted of ammonium salts of 2-^{13}C-trifluoroacetate, 2-^{13}C-bromodifluoroacetate and 2-^{13}C-dibromofluoroacetate.

points [49]. In recent work, boron has been enriched to 93.21% in ^{10}B and 99.01% in ^{11}B by zone refining [50]. It is known that isotope effects on heat capacity and crystal transition temperature can be detected by differential scanning calorimetry [51]. Zone refining of low melting ammonium trifluoroacetate (M.P 123°C) would be more energy efficient than that of high melting boron (M.P 2079°C). In future, we plan to develop an analytical HPLC method and to use preparative HPLC for the purification of ammonium 2-^{13}C-trifluoroacetate. Infrared difference spectroscopy may be able to resolve the carbon isotope effect for ammonium 2-^{13}C-trifluoroacetate in the condensed phase. Differential scanning calorimetry of ammonium 2-^{13}C-trifluoroacetate will provide the thermodynamic information needed to plan and develop a zone refining method for the extraction of ammonium 2-^{13}C-trifluoroacetate from an inexpensive natural abundance melt.

Another promising route to ^{13}C enriched TFA that we may explore is that of fluorodeoxygenation [52], starting with glycine. Recently a new and better synthesis of arylsulfur trifluorides has been reported for reagents that may provide a convenient route to fluorodeoxygenation of carboxylic acids [53].

Conclusion

We have presented a critical scientific narrative of a promising technology. The advantage of the $^{13}CF_3$ NMR spin label in protein NMR has been explained. Progress in incorporation of this spin label in a perdeuterated amino acid and in an important protein is reported. We comment on some of the synthetic subtleties encountered.

Whereas incorporation of monofluorinated aromatic amino acids in proteins using *in vivo* or *in vitro* expression systems is now routine, incorporation of trifluoromethyl analogs of branched chain amino acids is at present a challenging technology. A biochemical hurdle is the specificity of an aminoacyl tRNA synthetase. Wang et al. determined that the specificity constant, kcat/KM of 5,5,5-trifluoroisoleucine is only 1/134 that of isoleucine for *E. coli* isoleucyl tRNA synthetase [27]. Considerations of enzyme kinetics and competitive inhibition dictate that the background concentration of isoleucine needs to be effectively zero for $tRNA^{Ile}$ to be charged with 5,5,5-trifluoroisoleucine by isoleucyl tRNA synthetase. In addition to the catalytic domain, where the amino acid and $tRNA^{aa}$ are specifically ligated, aminoacyl tRNA synthetase also has an editing domain for the hydrolysis of mischarged tRNA. In the case of isoleucyl tRNA synthetase, the editing domain has been evolved most specifically for the hydrolysis of val· $tRNA^{Ile}$. The successful incorporation of 5,5,5-trifluoroisoleucine in mDHFR, mIL II (27) and in calmodulin [54] (Kainosho M: private communication) implies that 5TFI· $tRNA^{Ile}$ is too large for the editing site. Practical questions remain. Is an expression system based upon *E. coli* more advantageous than one based upon a eukaryotic organism? Is a cell free lysate system preferable to *in vivo* expression system?

This paper covers many areas of synthetic chemistry, organic, fluoro, isotopic, and biochemical. We have identified ammonium 2-^{13}C-trifluoroacetate as an important synthon for introduction of the $^{13}CF_3$ group into amino acids. Completed synthesis of methallylcyanide-d_7 and conceptual synthesis of 5-^{13}C-5,5,5-trifluoroisoleucine-d_7, provide an important building block for the exploitation of $^{13}CF_3$ in protein NMR.

5,5,5-TFI is an unnatural AA, more so than monofluorinated amino acids. Either as uniform, or site specific isotopic labels, deuterium and ^{13}C amino acids are readily synthesized and incorporated. Fluorine is the 13th most abundant isotope in the earth's crust, yet even after 3.5 billion years of biology only about a dozen fluorinated natural products have been evolved, attributed to fluorine's chemistry as a "superhalogen" [55]. Organofluorine compounds as polymers or as drugs have proven useful in material science and pharmacology. The target spin $^{13}CF_3$ label should prove useful in multidimensional heteronuclear NMR structure dynamics studies of proteins [56]. Synthesis of 5,5,5-TFI has been proven by its incorporation in the calcium binding protein calmodulin. Methallylcyanide-d_7 has been produced with military grade deuterium isotope purity, 99.4%. Trace quantities of 2-^{13}C-trifluoroacetate have been characterized by ^{13}C-^{19}F NMR coupling. This contribution may pave the way to future study [57-59].

Acknowledgements

This work is part of a project in heteronuclear multidimensional NMR.

References

1. Danielson MA, Falke JJ (1996) Use of 19F NMR to probe protein structure and conformational changes. Annu Rev Biophys Biomol Struct 25: 163-195.
2. Gerig JT (1994) Fluorine NMR of proteins. Progress in Nuclear Magnetic Resonance Spectroscopy 26: 293-370.
3. Gakh YG, Gakh AA, Gronenborn AM (2000) Fluorine as an NMR probe for structural studies of chemical and biological systems. Magnetic Resonance in Chemistry 38: 551-558.
4. Frieden C, Hoeltzli SD, Bann JG (2004) The preparation of 19F-labeled proteins for NMR studies. Methods in enzymology 380: 400-415.
5. Linda A Luck and Joseph J Falke (1991) Fluorine ^{19}F NMR Studies of the d-Galactose Chemosensory Receptor. 1. Sugar Binding Yields a Global Structural Change. Biochemistry 30: 4248-4256.
6. Luck LA and Falke JJ (1991) Fluorine-19 NMR studies of the D-galactose chemosensory receptor. 2. Calcium binding yields a local structural change. Biochemistry 30: 4257-4261.
7. Salopek-Sondi B, Vaughan MD, Skeels MC, Honek JF, Luck LA (2003) ^{19}F NMR studies of the leucine-isoleucine-valine binding protein: Evidence that a closed conformation exists in solution. Journal of Biomolecular Structure and Dynamics 21: 235-246.
8. Li H, Frieden C (2005) NMR studies of 4-19F-phenylalanine-labeled intestinal fatty acid binding protein: evidence for conformational heterogeneity in the native state. Biochemistry 44: 2369-2377.
9. Bann JG, Pinkner J, Hultgren SJ, Frieden C (2002) Real-time and equilibrium 19F-NMR studies reveal the role of domain–domain interactions in the folding of the chaperone PapD. Proceedings of the National Academy of Sciences 99: 709-714.
10. Naugler D, Prosser RS (2017) Notes on Synthesis of Perdeutero-5-13C, 5, 5, 5-Trifluoroisoleucine VI. bioRxiv.
11. Shu Q, Frieden C (2004) Urea-dependent unfolding of murine adenosine deaminase: Sequential destabilization as measured by ^{19}F NMR. Biochemistry 43: 1432-1439.
12. Shu Q, Frieden C (2005) Relation of enzyme activity to local/global stability of murine adenosine deaminase: 19 F NMR studies. Journal of molecular biology 345: 599-610.
13. Cistola DP, Hall KB (1995) Probing internal water molecules in proteins using two-dimensional ^{19}F-^{1}H NMR. Journal of biomolecular NMR 5: 415-419.
14. Thu NTH, Pratt EA, Ho C (1991) Interaction of the membrane-bound D-lactate dehydrogenase of Escherichia coli with phospholipid vesicles and reconstitution of activity using a spin-labeled fatty acid as an electron acceptor: a magnetic resonance and biochemical study. Biochemistry 30: 3893-3898.
15. Prosser RS, Luchette PA, Westerman PW (2000) Using O2 to probe membrane immersion depth by 19F NMR. Proc Natl Acad Sci USA 29: 9967-9971.
16. Luchette PA, Prosser RS, Sanders CR (2002) Oxygen as a paramagnetic probe of membrane protein structure by cysteine mutagenesis and 19F NMR spectroscopy. Journal of the American Chemical Society 124: 1778-1781.
17. Hull WE and Sykes BD (1975) Dipolar nuclear spin relaxation of 19F in multispin systems. Application to ^{19}F labeled proteins. The Journal of Chemical Physics 63: 867-880.
18. Hull WE, Sykes BD (1975) Fluorotyrosine alkaline phosphatase: internal mobility of individual tyrosines and the role of chemical shift anisotropy as a ^{19}F nuclear spin relaxation mechanism in proteins. Journal of molecular biology

98: 121-153.

19. Sykes BD, Hull WE, Snyder GH (1978) Experimental evidence for the role of cross-relaxation in proton nuclear magnetic resonance spin lattice relaxation time measurements in proteins. Biophysical journal 21: 137-146.

20. Rozovsky S, Jogl G, Tong L, McDermott AE (2001) Solution-state NMR investigations of triosephosphate isomerase active site loop motion: ligand release in relation to active site loop dynamics. Journal of molecular biology 310: 271-280.

21. Vaughan MD, Cleve P, Robinson V, Duewel HS, Honek JF (1999) Difluoromethionine as a novel 19F NMR structural probe for internal amino acid packing in proteins. Journal of the American Chemical Society 121: 8475-8478.

22. Renner C, Alefelder S, Bae JH, Budisa N, Huber R, et al. (2001) Fluoroprolines as Tools for Protein Design and Engineering. Chem Intl Ed 40: 923-925.

23. Feeney J, McCormick JE, Bauer CJ, Birdsall B, Moody CM, et al. (1996) ^{19}F Nuclear Magnetic Resonance Chemical Shifts of Fluorine Containing Aliphatic Amino Acids in Proteins: Studies on Lactobacillus casei Dihydrofolate Reductase Containing (2 S, 4 S)-5-Fluoroleucine. Journal of the American Chemical Society 118: 8700-8706.

24. Wang P, Tang Y, Tirrell DA (2003) Incorporation of trifluoroisoleucine into proteins in vivo. Journal of the American Chemical Society 125: 6900-6906.

25. Palmer III AG (2004) NMR characterization of the dynamics of biomacromolecules. Chemical reviews 104: 3623-3640.

26. Kay LE (2005) NMR studies of protein structure and dynamics. Journal of Magnetic Resonance 173: 193-207.

27. Pervushin K, Riek R, Wider G, Wüthrich K (1997) Attenuated T2 relaxation by mutual cancellation of dipole–dipole coupling and chemical shift anisotropy indicates an avenue to NMR structures of very large biological macromolecules in solution. Proceedings of the National Academy of Sciences 94: 12366-12371.

28. Tugarinov V, Hwang PM, Ollerenshaw JE, Kay LE (2003) Cross-correlated relaxation enhanced 1H- 13C NMR spectroscopy of methyl groups in very high molecular weight proteins and protein complexes. Journal of the American Chemical Society 125: 10420-10428.

29. Ohki SY, Tsuda S, Joko S, Yazawa M, Yagi K, et al. (1991) ^{1}H NMR study on amide proton exchange of calmodulin-mastoparan complex. The Journal of Biochemistry 109: 234-237.

30. Hagan D (1994) The fluorinated Natural products. Natu Prod Reports 11: 123-133.

31. He B, Kemppainen JA, Voegel JJ, Gronemeyer H, Wilson EM (1999) Activation function 2 in the human androgen receptor ligand binding domain mediates interdomain communication with the NH2-terminal domain. Journal of Biological Chemistry 274: 37219-37225.

32. Liu Y, Engelman DM, Gerstein M (2002) Genomic analysis of membrane protein families: abundance and conserved motifs. Genome boil 3: 51-54.

33. Pashinnik VE, Martyniuk EG, Tabachuk MR, Shermolovich YG, Yagupolskii LM (2003) A new method for the synthesis of organosulfur trifluorides. Synthe commun 33: 2505-2509.

34. Ulmschneider MB, Sansom MS (2001) Amino acid distributions in integral membrane protein structures. Biochim Biophys Acta 1512: 1-14.

35. Muller N (1987) Synthesis of 5, 5, 5-trifluoro-DL-isoleucine and 5, 5, 5-trifluoro-DL-alloisoleucine. J Fluor chem 36: 163-170.

36. Marson CM, Grabowska U, Walsgrove T, Eggleston DS, Baures PW (1994) Stereocontrolled construction of condensed, gamma-lactam ring systems by cationic cyclizations-Rearrangement of a gamma-lactam to a delta-lactam. The J Org Chem 59: 284-290.

37. Compton DA, Murphy WF, Mantsch HH (1981) The IR spectra of 3-methyl-2-butenenitrile and 3-methyl-3-butenenitrile-Spectrochimica Acta Part A. Molec Spectro 37: 453-455.

38. Naugler D, Prosser RS (2017) Notes on Synthesis of Perdeutero-5-13C, 5, 5, 5-Trifluoroisoleucine VI. bioRxiv, pp: 140681.

39. Dmowski W, Biernacki A (1996) Electrochemical trifluoromethylation of dimethyl acetylenedicarboxylate. J Fluorine Chem 78: 193-194.

40. Kukhar VP (1994) Fluorine-containing amino acids. J of Fluorine Chem 69: 199-205.

41. Nihei T, Kurata S, Kondo Y, Umemoto K, Yoshino N, et al. (2002) Enhanced hydrolytic stability of dental composites by use of fluoroalkyltrimethoxysilanes. J Dental Research 81: 482-486.

42. Curran DP (2001) Fluorous reverse phase silia gel-A new tool for preparative separations in synthetic organic and organofluorine chemistry. Synlett 09: 1488-1496.

43. Wasser DJ, Johnson PS, Klink FW, Kucera F, Liu CC (1987) The electrochemical fluorination of acetyl fluoride. J Fluorine Chem 35: 557-569.

44. Curran DP, Amatore M, Guthrie D, Campbell M, Go E, et al (2003) Synthesis and reactions of fluorous carbobenzyloxy (FCbz) derivatives of α-amino acids. J Org Chem 68: 4643-4647.

45. Luo Z, Williams J, Read RW, Curran DP (2001) Fluorous Boc (FBoc) carbamates: new amine protecting groups for use in fluorous synthesis. The J Organic Chem 66: 4261-4266.

46. Glavin DP, Bada JL (1998) Isolation of amino acids from natural samples using sublimation. Anal Chem 70: 3119-3122.

47. Bloodworth AJ, Bowyer KJ, Mitchell JC (1987) A mild, convenient, halogen-exchange route to gem-difluorides and trifluorides. Tetrahed lett 28: 5347-5350.

48. Rhee KH, Basila MR (1968) Chemisorption of BF3 on catalytic oxide surfaces: Infrared spectroscopic studies. J Cataly 10: 243-251.

49. Yoshidome T, Dang Z, Morrow BA (2003) Infrared spectroscopic analyses of transformations of chemical species on the silica highly-reacted with gaseous BF3. Analy Sci 19: 429-435.

50. Bittenson S, Houston PL (1977) Carbon isotope separation by multiphoton dissociation of CF3I. The J Chem Phy 67: 4819-4824.

51. Vanhook WA (1995) Thermodynamic Analysis of Isotope Effects on Triple Points and/or Melting Temperatures. Phys Sci 50: 337-346.

52. Nogi N, Hirano T, Honda K, Tanaka S, Noda T (1998) J Surf Anal 4: 280-283.

53. Takeda M, Kainosho M (2012) Cell-Free Protein Production for NMR Studies. In: Shekhtman A, Burz D (eds.), Protein NMR Techniques. Methods in Molecular Biology (Methods and Protocols). Humana Press, p: 831.

54. Shustov GV, Denisenko SN, Chervin II, Kostyanovskii RG (1988) Bul Acad Sci USSR, Division of Chemical Science 37: 1422-1427.

55. Shigematsu H, Nakadaira H, Matsui T, Tobo A, Ohoyama K (2002) Pb-and Ti-isotope Effects in Pb TiO3 by Heat Capacity Measurement and Neutron Powder Diffraction. J Nuclear Sci Technol 39: 395-398.

56. Cushley RJ, Naugler D, Ortiz C (1975) ^{13}C Fourier Transform Nuclear Magnetic Resonance. XI. Pyridine N-Oxide Derivatives. Canadian Journal of Chemistry 53: 3419-3424.

57. Naugler DG, Cushley RJ (2000) Spectral estimation of NMR relaxation. Journal of Magnetic Resonance 145: 209-215.

58. Naugler D (2017) Interim, Mendeley Data. Available from: https://data.mendeley.com/datasets/42n45mkstx/1

59. Naugler D, Cushley RJ, Clark-Lewis I (2017) NMR Analysis of the Interaction of Ethanol with the Nicotinic Acetylcholine Receptor. J Bioanal Biomed 9: 177-183.

The HPTLC Validated Method Development for the Quantification of Naringin from the Partially Purified Labisia Pumila Dichloromethane Extract

Stepfanie NS[1], Sareh K[1], Gabriel AA[2], Teo SS[1], Farahnaz A[3] and Patrick NO[1*]

[1]*Department of Applied Sciences, UCSI University, 56000 Cheras, Kuala Lumpur, Malaysia*
[2]*Department of Pharmaceutical Science, UCSI University, 56000 Cheras, Kuala Lumpur, Malaysia*
[3]*Department of Anti-aging and Regenerative medicine, UCSI University, 56000 Cheras, Kuala Lumpur, Malaysia*

Abstract

Partially purified fraction E, dichloromethane extract of leaves from *Labisia Pumila* has been shown to possess anti-ulcer, anti-inflammatory, anti-asthmatic activities among others. Naringin has been identified by LC-MS and HPLC as the major compound present in fraction E. The present study reports the development and validated TLC densitometric method for the quantification of Naringin present in Fraction E. ICH guidelines were followed to develop this method. CAMAG-HPLTC system comprising of a TLC Visualizer and Linomat 5 sample applicator was used in this study. The separation was performed using TLC aluminum plates pre-coated with silica gel 60 F254. Optimized mobile phase consisted of Methanol: Ethyl Acetate (60:40 v/v). Win-CATS-V 1.2.3 software and Video Scan were used to identify and quantify Naringin at 366nm in fluorescence mode. The peak heights at Rf 0.648 ± 0.01 gave a linear range from 200-1000µg/ml correlation coefficient R2± SD = 0.973 ± 0.024. The LOD and LOQ were found to be Naringin 0.74 ± 0.29 µg and 7.73 ± 0.26 µg, respectively. Repeatability gave CV % < 2.0. We recovered 92.56% of Naringin and 642.4 mg/g (64.24%) was found to be present in Fraction E. The HPTLC method was found to be reproducible, accurate and convenient for rapid screening of bioactive constituents present in Fraction E. This developed method will be used for analysis and quality control of drug formulations containing *Labisia Pumila*.

Keywords: HPTLC; *Labisia Pumila;* ICH guidelines

Introduction

Labisia Pumila (LP) (var. Alata) a member of the Myrsinaceae family is a traditional herb found widely throughout the rainforest of Indochina [1]. Its other names include Kacip Fatimah, Sangkoh (Iban), Mata Pelanduk Rimba, Selusoh Fatimah and Tadah Matahari [1]. Indigenous women of the Malay Archipelago use this herb to increase libido, improve post-partum health and ease menstrual problems [1, 2]. This plant has been reported by many authors to possess; anti-bacterial, oestrogenicity, anti-inflammatory, anti-photoaging, antioxidant and gastro-protective with positive results [3-6]. The partially purified extracts of LP have been extensively studied [7-9] and Fraction E has been shown to possess antioxidant and anti-inflammatory activity. Different studies on the constituents of the different varieties of LP have reported them to contain variable patterns of flavonoids, phenolic and various bioactive volatile compounds [10, 11]. Phytochemical compounds like naringin, quercertin, rutin, kaempferol, vanillic acid and hesperetin among others have been reported to be constituted in LP extract [11]. Naringin (Cas: 10236-47-2) (4′,5,7-Trihydroxyflavanone 7-rhamnoglucoside) is a flavanone glycoside that has been reported to possess *in vitro* [12] and *in vivo* [13, 14] antioxidant properties. It was also reported to improve diabetic foot ulcers its down-regulation of anti-inflammatory, inhibition of oxidative stress and hyperglycemia and up regulation of growth factors (IFG-1, TGF-β and VEGF-c) expression. Preliminary phytochemical analysis on the partially purified extracts showed Fraction E tested positive for phenolic compounds while LCMS and HPLC (unpublished data) confirmed the presence of Naringin its major constituent.

HPTLC is quickly becoming the leading chromatographic technique used to quantify the amount of different compounds in complex samples [15]. The objective of this experiment was to report a new high performance-thin layer chromatography (HPTLC)-densitometric procedure for the separation and quantitative determination of Naringin in fraction E of partially purified DCM leaf extract of *Labisia Pumila*. The Validation of this chromatographic procedure was according to the International Conference on Harmonisation of Technical Requirements For Registration of Pharmaceuticals For Human Use guideline (ICH guideline) [16].

Materials and Methods

Plant samples and chemicals

The leaves of LP var. alata (Kacip Fatimah) were obtained and identified from a forest in Sungai Perak, by Dr. Shamsul Khamis, a research officer (plant taxonomy) from the Laboratory of Natural Products (NATPRO), Institute of Bioscience in University Putra Malaysia. Naringin (Cas: 10236-47-2) (>90%), Pre-coated aluminum silica gel TLC plates (F_{254} 20 x 20 cm) and 2-amino-ethyldiphenylborinate (Neu's reagent) were obtained from (Merck, Germany). All solvents used were of analytical grade.

Extraction of plant materials

The fresh leaves were air-dried for 7 days before being powdered by dry mill. About 500 g of plant material was kept in 6 L of

***Corresponding author:** Patrick Nwabueze Okechukwu, Faculty of Applied Sciences, UCSI University, UCSI Heights No. 1, Jalan Menara Gading, 56000 Kuala Lumpur, Malaysia
E-mail: patrickn@ucsiuniversity.edu.my

dichloromethane for 2 days. This extraction process was repeated six times before filtering with No. 1 Whatman filter paper. The filtrate was rotary evaporated at 40°C and kept to dry in a fume cupboard. The extract was kept in a desiccator for further separation by column chromatography. The partial purification of the DCM extract by column chromatography used a 42 x 2.5 cm vertical column equipped with a stop cock and glass frit to support the silica gel, pore size 60 Å, 200-400 mesh. 100% hexane was used to pre-elute the column. A decreasing polarity solvent ratio (100% hexane (Hx) – Hx: Ethyl acetate (EA) – EA: Methanol (MeOH) – 100% MeOH) was used to obtain Fraction E (60:40 Hx:EA to 100% MeOH). The fractions were collected based on their chromatogram profiles analyzed on (TLC plates Silica gel 60 F_{254}) and mobile phase 60:40 Hx:EA.

Preparation of standard and extract solution

Naringin and the Fraction E sample (5 mg/ml) were prepared by transferring accurately weighed 5mg into a 1.5ml centrifuge tube; the extract was dissolved in 1 mL of HPLC analytical grade methanol and ultra-sonicated for 20 minutes and filtered with 0.45 µm millipore sterile syringe filter. Standards and sample were prepared daily immediately before use.

HPTLC Instrumentation and chromatography conditions

20 µl of the extracts were separately applied (Samples and standard) onto the TLC plate with 6 mm wide band or spotted with an automatic TLC applicator Linomat-V with N_2 flow (Camag, Switzerland), 8 mm from the bottom with instruction input defined from win-CATS-V 1.2.3 software. After sample application, the plates were developed in a 10 x 20 cm horizontal Camag twin glass chamber pre-saturated with the mobile phase (10 ml each side) for 20 minutes at room temperature (25 -27°C). The mobile phase consisted of MeOH: EA (60:40 v/v). Linear ascending development was carried out until the 8 cm mark. The plates were observed after 30 minutes air drying under the Camag UV visualizer (366 nm) as shown in Figure 1. The plates were sprayed with Neu's reagent spray. Video Scan software in fluorescence mode was used to quantify the plates post derivation.

Validation method

ICH guidelines were followed for the validation of the quantitative method developed for precision, repeatability and accuracy of naringin in Fraction E [16]. A concentration range of 200 to 5000 µg/ml was spotted on the TLC plate in triplicates. Linearity, LOD and LOQ was done according to methods described by [17]. Precision was evaluated by the analysis of replicate (n = 3) applications of freshly prepared standard solution at concentrations (200, 600 and 1000 µg/ml) by methods described by [18]. The repeatability of sample application and measurement of peak height was expressed in terms of coefficient of variation (CV%). Accuracy was measured by performing recovery experiments by spiking (200, 600, 1000 µg/ml) of Naringin with a single known concentration of Fraction E (600 µg/ml) and percentage recovery was calculated according to the formula below Figure 2:

% Recovery = (Peak height of Spiked sample - Peak height of Known Sample)

(Peak height of known standard concentration)

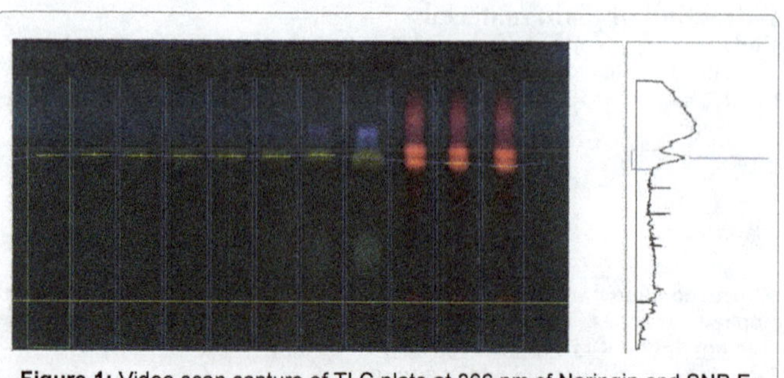

Figure 1: Video scan capture of TLC plate at 366 nm of Naringin and SNP E.

Results and Discussion

The amount of herbal drugs used worldwide has risen dramatically in the recent decade. Natural health products and Chinese herbal remedies are two examples from among the variety of plant drugs in the market [15]. High performance thin layer chromatography has been reported to be a universally accepted method for evaluating the chemical composition of natural products [19], specifically in phytochemical analysis [15]. This is due to advantages HPTLC offers like; image result presentation, it's simple, cost efficiency, rapid obtainable results and high sample capacity [15].

Standard curve, LOD and LOQ

The methods linearity was determined with a specified range to obtain test results in direct proportion to the concentrations of the analyte. Naringin had linear range from 200-1000µg/ml with correlation coefficient $R^2 \pm$ SD = 0.973 ± 0.024 as reported in Table 1. The limit of quantification (LOQ) is the lowest amount of the analyte that can be quantitatively determined in sample with defined precision and accuracy under standard conditions. LOQ is the amount of loaded sample producing a peak area that is equal to the sum mean blank area and ten times its standard deviation [20]. Sample concentration at 900 µg/ml was interpolated with the linear regression curve using graphpad prism 6 Figure 3. The amount of naringin found in Fraction E was found to be 642.39 ± 22.9 mg/g (64.2%) The LOD and LOQ have been reported in Table 1.

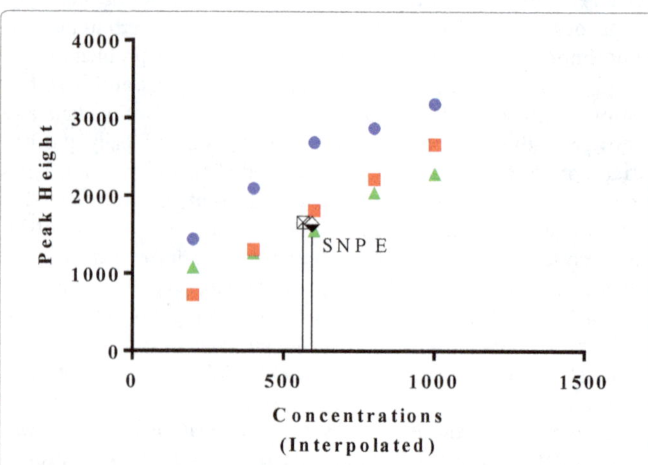

Figure 2: Interpolation of amount of naringin in SNP E from the standard curve.

Parameters	Results (mean ± SD) (n=3)
Regression equation	y = 2.037x + 725.8
Correlation coefficient	0.973 ± 0.024
Linearity range	200 to 1000 µg/ml
Limit of detection	0.75 ± 0.29 µg
Limit of quantification	7.73 ± 0.26 µg
Amount naringin (mg/g)	642.39 ± 22.9 mg/g (64%)

Table 1: Amount of Calibration Parameters.

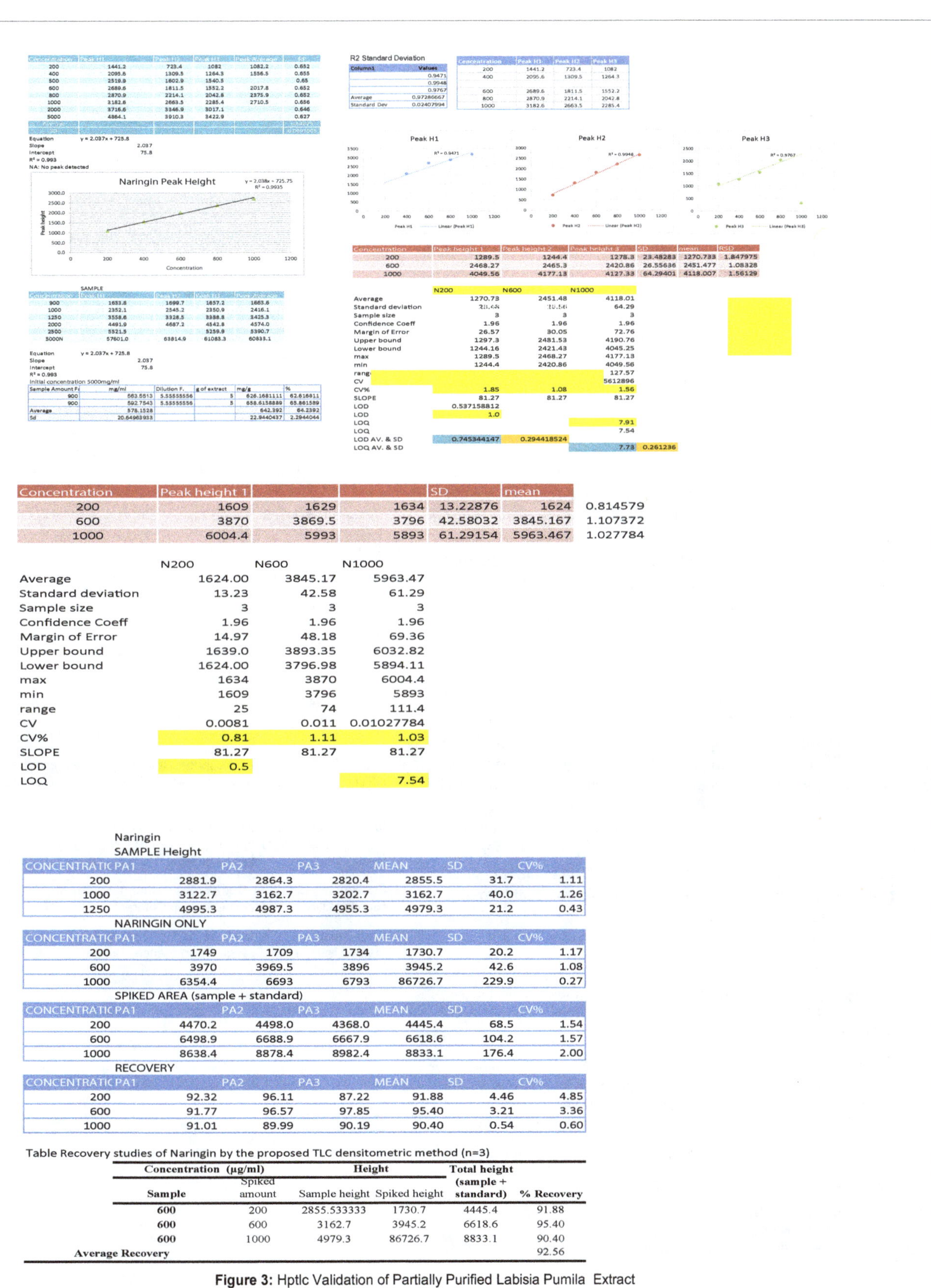

Concentration	Peak H1	Peak H2	Peak H3	Peak Average	Rf
200	1441.2	723.4	1082	1082.2	0.652
400	2095.6	1309.5	1264.3	1556.5	0.655
500	2519.9	1602.9	1540.5		0.65
600	2689.6	1811.5	1552.2	2017.8	0.652
800	2870.9	2214.1	2042.8	2375.9	0.652
1000	3182.6	2663.5	2285.4	2710.5	0.656
2000	3716.6	3346.9	3017.1		0.646
5000	4864.1	3910.3	3422.9		0.627
[illegible]					[illegible]
[illegible]					[illegible]

Equation y = 2.037x + 725.8
Slope 2.037
Intercept 75.8
$R^2 = 0.993$
NA: No peak detected

SAMPLE

Concentration	Peak H1	Peak H2	Peak H3	Peak Average
900	1633.8	1699.7	1657.2	1663.6
1000	2352.1	2545.2	2350.9	2416.1
1250	3558.6	3328.5	3388.8	3425.3
2000	4491.9	4687.2	4542.8	4574.0
2500	5521.5	5259.9	5259.9	5390.7
5000N	57601.0	63814.9	61083.3	60833.1

Equation y = 2.037x + 725.8
Slope 2.037
Intercept 75.8
$R^2 = 0.993$
Initial concentration 5000mg/ml

Sample Amount F	mg/ml	Dilution F.	g of extract	mg/g	%
900	563.5513	5.55555556	5	626.1681111	62.616811
900	592.7543	5.55555556	5	658.6158889	65.861589
Average	578.1528			642.392	64.2392
Sd	20.64963933			22.9440437	2.2944044

R2 Standard Deviation

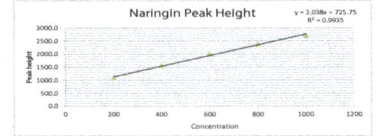

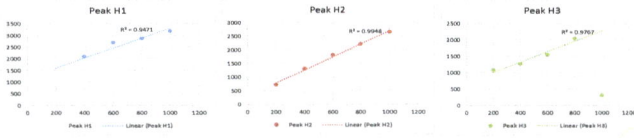

Column1	Values
	0.9471
	0.9948
	0.9767
Average	0.97286667
Standard Dev	0.02407994

Concentration	Peak H1	Peak H2	Peak H3
200	1441.2	723.4	1082
400	2095.6	1309.5	1264.3
600	2689.6	1811.5	1552.2
800	2870.9	2214.1	2042.8
1000	3182.6	2663.5	2285.4

Concentration	Peak height 1	Peak height 2	Peak height 3	SD	mean	RSD
200	1289.5	1244.4	1278.3	23.48283	1270.733	1.847975
600	2468.27	2465.3	2420.86	26.55636	2451.477	1.08328
1000	4049.56	4177.13	4127.33	64.29401	4118.007	1.56129

	N200	N600	N1000	
Average	1270.73	2451.48	4118.01	
Standard deviation	[illegible]	[illegible]	64.29	
Sample size	3	3	3	
Confidence Coeff	1.96	1.96	1.96	
Margin of Error	26.57	30.05	72.76	
Upper bound	1297.3	2481.53	4190.76	
Lower bound	1244.16	2421.43	4045.25	
max	1289.5	2468.27	4177.13	
min	1244.4	2420.86	4049.56	
rang			127.57	
CV			5612896	
CV%	1.85	1.08	1.56	
SLOPE	81.27	81.27	81.27	
LOD	0.537158812			
LOD	1.0			
LOQ			7.91	
LOQ			7.54	
LOD AV. & SD	0.745344147	0.294418524		
LOQ AV. & SD			7.73	0.261236

Concentration	Peak height 1			SD	mean	
200	1609	1629	1634	13.22876	1624	0.814579
600	3870	3869.5	3796	42.58032	3845.167	1.107372
1000	6004.4	5993	5893	61.29154	5963.467	1.027784

	N200	N600	N1000
Average	1624.00	3845.17	5963.47
Standard deviation	13.23	42.58	61.29
Sample size	3	3	3
Confidence Coeff	1.96	1.96	1.96
Margin of Error	14.97	48.18	69.36
Upper bound	1639.0	3893.35	6032.82
Lower bound	1624.00	3796.98	5894.11
max	1634	3870	6004.4
min	1609	3796	5893
range	25	74	111.4
CV	0.0081	0.011	0.01027784
CV%	0.81	1.11	1.03
SLOPE	81.27	81.27	81.27
LOD	0.5		
LOQ			7.54

Naringin
SAMPLE Height

CONCENTRATIC	PA1	PA2	PA3	MEAN	SD	CV%
200	2881.9	2864.3	2820.4	2855.5	31.7	1.11
1000	3122.7	3162.7	3202.7	3162.7	40.0	1.26
1250	4995.3	4987.3	4955.3	4979.3	21.2	0.43

NARINGIN ONLY

CONCENTRATIC	PA1	PA2	PA3	MEAN	SD	CV%
200	1749	1709	1734	1730.7	20.2	1.17
600	3970	3969.5	3896	3945.2	42.6	1.08
1000	6354.4	6693	6793	86726.7	229.9	0.27

SPIKED AREA (sample + standard)

CONCENTRATIC	PA1	PA2	PA3	MEAN	SD	CV%
200	4470.2	4498.0	4368.0	4445.4	68.5	1.54
600	6498.9	6688.9	6667.9	6618.6	104.2	1.57
1000	8638.4	8878.4	8982.4	8833.1	176.4	2.00

RECOVERY

CONCENTRATIC	PA1	PA2	PA3	MEAN	SD	CV%
200	92.32	96.11	87.22	91.88	4.46	4.85
600	91.77	96.57	97.85	95.40	3.21	3.36
1000	91.01	89.99	90.19	90.40	0.54	0.60

Table Recovery studies of Naringin by the proposed TLC densitometric method (n=3)

Concentration (µg/ml)		Height		Total height	
Sample	Spiked amount	Sample height	Spiked height	(sample + standard)	% Recovery
600	200	2855.533333	1730.7	4445.4	91.88
600	600	3162.7	3945.2	6618.6	95.40
600	1000	4979.3	86726.7	8833.1	90.40
Average Recovery					92.56

Figure 3: Hptlc Validation of Partially Purified Labisia Pumila Extract

Amount (µg/ml)	Intraday Precision		Interday Precision	
	SD in height	%CV	SD in height	%CV
200	13.23	0.81	23.48	1.85
600	42.58	1.11	26.56	1.08
1000	61.29	1.03	64.29	1.56

Table 2: Intra and Inter day precision of naringin (n=3).

Concentration (µg/ml)		Sample height		Total height (sample + standard)	% Recovery
Sample	Spiked amount	Sample height	Spiked height		
600	200	2855.5	1730.7	4445.4	91.88
600	600	3162.7	3945.2	6618.6	95.40
600	1000	4979.3	86726.7	8833.1	90.40
Average					92.56

Table 3: Recovery studies of naringin by the proposed TLC densitometric method (n=3).

Precision and accuracy

Precision of comparison (intra-day and inter-day) were determined under different conditions, different day, different reagents, on the same sample. Intra and Intra-day was carried out using the standard on three different days where results were expressed in CV %. In terms of repeatability of the measurement peak area this method obtained a coefficient of variation not more than 2% (Table 2). The percent recovery was found to be 92.56%. The results are shown in Table 3.

Conclusion

A Quantitative HPTLC method for estimating the amount of Naringin in Fraction E from the partially purified DCM leaf extract from *Labisia Pumila* has been described in this paper. This method was able to obtain precise and accurate results. The data could be used as a quality control technique for the evaluation of Naringin in Fraction E. The method gave good peak resolution in the analysis of bioactive constituents present in the sample. Linearity gave an R_2 value of more than 0.9. Precision and recovery was able to give CV% less than 2%. The proposed HPTLC method for the analysis of fraction E from the partially purified leaf DCM extract of *Labisia Pumila* reported here is simple, sensitive, economic and suitable for rapid routine quality control analysis and quantification of Naringin in herbal drug preparation and may be useful for standardization purposes.

Acknowledgement

The authors are grateful to FRGS (MoHE), Reference code: 2/2013/SGO5/UCSI/02/1 and the Faculty of Applied Sciences, UCSI University for providing the facility for this project.

References

1. Abdullah N, Chermahini SH, Suan CL, Sarmidi MR (2013) *Labisia pumila*: A review on its traditional, phytochemical and biological uses. World Appl Sci J 27:1297- 1306.
2. Al-Wahaibi A, Wan Nazaimoon WM, Norsyam WN, Farihah HS, Azian AL (2008) Effect of water extract of *Labisia pumila* Var Alata on aorta of ovariectomized Sprague Dawley rats. Pakistan J Nutr 7:208- 213.
3. Nadia ME, Nazrun AS, Norazlina M, Isa NM, Norliza M, et al. (2012) The Anti-Inflammatory, Phytoestrogenic, and Antioxidative Role of *Labisia pumila* in Prevention of Postmenopausal Osteoporosis. Adv Pharmacol Sci 2012: 706905.
4. Okechukwu NP, Marunga J (2011) Gastro-protective effect of dichloromethane crude extract of *labisia pumila*. Int J Pharmacol Toxicol Sci 2:1- 7.
5. Choi HK, Kim DH, Kim JW, Ngadiran S, Sarmidi MR, et al. (2010) *Labisia pumila* extract protects skin cells from photoaging caused by UVB irradiation. See comment in PubMed Commons below J Biosci Bioeng 109: 291-296.
6. Sanusi RAM, Shukor NAA, Sulaiman MR (2013) Anti-inflammatory Effects of *Labisia pumila* (Blume) F . Vill-Naves . Aqueous Extract. Sains Malaysiana 42:1511- 1516.
7. Samuagam L, Akowuah GA, Okechukwu PN (2011) Partial purification and antinociceptive investigation of leaves of *Labisia Pumila*. Asian J Pharm Clin Res 4: 44- 46.
8. Okechukwu PN, Ekeuku SO, Loshnie S, Akowuah GA (2014) Anti-inflammatory, Analgesic, Antinociceptive and Antipyretic Investigation of Bioactive Constituents from Partial Purified Dichloromethane Crude Extracts from Leaves of *Labisia Pumila*. Int J Pharm Res Sch 3: 743- 751.
9. Okechukwu NP, Lee JA, Akowuah AG (2011) *In-vitro* antihistaminergic, anticholinergic, antioxidant and mast cell stabilizing effects of partially purified *Labisia pumila* leaf extract. FASEB J 25: 1020- 1027.
10. Karimi E, Jaafar HZ (2011) HPLC and GC-MS determination of bioactive compounds in microwave obtained extracts of three varieties of *Labisia pumila* Benth. Molecules 16: 6791-6805.
11. Chua LS, Latiff NA, Lee SY, Lee CT, Sarmidi MR, et al. (2011) Flavonoids and phenolic acids from *Labisia pumila* (Kacip Fatimah). Food Chem 127: 1186-1192.
12. Das R, Dutta A, Bhattacharjee C (2013) Assessment on on the Antibacterial Potential of Phytochemical Naringin-An *in vitro* Evaluation. Int J Emerg Technol Adv Eng 3: 353- 357.
13. Thangavel P, Muthu R, Vaiyapuri M (2012) Antioxidant potential of naringin - a dietary flavonoid-in N-Nitrosodiethylamine induced rat liver carcinogenesis. Biomed Prev Nutr 2:193- 202.
14. Jeon SM, Bok SH, Jang MK, Kim YH, Nam KT, et al. (2002) Comparison of antioxidant effects of naringin and probucol in cholesterol-fed rabbits. Clin Chim Acta 317: 181- 190.
15. Ciesla L, Zaluski D, Smolarz HD, Hajnos M, Waksmundzka-hajnos M (2011) HPTLC-Densitometric Method for Determination of Eleutherosides B , E , and E 1 in Different Eleutherococcus Species. J Chromatogr Sci 49:182- 188.
16. Q2A I (1994) Text on validation of analytical procedures. In: Proceedings of the International Conference on Harmonization of Technical Requirements for Registration of Pharmaceuticals for Human Use. Japan.
17. Kumar P, Dwivedi SC, Kushnoor A (2011) Development and Validation of HPTLC Method for the Determination of Nelfinavir as Bulk Drug and in Tablet Dosage Form. 2:1783- 1788.
18. Maheshwari S, Khandhar AP, Jain A (2010) Quantitative Determination and Validation of Ivabradine HCL by Stability Indicating RP-HPLC Method and Spectrophotometric Method in Solid Dosage Form. EJAC 5:53- 62.
19. Schoenbart B (1999) Comparative Chromatographic Analysis of Dan Shen (Salvia miltiorrhiza) Roots Grown in Different Regions of the World. USA 1-14.
20. Dhandhukia PC, Thakker JN (2011) Quantitative Analysis and Validation of Method Using HPTLC. High-Performance Thin-Layer Chromatogr pp 203- 221.

Reliable and Sensitive SPE-HPLC-DAD Screening of Endocrine Disruptors Atrazine, Simazine and their Major Multiresidues in Natural Surface Waters: Analytical Validation and Robustness Study Perfomance

Juan José Berzas Nevado, Carmen Guiberteau-Cabanillas, María Jesus Villasenor Llerena* and Virginia Rodríguez-Robledo

Department of Analytical Chemistry and Food Technology, University of Castilla-La Mancha, 13071 Ciudad Real, Spain

Abstract

An analytical method combining off-line solid phase extraction (SPE) and reversed phase high performance liquid chromatography (HPLC) has been developed for simultaneous determination of six herbicides. The compounds analyzed in natural surface water samples, were: the herbicides atrazine (AT), simazine (SI) and their major dealkylated and hydroxylated degradation products such as deisopropyl-atrazine (DIA), deethyl-atrazine (DEA), hydroxyl-atrazine (H-AT) and hydroxyl-simazine (H-SI), all of them with herbicide activity too. As optimum conditions, a Kromasil C18 column as stationary phase, a mixture of acetonitrile and phosphate buffer pH 7.0 as mobile phase in gradient elution mode and 215 nm as measuring wavelength, were selected.

A polymeric cartridge (Bond-Elut ENV-1000 mg) was used for a SPE procedure. The optimized off-line SPE HPLC-DAD method was evaluated in terms of validation and robustness procedure, obtaining among other parameters, limits of quantification between 81 and 100 ng L^{-1}, considering the enrichment factor for all of the studied herbicides. The proposed method was successfully applied to the screening of herbicides in natural surface water samples from different locations in the 23th aquifer (Albacete, Spain).

Keywords: S-Triazines; Degradation products; Solid-phase extraction; High performance liquid chromatography; Natural surface waters from aquifer

Introduction

Pesticides such as s-triazines are widely used as selective pre and post-emergence herbicides for the control of broadleaved, grassy areas, in many agricultural crops like corn, wheat, as well as in railways, roadside verges and golf courses. After application of triazines, several degradation processes take place leading to dealkylation of amine groups (positions 4 and 6 of the triazinic structure), or hydrolysation of the substituent in position 2 according to the chemical structures shown in Table 1.

Atrazine (AT), (2chloro-N4-ethyl-N6-isopropyl-1,3,5-triazine-4,6-diamine), Simazine (SI), (2chloro-N4,N6-diethyl-1,3,5-triazine-4,6-diamine) and their main dealkylated and hydroxylated degradation products, Deethyl-atrazine (DEA), (dealkylated degradation product obtained by losing of ethyl group in N4 amine function of Atrazine structure) Deisopropyl-atrazine (DIA), (dealkylated degradation product obtained by losing of isopropyl group in N6 amine function of Atrazine structure) hydroxyl-atrazine (H-AT) (degradation product obtained by replacing the Cl substituent at position 2 by hydroxyl group in Atrazine structure) and hydroxyl-simazine (H-SI), (degradation product obtained by replacing the Cl substituent at position 2 by hydroxyl group in Simazine structure) are considered inside to the group of endocrine-disrupting chemicals [1], according to U.S. Environmental Protection Agency (EPA). The EPA defines an environmental endocrine or hormone disrupting chemical as an exogenous agent, which interferes with the synthesis, secretion, transport, binding, action or elimination of natural hormones in the body, and that are responsible for the maintenance of homeostasis, reproduction, development and/or behavior [2].

Both atrazine, simazine and their degradation products, are relatively persistent making necessary their analysis and quantification in environmental samples.

The continuous monitoring of pesticides and degradation products in natural waters, is a great importance matter to control the toxicity of the environment. Nowadays, water quality receives considerable attention for its obvious implication in public health, so, stringent regulations have been issued by legislation agencies. In this sense, current European Union (EU) directives [3], dictates that concentrations of individual pesticides must not exceed maximal admissible values of 0.1 μg L^{-1} in drinking water, whereas in natural surface waters, the alert and alarm threshold values are typically 1 and 3 μg L^{-1}, respectively.

In the literature, most of the reported methods for quality determination and quantification of s-triazines in water samples, involves separation by high performance liquid chromatography (HPLC) or gas chromatography (GC) after liquid-liquid (LLE) [4,5], solid phase extraction (SPE) [6,7], or either more currently using liquid-liquid microextraction or solid-phase microextraction [8,9]; being well-known the advantages of SPE technique over LLE. Therefore, SPE procedures using C18-silica, polymeric sorbents, (i.e. nonpolar polystyrene-divinylbenzene), or other similar matrixes [10], for the analysis of both organic contaminants in water samples [11], and determination of triazines in environmental samples [12-14], have been previously reported. Other kinds of materials as multiwalled

***Corresponding author:** M Jesus Villasenor Llerena, Departamento de Química Analíticay Tecnología de Alimentos, Universidad de Castilla-La Mancha, Spain
E-mail: mjesus.villasenor@uclm.es

Pesticides	Chemical Structures	pK_a
Atrazine (AT)	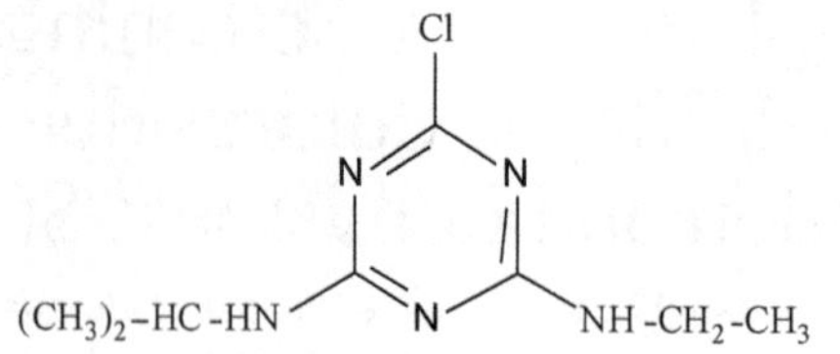 Cl; N N; $(CH_3)_2$-HC-HN N NH-CH_2-CH_3	1.68-1.85
Simazine (SI)	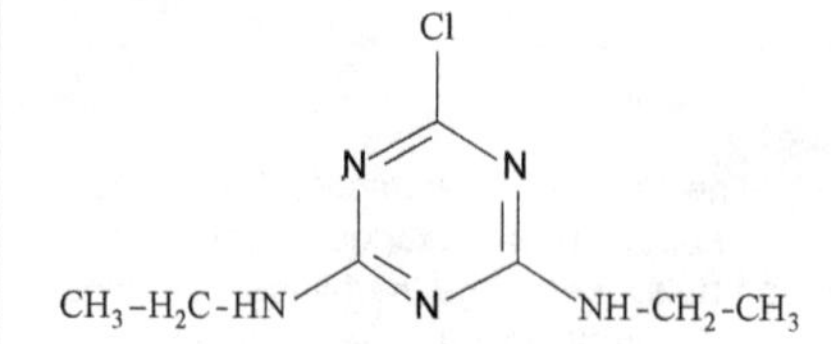 Cl; N N; CH_3-H_2C-HN N NH-CH_2-CH_3	1.65-1.80
Deethyl-atrazine (DEA)	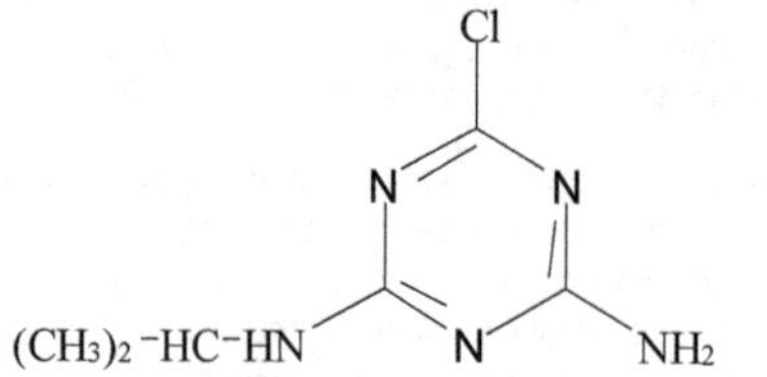 Cl; N N; $(CH_3)_2$-HC-HN N NH_2	1.65
Deisopropyl-atrazine (DIA)	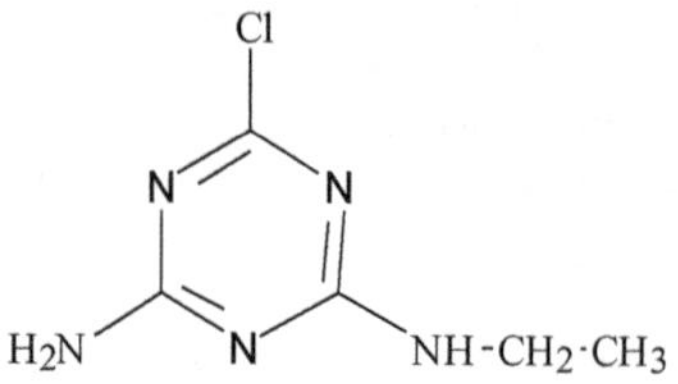 Cl; N N; H_2N N NH-CH_2-CH_3	1.58
Hydroxyl-atrazine (H-AT)	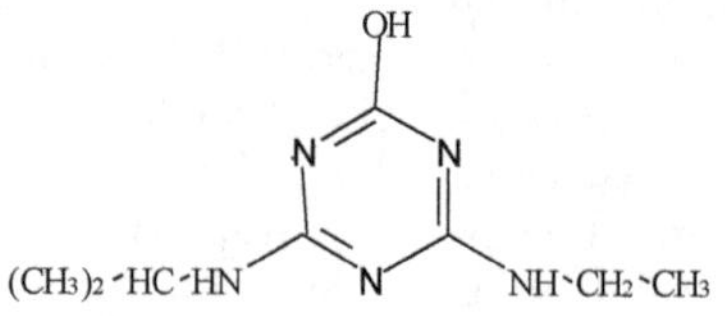 OH; N N; $(CH_3)_2$-HC-HN N NH-CH_2-CH_3	5.2
Hydroxyl-simazine (H-SI)	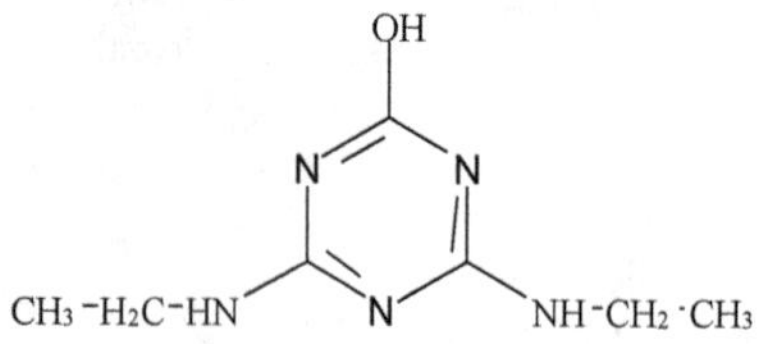 OH; N N; CH_3-H_2C-HN N NH-CH_2-CH_3	5.2

Table 1: Chemical structures and pKa values of studies analytes.

carbon nanotubes (MWCN) have also been successfully used with exceptional merit as SPE sorbents for enrichment of environmental pollutants [15]. Currently, Al-Degs et al. [16], developed a sensitive multiresidue procedure for the analysis of four common pesticides, where two of them were SI and AT, using off-line SPE-HPLC-UV and MWCN as adsorbent for the preconcentration of these herbicides, in water samples.

However, few papers have been published related to the screening of the main degradation products, which most of them have got herbicide activity too [17,18].

The parent s-triazines SI and AT and with some restriction the dealkylated degradation products too, have been analyzed by GC using a mass spectrometry detector (MSD), or a nitrogen-phosphorus detector (NPD) [19-22]. In this way, Berzas et al. [23], separated and determinated nine common herbicides as SI, AT and their dealkylated degradation products DEA, DIA among other related herbicides compounds, using a very sensitive capillary gas chromatographic mass spectrometric-selective ion monitored (GC-MS-SIM) as analytical technique. Although it is well-known, GC-MS still remains as the most powerful technique to separate and quantify the above-cited herbicides, but if among the analytes are included their hydroxyl-degradation products, they cannot be easily analyzed directly by GC, because of their high polarity and non-volatilily. Then, HPLC and capillary electrophoresis (CE) are other suitable alternative techniques, for the analysis of both s-triazines and their hydroxylated degradation products. In fact, a recent work by Carabias et al. [24] makes a comparison between non-aqueous capillary electrophoresis and HPLC methods, for the determination of several chloro-s-triazines, thiomethyl-s-triazines and three main degradation products of parent compound AT. This research concluded that both methods afford similar results for the analysis of surface and drinking water samples.

HPLC techniques present as great advantage to be directly

applicable to both s-triazines and their OH degradation products, reporting in the literature couplings with different detection modes like UV [25], MS [26-28] or even MS-MS tandem [7,29]. On a similar way, a currently paper developed by Dos Santos et al. [30], describes a sequential injection chromatographic procedure for separation and quantification of the herbicides SI, AT and propazine, exploring also the low backpressure of a 2.5 cm long monolithic C18 column. Most of these methods are mainly focused on parent triazines, even in their dealkylated residues [28], but not in the hidroxylated ones until now.

Therefore, the development of analytical methods for screening the presence and amounts of herbicides and their residues in natural waters, is an obvious interesting matter to prevent toxicological risks.

In the present work, an easy and fast off-line SPE HPLC-DAD procedure has been proposed for the isolation, separation and quantification of parents herbicides SI, AT and their main dealkylated (DEA, DIA) and hydroxylated (H-AT, H-SI) degradation products, commonly found in natural water samples.

An extensive validation study, which includes an exhaustive design robustness procedure as novelty, has been also carried out to assess the reliability of the obtained results. Finally, the whole method was successfully applied to the screening of SI, AT and their main dealkylated and hydroxylated degradation products, in natural surface water samples from different soundings at 23rd aquifer (Albacete, Spain).

This final analysis using natural water samples proved the usefulness of our proposed method, using a simple technique with optical detection to analyze the targeted compounds at relevant levels. This is especially significant for hydroxylated degradation products, very difficult to analyze by other methods without a previous derivatization, but nevertheless, very common in these surface natural waters.

As another value of the method, the LOQs reached for studied herbicides (between 81 and 100 ng L^{-1}), are from ten to thirty fold lower than the maximal allowed amounts for natural surface waters(1 and 3 µg L^{-1} as alert and alarm levels), according to European Union directives [3].

The added value of our method remains in its proved reliability, in opposite to others types of sorbents because of its solubility problems (MWCN), or the use of backpressure monolithic columns due to the difficult separated elution, which involves separation, quantification and detection problems. Besides, these recent approaches have been just used until now for the analysis of the parent compounds, but not for their degradation products yet.

Experimental

Chemicals

All solvents and reagents were of analytical grade unless indicated otherwise. Acetonitrile (HPLC-grade) was obtained from Scharlau S.L. (Barcelona, Spain) and methanol (HPLC-grade) from Panreac (Barcelona, Spain). Buffer solutions were prepared using the following reagents: sodium dihydrogenphosphate, disodium hydrogenphosphate and phosphoric acid (HPLC-grade). All of these compounds were obtained from Merck (Darmstadt, Germany).

Simazine (SI) and atrazine (AT) were obtained from Riedel-de Haën (Seelze, Germany), whereas hydroxyl-atrazine (H-AT) and hydroxyl-simazine (H-SI), deisopropyl-atrazine (DIA) and deethyl-atrazine (DEA), were purchased from Dr. Ehrenstorfer (Augsburg, Germany).

Apparatus

The HPLC system utilized was a Shimadzu (Kyoto, Japan) model LC-10AD with a solvent delivery pump, a Rheodyne injection valve (20 µL sample loop) and a SPD-M10A photodiode-array detector. The system was controlled by a computer equipped with the CLASS-LC 10 software, which was used for all of the measurements and treatment of data.

Separation was achieved at room temperature using a reversed-phase Kromasil C18 column (4.6 i.d. ×150 mm; particle size 5 µm), purchased from Waters (Milford, MA, USA). pH values were measured by using a Crison model 2001 pHmeter provided with a combined glass electrode.

Phosphate buffer solutions and water samples were filtered through a 0.45 µm MFTM-membrane filters, and acetonitrile was filtered through a 0.5 µm Fluoropore TM membrane filters. Both membrane filters were purchased from Millipore (Milford, MA, USA).

The extraction and pre-concentration processes were carried out using a device developed in our laboratories consisting of a water manifold, Visiprep TM Sep-Pack system, from Supelco (Madrid, Spain), coupled to a Vacuum pump XF 54 23050, from Millipore (Milford, MA, USA). The SPE procedure was performed using Bond Elut-ENV SPE Sep-Pack Plus cartridges, with styrene divinylbenzene co-polymer sorbent (1000 mg), from Water (Milford, MA, USA). Methanolic extracts were filtered with 0.45 µm×13 mm membrane filter for clarification of aqueous and mild organic solution (Millex-HV, Millipore).

A Centrifugal evaporator (JOUAN RC10.09), coupled to a refrigerated RCT90 Aspirator from Shimadzu (Madrid, Spain), was used to evaporate the extracts of the water samples.

Operating conditions

Standard and water samples: Stock standard solutions (200 mg L-1) were prepared by dissolving pure solid samples in acetonitrile:water (50:50 v/v) for SI and AT, and in methanol:water (50:50 v/v) for DIA, DEA, H-SI and H-AT. Three drops of formic acid were previously added to hydroxyl-triazines to get their complete dissolution.

Working standard solutions were daily prepared by dilution of suitable aliquots from the stock solutions in methanol. All solutions were stored at 4°C in the dark.

Standard water samples were prepared by spiking different amounts (from 0.1 µg L^{-1}) of every herbicide in tap or Milli-Q water samples.

Natural surface water samples were obtained from the 23th aquifer (Albacete, Spain), where they were collected in plastic containers, and directly transferred into 1 L glass topaz bottles. After their collection, these water samples were immediately filtered at laboratory using a 0.45 µm filter to remove the suspended matter, and finally they were kept in the dark at 4°C until their later analysis.

Both tap and natural surface water samples were previously adjusted to 7.0 pH value, before their analysis.

Off-line solid phase extraction and HPLC procedure:

SPE procedure: The extraction and isolation of triazinic herbicides and their degradation products from the spiked tap water and from natural surface water samples, were performed by a SPE process using the already cited 6-mL Bond Elut-ENV SPE-cartridges. Owing to the high level of complexity of the natural surface water matrix, where the analytes under investigation were present at very low levels, the SPE process was used for both pre-concentration and cleaning up stage.

Every step of this extraction procedure was carefully studied and the results were as follows: the cartridge was previously conditioned with 5 mL methanol and 10 mL phosphate buffer 10 mM, pH 7. After that, volumes of water samples (1000 mL) were passed through stationary phase (flow rate <20 mL min^{-1}), then the cartridge was washed with 10 mL phosphate buffer (10 mM, pH 7), and finally the analytes were eluted with methanol (5 mL, three times) into conical glass vessels.

To reach a high enrichment factor in the extraction-preconcentration procedure, the eluted volume (15 mL) was evaporated at 60°C until 200 μL as final volume. Besides, it was necessary to add 100 μL of acetonitrile to the final extract, since it provided a better peak shape and lower baseline noise, reaching so an enrichment factor of 3333. Those final extracts were filtered through a 0.45 μm×13 mm (Millex-HV) filter and immediately injected into the HPLC-DAD equipment.

HPLC-DAD procedure: A Khromasil C18 column as reversed stationary phase for liquid chromatography was selected for the development of analytical separation, whereas a binary gradient elution program combining phosphate buffer 5 mM, pH 7.0 and acetonitrile was chosen as mobile phase.

An elution gradient using a constant flow rate of 1 mL min^{-1} and an injection volume of 20 μL was used during the analytical separation.

The gradient program profile selected was as follows

Step	Acetonitrile (%)	Time (min)
1	20	1
2	20 to 40	1
3	40	10
4	40 to 60	1
5	60	1
6	60 to 20	1
7	20	5*

*Getting the initial conditions, reaching so the separation of the six analytes in a running time lower than 14 min, as it can be seen in Figure 1.

A wavelength of 215 nm was selected as optimal for detection and quantification of the six herbicides. Nevertheless, on the validation procedure, detection wavelengths were selected at the maximum of spectra for every herbicide, with the purpose to increase the selectivity and decreasing the baseline noise.

Results

Preliminary studies about SPE of s-triazines from natural water samples

As it has been mentioned in the introduction, several works using SPE as isolation procedure and different adsorbents, such as C18 or polymeric ones [10-14], have been published. Currently, innovative material has been successful used with exceptional merit as SPE adsorbents for enrichment of environmental pollutants in water samples [15,16].

In our work, to achieve as much as possible the isolation of s-triazines compounds and their major degradation products from natural water samples, different commercial available cartridges and extraction disks were tested.

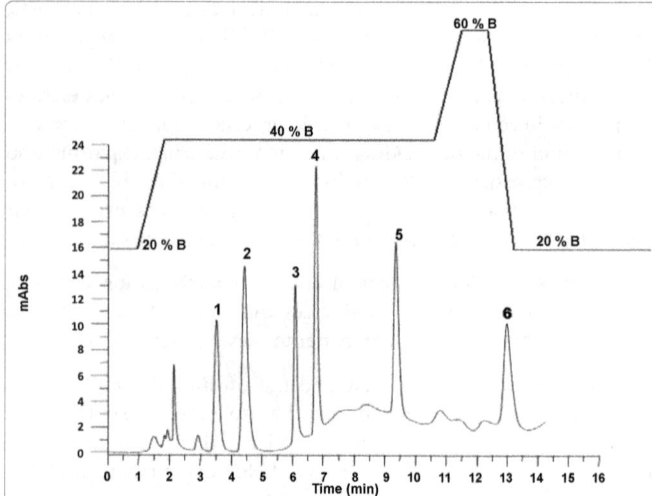

Figure 1: HPLC-DAD separation for a standard mixture (10 mg L^{-1}) of hydroxyl-simazine (1), deisopropyl-atrazine (2), hydroxyl-atrazine (3), deethyl-atrazine (4), simazine (5) and atrazine (6), obtained in the selected chromatographic conditions, and with the gradient profile also shown.

Several kinds of cartridges with different polarity, amounts of stationary phases, etc. were compared in order to select the most suitable one to achieve the SPE process. All this information is available as "Supplementary Material".

Once selected the most suitable cartridge (polarity degree and amount of stationary phase), every step of the SPE procedure, such as the appropriate organic and aqueous solvents, volume for washing stages, samples volume, elution volume and final volume of extracts, was evaluated to achieve a complete extraction. The best recoveries were reached with the optimized procedure already described in Experimental section.

High performance liquid chromatography procedure

The influence of chromatographic parameters like both nature of buffer, pH, ionic strength of mobile phase, percentage of organic modifier, and mobile phase flow rate, was studied on retention times (RT) and resolution between peaks (RS). The selection of these chromatographic parameters was performed attending to reach the best sensitivity and efficiency (N), a good RS between peaks and shorter analysis times.

Preliminary studies were carried out in isocratic mode using different percentages of acetonitrile or methanol in the mobile phase. Acetonitrile was chosen as the organic modifier because it provided higher sensitivity and better peak shape.

Due to the ionization groups of studied triazines (whose chemical structures are shown in Table 1), it was necessary to use a buffer solution as aqueous mobile phase to control the pH. Under knowledge of pKa values of compounds (Table 1), a pH range was studied between 3.0 and 7.0 using phosphate buffers. A 7.0 pH value was selected as optimum because it provided non-ionic forms for the analytes, getting better interaction with the stationary phase, increasing so the RTs, and improving thus the selectivity for the herbicides.

An ionic strength of 5 mM phosphate buffer and a flow rate of 1.0 mL min^{-1}, were chosen as a compromise solution between a good RS (>1.5), higher N, good peak shape and lower analysis times.

On the other hand, a gradient elution program was required to achieve a suitable separation since isocratic mode exhibited too long RTs for SI and AT, as well as RS values lower than 1.5 between H-AT and DEA peaks. The best separation was carried out with the gradient profile already reported in the Experimental section 2.3.2.

A diode array detector was selected taking into account the good spectrophotometric absorption properties (and therefore very good sensitivity) of these aromatic compounds. A wavelength of 215 nm was chosen as optimal one for detection and quantification of the six herbicides.

Figure 1 shows a chromatogram for a standard mixture of the six analytes (10 mg L^{-1}), obtained under the already selected chromatographic conditions. The selected gradient profile is also shown in this figure, which provides a very good resolution between peaks in a run time about 14 min.

Validation off-line SPE HPLC-UV procedure

To assess the reliability of the whole off-line SPE HPLC-DAD procedure, its analytical performance characteristics such as stability, precision, linearity and limits of quantitation among others, were evaluated on extracts obtained from spiked tap water samples. The use of spiked tap water as matrix validation was due to two reasons. The first one was because of tap water matrix was supposed to be free of targeted pesticides, and therefore, we considered it as the most appropriate one to performed precision and accuracy in a reliable way. The other reason was due to the high variability of the different surface water samples from aquifer, and obviously, the inherent difficulty to found any suitable and representative surface water sample of this variety, to carry out a valuable validation study on it. The extracts were obtained by the SPE procedure previously described in Section 2.3.2, which provided an enrichment factor of 3333 for every compound.

Figure 2a presents the off-line SPE-HPLC-DAD separation, corresponding to the extract obtained from a spiked tap water sample (0.30 µg L^{-1} for every analyte).

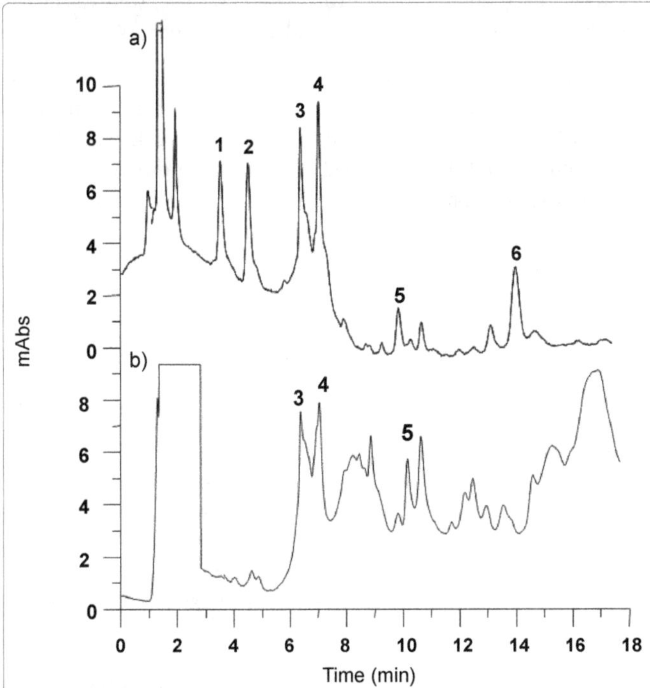

Figure 2: a) Off-line SPE HPLC-DAD separation corresponding to an extract from a spiked tap water (0.30 µg^{-1} L for every herbicide). **b)** Off-line SPE HPLC-DAD screening from natural surface water sample corresponding to sounding 5, where SI, H-AT and DEA were found.

Discussion

Stability

Although this aspect of the study is often considered to be related to the robustness of the procedure, this factor should be examined at the beginning of the validation process, because it determines the validity of the obtained data in the subsequent tests.

Stability of stock standard solutions was determined by comparing the response factors (concentration/average peak areas) of solutions stored in darkness at 5°C, with those ones of freshly prepared solutions, for triplicate injections each one. All of the studied herbicides were found stable at least for two months (with differences in response factors lower than 0.2%), whereas extracts of spiked tap water were just stable for ten hours.

Precision

The precision of the proposed method has been expressed in terms of Relative Standard Deviation (RSD). Repeatability and intermediate precision were evaluated in accordance with ICH (International Conference on Harmonisation) criteria [31].

To check the repeatability of the chromatographic proposed method, twelve extracts from spiked tap water samples containing 3 µg L^{-1} of every endocrine-disrupting compound were injected and analyzed. The obtained RSD values were below 3% for peak areas, about 1% for RS and lower than 1% for RTs.

The intermediate precision of the overall off-line SPE HPLC-DAD procedure was evaluated by submitting six spiked tap water samples (3.5 µg L^{-1} for each compound) to this overall extraction-liquid chromatographic process in triplicate injections in two different days. The Snedecor F-test was used to compare both series of data from the two different days [32]. Significant differences were not observed at a confidence level of 95% and 5 degrees of freedom, since the reached experimental F values were lower than their respective theoretical ones, as it can be seen in Table 2.

LODs and LOQs

Limits of detection and quantification (LODs and LOQs) were estimated upon IUPAC criteria, by evaluating the baseline noise on six blank tap water extracts using the maximal sensitivity allowed by the system, over a period ten times the peak width, around the specific retention times for every herbicide. They were found values between 90 and 100 µg L^{-1} for LODs and from 300 to 350 µg L^{-1} for LOQs. These LOQs values were subsequently validated by the analysis of five different blank tap water extracts, spiked with amounts of each analyte corresponding to their respective LOQs. The relative errors obtained in the verification of these LOQs were lower than 15% in all of the cases.

It is important to emphasize that according to the preconcentration factor reached in the SPE step, the LODs and LOQs allowed in natural water samples could be 3333 times lower than the HPLC-DAD calculated values (Table 2). Consequently, as it be can see in Table 2, the LOQs reached for studied herbicides (between 81 and 100 ng L^{-1}), are from ten to thirty fold lower than the maximal allowed amounts for natural surface waters (1 and 3 µg L-1 as alert and alarm levels), according to European Union directives [3].

Pesticide	Intermediate precision[a]		LOQ[b] (ng L-1)	Equation[c]	Standard Deviation (SD) Slope Intercept		σ^2 [d]	Tap Water[e]		Deionised Water[e]	
	F_{exp} for Area	F_{exp} for RS						Recoveries		Recoveries	
								%	SD[f]	%	SD[f]
H-SI	1.18	2.92	96	y=120205 CHSI-23728	14003	18197	0.9985	71.50	3.20	69.19	2.06
DIA	1.39	4.46	88	y=143618 CDIA-37688	36842	14189	0.9929	86.32	5.63	89.24	2.65
H-AT	1.33	3.12	90	y=108686 CHAT-9760	18657	3569	0.9968	77.40	1.13	71.62	0.76
DEA	1.71	3.20	81	y=147826 CDEA-31214	26269	2898	0.9965	91.67	4.22	94.29	2.99
SI	1.05	1.47	95	y=70148 CSI-16736	13935	8061	0.9957	74.47	4.15	78.00	3.72
AT	1.28	4.93	100	y=112109 CAT-5123	25808	1970	0.9943	71.20	3.82	74.14	0.56

[a]F(5, 5) theorical=5.05

[b]LOQ including enrichment factor (3333)

[c]Peak area (y, no units) versus concentrations (μg L-1).

[d]Regression coefficients

[e]Mean of recoveries in spiked tap and deionized water samples at three different levels in triplicate.

[f]Mean of standard deviations (SD) of recoveries in spiked tap and deionized water samples at three different levels in triplicate

Table 2: Analytical performance characteristics for the developed off- line SPE- HPLC-DAD method.

Factor	Units	Limits	Level (-1)	Level (+1)	Nominal (0)
A. pH phosphate buffer		± 0.5	6.5	7.5	7
B. Buffer concentration	mM	± 2	3	7	5
C. Flow rate	mL min-1	± 0.2	1.2	0.8	1.00
D. Initial % solvent B (step1)	%	± 2	18	22	20
E. First gradient program time (step 2)	Min	± 0.2	0.8	1.2	1
F. % solvent B (step 3)	%	± 2	38	42	40
G. Second held time (step 3)	Min	± 1	9	11	10

Level 0: optimal values
Level +1: upper values
Level -1: lower values

Table 3: Variables selected as factors and values chosen as levels.

Linearity

The linear behaviour was determined from triplicate injections of spiked tap water samples, previously submitted to the SPE treatment, at six different concentration levels between 0.1 and 1.0 μg L-1 for every compound. The satisfactory linear regression equations, their standard deviations for slope and intercept and their regression coefficients are also summarized in Table 2. The obtained results can indicate that SI, AT, DIA, DEA, H-SI and H-AT responses are linear over the studied concentration range.

However, according to the Analytical Methods Committee (AMC) [33], a regression coefficient close to unity is not necessarily the outcome of a linear relationship, and as consequence, the lack of fit test was carried out by plotting the residuals (distances of the experimental points from the fitted regression lines) against concentration for every analyte. If, there is not lack of fit, the plot will look like a random sample from a normal distribution with zero-mean that is the calibration results inherently linear. This one was the observed situation on applying this test to our calibration graphs, and therefore the linearity of these relationships was thus confirmed.

Accuracy

To test the accuracy of the proposed procedure, three tap and three deionized water samples spiked with the six herbicides in variable amounts between 0.1-2 μg L-1 (in triplicate), were submitted to the off-line SPE HPLC-UV procedure. The results (average recoveries), summarized in Table 2, were between 85-95% for DIA and DEA and between 70-80% for the other herbicides. In this table, it can also be noted that recoveries are quite similar for deionised and tap water samples. Therefore, we can say that the proposed SPE procedure reach to remove the presence of salts, suspension and dissolving matter in these tap water samples.

Robustness

Robustness evaluates the influence of small changes on the internal factors of an analytical procedure, providing an indication of its reliability during normal usage [34]. The purpose of a robustness test is to identify possible sources of errors, when changes occur in the specified method conditions [35-38]. In this paper we have tested the influence of small variations on internal parameters of the method, whose influence has been studied at three different levels (Table 3).

The Plackett-Burman fractional factorial model, based on balanced incomplete blocks has been employed for this evaluation, using a design for seven factors and fifteen experiments (N=15) [39].

The mean effect of each variable is the average difference between observations made at the extreme levels and those made at the optimal one. Mean effects and standard errors (DA, DB, DC...) were calculated using the procedures described by Youden and Steiner [34].

Results of the levels variation effects were checked on the most critical chromatographic responses of the method: resolution between H-SI and DIA, between H-AT and DEA, and AT peak area efficacy. Taking into account these obtained deviations, the proposed method has proved to be robust for all of the variations tested in this study, since in all of the cases for the operating factors (A-G), on every studied chromatographic response, the following rule was observed:

being

$$S=\sqrt{\frac{2}{7}\left(DA^2+DB^2+DC^2+...+DG^2\right)}$$

Figure 3 shows the effects of negative (-1) and positive (+1) levels for the seven selected factors on RS between H-AT and DEA, where as it can be seen the most critical factors resulted to be the % acetonitrile at positive level in the four step, and also the influence of buffer concentration at negative level. Anyway, the proposed analytical method has proved to be enough robust, when the influence of these seven factors was evaluated on this critical chromatographic response.

Peak homogeneity

It is useful to investigate the purity of separated peaks because co-migration of peaks is also possible in HPLC, as in any other separation technique. Several techniques for validating peak purity have been proposed in the literature [40,41].

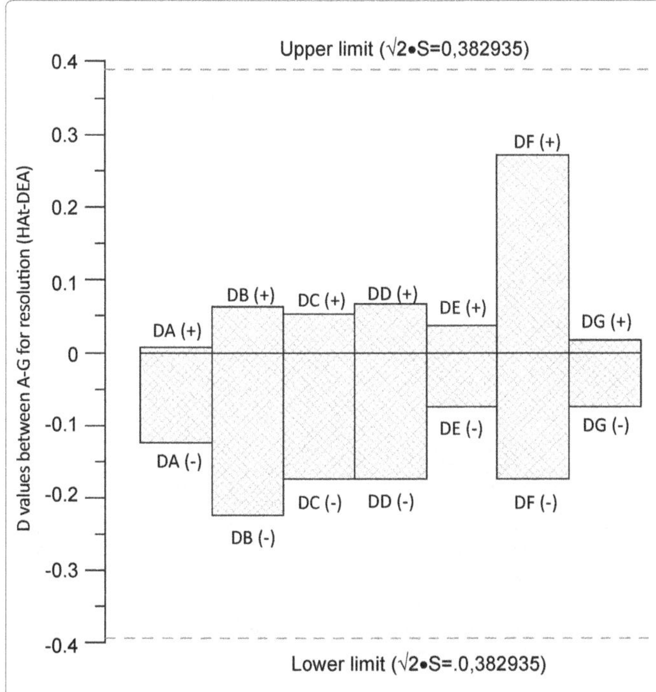

Figure 3: Variation effects of the levels (-1) and (+1) for the seven selected operating factors on RS between H-AT and DEA.

In our work two techniques have been used to validate the peak purity of the compounds under study.

One of them, called "normalization and comparison of spectra at different peak sections", where the absorption spectra acquired at the apex, at the ascending and at the descending parts of the corresponding peak for every herbicide present in each sample, were compared.

The second one, named "absorbance at two wavelengths" where a compound can be define as "pure", when the molar absorptivity (A) at a wavelength λ_1 is directly proportional to that one at any other wavelength λ_2, that means, A (λ_1)=KA (λ_2); where K characterizes the pure compound at the selected wavelength. The absorbance ratio for a pure compound is constant. A plot of this ratio against RTs will give a square wave function of K amplitude value. Any co-eluting impurity peak leads to a deviation from the flat ratio plot. The wavelengths have to be selected at the highest absorbance difference. An obtained square wave function demonstrates the purity of the chromatographic peaks. In all of our applications analyzed by the proposed method, similar results to describe above were obtained. Both methods were satisfactorily applied to check the lack of interferences from the matrix in natural surface water samples from aquifer, on the peaks assigned as focused compounds.

Application to natural water samples

Natural surface water samples from different soundings at several locations in the 23th aquifer (Albacete, Spain), have been analyzed using the proposed off-line SPE HPLC-DAD procedure. In Table 4 has been summarized the results obtained for quantitative analyses.

As it can be observed, AT has been detected in two samples, whereas SI was just quantitated in one sounding corresponding to the fifth one. The studied degradation products H-SI, H-AT, DIA and DEA were found between concentration values of 0.10 and 0.31 μg L^{-1} in the different soundings (Table 4). H-AT was present in all of the analyzed samples, whereas DIA was found in just two of them. These results support the presence of main degradation products of parents SI and AT, proving so the great importance to include them as aim of study for the analysis of environmental samples as natural surface water samples.

The Figure 2b shows an off-line SPE HPLC-DAD screening from sounding 5, where amounts of SI and two degradation products (DEA an H-AT) were found.

The specificity of targeted analytes in the natural surface water samples, was checked by the two methods already described in "peak homogeneity" section, in order to perform a reliable assignment of quantified analytes.

Previously the authors proposed an off-line SPE GC-MS-SIM method, to perform the qualitative and quantitative determination of other s-triazine herbicides and related degradation products, where AT, SI, DIA and DEA were satisfactorily analyzed [23]. However, this method did not permit to determine OH-triazine degradation products, H-AT and H-SI, without a previous derivatization step. For this reason, it was necessary to develop an alternative and simple method, which could do it with LOQs at similar ranges, allowing us in this way to quantify these hydroxylated degradation products, which are very often found at relevant levels in the natural surface water samples, as we can see in Table 4.

Conclusion

In this work, a sensitive, easy and simple chromatographic method has been described for the separation and quantification of endocrine-disrupting chemicals SI, AT and their major degradation products as hydroxylated and dealkylated ones, H-SI, H-AT and DIA and DEA, respectively. A SPE procedure for the isolation and preconcentration of these analytes from water matrix was developed. The SPE procedure was extensively evaluated to obtain a high enrichment factor, for achieving a complete extraction and reaching a sensitivity increase. The instrumental conditions for this off-line SPE HPLC-DAD method were established for determination at trace levels of these six herbicides, widely found in Spain, in natural surface and ground water samples.

The reliability of the off-line SPE HPLC-DAD method was successfully proved by a validation procedure including stability of solutions, precision, linearity, determination and quantification limits, accuracy, statistical robustness evaluation and specificity studies.

Since our method provides enough low LOQs to quantify all of the analytes (from ten to thirty fold lower) than the relevant maximal amounts permitted in natural surface waters (1 and 3 μg L^{-1} as alert and alarm levels), upon European Union directives [3], it was also applied to qualitative and quantitative analysis of endocrine-disrupting

Sounding	H-SI	DIA	H-AT (μgl^{-1} ± SD	DEA	SI	AT
1	0.24 ± 0.06	n.d	0.31 ± 0.09	0.12 ± 0.03	n.d	n.d
2	0.21 ± 0.07	n.d	0.17 ± 0.05	0.12 ± 0.03	n.d	n.d
5	n.d	n.d	0.28 ± 0.05	0.24 ± 0.08	0.20 ± 0.06	n.q
6	n.d	0.12 ± 0.03	n.q	n.q	n.d	0.13 ± 0.03
7	n.d	n.q	n.q	n.d	n.d	n.d
Location 1	n.q	n.d	n.q	n.d	n.d	n.d
Location 2	n.d	n.d	0.23 ± 0.05	n.d	n.d	n.d
Location 3	n.q	n.d	0.10 ± 0.03	n.d	n.d	n.d

n.d=no detected
n.q=no quantificated, <LOQ

Table 4: Concentrations (μg L^{-1}) and Standard Desviation (n=3) found for every herbicide in the different soundings from the 23rd aquifer (Albacete, Spain).

herbicides in different natural surface water samples from the 23rd aquifer.

The results confirmed mainly the presence of degradation products such as H-AT, H-SI and DEA, at concentration levels of µg L^{-1}, proving so the importance to include the mentioned dealkylated and hydroxylated-triazines as targeted analytes. Among the different soundings, AT, SI and DIA were quantified in just one sample (different among them), whereas H-AT was present in all of the analyzed surface waters from aquifer. The degradation products H-SI, H-AT, DIA and DEA were found between concentration values of 0.10 and 0.31 µg L^{-1}.

As added advantage, this method allows the analysis of hydroxyl-degradation products, analytes which are difficult to analyse by GC methods without previous derivatization, but nevertheless, they are very common in natural surface waters from aquifer, and consequently, they are very important matter of analysis, as we can see from the results shown in Table 4.

Another main value of our method remains on its proved reliability, even for the analysis of the degradation products, whose study has been reported by few papers until now, despite their herbicide activity too. This is in opposite to other more recent works previously published, like the use of MWCN as sorbent in SPE process, or either the use of back pressure monolithic columns. Both of them have been used until now, just for the analysis of the parent compounds, but not for their degradation products yet. The results confirmed mainly the presence of degradation products such as H-AT, H-SI and DEA, at concentration levels of µg L^{-1}, proving so the importance to include the mentioned dealkylated and hydroxylated-triazines as targeted analytes.

Acknowledgements

The authors would like to thank"Convenio específico de colaboración entre Consejería de Obras Públicas de la Junta de Comunidades de Castilla –La Mancha y la Universidad de Castilla la Mancha en materia de calidad de aguas" and"Centro Regional de Estudios del Agua (CREA)".

Supplementary Information

Several types of cartridges, changing polarity of column, amounts of stationary phases, i.e. C18 cartridges (500, 1000 and 2000 mg), styrene divinylbenzene cartridges (500 and 1000 mg) and C18 disks (diameter 47 mm) were compared. It was observed the co-polymer styrene divinylbenzene-1000 mg cartridges provided higher recoveries (70-90%), than the recoveries (<50%) obtained using reversed phases (C18 cartridge or disks). This can be probably due to the non-ionic character of studied compounds, so this stationary phase was selected for later studies.

References

1. Jiang H, Adams C, Graziano N, Roberson A, McGuire M, et al. (2006) Occurrence and removal of chloro-s-triazines in water treatment plants. Environ Sci Technol 40: 3609-3616.
2. Special Report on Environmental Endocrine Disruption: An Effects Assessment and Analysis (1997) Environmental Protection Agency, Risk Assessment Forum, 630/R-96/012, Washington, USA.
3. EEC Drinking Water Guidelines, 80/778/EEC, EEC No L229/11-29, Brussels, 1980.
4. Sabik H, Jeannot R (1998) Determination of organonitrogen pesticides in large volumes of surface water by liquid-liquid and solid-phase extraction using gas chromatography with nitrogen-phosphorus detection and liquid chromatography with atmospheric pressure chemical ionization mass spectrometry. J Chromatogr A 818: 197-207.
5. Vandecasteele K, Gaus I, Debreuck W, Walraevens K (2000) Identification and quantification of 77 pesticides in groundwater using solid phase coupled to liquid-liquid microextraction and reversed-phase liquid chromatography. Anal Chem 72: 3093-3101.
6. Saez A, Gómez de Barreda D, Gamon M, Garcia de la Cuadra J, Lorenzo E, et al. (1996) UV detection of triazine herbicides and their hydroxylated and dealkylated degradation products in well water. J Chromatogr A 721: 107-112.
7. Koal T, Asperger A, Efer J, Engewald W (2003) Chromatographia 57: 93-101.
8. Zhou Q, Pang L, Xie G, Xiao J, Bai H (2009) Determination of atrazine and simazine in environmental water samples by dispersive liquid-liquid microextraction with high performance liquid chromatography. Anal Sci 25: 73-76.
9. Djozan D, Mahkam M, Ebrahimi B (2009) Preparation and binding study of solid-phase microextraction fiber on the basis of ametryn-imprinted polymer: application to the selective extraction of persistent triazine herbicides in tap water, rice, maize and onion. J Chromatogr A 1216: 2211-2219.
10. Tolosa I, Douy B, Carvalho FP (1999) Comparison of the performance of graphitized carbon black and poly(styrene-divinylbenzene) cartridges for the determination of pesticides and industrial phosphates in environmental waters. J Chromatogr A 864: 121-136.
11. Topuz S, Alpertunga B (2003) Determination of Triazines and N-Methylcarbamate Pesticides in Water by High-Performance Liquid Chromatography-Diode Array Detection Int J Environ Anal Chem 83: 787-795.
12. Nogueira JMF, Sandra T, Sandra P (2004) Multiresidue screening of neutral pesticides in water samples by high performance liquid chromatography–electrospray mass spectrometry. Anal Chim Acta 505: 209-215.
13. Pichon V, Charpak M, Hennion MC (1998) Multiresidue analysis of pesticides using new laminar extraction disks and liquid chromatography and application to the French priority list. J Chromatogr A 795: 83-92.
14. Brossa L, Marcé RM, Borrull F, Pocurull E (2003) Determination of endocrine-disrupting compounds in water samples by on-line solid-phase extraction-programmed-temperature vaporisation-gas chromatography-mass spectrometry. J Chromatogr A 998: 41-50.
15. Zhou Q, Xiao J, Wang W, Liu G, Shi Q, et al. (2006) Determination of atrazine and simazine in environmental water samples using multiwalled carbon nanotubes as the adsorbents for preconcentration prior to high performance liquid chromatography with diode array detector. Talanta 68: 1309-1315.
16. Al-Degsa YS, Mohammad A Al-Ghoutib (2008) Preconcentration and determination of high leachable pesticides residues in water using solid-phase extraction coupled with high-performance liquid chromatography. 88: 487-498.
17. Carabias-Martínez R, Rodríguez-Gonzalo E, Herrero-Hernández E (2005) Determination of triazines and dealkylated and hydroxylated metabolites in river water using a propazine-imprinted polymer. J Chromatogr A 1085: 199-206.
18. Lerch, RN, Li YX (2001) Analysis of Hydroxylated Atrazine Degradation Products in Soils. Int J Environ Anal Chem 79: 167-183.
19. Loos R, Niessner R (1999) Analysis of atrazine, terbutylazine and their N-dealkylated chloro and hydroxy metabolites by solid-phase extraction and gas chromatography-mass spectrometry and capillary electrophoresis-ultraviolet detection. J Chromatogr A 835: 217-229.
20. Bagheri H, Khalilian F (2005) Immersed solvent microextraction and gas chromatography–mass spectrometric detection of s-triazine herbicides in aquatic media. Anal Chim Acta 537: 81-87.
21. Sandra P, Beltran J, David F (1995) Enhanced selectivity in the determination of trianzines in environmental samples by Benchtop CGC-MS-MS. J High Result Chromatogr 18: 545-548.
22. Brossa L, Marcé RM, Borrull F, Pocurull E (2003) Determination of endocrine-disrupting compounds in water samples by on-line solid-phase extraction-programmed-temperature vaporisation-gas chromatography-mass spectrometry. J Chromatogr A 998: 41-50.
23. Berzas JJ, Guiberteau C, Villasenor MJ, Rodriguez V (2007) Sensitive SPE GC-MS-SIM screening of endocrine-disrupting herbicides and related degradation products in natural surface waters and robustness study. Microchemical Journal 87: 62-71.
24. Carabias-Martínez R, Rodríguez-Gonzalo E, Miranda-Cruz E, Domínguez-Alvarez J, Hernández-Méndez J (2006) Comparison of a non-aqueous capillary electrophoresis method with high performance liquid chromatography for the determination of herbicides and metabolites in water samples. J Chromatogr A 1122: 194-201.
25. Lerch RN, Blanchard PE, Thurman EM (1998) Contribution of Hydroxylated Atrazine Degradation Products to the Total Atrazine Load in Midwestern Streams. Environ Sci Technol 32: 40-48.
26. Corcia DA, Crescenzi C, Guerriero E, Samperi R (1997) Ultratrace

Determination of Atrazine and Its Six Major Degradation Products in Water by Solid-Phase Extraction and Liquid Chromatography-Electrospray/Mass Spectrometry Environ Sci Technol 31: 1658-1663.

27. García MÁ, Santaeufemia M, Melgar MJ (2012) Triazine residues in raw milk and infant formulas from Spanish northwest, by a diphasic dialysis extraction. Food Chem Toxicol 50: 503-510.

28. Smith GA, Pepich BV, Munch DJ (2008) Preservation and analytical procedures for the analysis of chloro-s-triazines and their chlorodegradate products in drinking waters using direct injection liquid chromatography tandem mass spectrometry. J Chromatogr A 1202: 138-144.

29. Kock-Schulmeyer M, Villagrasa M, López de Alda M, Céspedes-Sánchez R, Ventura F, et al. (2013) Occurrence and behavior of pesticides in wastewater treatment plants and their environmental impact. Sci Total Environ 458-460: 466-476.

30. Dos Santos LB, Infante CM, Masini JC (2009) Development of a sequential injection chromatography (SIC) method for determination of simazine, atrazine, and propazine. J Sep Sci 32: 494-500.

31. http://www.fda.gov/downloads/RegulatoryInformation/Guidances/UCM128049.pdf

32. Massart DL, Vandeginste BGM, Deming SN, Kaufmann L, Chemometrics: A Text book (1st edn) Oxford University Press, Oxford 1988.

33. AMC technical brief, Analytical Methods Committee. The Royal Society of Chemistry, 2000.

34. Youden WJ, Steiner EH (1975) The Association of Official Analytical Chemistry, Washington DC, 33.

35. Van Leeuwen JA, Buydens LMC, Vandeginste BGM, Kateman G, Schoenmarkers P, et al. (1991) The ruggedness test in HPLC method validation. Chemometric and Intelligent Laboratory Systems 10: 337-347.

36. Massart DL, Vandeginste BGM, Buydens LMC, De Jong S, Lewi PJ, et al. (1997) J Handbook of Chemometrics and Qualimetrics. Part, Elsevier, Amsterdam.

37. Miller JC, Miller JN (1993) Statistics for Analytical Chemistry, (edn) Ellis Horwood, New York, USA.

38. Mulholland M (1987) Development and evaluation of an automated procedure for the ruggedness testing of chromatographic conditions in high-performance liquid chromatography. J Chromatogr 395: 551-593.

39. Plackett L, Burman JP, Biometrika (1946) The Design of Optimum Multifactorial Experiments. Biometrika Trust 33: 305-325.

40. Fabre H, Le Bris A, Balchin MD (1995) Evaluation of different techniques for peak purity assessment on a diode-array detector in liquid chromatography. J Chromatogr A 697: 81-88.

41. Huber L (1989) Applications of Diode-Array Experience. Hewlett-Packard 12: 5953-2330.

Selectivity, Thermodynamic and Anisotropic Properties of Substituted Liquid-Crystal Cyanoazoxybenzenes as Stationary Phases for Gas Chromatography

Kuvshinova SA[1]*, Burmistrov VA[1], Novikov IV[1], Alexandriysky VV[1] and Koifman OI[1,2]

[1]*Research Institute of Macroheterocyclic Compounds, Ivanovo State University of Chemistry and Technology, Sheremetevskii pr. 7, Ivanovo, 153000 Russia*
[2]*Institute of Solutions Chemistry, Russian Academy of Sciences, Russia*

Abstract

Herein we discuss physical properties of 4-(ω-hydroxyalkoxy)-4'-cyanoazoxybenzene homologs. 1D and 2D correlation NMR spectroscopy (in particular, ^{1}H, ^{15}N-HMBC experiment) have allowed elucidation of structure of the prepared rod-like supramolecular cyanoazoxybenzenes. Mesomorphic properties of the compounds have been studied by means of polarization thermomicroscopy and differential scanning calorimetry. All the studied cyanoazoxybenzenes have revealed enantiotropic nematic mesomorphism over wide temperature range. Nematic mesophase of the eighth homolog has possessed large positive dielectric anisotropy. Introduction of small amounts of the prepared cyanoazoxybenzenes as additive has stabilized the mesophase and has increased the dielectric anisotropy of 4-pentyloxy-4'-cyanobiphenyl. Gas-liquid chromatography studies have shown that sorbents based on 4-(2-hydroxyethyloxy)-4'-cyanoazoxybenzene are highly selective towards various structural isomers; that cannot be achieved using conventional nematic liquid crystals. Thermodynamic evidence of specific interactions between the mesogen and the non-mesomorphic sorbate has been discovered.

Keywords: Mesogen; Azoxybenzene; Hydrogen bond; Position isomerism; NMR spectroscopy; Nematic phase; Supramolecule; Dielectric anisotropy; Double refraction; Sorbent; Structural selectivity

Introduction

Modern supramolecular chemistry is among most dynamically developing branches of science [1-5]. It includes study of chemical, physical, biological, and other aspects of complex chemical systems linked together via intermolecular (non-covalent) interactions. Hydrogen bond occupies a special place among diverse non-covalent interactions (van der Waals and donor-acceptor ones, coordination bonding involving metal ions, etc.). Being stereospecific, relatively strong, and dynamic, hydrogen bond has been recognized as key interaction in supramolecular systems [6].

Intermolecular hydrogen bonds are crucially important in formation of liquid-crystalline materials as well, as supramolecular self-assembly may result in emergence of novel properties: phase transitions, photoinduced effects, conductivity, proton transport, etc. In view of this, liquid crystal systems linked via hydrogen bonds are considered typical objects of supramolecular chemistry, as mesomorphism can be discussed regarding a sufficiently populated supramolecular assembly [7-9].

Liquid-crystalline molecular associates may be formed either of identical molecules (Figures 1a-c) or of chemically different compounds (Figure 1d).

Studies of Bennet [10] and Gray [11] pioneered in study of liquid crystals based on aromatic carboxylic acids and cinnamic acid; such materials exhibit mesomorphism due to formation of cyclic dimers linked via hydrogen bonds (Figure 1a). Mesogenic dimer is the simplest assembled form of supramolecular liquid crystals, the most widespread, and the best studied [12-15]. Amides are known to form supramolecular chains via hydrogen bonding [16] (Figure 1b), whereas polyhydric alcohols (sugars [17-20], hexahydroxycyclohexane [21], and diols [22-24]) may form layered structures (Figure 1c).

Supramolecular mesogens formed via the molecular recognition of involving various compounds are by far more numerous [25,26] (Figure 1d). Such binary systems may consist of two liquid-crystalline [27,28], two non-mesomorphic [29], or a liquid-crystalline and a non-mesomorphic [30,31] components. Noteworthily, azaheterocyclic compounds (derivatives of pyridine, azopyridine, and 4,4'-bipyridine) have been recognized as the most promising non-mesogenic building blocks [32-38]. In all the cases formation of strong hydrogen-bound complex are reflected by special properties of a liquid-crystalline product, distinct from these of the starting components.

Our group has contributed to investigation of liquid-crystalline systems involving hydrogen bonds as well [39,40]. We have extended the range of conventional reactive polar substituents constituting the discussed systems by synthesis study of mesogenic derivatives of azobenzene, benzylideneaniline, phenylbenzoate, and biphenyls containing aldehyde, aldoxime, epoxy, and other terminal groups [41,42]. The self-assembly via hydrogen bonding to form polymeric supramolecular assemblies was aided by the prepared bifunctional mesogenic compounds bearing at least two specifically interacting complementary fragments in the molecule [43-45]. We ruled out certain regularities of the self-assembly effect on mesomorphic, volume, rheological, dielectric, orientation, and sorption properties of the functional liquid crystals [46-49]. We generalized the most significant results of our studies and denoted the promising fields of practical applications of supramolecular mesogenic structures [50].

***Corresponding author:** Sofya A Kuvshinova, Research Institute of Macroheterocyclic Compounds, Ivanovo State University of Chemistry and Technology, Sheremetevskii pr. 7, Ivanovo, 153000 Russia
E-mail: sofya.kuv@yandex.ru

Analysis of the available reference literature on supramolecular liquid crystals has revealed that majority of the published reports have discussed preparation and formation conditions of supramolecular mesophases as well as phase diagrams of binary mixtures of the complementary components, identification of the mesophases, and variation of phase transition temperature. However, the information on the effect of specific interactions on physical properties of supramolecular liquid crystals has been scarce so far.

The azoxybenzenes class is of significant fundamental and applied importance among organic mesomorphic substances. Some of their properties, including high thermal stability, wide temperature range of mesomorphism, good miscibility with other mesogens, and sufficiently low viscosity make them promising materials for nonlinear optics [51-53] and gas chromatography [54] applications.

In this work we present the studies of physical properties of compounds containing two benzene rings bridged by azoxy group as a molecule core, the end groups being -CN and -ROH (R stands for hydrocarbon spacer) (Figure 2); possible practical applications of these compounds are discussed as well.

A special feature of the discussed compounds is the presence of two polar reactive substituents, allowing for further chemical modification to give functional materials with desired mesogenic properties as well as those based on macroheterocycles suitable for advanced technical and scientific applications.

In particular, the interaction of epichlorohydrin and epoxy function allows introduction of the oxirane heterocycle at the aliphatic substituent of cyanoazoxybenzenes **Ia-e**. Such mesogens may be efficient stabilizers in compositions based on thermoplastic polymers.

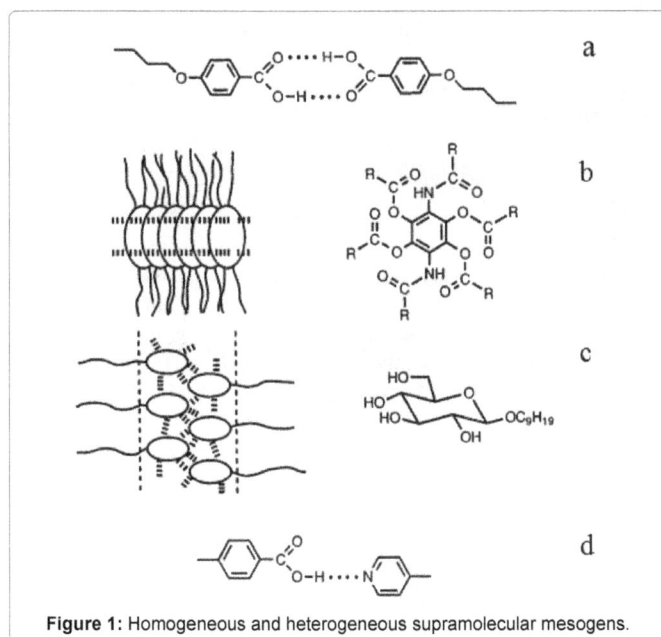

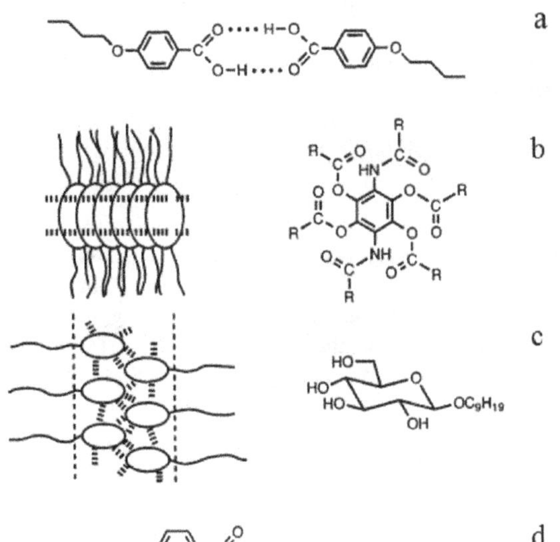

Figure 1: Homogeneous and heterogeneous supramolecular mesogens.

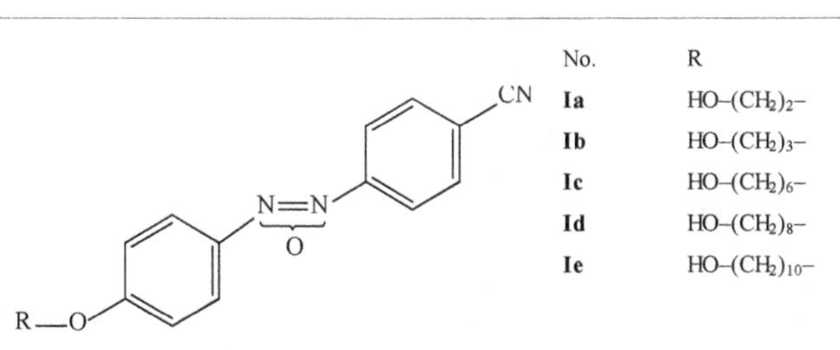

No.	R
Ia	HO–(CH$_2$)$_2$–
Ib	HO–(CH$_2$)$_3$–
Ic	HO–(CH$_2$)$_6$–
Id	HO–(CH$_2$)$_8$–
Ie	HO–(CH$_2$)$_{10}$–

Figure 2: Structure of 4-(ω-hydroxyalkyloxy)-4'-cyanoazoxybenzenes.

Introduction of the terminal acryloyl moiety via interaction of the studied azoxybenzenes **Ia-e** with acryloyl chloride results in formation of fairly promising monomers suitable for development of smart light-controlled liquid-crystalline polymers that may be used for reversible or irreversible black-and-white or color information recording and storage, optical memory systems, display technology, optoelectronics, holography, etc.

Crownether-containingmonomersbasedonreactiveazoxybenzenes **Ia-e** allow preparation of multi-functional photochromic-ionophore liquid-crystalline copolymers for development of self-assembled photo-controlled sensor devices. On the other hand, hydrolysis of the -CN group yields carboxylic acids, the corresponding acyl chlorides being capable of the reactions with reactive substituents of porphyrins and phthalocyanines, thus opening vast opportunities for structure modification of macroheterocyclic compounds to provide various functional materials.

Experimental

Preparation, modification and mesomorphic properties

Preparation of 4-(ω-hydroxyalkyloxy)-4'-cyanoazoxybenzenes was described elsewhere [55]. The mesogens were purified by re-crystallization from ethanol followed by incubation under residual pressure of 200 Pa at the temperature of mesophase existence during 12 h. The sufficient purity of the prepared specimens was judged by the absence of the impurities signals in the NMR spectra, the constant temperature of nematic-isotropic phase transition (T_{NI}) after repeated purification steps, and no disintegration into the nematic (N) and isotropic (I) phases in the course of the phase transition.

^{1}H, ^{13}C, and ^{15}N NMR spectra were registered using a Bruker Avance III-500 instrument in $CDCl_3$ at 35°C. The carbon (δ_C=77.00 ppm) and residual proton (δ_H=7.27 ppm) signals of the solvent and liquid ammonia (δ_N=0.0 ppm) were used as internal and external references, respectively. All the experiments were run according to the manufacturer recommendation. The evolution time in HMBC ^{1}H-^{13}C and ^{1}H-^{15}N experiments was of 60 and 125 ms, respectively.

Phase transition temperatures of the individual liquid crystals and the binary mesogens systems were measured using polarization thermomicroscopy (PTM) and differential scanning calorimetry (DSC) techniques. The determined temperatures were further checked during gas-liquid chromatography and dielectric constant measurement experiments.

The PTM studies were performed using a Polam L211 polarization microscope equipped with a temperature stage allowing heating at 0.1-3.5 deg/min rate over a wide temperature range of 0-500°C as well as prolonged incubation of a specimen at a desired temperature. The accuracy of temperature reading was of ± 0.1°C.

DSC curves were recorded using a DSC 204 F1 Phoenix calorimeter; the heating and cooling experiments were performed at 30-350°C at the rate of 5 deg/min.

The mixtures of 4-pentyloxy-4'-cyanobiphenyls (5OCB) with **Ia-e** were prepared by weighing using a MP 20 (accuracy of 0.05 mg) and Sartorius Genius ME215 P (0.01 mg) balances. Liquid-crystalline substance 5OCB (Aldrich) was used as received.

Dielectric properties and birefringence

Static dielectric constants of compound **Ic** and nematic binary mixtures in the mesomorphic and isotropic liquid states was measured taking advantage of dielcometric technique using a constant-temperature cell consisting of two parallel 19.6 mm^2 plates separated by a 0.25 mm gap filled with a tested substance. The specimen orientation was aided by a 2000 G electromagnet. The cell capacitance was measured with a LCR-817 (INSTEK) instrument at 1000 Hz and 1.2 V. The cell was calibrated using reference compounds with known dielectric constant. The dielectric constant was calculated as follows:

$$\varepsilon = \frac{C_X - C_M}{C_W}$$

with C_X being capacitance of the cell with the studied specimen; C_M being wiring capacitance as determined from the cell calibration; and C_W being capacitance of the cell working distance from the calibration.

In order to estimate the dielectric anisotropy $\Delta\varepsilon = \varepsilon_I - \varepsilon_\perp$ we performed the measurements parallel (ε_I) and perpendicular ($\varepsilon_\perp$) to the nematic liquid crystal director, over a range of temperatures set up using an UH-16 thermostat. The relative error of dielectric constant measurement was of 1%, Δε calculated using statistics methods was of 3%.

Birefringence of a liquid-crystalline materials was expressed as $\Delta n = n_e - n_0 = n_I - n_\perp$ Refractive indexes of ordinary ray in the mesomorphic state $n_0 = n_\perp$ and in the isotropic phase n_{is} were determined using a constant-temperature Abbe refractometer at 589 nm with accuracy of ± 0.0005. Surface of the refractometer prisms was rubbed in order to facilitate the specimen orientation. Refractive index of extraordinary ray $n_e = n_I$ was calculated using the equation for the average value $n^2 = \frac{n_e^2 + 2n_0^2}{3}$, the latter being determined via n_{is} extrapolation to the range of nematic phase. The error of birefringence determination was of 0.5% (n) and 1.0% (Дn).

Gas-liquid chromatography

4,4'-Dimethoxyazoxybenzene **LC-1** (Aldrich) was used as received. 4-Propyloxy-4'-cyanoazoxybenzene **LC-2** was prepared as described elsewhere [56].

Mesogens **LC-1**, **LC-2**, and **Ia** (9.95 wt.%) were applied onto a solid stationary phase Chromaton N-AW (0.40-0.63) via evaporation of the chloroform solution. The so prepared sorbent was introduced into metal 1 m × 3 mm columns under vacuum. Each column was conditioned during 6 hr at the highest operating temperature.

Retention time of the sorbates (p- and m-xylene and 3- and 4-methylanisole as well as isomeric изомерные lutidins and picolines) was measured using a Shimadzu GC-2014 gas chromatograph equipped with a flame ionization detector, helium being the carrier gas. Shimadzu GC-solution Chromatography Data System Version 2.4 software allowed setting the column, evaporator, and detector temperatures (at 0-400°C with accuracy of ± 0.1°C) as well as feed and pressure of the carrier gas at the column input and output; the retention time was determined with accuracy of 0.5 s. An autosampler Shimadzu AOC-20i syringe Shimadzu (10 μL) was used to apply small volume of the sorbates (no more than 0.1 μL) to the column so that the experiment conditions matched the limiting dilution and the sorbate concentration was within the linear range of the dissolution isotherm. The dead (void) time was determined using propane as reference.

For physico-chemical parameters of the sorbates, the corresponding A, B, and C coefficients in the Antoine equation, and procedures of calculation of the sorbates saturated vapor pressure and thermodynamic sorption parameters [57].

Results and Discussion

Identification and mesomorphic properties

Structures of azoxybenzenes **Ia-e** were elucidated by means of NMR spectroscopy. The ^{1}H NMR spectra contained the duplicated set of signals of *para*-disubstituted benzene rings (the AA'BB' system) pointing at formation of the isomers A and B differing in oxygen position under the preparation conditions. Integration of the ^{1}H NMR spectra of samples **Ia-e** revealed that the A/B ratio was almost equimolar (Figure 3).

The presence of nitrogen atom in the compounds **Ia-e** allowed elucidation of their structure by means of the ^{1}H,^{15}N-HMBC experiment optimized for the spin-spin coupling constant of 4 Hz revealing the ^{1}H-^{15}N interactions via several bonds. Noteworthily, the signals of cyano group nitrogen were not observed in the spectra when using the indirect detection of ^{15}N, likely due to the low value of heteronuclear spin-spin interaction constant. However, the presence of -C≡N groups was confirmed by ^{13}C NMR spectra containing characteristic signals at the 118 ppm region. On the other hand, appearance of the ^{15}N signals at 310-330 Hz evidenced about formation of azoxy derivatives. It is known that the signal of the oxidized nitrogen atom (=NO-) of azoxybenzenes experienced an upfield shift in the ^{15}N NMR spectra [58,59]. Hence, the structure of isomers A and B as well as their ratio could be elucidated from the available spectral data (Figure 4).

We attempted separation of isomers of supramolecular azoxybenzenes **Ia-e** via recrystallization from ethanol or benzene as well as via column chromatography on alumina (methylene chloride as eluent) or on silica gel (diethyl ether-chloroform-benzene-ethanol-acetic acid 1:1:1:0.25:0.25 as eluent). However, the isomers ratio of compounds **I** was not changed upon the purification, likely due to the structural features. The presence of hydroxyl groups capable of strong specific interactions in compounds **I** seemed to level off the differences in the solvation and elution behavior originating from structure of the bridging groups Table 1.

Phase transitions temperature of 4-(ω-hydroxyalkyloxy)-4'-cyanoazoxybenzenes **Ia-e** and their structural analogs 4-alkyloxy-4'-cyanoazoxybenzenes as determined by the PTM method are collected in Table 1. The tabulated data revealed significant influence of the terminal groups on the phase transitions temperature. First, the active substituent stabilized the nematic phase. In particular, whereas the higher homologs of "conventional" cyanoazoxybenzenes were smectic-nematic, the bifunctional azoxybenzenes **Ia-e** exhibited nematic mesomorphism even in the case of compound **Ie** with 10 carbon atoms in the aliphatic fragment. Moreover, introduction of the terminal hydroxyl group resulted in increase of the phase transitions temperature, the increase being more prominent for the nematic-isotropic transition; hence, the temperature range of existence of the anisotropic phase was expanded. That could be due to appearance of sufficiently strong hydrogen bonds. Two types of interaction were possible: the chain-like "head to tail" association and the "tail to tail" dimerization (Figure 5). In the both cases the effective anisotropy of molecular polarizability was expected to increase, and the mesophase thermal stability should have been enhanced. Additionally, that should have resulted in significant restriction of the aliphatic fragments mobility and their higher orientation ordering.

DSC data demonstrated that energy of the nematic-isotropic phase transition of supramolecular azoxybenzenes **Ia-e** was of 2.64-2.87 kJ/mol, significantly higher than that of the azoxybenzenes not containing terminal OH groups in the aliphatic fragment and hence not capable of

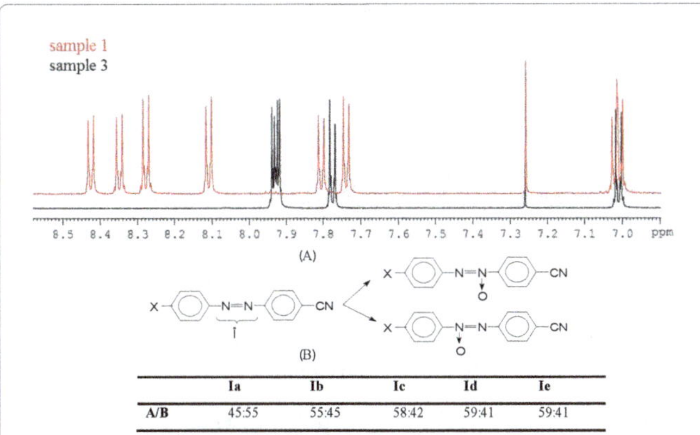

	Ia	Ib	Ic	Id	Ie
A/B	45:55	55:45	58:42	59:41	59:41

Figure 3: Parts of ¹H NMR spectra (region of resonance of aromatic protons) of azo- (sample 3) and azoxybenzene (sample 1); the ratio of isomers in the cases of **Ia-e** as derived from the NMR data.

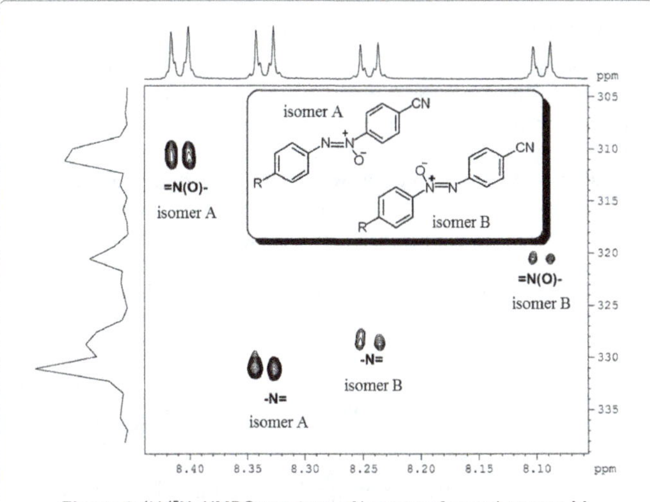

Figure 4: ¹H,¹⁵N–HMBC spectrum of isomers of azoxybenzene **Id**.

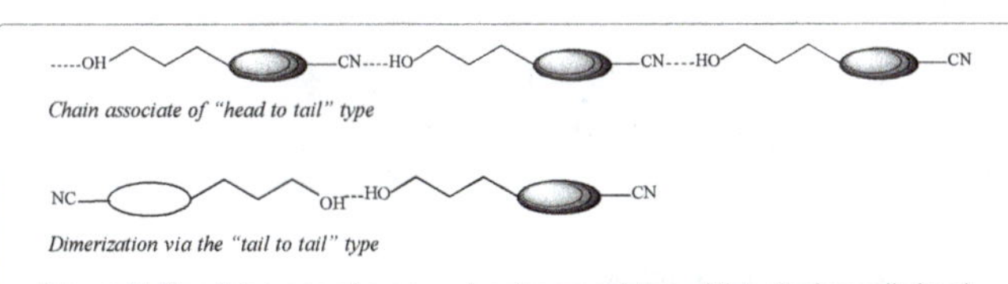

Figure 5: Possible types of supramolecular association of 4-(ω-hydroxyalkyloxy)-4'-cyanoazoxybenzenes **Ia-e**.

Compound	Isomers ratio		Phase transitions temperature, °C		Compound	Phase transitions temperature, °C	
	I-A	I-B	C→N	N→I		C→N	N→I
HO—(CH$_2$)$_n$—O—C$_6$H$_4$—N=N(O)—C$_6$H$_4$—					C$_n$H$_{2n+1}$O—C$_6$H$_4$—N=N(O)—C$_6$H$_4$—C		
Ia(n=2)	45	55	136.6	188.1	n=2	169.5	179.0
Ib (n=3)	55	45	124.5	167.0	n=3	127.5	156.5
Ic (n=6)	58	42	127.5	166.4	n=6	114.0	
Id (n=8)	59	41	117.6	150.3	n=8	C 70.8 S 120.4 N 133.7 I	
Ie (n=10)	59	41	111.3	138.0	n=9	C 73.5 S 119.5 N 127.7 I	

Table 1: Temperature of phase transitions of azoxybenzenes **I** and their structural analogs (accuracy of ± 0.1°C).

the self-assembly via hydrogen bonding (0.85-1.37 kJ/mol). The high enthalpy of the phase transition from the ordered liquid-crystalline state into the disordered isotropic liquid was likely due to disruption of the intermolecular hydrogen bonds.

Anisotropic properties

Operation parameters of visual indication devices taking advantage of electrooptical effects of liquid crystals depend on anisotropy of dielectric properties of the mesogens. Therefore, measurement of components of dielectric permittivity tensor is of primary importance for development of new materials and estimation of their potential. In view of that, we studied static dielectric permittivity of 4-(8-hydroxyoctyloxy)-4'-cyanoazoxybenzene **Id** (Figure 6).

The studied compound **Id** revealed positive dielectric permittivity anisotropy $\varepsilon_I > \varepsilon_\perp$ and sufficiently high dielectric constant value, due to the presence of highly polar cyano group in the terminal fragment. Phase transitions temperature determined using dielcometric method coincided well with those determined by thermomicroscopy.

In the case of conventional 4-octyloxy-4'-cyanoazoxybenzene Δε=9.5 [56] at $T_{red}=T-T_{NI}=-7$°C. Introduction of terminal hydroxyl group increased the Дε value up to 11.8 in the case of compound **Id**. That could be due to both the increase of overall dipole moment of the molecule upon addition of a polar hydroxyl group and significant changes of nature of the association processes in the nematic and isotropic phases of supramolecular azoxybenzenes as compared to their analogs not capable of self-assembly.

The high temperature of mesophase existence complicates the application of compounds **I** in pure form. In view of that, investigation of efficiency of bifunctional azoxybenzenes **Ia-e** as dopants of low-temperature cyanobiphenyls is of definite interest. In particular, we studied the influence of compounds **Ia-e** on mesomorphic and anisotropic properties of their mixtures with 4-pentyloxy-4'-cyanobiphenyl (5OCB).

Figure 7 displays a part of phase diagram of the 5OCB + **Ia** mixture and the values of slope of the phase boundary of the 5OCB + 4-(ω-hydroxyalkyloxy)-4'-cyanoazoxybenzene mixtures. We found that compounds **Ia-e** were fully miscible with 5OCB and stabilized the mesophase. The highest stabilizing effect was revealed in the case of compound **Ia**, and compound **Ie** was the least efficient stabilizer; seemingly, the effect was determined by the azoxybenzenes geometry parameters.

We obtained temperature dependences of static dielectric permittivity and refractive indexes of the studied systems; they were recalculated into the data on dielectric anisotropy Δε and birefringence Δ*n*, respectively (Figures 8 and 9). The temperature dependences of *e*, *n*, Δε, and Δ*n* were typical of mesomorphic materials; addition of hydroxyl-containing components **Ic-e** resulted in enhancement of dielectric anisotropy and birefringence.

Selectivity and thermodynamics of dissolution of structural isomers in liquid-crystalline azoxybenzenes

Doping of composite liquid-crystalline materials is not the only possible practical application of 4-(ω-hydroxyalkyloxy)-4'-cyanoazoxybenzenes **Ia-e**. Due to wide range of the mesophase existence and low viscosity, liquid-crystalline azoxybenzenes are widely used as stationary phases in gas chromatography for analytical separation of structural isomers of organic compounds [60]. 4,4'-Methoxyethoxyazoxybenzene (MEAB) exhibiting the factor of structural selectivity towards meta-para xylenes separation (1.13

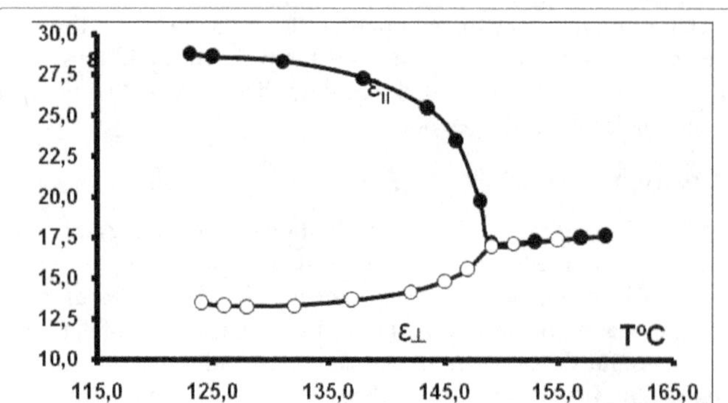

Figure 6: Dielectric permittivity of 4-(8-hydroxyoctyloxy)-4'-cyanoazoxybenzene **Id** as function of temperature.

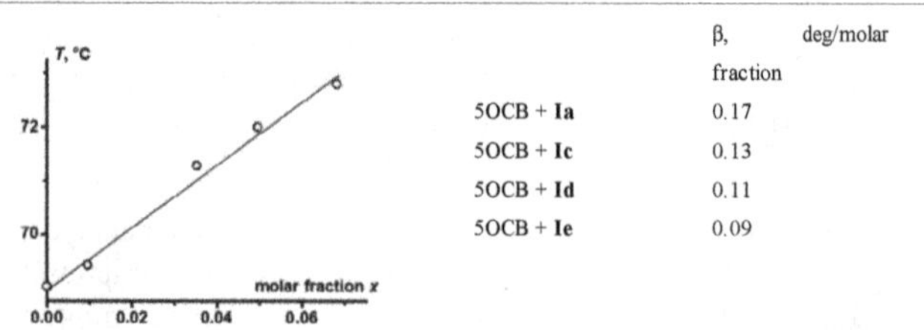

Figure 7: Fragment of the phase diagram of 5OCB–**Ia** mixture and slopes of the phase boundary of 5OCB–4-(ω-hydroxyalkyloxy)-4'-cyanoazoxybenzene mixtures.

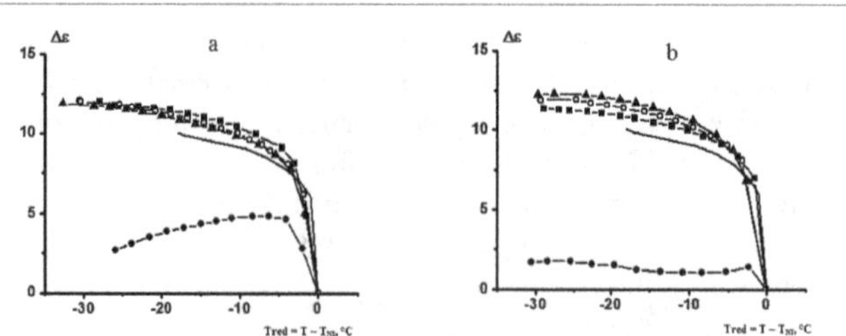

Figure 8: Anisotropy of dielectric permittivity of 5OCB (*solid line*) and of its mixtures with **Ia** (•), **Ic** (■), **Id** (▲), and **Ie** (○) with the additive fraction of 3 wt% (a) and 5 wt% (b) as function of temperature.

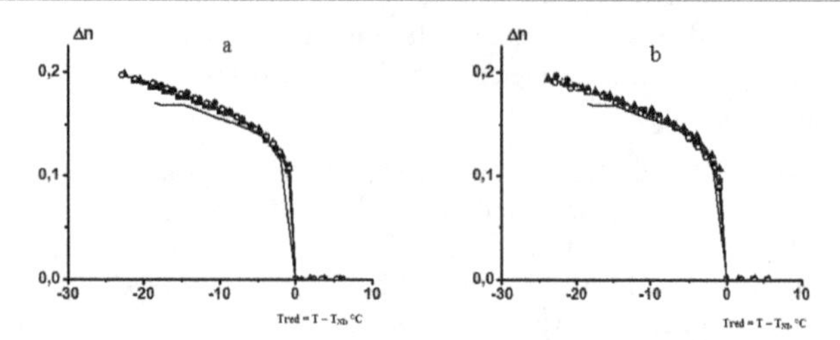

Figure 9: Anisotropy of birefringence of 5OCB (*solid line*) and of its mixtures with **Ia** (•), **Ic** (■), **Id** (▲), and **Ie** (○) with the additive fraction of 3 wt% (a) and 5 wt% (b) as function of temperature.

in the nematic phase) is among the most selective liquid crystals for chromatography applications [61]. That allows separation of xylenes using a 1 m column, the operation being often impossible using 10-15 m conventional capillary columns. Such excellent selectivity is due to the absence of extensive regions of low ordering in the MEAB mesophase, the reason for the perfect ordering being the short and limitedly mobile aliphatic substituents (methyloxy and ethyloxy) of this mesogen. Hence, the orientation ordering over the liquid-crystalline sample volume and steric limitations for the separated sorbates are leveled off.

At the same time, the MEAB selectivity is not sufficient for gas chromatography separation of certain multicomponent mixtures. We have earlier demonstrated that taking advantage of supramolecular factor of structural selectivity allows for efficient analytical separation of xylene isomers as well as higher-boiling compounds: *p*- and m-cresols, p- and m-methylanisoles, 3,4- and 3,5-lutidins, etc. [62]. In view of that, MEAB analogs capable of self-assembly via the active terminal groups are of definite fundamental and applied interest.

Using the gas chromatography experiment data (retention times of pairs of structural isomers) we determined the highest coefficients of structural selectivity 6 of a sorbent containing 4-(2-hydroxyethyloxy)-4'-cyanoazoxybenzene **Ia** as stationary phase (Table 2).

The experimental results demonstrated excellent selectivity of the sorbent based on 4-(2-hydroxyethyloxy)-4'-cyanoazoxybenzene **Ia** towards isomers of high-boiling organic compounds (α=1.75 in the case of 3,4- and 3,5-lutidins) and high selectivity towards separation of para- and meta-xylenes (α=1.12). Hence, mesogen **Ia** is a promising liquid-crystalline stationary phase to be used for quantitative analysis of mixtures of organic compounds [55].

Noteworthily, sorption behavior of isomeric lutidins was remarkably different from that of xylenes, methylanisoles, and cresols. In particular, the less anisotropic 3,4-lutidin revealed higher retention time as compared to the more anisotropic 3,5- isomer. Moreover, the separation coefficient of lutidins was higher than that of other sorbate pairs. Those peculiarities evidenced that specific interactions of the electron-donating sorbate and the proton-donating terminal group of the mesogen should be accounted for along with the steric separation factor in the case of sufficiently basic sorbates. Indeed, 3,4-lutidin incorporation in the stationary phase was favorable over the interaction of more anisotropic 3,5-lutidin:

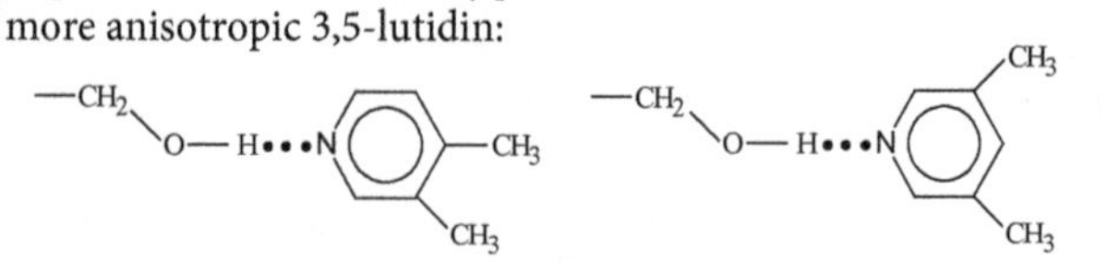

At the same time, the integral structural selectivity towards separation of pyridine derivatives could be affected by the following factors:

1. purely steric limitations;
2. selective dipole-dipole interactions;
3. selective specific interactions;
4. supramolecular effect.

In order to elucidate contributions of the above-listed factors, we studied dissolution selectivity and thermodynamics of electron-donating isomeric lutidins and picolines in the stationary phases based on the following nematic azoxybenzenes:

- low-polar "conventional" 4,4'-dimethoxyazoxybenzene **LC-1**

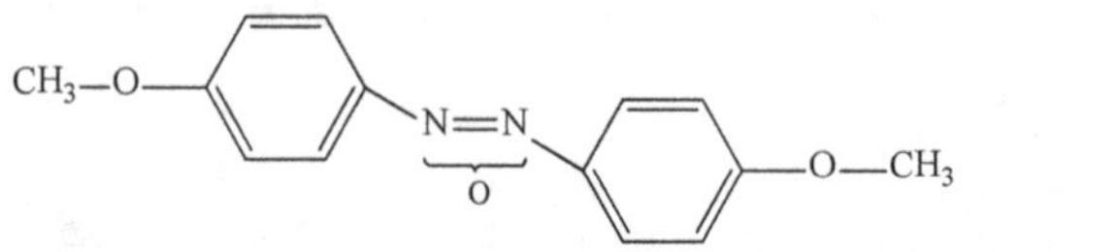

- high-polar "conventional" 4-propyloxy-4'-cyanoazoxybenzene **LC-2**

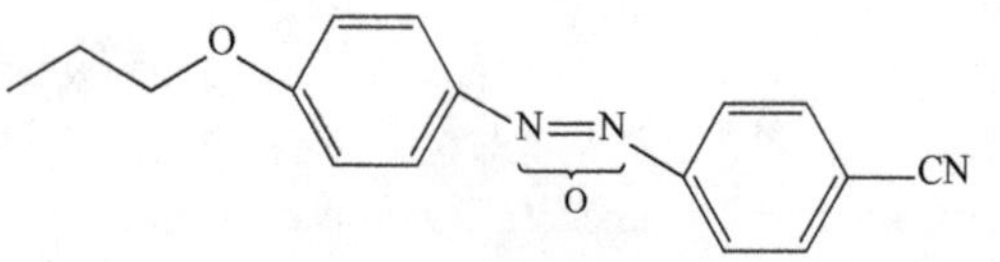

- supramolecular 4-(2-hydroxyethyloxy)-4'-cyanoazoxybenzene **Ia**

We calculated the Herington coefficients of structural selectivity from the gas chromatography experiment data (Table 3).

Supramolecular liquid-crystalline material **Ia** exhibited the highest selectivity. Decrease of the overall structural selectivity of **LC-2** as compared to **LC-1** was likely due to the increase of the terminal alkyl substituent length resulting in appearance of the regions of deteriorated orientation ordering.

We recalculated the retention experimental data into thermodynamic parameters of dissolution of isomeric lutidins and picolines (Tables 4-6).

Peculiarities of thermodynamic compensation effect were analyzed by plotting the dataset in the $-\Delta\bar{H}_2^\infty = f(-T\Delta\bar{S}_2^\infty)$ coordinates (Figure 10). The data unequivocally evidenced about prevailing of the entropy factor in dissolution of isomeric pyridine derivatives, due to steric limitations imposed by a liquid-crystalline matrix.

Contribution of the hydrogen bond strength between the liquid crystal and the non-mesogen into the retention selectivity we performed quantum-chemical simulation of mesogens, sorbates, and their complexes (Figure 11). Geometry optimization as well as computation of the force field and the vibrations frequency was performed taking advantage of the DFT method (B3LYP hydrid functional) with the Dunning split-valence basis set cc-pVTZ. The simulation consisted of two stages: first, the starting geometry was optimized using the Hartree-Fock method; second, the configurations corresponding to the minima at the potential energy surface were used as starting ones for DFT analysis. The simulation was performed using PC GAMESS software [63]. Data input preparation and results processing were carried out using Chem Craft software [64].

The simulation demonstrated that the bond energy was almost the same in all the complexes. That confirmed the previously discussed prevailing of the steric factor over the energy one in dissolution of the isomeric lutidins and picolines. Hence, the factor of steric limitations on the sorbate incorporation into the liquid-crystalline stationary phase was crucial to determine efficacy of separation of the electron-donating pyridine derivatives. The prominent selectivity was due to the high orientation ordering of the terminal substituents resulting from either their short length or the supramolecular hydrogen bonding interaction of the complementary groups.

Isomers	Column temperature, °C	α
p- and *m*-xylene	103.9	1.12
p- and *m*-methylanisole	105.6	1.1
3,4- and 3,5-lutidin	101.5	1.75
p- and *m*-cresol	139.8	1.03

Table 2: Highest coefficients of structural selectivity of compound **Ia**.

Mesogen	Lutidins					Picolines	
	α(2,5/2,6)	α(2,4/2,5)	α(2,3/2,4)	α(3,5/2,3)	α(3,4/3,5)	α(2/3)	α(3/4)
LC-1	1.80	1.07	1.04	1.54	1.34	1.70	1.11
LC-2	1.89	1.00	1.09	1.49	1.35	1.68	1.05
Ia	1.96	1.12	1.03	1.49	1.51	1.74	1.17

Table 3: Coefficients of structural selectivity of azoxybenzenes.

Sorbate	γ^∞	$-\Delta H^\infty$, kJ/mol	$-S^\infty$, J/(mol·K)	H^E, kJ/mol	S^E, J/(mol·K)	G^E, kJ/mol
T=401 K						
2,6-lutidin	1.76	18.0	49.3	16.5	36.8	2.0
2,5-lutidin	1.36	22.8	59.3	15.5	36.3	1.1
2,4-lutidin	1.38	22.0	57.3	14.4	33.6	1.1
2,3-lutidin	1.37	21.6	56.3	16.1	37.8	1.1
3,5-lutidin	1.23	26.2	66.8	13.2	31.3	0.7
3,4-lutidin	1.14	30.3	76.4	10.3	24.7	0.4
2-пиколин	1.29	21.3	55.3	12.4	29.0	0.8
3-пиколин	1.15	20.0	51.0	14.5	35.0	0.5
4-пиколин	1.09	23.1	58.2	10.6	25.6	0.3
T=413 K						
2,6-lutidin	1.30	30.4	75.8	3.3	5.7	0.9
2,5-lutidin	1.05	34.4	83.7	2.3	5.2	0.2
2,4-lutidin	1.06	35.2	85.6	2.3	5.2	0.2
2,3-lutidin	1.02	34.6	84.0	2.5	5.8	0.1
3,5-lutidin	0.93	36.1	86.8	2.5	6.6	-0.2
3,4-lutidin	0.88	38.9	93.0	0.3	1.9	-0.4
2-пиколин	0.97	29.9	72.2	1.7	4.2	-0.1
3-пиколин	0.88	31.7	75.6	1.9	5.7	-0.4
4-пиколин	0.84	32.5	77.3	1.2	4.4	-0.6

Table 4: Thermodynamic parameters of dissolution of substituted pyridines in the nematic and isotropic phases of **LC-1** at infinite dilution.

Conclusion

In this work we have presented the studies of structure and physical properties of supramolecular 4-(ω-hydroxyalkyloxy)-4'-cyanoazoxybenzenes. The prepared compounds contain terminal -CN and -OH groups in a single molecule. These groups are highly reactive, allowing for the structure modification to afford synthons for targeted synthesis of novel macroheterocyclic and mesogenic materials. Furthermore, these terminal substituents are capable of specific intermolecular interactions.

We have demonstrated that self-assembly of azoxybenzenes **Ia-e** containing active functional substituents decreases the tendency of the material to form smectic phase, enhances thermal stability of the nematic phase, and increases the dielectric anisotropy. Doping of liquid-crystalline 4-pentyloxy-4'-cyanobiphenyl with small amount of the prepared supramolecular azoxybenzenes **Ia-e** has resulted in the increase of dielectric permittivity anisotropy and birefringence. The effect is likely caused by formation of either supramolecules linked via the ~OH···HO~ hydrogen bonds or linear supramolecular assemblies containing the ~OH···NC~ linkage.

We have analyzed sorption and selective properties of the stationary phase based on supramolecular 4-(2-hydroxyethyloxy)-4'-cyanoazoxybenzene **Ia**. The mesogen exhibits high structural selectivity with respect to various isomers, unachievable when using "conventional" nematic azoxybenzenes; that is due to self-assembly into the chain associates. The prevailing factor of the high structural

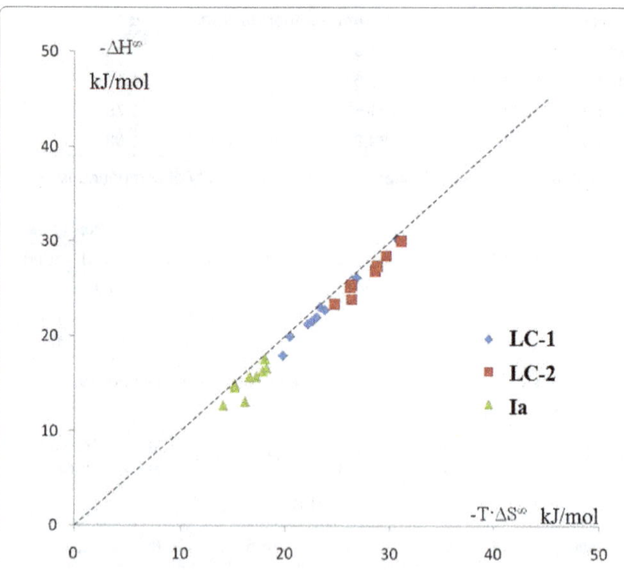

Figure 10: Graphical representation of thermodymanic compensation effect.

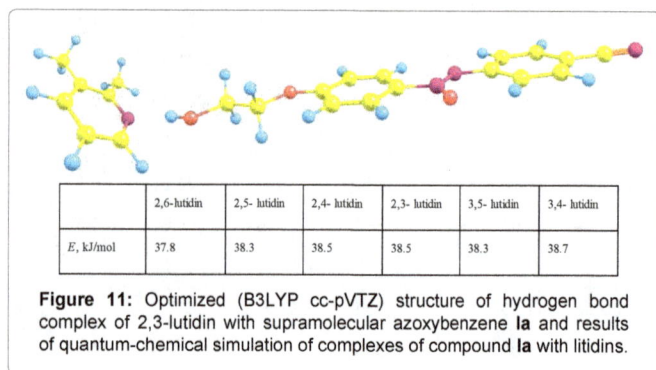

	2,6-lutidin	2,5- lutidin	2,4- lutidin	2,3- lutidin	3,5- lutidin	3,4- lutidin
E, kJ/mol	37.8	38.3	38.5	38.5	38.3	38.7

Figure 11: Optimized (B3LYP cc-pVTZ) structure of hydrogen bond complex of 2,3-lutidin with supramolecular azoxybenzene **Ia** and results of quantum-chemical simulation of complexes of compound **Ia** with litidins.

Sorbate	γ^{∞}	$-\Delta H^{\infty}$, kJ/mol	$-\Delta S^{\infty}$, J/(mol·K)	H^E, kJ/mol	S^E, J/(mol·K)	G^E, kJ/mol
T=408 K						
2,6-lutidin	2.04	23.9	64.5	10.3	19.2	2.4
2,5-lutidin	1.52	27.4	70.5	10.0	21.0	1.4
2,4-lutidin	1.64	26.9	70.0	9.2	18.4	1.7
2,3-lutidin	1.55	27.1	70.1	10.4	21.8	1.5
3,5-lutidin	1.43	28.5	72.6	11.3	24.8	1.2
3,4-lutidin	1.36	30.1	76.2	11.2	25.1	1.0
2-picoline	1.47	23.4	60.5	8.8	18.3	1.3
3-picoline	1.33	25.2	64.2	9.0	19.5	1.0
4-picoline	1.32	25.4	64.5	9.0	19.7	0.9
T=433.2 K						
2,6-lutidin	1.56	31.4	77.1	1.2	-1.7	1.6
2,5-lutidin	1.20	32.2	75.9	2.8	5.1	0.7
2,4-lutidin	1.30	32.5	76.6	2.8	5.1	0.9
2,3-lutidin	1.20	32.0	75.5	4.0	7.7	0.6
3,5-lutidin	1.12	35.4	82.8	2.0	3.6	0.4
3,4-lutidin	1.10	34.4	80.2	-1.2	-3.6	0.4
2-picoline	1.18	31.9	75.0	-1.4	-4.6	0.6
3-picoline	1.08	33.1	77.0	-0.6	-2.0	0.3
4-picoline	1.08	32.0	74.6	0.7	0.9	0.3

Table 5: Thermodynamic parameters of dissolution of substituted pyridines in the nematic and isotropic phases of **LC-2** at infinite dilution.

Sorbate	γ^{∞}	$-\Delta H^{\infty}$, kJ/mol	$-S^{\infty}$, J/(mol·K)	H^E, kJ/mol	S^E, J/(mol·K)	G^E, kJ/mol
T=443 K						
2,6-lutidin	2.29	13.1	36.4	19.0	35.9	3.0
2,5-lutidin	1.50	16.3	40.2	18.7	38.8	1.5
2,4-lutidin	1.54	16.6	41.0	17.4	35.6	1.6
2,3-lutidin	1.47	15.7	38.8	20.1	42.1	1.4
3,5-lutidin	1.24	15.7	37.4	21.5	46.6	0.8
3,4-lutidin	1.14	17.6	40.7	13.5	29.4	0.5
2-picoline	1.41	12.7	31.7	17.4	36.3	1.3
3-picoline	1.16	14.6	34.2	17.6	38.4	0.6
4-picoline	1.05	14.9	34.2	17.4	38.8	0.2

Table 6: Thermodynamic parameters of dissolution of substituted pyridines in the nematic phase of mesogen **Ia** at infinite dilution.

selectivity of the supramolecular mesogen is the entropy contribution resulting from the limited mobility of terminal groups after the mesogen self-assembly.

Acknowledgements

This work was financially supported by Russian Scientific Foundation (Agreement No. 14-23-00204).

References

1. Lehn JM (1985) Supramolecular chemistry: receptors, catalysts, and carriers. Science 227: 849-856.
2. Schneider HJ, Yatsimirsky AK (2000) Principles and methods in supramolecular chemistry. Chichester, New York Wiley. p: 349.
3. Lenh JM (1990) Perspectives in supramolecular chemistry - from molecular recognition towards molecular information processing and self-organization. Angew Chem Int Ed Engl 29: 1304-1319.
4. Mamardashvili GM, Mamardashvili NZH, Koifman OI (2005) Supramolecular porphyrin complexes. Russ Chem Rev 74: 765.
5. Zaitsev SYu (2006) Supramolecular systems at the phase boundary as models of biomembranes and nanomaterials (Supramolekulyarnye sistemy na granites razdela faz kak modeli biomembran i nanomaterialy). Zaitsev. Moscow: Nord Kompyuter. p: 189.
6. Steed JW, Atwood JL (2000) Supramolecular Chemistry: An Introduction (textbook). NY Wiley. p: 745.
7. Kato T (1998) Hydrogen bond systems. In: Demus D, Goodby J, Spiess HW, Vill V, editors. Handbook of liquid crystals, 2b Weinheim: Wiley-VCH.
8. Biginn U (2003) Thermotropic columnar mesophases from N-H O, and N H-O hydrogen bond supramolecular mesogens. Prog Polym Sci 28: 1049-1105.
9. Paleos CM, Tsiourvas D (2001) Supramolecular hydrogen-bonded liquid crystals. Liq Cryst 28: 1127-1161.
10. Bennet GM, Jones B (1939) Mesomorphism and polymorphism of some p-alkoxybenzoic and p-alkoxycinnamic acids. J Chem Soc 420-425.
11. Gray GW, Hartley JB, Jones B (1955) Mesomorphism and chemical constitution. V. The mesomorphic properties of the 4-n-alkoxydiphenyl-4'-carboxylic acids. J Chem Soc 4: 1412-1420.
12. Imrie CT (1998) Liquid Crystal Dimers and Oligomers. Imrie CT and Luckhurst GR in the Handbook of Liquid Crystals. Vol. 2B: Low Molecular Weight Liquid Crystals. edition Demus D, Goodby JW, Gray GW, Spiess HW and Vill V, Wiley-VCH, Weinheim. pp: 801-834.
13. Kavitha C, Pongali SPN, Madhu MMLN (2013) Thermal Analysis of Supramolecular Hydrogen-Bonded Liquid Crystals Formed by Nonyloxy and Alkyl Benzoic Acids. Mol Cryst Liq Cryst 574: 96-113.
14. Subhapriya P, Vijayanand PS, Madhu Mohan MLN (2013) Synthesis and Characterization of Supramolecular Hydrogen-Bonded Liquid Crystals Comprising of p-n-Alkoxy Benzoic Acids with Suberic Acid and Pimelic Acid. Mol Cryst Liq Cryst 571: 40-56.

15. Pandey A, Singh B (2012) Synthesis, Characterization and Mesomorphic Properties of Aromatic Acid Dimers. Mol Cryst Liq Cryst 552: 43-52.

16. Matsunaga Y (1986) Liquid Crystal Phases Exhibited by N,N'-Dialkanoyldiaminomesitylenes Matsunaga Y, Terada M. Mol Cryst Liq Cryst 141: 321-326.

17. Noller CR, Rockwell WC (1936) The Preparation of Some Higher Alkylglucosides. J Am Chem Soc 60: 2076-2077.

18. Fischer E, Helferich B (1911) Über neue synthetische Glucoside. Justus Liebigs Ann Chem 383: 68-91.

19. Jeffrey GA, Bhattacharjee S (1984) Carbohydrate Liquid Crystals. Mol Cryst Liq Cryst 110: 221-237.

20. Goodby JW (1984) Liquid Crystal Phases Exhibited by Some Monosaccharides. Mol Cryst Liq Cryst 110: 205-219.

21. Praefcke K, Marquardt P, Kohne B, Stephan W (1991) Liquid-Crystalline Inositol Ethers, their Syntheses and Columnar Mesophases. J Carbohydr Chem 10: 539-548.

22. Diele S, Madicke A, Geissler E, Meinel D, Demus D, et al. (1989) Liquid Crystalline Structures of the 2-(trans-4-n-alkylcyclohexyl)propan-1,3-diols. Mol Cryst Liq Cryst 166: 131-142.

23. Hentrich F, Diele S, Tschierske C (1994) Thermotropic liquid crystalline properties of 1,2-n-alkan-tetraols. Cryst 17: 827-839.

24. Patil VS, Padalkar VS, Tathe AB, Gupta VD, Sekar N (2013) Synthesis, Photo-physical and DFT Studies of ESIPT Inspired Novel 2-(2',4'-Dihydroxyphenyl) Benzimidazole, Benzoxazole and Benzothiazole. J Fluorescence 23: 1019-1029.

25. Kato T (2000) Hydrogen-Bonded Liquid Crystals: Molecular Self-Assembly for Dynamically Functional Materials. Structure & Bonding - Springer Berlin, Heidelberg 96: 95-146.

26. Naoum MM, Fahmi AA, Refaie AA, Alaasar MA (2012) Novel hydrogen-bonded angular supramolecular liquid crystals. Liq Cryst 39: 47-61.

27. Naoum MM, Fahmi AA, Alaasar MA (2008) Mol Cryst Liq Cryst 487: 74-91.

28. Naoum MM, Fahmi AA, Mohammady SZ, Abaza AH (2010) Effect of lateral substitution on supramolecular liquid crystal associates induced by hydrogen-bonding interactions between 4- (4'-pyridylazo-3-methylphenyl)-4''-alkoxy benzoates and 4-substituted benzoic acids. Liq Cryst 37: 475-486.

29. Naoum MM, Fahmi AA, Alaasar MA (2009) Mol Cryst Liq Cryst 506: 22-33.

30. Naoum MM, Fahmi AA, Alaasar MA (2008) Mol Cryst Liq Cryst 482: 57-70.

31. Naoum MM, Fahmi AA, Almllal WA (2010) Mol Cryst Liq Cryst 518: 109-128.

32. Machida S (1997) Response of a Hydrogen-Bonded Liquid Crystal to an Applied Electric Field Accelerated by a Poly(ç-benzyl-L-glutamate) Chemical Reaction Alignment Film. Langmuir 13: 576-580.

33. Alaasar M, Tschierske C, Prehm M (2011) Hydrogen-bonded supramolecular complexes formed between isophthalic acid and pyridine-based derivatives. Liq Cryst 38: 925-934.

34. Lu H, Wang JX, Song XZ (2011) Supramolecular Liquid Crystals Induced by Intermolecular Hydrogen Bonding. Mol Cryst Liq Cryst 537: 93-102.

35. He WL, Wang L, Yang Z, Yang H, Xie MW (2011) Synthesis and optical behaviour of hydrogen-bonded liquid crystals based on a chiral pyridine derivative. Liq Cryst 38: 1217-1225.

36. Lee JH, Jang I, Hwang SH, Lee SJ, Yoo SH, et al. (2012) Self-assembled discotic nematic liquid crystals formed by simple hydrogen bonding between phenol and pyridine moieties. Liq Cryst 39: 973-981.

37. Wei Q, Guo X, Yang H (2012) Synthesis and Mesomorphic Properties of Two Series of Hydrogen-Bonded Liquid Crystals Based on Laterally Fluorinated Benzoic Acid and 4,4'-Bipyridine With a Molar Ratio of 2:1. Mol Cryst Liq Cryst 557: 1-10.

38. Roohnikan M, Ebrahimi M, Ghaffarian SR, Tamaoki N (2013) Supramolecular self-assembly of novel hydrogen-bonded cholesteric liquid crystal exhibiting macromolecular behaviour. Liq Cryst 40: 314-320.

39. Burmistrov VA, Alexandriysky VV, Koifman OI (1992) Orientational effects of hydrogen bonding in liquid crystalline in solutions containing Shiff bases. Liq Cryst 12: 403-415.

40. Burmistrov VA, Alexandriysky VV, Koifman OI (1995) Influence of the molecular structure of a nematic solvent on hydrogen bonding with non-mesomorphic proton-donors. Liq cryst 18: 657-664.

41. Burmistrov VA, Yu V, Kareev O, Koifman I (1988) Mesogenic p-[4-(4-alkoxyphenylazo)benzylidene]-4-cyanoanilines. Zh Org Khim 24: 1742-1746.

42. Kuvshinova SA, Burmistrov BA, Novikov IV, Litov KM, Aleksandriiskii VV, et al. (2014) Synthesis, Mesomorphic and Dielectric Properties of 4-(Cyanomethoxy) phenyl 4-Alkoxybenzoates, 4-(Cyanomethoxy)-4'-alkoxyazo- and -azoxybenzenes. Russ J Org Chem 50: 615-620.

43. Zugenmaier P, Heiske A (1993) The molecular and crystal structure of a homologous series of bipolar mesogenic biphenyls - HO(CH2)nOC6H4-C6H4CN. Liq Cryst 15: 835-849.

44. Kuvshinova SA, Fokin DS, Burmistrov VA, Koifman OI (2009) Mesogenic 4-[4-(ω-hydroxyalkoxy) phenyl]diazenylcinnamic acids and their 4-cyanophenyl esters. Russ J Org Chem 45: 182-184.

45. Kuvshinova SA, Zav'yalov AV, Koifman OI, Aleksandriiskii VV, Burmistrov VA (2004) Mesogenic 4-(ω-Hydroxyalkoxy)-4'-formylazobenzenes. Russ J Org Chem 40: 1161-1164.

46. Burmistrov VA, Zavyalov AV, Novikov IV, Kuvshinova SA, Aleksandriisii VV (2005) Rheological properties of liquid-crystalline 4-(ω-hydroxyalkyloxy)-4'-cyanobiphenyls. Zh Fiz Khim 79: 142-145.

47. Burmistrov VA, Zavyalov AV, Novikov IV, Kuvshinova SA, Aleksandriisii VV (2005) Dielectric properties and orientation ordering of 4-(ω-hydroxyalkyloxy)-4'-cyanobiphenyls. Zh Fiz Khim 79: 1709-1712.

48. Kuvshinova SA (2010) Selectivity and thermodynamics of the dissolution of structural isomers in the stationary phases based on nematic 4-ethyloxy-4'-(ω-hydroxyalkoxy)azo- and azoxybenzenes. Russ J Phys Chem 84: 1956-1961.

49. Kuvshinova SA, Burmistrov VA, Fokin DS, Blokhina SV, Koifman OI (2009) Thermodynamic properties and selectivity of substituted liquid-crystal formylazobenzenes as stationary phases for gas chromatography. Russ J Anal Chem 64: 505-508.

50. Burmistrov VA, Aleksandriiskii VV, Koifman OI (2013) Hydrogen bond in thermotropic liquid crystals (Vodorodnaya svyaz' v termotropnykh zhidkikh kristallakh). Moscow: KRASAND 352.

51. Aronzon D, Levy EP, Collings PJ (2007) Trans-cis isomerization of an azoxybenzene liquid crystal. Liq Cryst 34: 707-718.

52. Kim MS, Song MY, Jeon B, Lee JY (2012) Synthesis and Nonlinear Optical Properties of Novel Y-type Polyurethane Containing Nitroazobenzene Group. Mol Cryst Liq Cryst 568: 111-116.

53. Kim MS, Song MY, Jeon B, Lee JY (2012) Synthesis and Properties of Novel Nonlinear Optical Polyimide Containing Nitrophenylazoresorcinoxy Group. Mol Cryst Liq Cryst 568: 105-110.

54. Witkiewicz Z, Suprynowicz Z, Dabrowski R (1979) Liquid crystalline cyanoazoxybenzene alkyls carbonates as stationary phases in small-bore packed micro-columns. J Chromatogr 175: 37-49.

55. Kuvshinova SA. Patent 2381214 (Russian Federation) 4-(2-Hydroxyalkyloxy)-4'-cyanoazoxybenzene revealing properties of liquid-crystalline stationary phase for gas chromatography.

56. Lazareva VT (1976) Liquid-crystalline compounds. IV. Para-substituted cyanoazobenzenes and cyanoazoxybenzenes 12: 149-156.

57. Reid RC, Prausnitz JM, Sherwood TK (1977) The properties of gases and liquids. McGraw-Hill.

58. Sawada M, Takai Y, Tanaka T, Hanafusa T, Okubo M, et al. (1990) Substituent effects on 15N and 17O NMR chemical shifts in 4'-Substituted trans-NNO-azoxybenzenes. Bull Chem Soc Japan 63: 702-707.

59. Marek R, Lycka A (2002) 15N NMR spectroscopy in structural analysis. Curr Org Chem 6: 35-66.

60. Witkiewicz Z, Oszczucłowski J, Repelewicz M (2005) Liquid-crystalline stationary phases for gas chromatography. J Chromatogr A 1062: 155-174.

61. Egorova KV, Vigdergauz MS (1985) Sorption and chromatography properties of liquid-crystalline p,p'-methoxyethoxyazoxybenzene. Zh Fiz Khim 59: 2774-2777.

62. Burmistrov VA, Kuvshinova SA, Fokin DS, Koifman OI (2009) Supramolecular liquid crystalline stationary phases for gas chromatography. The V China-Russia-Korea International Symposium on Chemical Engineering and New Materials Science. Daejeon, Korea. pp: 71-76.

63. Granovsky AA PC GAMESS version 7.1.E (FireFly), build number 5190.

64. Zhurko GA. Chemcraft software.

21

Supercritical-Fluid Chromatography with Diode-Array Detection for Emerging Contaminants Determination in Water Samples. Method Validation and Estimation of the Uncertainty

Vilma del C Salvatierra-Stamp[1], Norma S Pano-Farias[1], Silvia G Ceballos-Magaña[2], Jorge Gonzalez[1], Valentin Ibarra-Galván[1] and Roberto Muñiz-Valencia[1*]

[1]*Facultad de Ciencias Químicas, Universidad de Colima, Carretera Colima-Coquimatlán, Coquimatlán, Colima, Mexico*
[2]*Facultad de Ciencias, Universidad de Colima, c/Bernal Díaz del Castillo 340, Colima, Mexico*

Abstract

Here we present a communication about the article "Salvatierra-Stamp VC, Ceballos-Magaña SG, Gonzalez J, Ibarra-Galván V, Muñiz-Valencia R (2015) Analytical method development for the determination of emerging contaminants in water using supercritical-fluid chromatography coupled with diode-array detection. Analytical and Bioanalytical Chemistry 407:4219-4226". In this paper, a selective, linear, accurate and precise supercritical-fluid chromatography coupled with diode-array detection method was developed and validated for the determination of seven emerging contaminants: two pharmaceuticals, three endocrine disruptors, one bactericide and one pesticide. The compounds were base-line separated in around 10 minutes. Also, the method involved a sample treatment optimization by means of C18-OH solid phase extraction cartridges. The developed method was validated. In this sense, the correlation coefficient and recovery was higher than 0.9997 and 94%, respectively. Limit of detection and quantification was in the range of 0.10-1.59 µg/L and 0.31-4.83 µg/L, respectively. The measurement uncertainty was evaluated using the top-down model considering six sources of uncertainty. For all compounds, the uncertainty associated with accuracy and linearity regression was the main contribution to the combined uncertainty. Expanded uncertainties for each compound in method analysis were lower than 10.8%. Finally the method was successfully applied to environmental water samples.

Keywords: Emerging contaminants; Supercritical-fluid chromatography; Solid phase extraction; Water samples; Uncertainty assessment

Introduction

Emerging contaminants (ECs) consist of a large and growing group of compounds of natural and anthropogenic origin, among which include pharmaceuticals, pesticides, personal care products, hormones, industrial chemicals, etc. [1]. The occurrence of emerging contaminants in the aquatic environment has become an environmental problem of global concern. Traditionally these organic compounds were not considered pollutants; however, today it is known that many of them have potential harmful effects in various organisms, causing toxic effects and disorders in the endocrine system and can cause irreversible effects [2].

Pharmaceuticals and daily personal care products (PCPs) are widely used all over the world; however their disposal and body-excretion are usually not controlled [3,4]. A similar situation occurs with household pesticides which are often used without heed and disposed of as household waste. These actions lead to the contamination of the environment with different classes of pesticides [5]. The presence of these pollutants in the environment is usually in a very low concentration range: from ng/L to µg/L.

Sometimes the analysis of these pollutants can be difficult due to the wide variety of compounds. For this reason various techniques are used for both cleaning up and pre-concentration. The most widely used sample preparation is the solid phase extraction (SPE) [4,6,7], while liquid chromatography coupled with mass spectrometry (LC-MS) is the preferred analytical technique due to its high selectivity and sensitivity [8,9]; however, this MS detector requires high operating costs and is not available in many laboratories.

The use of supercritical-fluid chromatography (SFC) is an alternative technique that improves the analysis of these compounds, which has stimulated the interest of many researchers. In this technique, a mixture of carbon dioxide (CO_2) and an organic solvent as mobile phase is commonly used, allowing the use of high flow-rates with low pressure falls through the column, leading short analysis time and decreasing the consumption of the organic solvent, making the use of SFC a faster and attractive technique. Despite these advantages there are few publications related to the determination of contaminants in environmental samples [10]. Furthermore there are much fewer publications employing SFC coupled with diode-array detector (DAD) for analyzing emerging contaminants in water samples [11].

Nowadays, the demonstration of reliability and quality of the data produced in chemical analysis is of great importance during method development, especially for accredited laboratories. To ensure the reliability of the results both traceability and estimation of measurement uncertainty are important. This reliability and comparability of the data are obtained from the method validation and uncertainty assessment. The uncertainty is a quantitative indicator associated with a level of

***Corresponding author:** Roberto Muñiz-Valencia, Facultad de Ciencias Químicas, Universidad de Colima, Carretera Colima-Coquimatlán, 28400, Coquimatlán, Colima, Mexico, E-mail: robemuva@yahoo.com

confidence, in which the different possible sources of uncertainty are evaluated, including the stock solution preparation, sample preparation, glassware, etc. The aim of this evaluation is to show critical stages of an analytical method where uncertainty could be reduced.

The uncertainty can be estimated using different models being "top-down" and "bottom-up" the most commonly used. The "bottom-up" model includes a decomposition process of all unit operations performed by the analyst, grouped into common activities and an estimate of their contribution to the combined uncertainty value of the measurement process. The "top-down" model is based on method validation and precision data derived from the results obtained in the laboratory [12]. This model evaluates each individual uncertainty for every single step.

The EURACHEM/CITAT Guide Quantifying Uncertainty in Analytical Measurement recommends the following steps for proper estimation of uncertainty: i) specify measurand, ii) identify uncertainty sources, iii) quantify uncertainty components and iv) calculate combined and expanded uncertainty. In general, there are few papers about CEs analysis in which the uncertainty is estimated.

The objective of the work reported in the article Analytical and Bioanalytical Chemistry (2015) 407:4219-4226 was to develop, validate and apply an easy and sensible SPE-SFC-DAD analytical method to quantify glyburide, carbamazepine, 17 α-ethinyl estradiol, 17 β-estradiol, bisphenol A, diuron and triclosan in water samples. Moreover to determine the critical stages of the proposed method an estimation of the uncertainty applying the top-down model was carried out. Finally, the combined and expanded uncertainty for each compound was calculated from the contribution of each stage.

Materials and Methods

Solvents and materials

All analytical standards: carbamazepine (CBZ), glyburide (GBD), 17 α-ethinyl estradiol (17EE), 17 β-estradiol (17E), bisphenol A (BPA), triclosan (TCS) and diuron (DIU) and LC-MS grade solvents (methanol (MeOH) and acetonitrile (AcN)) were purchased from Sigma-Aldrich (Saint Louis, MO, USA). Working standard solutions were prepared weekly, stored at 4°C and protected from light.

The cartridges used for solid phase extraction (SPE) were Bond Elut-C18OH (1 g per 6 mL) and HF Bond Elut-C18 (500 mg per 3 mL) from Varian, Agilent (Santa Clara, CA. USA) and Discovery DSC18 (500 mg per 6 mL) from Sigma-Aldrich (Saint Louis, MO, USA). A Visiprep SPE vacuum manifold from Sigma-Aldrich (Saint Louis, MO, USA) and a vacuum pump model EV-40, from EVAR (Guadalajara, Mexico) were also used.

Collection and treatment of water samples

Different water samples (WS) were collected in amber glass bottles previously rinsed with ultra-pure water at a depth of 0.5 to 1 meter, in rivers near Colima City, Mexico. These samples were kept in coolers and stored at 4°C, protecting them from daylight until analysis. All samples were filtered through 0.45 μm highly hydrophilic polyvinylidene fluoride membrane filters from Phenomenex (Torrance, CA, USA) to eliminate suspended matter.

The SPE protocol used Bond Elut-C18OH cartridges previously raised with 3 mL of methanol and 6 mL of ultrapure water. Then, 150 mL water sample was placed in a 250 mL flask and processed at 8 mL/min flow-rate at pH 5.5 through the cartridge. Subsequently, the elution was performed using 3 mL MeOH at 1 mL/min flow-rate and injected into the SFC-DAD system.

SFC-DAD analysis

Chromatographic separation was carried out on an Acquity Ultra Performance Convergence Chromatography (UPC^2) system equipped with a binary solvent pump, with refrigerated auto sampler with a 10 μL loop, back-pressure regulator (BPR), convergence manager and column oven. This equipment was coupled with a diode-array detector (DAD). The BPR was maintained at 2000 psi. The column Viridis BEH-2-EP (4.6 mm × 100 mm, 5 μm particle size) with 10 μL injection loop was used at 40°C. The temperature of the auto sampler was maintained at 15°C. Data collection and analyses were performed using Empower™ Pro 3 Software. All system components are from Waters (Milford, MA, USA). Pressurized CO_2 (99.999%) was purchased from Praxair (Colima, Mexico).

The separation used gradient elution mode at a flow-rate of 1.4 mL/min. The gradient elution was performed as follows: from 0 to 5 min increase from 5 to 30% AcN; and finally from 5 to 10 min increase from 30 to 40%. Column reequilibration was performed for 3 min at the initial conditions. The qualitative analysis was performed in the 190-360 nm range, using the UV-absorbance DAD detector. Quantitative analysis was performed at 215 nm. ECs peak identification and the evaluation of purity each EC peak was done by comparing their retention times and UV spectra.

Method validation

Method validation was performed following the ICH and European Commission Decision 2002/657/EC guidelines [13,14], assessing selectivity, linearity, precision, accuracy and limits (quantification and detection).

Measurement uncertainty

The uncertainty was determined according to the procedures recommended by EURACHEM/CITAT Guide Quantifying Uncertainty in Analytical Measurement.

In the estimation of the uncertainty it is necessary to take each source and treat it separately to assess their contribution. When this component is expressed as a standard deviation is known as standard uncertainty, and is indicated as u(y).

As can be seen in equation 1, the "y" value of a combined standard uncertainty, $U_C(y)$, is calculated based on the uncertainty of the independent parameters p, q, r,...; e.g., u(y=q)

$$U_C(y(p,q,r,...)) = \sqrt{u(p)^2 + u(q)^2 + u(r)^2 ...} \quad (1)$$

Where y(p, q, r...) is a function of the individual parameters that cause uncertainty such as stock and working solution preparation, sample preparation, precision, accuracy, calibration curve, etc.

The expanded uncertainty (U_e) provides an interval where the measured value is expected to lie with an appropriate level of confidence. As shown in equation 2, U_e value is calculated by multiplying the combined uncertainty $u_c(y)$ by a coverage factor "k". The value selection for "k" is based on the level of confidence desired, and in this paper the coverage factor chosen was 2, corresponding to a confidence level of 95%.

$$U_e = u_C(y)*k \quad (2)$$

During the estimation of the expanded uncertainty for each

ECs of the presented method six sources of uncertainty were taken into account: stock solution preparation (u_1) and working solution preparation (u_2), sample preparation (u_3), precision (u_4), accuracy (u_5) and calibration curve (u_6).

In the stage of stock solution preparation, the purity of analytes is a common source of uncertainty. All ECs standards have a purity of 98%, with a tolerance of ± 2%. To determine the individual uncertainties the value of tolerance established by the manufacturer (a) should be treated as a rectangular distribution so that "d=3" in equation 3. In micropipettes and flasks manipulation "d=6", uncertainty is obtained by dividing the manufacturer´s specification by 6 to convert triangular deviation limits to standard deviation, using equation 3. On the other hand, the weighing of analyte is another element of uncertainty. In this case the manufacturer provides data of u(x)=0.1 mg.

$$u(x) = \frac{a}{\sqrt{d}} \tag{3}$$

With respect to volumetric uncertainty associated with the glassware, used in the preparation of stock and working solution, and in sample treatment (SPE), was calculated considering the coefficient of volume expansion of the liquid being employed at working temperature (equation 4).

$$u(x) = V * (\Delta T) * \alpha \tag{4}$$

Where V=working volume, ΔT=difference between working temperature and calibration temperature and α=coefficient of volume expansion of the liquid employed.

In the estimation of combined uncertainty (u_c), the relationship between the individual uncertainties u(y) value "y" and the uncertainty of the individual parameters in each stage is evaluated with equation 5, where (u(x)/x) is the individual uncertainty (volumetric, weighing, analyte purity, etc) expressed as relative standard deviation.

$$u_C(y(x_1, x_2, x_3, \ldots)) = \sqrt{\left(\frac{u(x_1)}{x_1}\right)^2 + \left(\frac{u(x_2)}{x_2}\right)^2 + \left(\frac{u(x_3)}{x_3}\right)^2 \ldots} \tag{5}$$

In the estimation of the uncertainty associated with the precision and accuracy, the number of repetitions is an important factor. Both estimations are performed for each analyte in the same way. For these estimations the standard deviation of the normalized differences (SD_{ND}) for each analyte is divided by $\sqrt{2}$ at a determined concentration (Equation 6). In this study the concentration chosen was the middle concentration of the calibration curve (0.2 mg/L).

$$u(\text{precision or accuracy}) = \frac{SD_{ND}}{\sqrt{2}} \tag{6}$$

With regard to the uncertainty associated with the calibration curve, the curve equation for each analyte is an important factor for its evaluation. The signal average (absorbance measurements) of the replicates was calculated at 0.2 mg/L. Equations 7-9 shows the individual calculation of uncertainty for each analyte.

$$u(c_0) = \frac{S}{B_1}\sqrt{\frac{1}{p} + \frac{1}{n} + \frac{\left(c_0 - \bar{c}\right)}{S_{xx}}} \tag{7}$$

Where S represent the residual standard deviation (equation 8), B_1 represent the calibration curve slope, c_0 is the evaluated concentration (0.2 mg/L), p is the number of repetitions to determine c_0, n is the number of measurements for all the concentrations in calibration curve, $\bar{c}$ is the average concentration of the calibration point, and S_{xx} was calculated using Equation 9. B_0 is the intercept of calibration curve for each analyte, A is the signal of the concentration of the calibration point (j....n) and cj is the concentration of the calibration point (j....n).

$$S = \sqrt{\frac{\sum_{j=1}^{n}\left[A_j - \left(B_0 + \left(B_1 * c_j\right)\right)\right]^2}{n-2}} \tag{8}$$

$$S_{xx} = \sum_{j=1}^{n}\left(c_j - \bar{c}\right)^2 \tag{9}$$

Results and Discussion

SFC-DAD optimization

The chromatographic conditions were optimized to separate each EC with good resolution within a reasonable analysis time. To achieve this, various organic modifiers are usually added to the supercritical fluid CO_2, which increases the solubility of polar compounds, improving peak shape and sensitivity of the method. In this method AcN was used as organic modifier added to the mobile phase. Optimization of chromatographic separation was carried out on a Viridis BEH 2-EP column injecting a methanolic solution mixture of ECs at 1 mg/L. Using the conditions described in 2.3 section, the baseline separation was achieved in less than 10 minutes (Figure 1), with resolution (Rs) values higher than 2.3 and symmetry peaks values near 1, with exception of GBD.

The effect of temperature (35 and 40°C) and back-pressure regulator (BPR) pressure (1500 and 2000 psi) was evaluated using the optimum gradient separation. These parameters may affect chromatographic separation and sensitivity by changing the density and viscosity of the mobile phase. After establishing the optimal conditions of temperature and BPR pressure at 40°C and 2000 psi, respectively, the effect of flow-rate at 1 and 1.4 mL/min were tested. At a flow-rate of 1 mL/min, a significant decrease resolution of 17EE and BPA was observed. Therefore, the flow-rate was set at 1.4 mL/min. The wavelength for ECs detection was set at 215 nm.

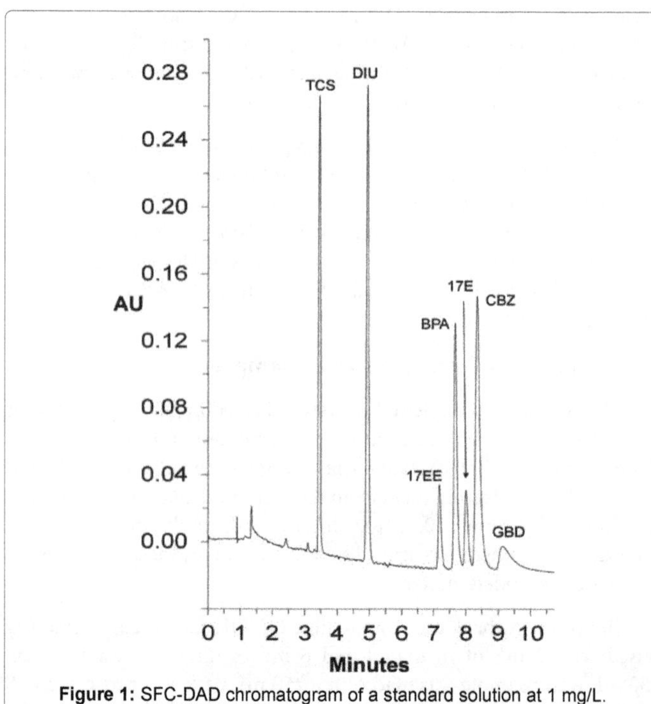

Figure 1: SFC-DAD chromatogram of a standard solution at 1 mg/L.

SPE optimization

For optimization of SPE three different SPE cartridges (HF Bond Elut-C18, Bond Elut-C18OH and Discovery DSC-18) were compared. The working protocol was as follows: The cartridge was conditioned with 3 mL of MeOH and then 6 mL of ultra-pure water. Subsequently, a volume of 150 mL of water sample (WS) spiked with a ECs standard mixture of 5 mg/L was placed in a 250 mL flask and passed through the cartridge at 4 mL/min flow-rate. ECs were eluted with 3 mL of MeOH at 1 mL/min flow-rate, and injected into the SFC-DAD system. As can be seen in Table 1, the best results were obtained using Bond Elut-C18OH cartridges and consequently selected for further optimization.

The influence of sample flow-rate (4, 6 and 8 mL/min) and pH (4.0, 5.5 and 7.0), washing solution (0, 5 and 15% MeOH, v/v) and elution solvent volume (4, 3, 2, and 1 mL MeOH) on the ECs extraction in WS was evaluated. There were no significant influence when varying sample flow-rate (RSD<1.5%) and sample pH (RSD<3.7%). The effect of washing solution was evaluated by visual inspection and no change in signals was observed. Respect to the elution solvent volume, it was observed that by using of 3 mL MeOH higher responses were obtained comparing with 4, 2 and 1 mL MeOH.

Therefore, the optimal conditions for SPE step were: sample flow-rate of 8 mL/min at pH 5.5, no washing step was applied and elution solvent volume was 3 mL MeOH.

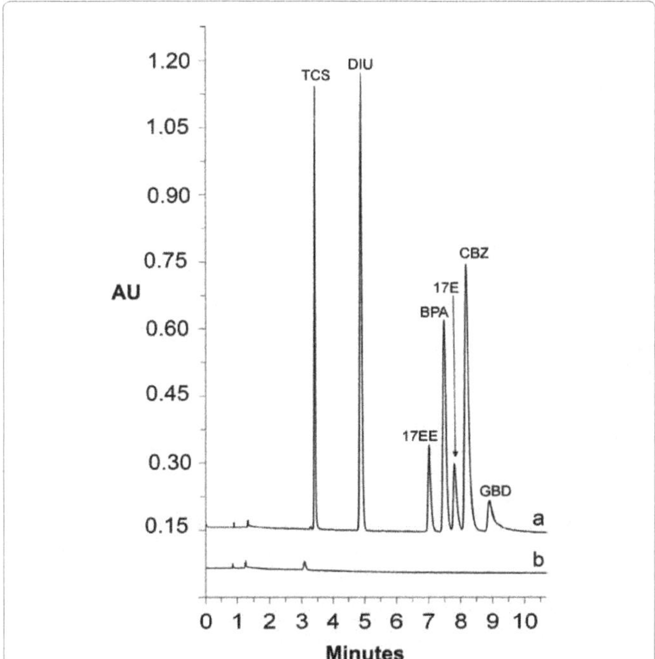

Figure 2: SFC-DAD chromatogram of a (a) WS spiked at 5 mg/L and (b) blank WS.

Method validation

The method selectivity was evaluated by qualitative comparison of the chromatograms obtained from the standards to 1 mg/L (Figure 1) and WS added with 5 mg/L (Figure 2a) by performing the procedure described in 2.3 and 2.2 section. UV spectrum of each peak detected was compared with the previously stored of the respective standard. As can be seen in Figure 2b, blank sample chromatogram has no signals caused by impurities. Thus, this method is suitable for analysis of ECs in WS.

The method linearity was performed using a calibration curve of 8 concentrations: 0.014, 0.02, 0.04, 0.1, 0.20, 0.40, 0.60 and 2 mg/L, according to the methodology described in the 2.3 section. The results were evaluated by linear regression where the correlation coefficients (r) for all ECs were higher than 0.9997.

The precision and accuracy of the method were evaluated by triplicate analysis at a concentration of 0.2 mg/L. With regard to precision, evaluated by intra-day repeatability, triplicate determinations were carried out on three consecutive days, with relative standard deviations (RSD) less than 9.2%, as shown in Table 2. Referring to accuracy, expressed as recovery percentages (%R), were found in a range of 94.5 to 103.1% (Table 2), which are within the acceptance criteria.

For the evaluation of LOD and LOQ 20 blank WS were analyzed and the standard deviation (SD) of responses in the retention time of each analyte was calculated. These values of SD were divided by the slope of the calibration curve of each analyte and multiplied by 3.3 and 10 to calculate LOD and LOQ, respectively. The values obtained are shown in Table 2.

Compound	Recovery (%)		
	Bond Elut-C18OH	HF Bond Elut-C18	Discovery DSC-18
TCS	87.8	87.9	12.8
DIU	74.0	62.4	37.6
17EE	81.2	67.6	24.2
BPA	60.4	51.2	16.0
17E	86.0	73.4	0
CBM	64.7	55.3	26.3
GBD	65.1	58.6	21.2

Table 1: Recoveries using three different SPE cartridges.

Compound	Intra-day Precision [a] (% RSD)	Accuracy [a] % Recovery (RSD)	LOD[b] (μg L^{-1})	LOQ[b] (μg L^{-1})
TCS	5.5	101.0 (6.0)	0.40	1.22
DIU	5.5	94.7 (5.8)	0.43	1.31
17EE	7.0	103.1 (7.2)	1.59	4.83
BPA	6.1	102.6 (6.2)	0.23	0.69
17E	6.7	95.6 (7.0)	0.71	2.15
CBZ	5.5	101.9 (5.7)	0.10	0.31
GBD	6.0	96.0 (6.1)	1.41	4.27

[a]Intra-day precision and recovery were assessed at 0.20 mg/L (n=3)
[b]Limit of detection (LOD) and quantification (LOQ)

Table 2: Validation parameters: Intra-day precision, accuracy, and limit of detection (LOD) and quantification (LOQ) for ECs using WS.

Measurement uncertainty

Six sources of uncertainty were taken into account: the preparation of the stock solution at 1000 mg/L (u_1), the preparation of working solution at 100 mg/L (u_2), the sample preparation step using SPE (u_3), uncertainty associated with precision (u_4), uncertainty associated with accuracy (u_5) and uncertainty associated with the linearity regression (u_6). Uncertainty of volumetric flasks and automatic pipettes (associated with u_1, u_2 and u_3) were calculated using the tolerance values provided by the manufacturer. Uncertainty associated with precision (u_4) and accuracy (u_5) were evaluated taking into account the middle point of the calibration curve (0.2 mg/L, n=20). For linearity regression uncertainty (u_6) signals, calibration curve data and number of repetitions were used. The combined uncertainty (Uc) of this method was calculated using the Equation 1. The expanded uncertainty (Ue) is obtained by

Compound	Precision, % (u_4)	Accuracy, % (u_5)	Linearity regression, % (u_6)	Uc (%)	Ue, k=2, (%)
TCS	1.88	-3.13	3.46	5.40	10.80
DIU	1.66	-3.14	2.92	5.00	10.00
17EE	1.53	-2.18	2.99	4.46	8.92
BPA	0.19	-2.30	2.86	4.17	8.34
17E	0.28	-2.35	2.78	4.15	8.30
CBZ	2.43	-2.50	2.67	4.81	9.62
GBD	2.03	-2.41	2.63	4.55	9.10

Table 3: Uncertainties estimated in each step of ECs analysis for each compound. Uncertainty associated with stock solution preparation at 1000 mg/L (u_1), working solution preparation at 100 mg/L (u_2) and sample treatment (u_3) is 1.20, 0.59 and 1.45 %, respectively.

applying a coverage factor of k=2, corresponding to a 95% confidence level around the results where the measurand may lie.

As shown in Table 3, method expanded uncertainty (U_e) were in the 8.30-10.80% range, being the highest to TCS and the lowest to 17E. The individual expanded uncertainties values for preparation of stock (u_1) and working (u_2) solution for all analytes, u_1 and u_2 were 1.20 and 0.59 respectively due to their were obtained based on the manufacturer specifications data. In similar way, the uncertainty for all ECs in the stage of sample preparation (u_3) was 1.45. During this stage, the use of volumetric glassware had the higher contribution. Uncertainties associated with accuracy (u_5) and linearity regression (u_6) had the major contribution to U_c and U_e for all compounds.

As can be seen in Table 3, precision uncertainties are in the 1.5-2.5% range, with exception of BPA (0.19%) and 17E (0.28%). Accuracy uncertainties are around-2.4%, with exception of TCS (3.13%) and DIU (3.14%). Linearity regression uncertainties are in the 2.6-3.0%, with exception of TCS (3.46%). Expanded uncertainty values demonstrate that with a confidence level of 95%, this method is efficient and suitable for the determination of the target ECs in water samples.

Method feasibility

It was performed in 7 different surface WS. For determination, UV spectra of the peaks found in the sample were compared with the corresponding previously determined standards. The results obtained shows that TCS could be quantified at concentrations of 1.3 and 1.2 µg/L in WS4 and WS5, respectively, but in WS1, WS3, WS6, and WS7 were lower than LOQ. BPA concentration in WS4 was 0.70 µg/L, but lower than LOQ in WS2. DIU, 17E, 17EE, CBZ and GBD were not detected in any sample.

Conclusions

A sensitive and rapid SPE-SFC-DAD method for the determination of TCS, DIU, 17EE, 17E, CBZ and GBD in WS was developed. Chromatographic separation was performed on a Viridis BEH-2-EP column, using CO_2 and AcN as the mobile phase in gradient mode. SPE was carried out using Bond Elut-C18OH cartridges. The method was validated according to international validation guidelines, obtaining recoveries higher than 94%, with LOD and LOQ in a range of 0.10-1.59 mg/L and 0.31-4.83 mg/L, respectively. For all compounds, the uncertainty associated with accuracy and linearity regression was the main contribution to the Uc and Ue. Expanded uncertainty (Ue) values were lower than 10.80% this implies that the proposed method is reliable for the determination of the target ECs in water samples.

Acknowledgments

Salvatierra-Stamp thanks Consejo Nacional de Ciencia y Tecnologia (CONACyT) Mexico for the grant provided. This study was funded by CONACyT Mexico grand project 254184.

References

1. Gavrilescu M, Demnerová K, Aamand J, Agathos S, Fava F (2015) Emerging pollutants in the environment: present and future challenges in biomonitoring, ecological risks and bioremediation. N Biotechnol 32: 147-156.
2. Eertmans F, Dhooge W, Stuyvaert S, Comhaire F (2003) Endocrine disruptors: effects on male fertility and screening tools for their assessment. Toxicol In Vitro 17: 515-524.
3. Liu JL, Wong MH (2013) Pharmaceuticals and personal care products (PPCPs): a review on environmental contamination in China. Environ Int 59: 208-224.
4. Kostopoulou M, Nikolaou A (2008). Analytical problems and the need for sample preparation in the determination of pharmaceuticals and their metabolites in aqueous environmental matrices. TrAC 27: 1023-1034.
5. Rodriguez-Mozaz S, Ricart M, Köck-Schulmeyer M, Guasch H, Bonnineau C, et al. (2015) Pharmaceuticals and pesticides in reclaimed water: Efficiency assessment of a microfiltration-reverse osmosis (MF-RO) pilot plant. J Hazard Mater 282: 165-173.
6. Jiang JQ, Zhou Z, Sharma VK (2013) Occurrence, transportation, monitoring and treatment of emerging micro-pollutants in wastewater – A review from global views. Microchem J 110: 292-300.
7. Li X, Zheng W, Kelly WR (2013) Occurrence and removal of pharmaceutical and hormone contaminants in rural wastewater treatment lagoons. Sci Total Environ 445-446: 22-8.
8. Shi Z, Hu J, Li Q, Zhang S, Liang Y, et al. (2014) Graphene based solid phase extraction combined with ultra-high performance liquid chromatography-tandem mass spectrometry for carbamate pesticides analysis in environmental water samples. J Chromatogr A 1355: 219-227.
9. Paíga P, Lolić A, Hellebuyck F, Santos LH, Correia M, et al. (2015) Development of a SPE-UHPLC-MS/MS methodology for the determination of non-steroidal anti-inflammatory and analgesic pharmaceuticals in seawater. J Pharm Biomed Anal 106: 61-70.
10. Bernal JL, Martín MT, Toribio L (2013) Supercritical fluid chromatography in food analysis. J Chromatogr A 1313: 24-36.
11. Salvatierra-Stamp VC, Ceballos-Magaña SG, Gonzalez J, Ibarra-Galván V, Muñiz-Valencia R (2015) Analytical method development for the determination of emerging contaminants in water using supercritical-fluid chromatography coupled with diode-array detection. Anal Bioanal Chem 407: 4219-4226.
12. (2012) Quantifying Uncertainty in Analytical Measurement. 3rd edn. EURACHEM/CITAC Guide.
13. ICH (1996) Guidance for Industry Q2B Validation of Analytical Procedures: Methodology.
14. Rodríguez N, Ortiz MC, Sarabia LA (2009) Study of robustness based on n-way models in the spectrofluorimetric determination of tetracyclines in milk when quenching exists. Anal Chim Acta 651: 149-158.

Quantification of Selected Antiretroviral Drugs in a Wastewater Treatment Works in South Africa Using GC-TOFMS

Schoeman C[1*], Mashiane M[1], Dlamini M[1] and Okonkwo OJ[2]

[1]*Department of Environmental, Water & Earth Sciences, Faculty of Science, Tshwane University of Technology, Rand Water, 2 Barrage Road, Vereeniging, South Africa*
[2]*Department of Environmental, Water & Earth Sciences, Faculty of Science, Tshwane University of Technology, 175 Mandela Drive, Pretoria, South Africa*

Abstract

Increasing amounts of Pharmaceutical Personal Care Products (PPCPs) have been detected in the water cycle in recent years. Of all the PPCPs, very little information regarding the determination of Antiretroviral Drugs (ARVDs) is available. The aim of this study was to monitor the concentrations of two ARVDs, nevirapine and efavirenz in influent and effluent points at a Wastewater Treatment Works in Gauteng, South Africa. Treated wastewater, before and after chlorination, was also examined to determine if the target ARVDs were removed by chlorination. The target ARVDs were extracted from wastewater using Solid Phase Extraction (SPE) and the extracts were subsequently analysed using Gas Chromatography-Time of Flight Mass Spectrometry (GC-TOFMS).The method (extraction plus instrumental) was validated to determine limits of detection and quantification; accuracy; precision and uncertainties (at 40 ng/L) and all were found to be well within requirements for part per trillion analyses. The robustness of the method was also determined by analysing 10 quality control replicates on three non-consecutive days and found to be fit for purpose. The concentrations of nevirapine and efavirenz in wastewater influent were found to be as high as 2100 and 17400 ng/L respectively. As much as 50% of the ARVDs were removed by the wastewater treatment plant and resulted in treated effluent concentrations of nevirapine and efavirenz as high as 350 and 7100 ng/L respectively. Chlorination was not found to affect the ARVDs significantly. The findings from two other investigations, one in Germany and the other from South Africa, that have investigated ARVDsin surface water and wastewater are compared with those of this study.

Keywords: Antiretroviral drugs; Wastewater; Chlorination; Solid phase extraction; Gas chromatography-time of flight mass spectrometry; Method validation

Introduction

Pharmaceuticals are synthetic or natural chemicals that can be found in prescription medicines, over-the-counter therapeutic drugs and veterinary drugs [1]. These compounds contain active ingredients that are of benefit to society but can ultimately end up in the water cycle at trace concentrations (nanograms to low micrograms per litre). The occurrence of these PPCPs in the environment has been widely discussed and published in the literature for the last ten years. These compounds could well have been present in the water prior to this time, however, advances in analytical techniques and instrumentation have only allowed for their detection in the last decade. Many surveys have indicated that PPCPs are present in wastewater and effluents and, as such, can be expected to be a source of PPCPs in drinking water. Routine monitoring programmes exist for regulated chemical and microbiological parameters; however, this is not the case for PPCPs. Ad hoc surveys for particular PPCPs have generated data that is available in the literature. Available studies have reported that PPCP concentrations in surface waters, groundwater and partially treated waters are typically less than 100 ng/l and those in treated water are generally less than 50 ng/l [1]. Because domestic wastewater can be expected to be a source of PPCPs in drinking water, the removal of these compounds by wastewater treatment processes is important. Although such processes are not designed to remove PPCPs, they do so to varying degrees [1]. PPCPs removal during wastewater purification is dependent on their physical and chemical properties. Wastewater treatment plants that have biological treatment such as activated sludge processes or bio filtration have been shown to remove PPCPs at varying rates, ranging from less than 20% to in excess of 90%. Efficiencies have been shown to vary depending on the operational configuration of the treatment plant. Such a plant was the subject of this study. Factors influencing removal include sludge age, activated sludge tank temperature and hydraulic retention time. Advanced processes that include reverse osmosis, ozonation and advanced oxidation technologies can result in higher removal of PPCPs [1]. Traditional drinking water treatment processes such as coagulation do not remove many of the PPCPs. Free chlorine can remove approximately 50% of PPCPs, chloramines are less effective [1]. Advanced drinking water purification processes (ozonation, oxidation, activated carbon and membranes) result in removal rates of over 90% of PPCPs [1]. Literature indicates that concentrations of PPCPs in drinking water are usually more than 1000-fold below the minimum therapeutic dose, the lowest clinically active dose [1]. It was estimated that 2500 000 people in South Africa required Antiretroviral Therapy in 2012 [2]. A daily dose of combination therapy of HIV-ARVs (mean of 991 mg/day/person, range 590 - 1996) equates to a total of 542 944 kg of ARVD compounds ingested per year (assuming 1.5 million people are on ARVDs). Excretion of ARVDs varies depending on compound, though some, such as tipranavir are excreted at 80% and nevirapine at 2.7% via urine [3]. Assuming a mean of 30% excretion to sewage via urine and faeces, we estimate that about 162 883 kg of ARVDs could reach the aquatic systems of South Africa every year [3]. The large amounts of ARVDs that are potentially being discharged into the South African wastewater system

***Corresponding author:** Schoeman C, Department of Environmental, Water & Earth Sciences, Faculty of Science, Tshwane University of Technology, Rand Water, 2 Barrage Road, Vereeniging, South Africa
E-mail: cschoema@randwater.co.za

necessitate that they be monitored. Much work has been carried out on various matrices. For example, HPLC analysis of ARVDs in biological matrices [4]; tenofovirin agricultural soil [5]; tipranavir in human plasma [6]; antiretroviral drugs abacavir and tenofovir in human hair [7]; and nevirapine in plasma [8]. Studies that have been carried out on water samples include the examination of surface and wastewater in Germany [9]. Treated and raw wastewater and rivers were found to contain ARVDs, using Liquid Chromatography-Tandem Mass Spectrometry (LC-MS/MS). Relative recoveries exceeded 80% and limits of quantification ranged between 0.2 and 10 ng L^{-1}. Acyclovir, abacavir, lamivudine, nevirapine oseltamivir, penciclovir, stavudine, zidovudine were all detected. Further work carried out in South Africa [10] the simultaneous quantification of 12 antiretroviral compounds (zalcitabine, tenofovir, abacavir, efavirenz, lamivudine, didanosine, stavudine, zidovudine, nevirapine, indinavir, ritonavir and lopinavir) in surface water using the standard addition method is described. Water samples were concentrated by a generic automated solid phase extraction method and analysed by LC-MS/MS). Substantial matrix effect was encountered in the samples; an average method detection limit of 90.4 ng/L was reported. zalcitabine, tenofovir, abacavir, efavirenz, lamivudine, didanosine, stavudine, zidovudine, nevirapine, indinavir, ritonavir and lopinavir were all detected in the water samples analysed.

In the present study, influents and effluents water samples obtained from a Wastewater Treatment Works were extracted using SPE and ARVDs identified using GC-TOFMS. MS detection systems were used to identify PPCPs as described by [11]. GC-TOFMS has been used for the analysis of organic UV filters and insect repellents in wastewater [12]. Also, two-dimensional GC-TOFMS has been used to determine a number of PPCPs in river water [13]. TOFMS facilitates very rapid spectral scanning and this, along with sophisticated deconvolution software, allows the tentative identification of unknown compounds in the wastewater. Two ARVDs identified in this manner were nevirapine and efavirenz. These compounds are both non-nucleoside reverse transcriptase inhibitors, are soluble in methanol and could be expected to be extracted using SPE techniques. Reports of analysis using GC-MS [8] also indicated that this class of compounds can be analysed using GC-MS without derivatization. Details of the ARVDs [10] investigated and parathion [11] is shown in Table 1.

Materials and Methods

Sample collection

Sewage samples were collected from the Wastewater Treatment Works in Gauteng, South Africa. The samples that were collected were the influent to the works, the purified sewage effluent prior to chlorination and lastly, the chlorinated effluent discharged from the Works. Samples were taken in 1 litre clear Schott bottles and were stored at 4°C and analysed within 24 h of receipt.

Materials

All organic solvents, including reagent water, were manufactured by Honeywell Burdick and Jackson (USA) and of HPLC grade. Phosphoric acid was purchased from Merck SA, univ AR. SPE extraction cartridges used were Agela Cleanert PEP 60 mg 3 ml; the SPE drying cartridges used were Bond Elute Sodium Sulphate cartridges. Sample extracts were concentrated using a Techne dry block, DB 3 (nitrogen 99.999% purity). Analytical standards of nevirapine and efavirenz were supplied by USP, USA.

Methods

Extraction: Use was made of a manual SPE apparatus (Agilent, 20

Name (CAS No.)	Molecular Mass, g/mol	Structure	Log K_{ow}	pKa Strongest Acid/Base
Nevirapine (129618-40-2)	266.89	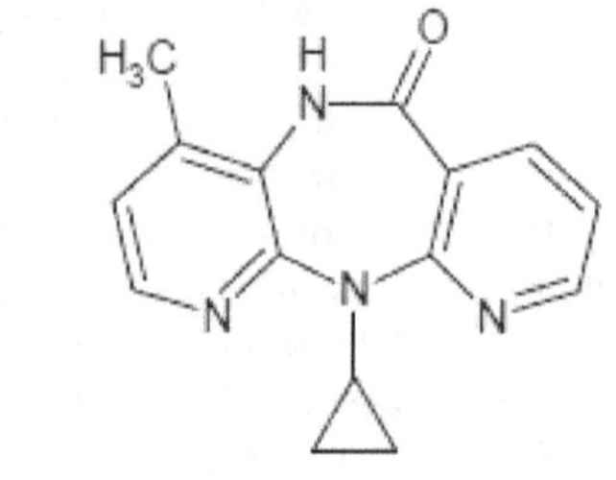	3.89	10.37/5.06
Efavirenz (154598-52-4)	315.68	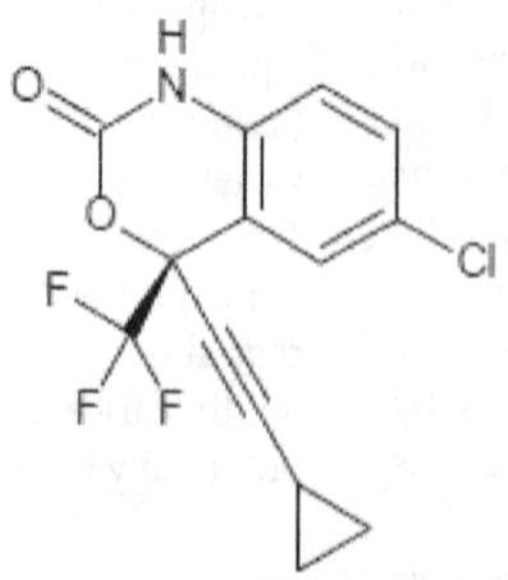	4.15	12.52/-1.5
Parathion, ISTD (56-38-2)	291.26	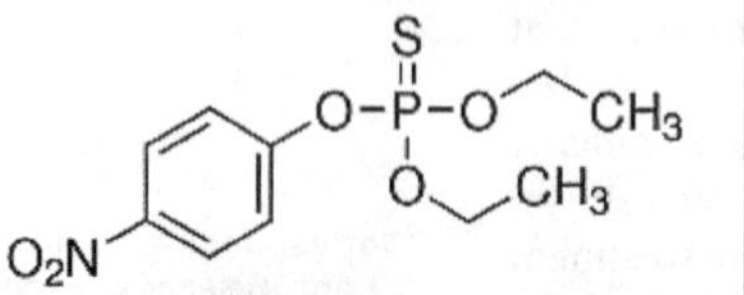	3.81	7.14

Table 1: Nevirapine, efavirenz and ISTD (parathion) CAS numbers, molecular masses, structures, Log Kow and pKa values.

position manifolds), SPE extraction cartridges were conditioned sequentially with one column volume of dichloromethane, ethyl acetate, methanol and de-ionized water containing 0.1% phosphoric acid (v/v). Samples (1 litre), calibration standards and the method blank were eluted through the cartridges at a flow rate not exceeding 10 ml/min (individual drops just visible). Once the sample had eluted through the cartridge, they were allowed to dry under vacuum for about 10 min. Bond Elute sodium sulphate SPE cartridges were washed (1 column volume of dichloromethane and 1 column volume of ethyl acetate) and placed below the extraction SPE cartridges to remove any water during the desorption step. Adsorbed compounds were desorbed from the cartridges by passing 500 µl of ethyl acetate (2x) and 600 µl dichloromethane through the cartridges under a gentle vacuum. The solvents were allowed to soak the SPE cartridges for 60 sec to enhance desorption. The solvents were pooled in a GC vial down blown with nitrogen (manifold temperature of 40°C) to about 150 µl and then made

up to 200 µl with ethyl acetate. The extract was analysed using GC-TOFMS. The use of matrix spiked calibration standards precluded the need for analyte recoveries to be determined as these were automatically accounted for when quantifying using these calibration standards.

Instrument parameters: Helium carrier gas used for the GC-TOFMS was supplied by Air Liquide and was 99.9999% pure. The GC inlet liner used was an SGE tapered focussing liner and the GC capillary column was a Phenomenex Zebron ZB-Semivolatiles GUARDIAN column (5m guard column, 30 m x 0.25 mm x 0.25 µm). The analytical instrumentation used was an Agilent 7890A GC (incorporating a Gerstel MPS 2 liquid auto sampler) coupled to a LECO Pegasus® HT TOF. LECO ChromaTOF® software version 4.24 was used for the identification of target compounds and for the quantificationof Nevirapine and efavirenz.The injection volume was 3 µl, pulsed splitless (2 min @ 50 psi) injection mode at 275°C with a purge delay of 0.5 min. The column flow was 1.2 ml/min, constant flow. The GC initial oven temperature was held at 80°C for 1 min. The temperature was then ramped to 320°C at 12.5°C/min. The transfer line between the GC and the TOFMS was maintained at 280°C. The TOFMS analysis was performed in the electron impact mode at 20 scans/sec and at a source temperature of 250°C. Detector voltage was boosted by 250 volts to maximize instrument sensitivity. LECO ChromaTOF® software version 4.24 was used to integrate and identify the target compounds in the wastewater extracts. The data processing method used for integration of the compounds incorporated a baseline offset of 0.5; auto smoothing; peak broadening calculated as the run progressed; and a signal/noise ratio of 5. The library searching was limited to forward searching of compounds of 1000 a.m.u. and less and the minimum mass spectral similarity match was set to 300. Such a low match factor was necessary to detect trace contaminants in the complex wastewater matrix. Resultant false positive matches were manually excluded from the results.

Nevirapine and Efavirenz quantification: The stock standard was diluted in ethyl acetate to give a working standard concentration of 1 ng/µl. This working standard was used to prepare the spiked matrix matched calibration standards. 20; 40; 60; 80; 100; and 200 µl were added to 1 litre of water deionized to prepare 20; 40; 60; 80; 100; and 200 ng/L solutions that were extracted in the same manner as the method blank and samples. These are then used to prepare the calibration curves. An Internal Standard (ISTD), parathion, was added to all calibration standards and samples (60 µl of a 1 ng/µl stock solution made up in ethyl acetate). ISTD was added before extraction to determine performance of the sample prep method. Quality control samples, waters spiked with ISTD and ARVD target compounds to 40 ng/L, were extracted and analysed with every batch of samples analysed. These quality control samples were analysed after the calibration standards and again at the end of the run sequence, or after 10 samples. Three sets of samples were analysed, the last of these sets were run using ten replicates for each of the influent; the pre- and post-chlorinated effluents and spiked quality control samples (on three non-consecutive days) to determine method robustness, uncertainties and method detection and quantitation limits of nevirapine and efavirenz. Wastewater influents and effluents samples were diluted with deionised water to ensure that they fell within the calibration range.

Statistical calculations: Random uncertainties of the calibration curves of nevirapine and efavirenz were used to calculate method Limits of Detection and Quantification (LODs and LOQs) as shown below,

- In the absence of meaningful blanks, the LOD was determined using the slope and regression uncertainties (as determined by Regression Analysis), $X_{LOD}=3S_b/b$. S_b is the Slope Uncertainty and b is the slope, the Slope and the Random Uncertainties are generally assumed to be equivalent and the Slope Uncertainty has been used in these calculations.
- `In the absence of meaningful blanks, the LOQ was determined using the slope and regression uncertainties (as determined by Regression Analysis), $X_{LOQ}=10S_b/b$. S_b is the Slope Uncertainty and b is the slope, the Slope and the Random Uncertainties are generally assumed to be equivalent and the Slope Uncertainty has been used in these calculations.

Method precision, accuracy and uncertainties (95% confidence) were calculated from the results obtained from 10 quality control samples extracted and analysed on the same day and were determined as follows.

- Precision was calculated using % RSD = (mean of SDEV of QCs/mean of QCs) x100.
- Accuracy was determined as follows, % Accuracy = (mean value/true value).
- Uncertainties were determined at 95% levels of confidence. Included in the Uncertainty Budget are the contributions made by the uncertainties of % standard purity; uncertainties of volume; uncertainty of mass; uncertainty of regression; and uncertainty of repeatability.

The robustness for this method was determined by comparing the results for quality control samples over 3 non-consecutive days. The statistical tool used in this case was the F-Test (at the 95% confidence level), which examines whether the standard deviation of two sets of data are similar or dissimilar from each other.

- No. of Replicates – minimum 8
- No. of sets of data – 3

The F_{calc} was calculated as follows for each compound:

$$F_{calc} = SD_1^2 / SD_2^2$$

Where SD 2 > SD 1

Fcrit = 3.18 with $(9_x, 9_y)$

Where: x = degrees of freedom for SD_2

y = degrees of freedom for SD_1

9 = degrees of freedom (number of replicates)

Results and Discussion

The determination of nevirapine and efavirenz in wastewater and treated wastewater samples is robust. Method statistics are shown in Table 2; calibration curves for nevirapine and efavirenz are shown in Figure 1 and Figure 2. Regression statistics of the calibration data yield excellent r^2 values and linearity (F factors of 57334 and 3027 for nevirapine and efavirenz respectively). Low random uncertainties for both curves resulted in low LODs and LOQs. Method precision, accuracy and uncertainties (95% confidence) were calculated from the results obtained from 10 quality control samples extracted and analysed on the same day and were found to be acceptable for parts per trillion analytical determinations [14]. The method was deemed robust and fit for purpose [15].

GC-TOFMS has been compared to quadrupole GC-MS using synthetic drug standards [16]. Lower LODs were evident with GC-TOFMS and were comparable to GC-ECD, with the added advantage of high quality full scan mass spectra. Peak deconvolution in the ChromaTOF® software has been shown to be poor with higher concentrations of sample extracts and this tends to lead to compromised chromatography [17]. Wastewater influents were diluted 100 times and the effluents 50 times respectively with deionised water for the observed concentrations to be within the calibration range. A further advantage of the dilution of the samples is that the chromatographic integrity of the system was less compromised and greater numbers of samples could be analysed before instrument maintenance was required.

A typical reconstructed ion chromatogram, showing ions characteristic of the compounds investigated, is shown in Figure 3. Although all ions were detected all the time, however, only selected ions (reconstructed ion chromatogram) are displayed for clarity (Figure 3). As can be seen, both nevirapine and efavirenz chromatograph well. Peak shapes are good (no tailing is evident) unless the column is overloaded. In the calibration range used in this work (20 to 200 ng/l) no overloading of the capillary column was observed. Retention times were not excessive (<1000 seconds) and were in the same region of the chromatogram as the ISTD, Figure 3 below. The mass spectra are shown in Figures 4a and 4b (nevirapine) and Figures 5a and 5b (efavirenz). For clarity mass spectra from both matrix spiked calibration samples and an analytical standard are included. Mass spectra for both nevirapine and efavirenz included ions of >200 a.m.u. of high abundances and facilitated accurate deconvolution and quantification. The mass spectra for the matrix spike and the analytical standard agreed well, ions characteristic of hydrocarbons (57, 71, 85, and 99 a.m.u. are indicative of alkanes) are additional in the in the matrix spike and were expected. The ions used for quantification of for efavirenz and nevirapine were 246 and 265 a.m.u. respectively, the ion used for the ISTD was 291 a.m.u. The nevirapine and efavirenz concentrations that were detected in the samples are shown in Figure 6. As can be seen in Figure 6, the concentrations of nevirapine and efavirenz vary from one sampling event to another. The concentrations of nevirapine and efavirenz in wastewater influent were found to be as high as 2100 and 17400 ng/L respectively. As much as 50% of the ARVDs were removed by the Wastewater Treatment Works and resulted in treated effluent concentrations of nevirapine and efavirenz as high as 350 and 7100 ng/L respectively. Also, the amounts removed by the Wastewater Treatment Works varied, and this is most likely related to both the nature of influent and the operation of the works. What is consistent is that the Wastewater Treatment Works does remove both compounds to some degree. This is in agreement with findings described earlier [1]. The effect of the chlorination of the treated wastewater on ARVDs concentrations was inconclusive.

Although efavirenz was detected at various surface water sampling points in South Africa in the study by Wood et al [10], concentrations were too low for quantification. Nevirapine was detected in all of the samples in this study but was only quantitated in 9 of the 24 sampling stations; the highest reported concentration was 1480 ng/L. Nevirapine is widely used for the treatment of HIV and for the prevention of mother-to-child transmission and its environmental persistence [18] make it likely to be found in the environment. Work carried out in Germany [9] indicated that nevirapine and was not removed by a wastewater treatment works whilst other ARVDs were removed. Both of these studies used LC-MS/MS to determine the ARVDs.

Conclusion

Parameter	Nevirapine	Efavirenz
LOD, ng/L	1.8	7.8
LOQ, ng/L	6.0	25.9
Precision, %	10.2%	3.5%
Accuracy, %	106%	109%
Uncertainty, ng/L	1.6	6.4

Table 2: Method statistics.

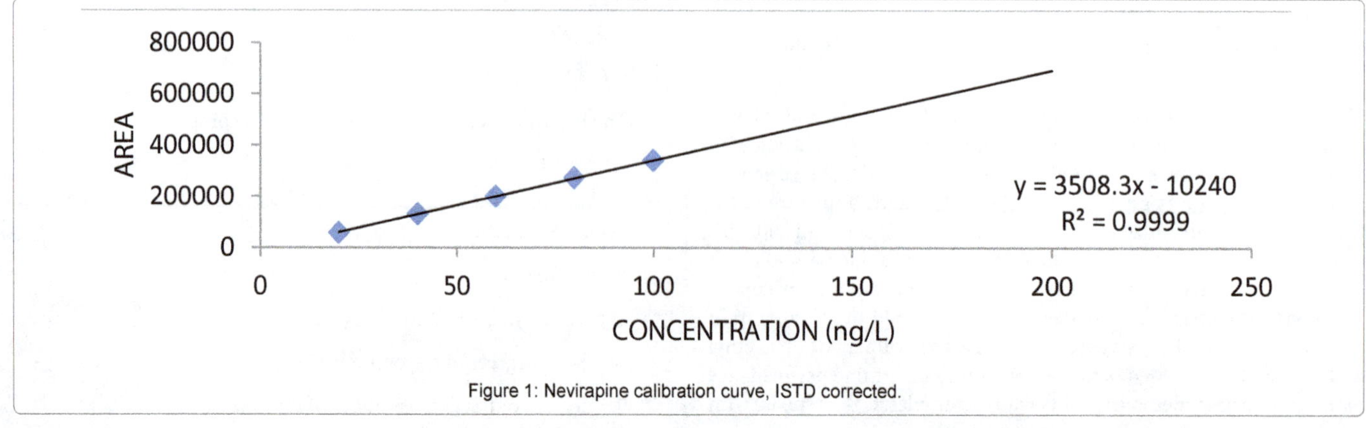

Figure 1: Nevirapine calibration curve, ISTD corrected.

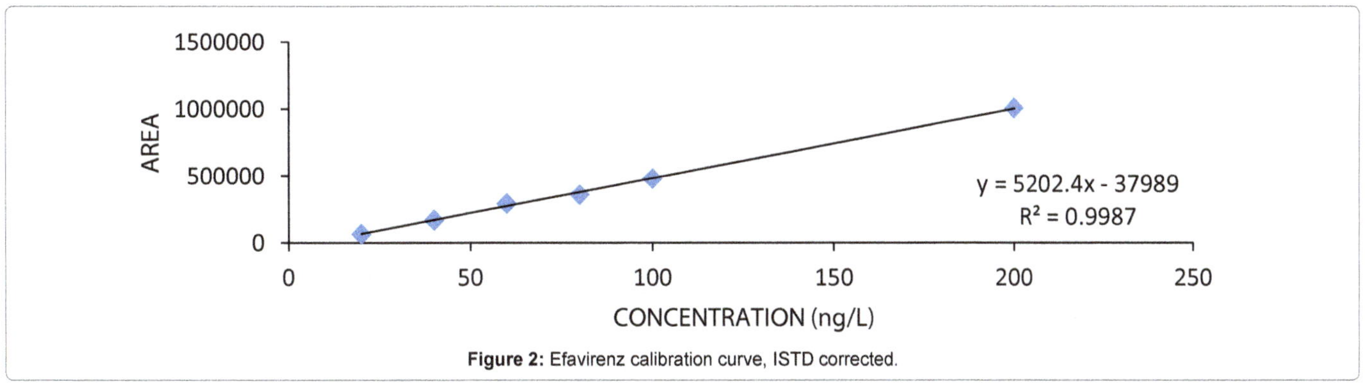

Figure 2: Efavirenz calibration curve, ISTD corrected.

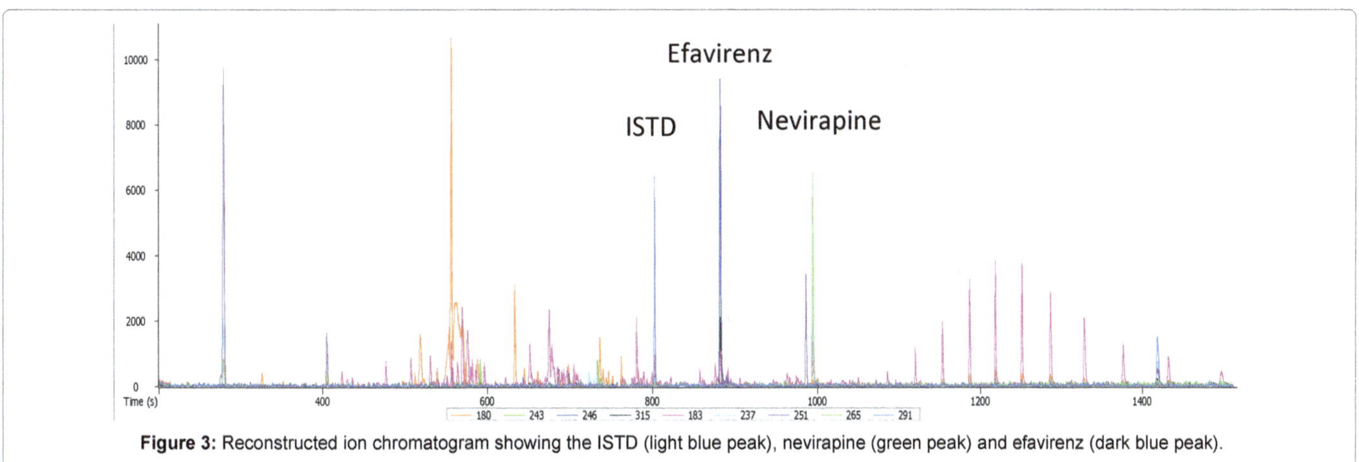

Figure 3: Reconstructed ion chromatogram showing the ISTD (light blue peak), nevirapine (green peak) and efavirenz (dark blue peak).

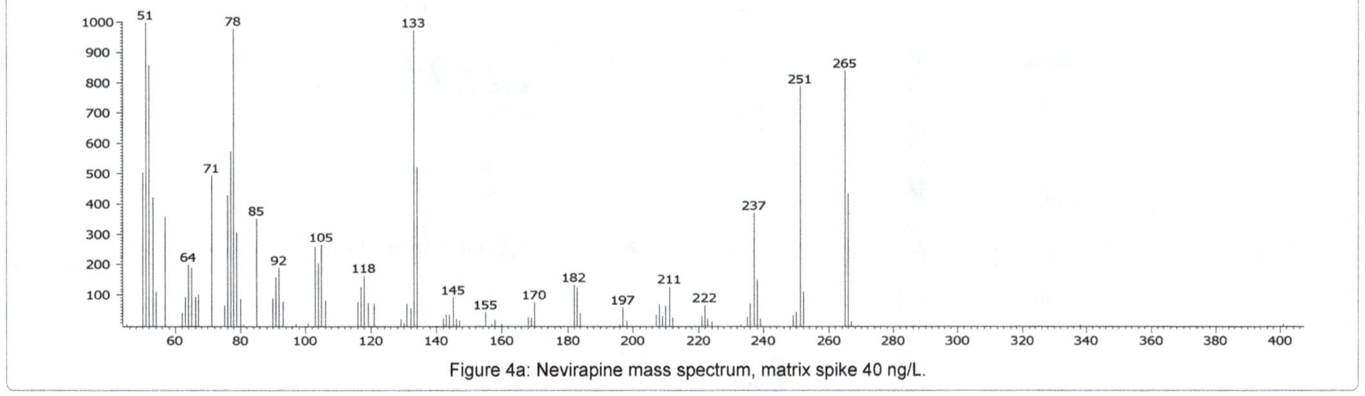

Figure 4a: Nevirapine mass spectrum, matrix spike 40 ng/L.

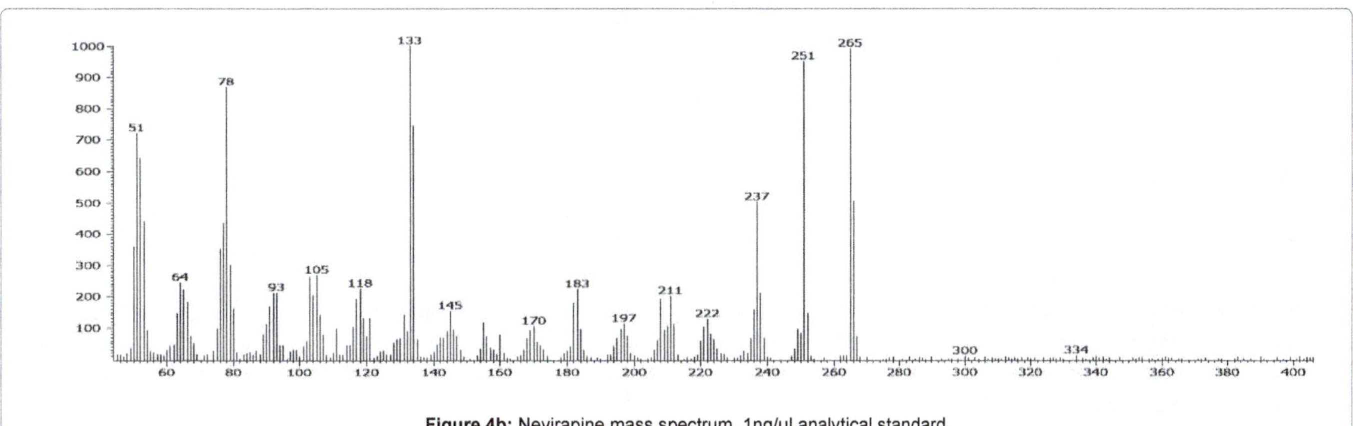

Figure 4b: Nevirapine mass spectrum, 1ng/µl analytical standard.

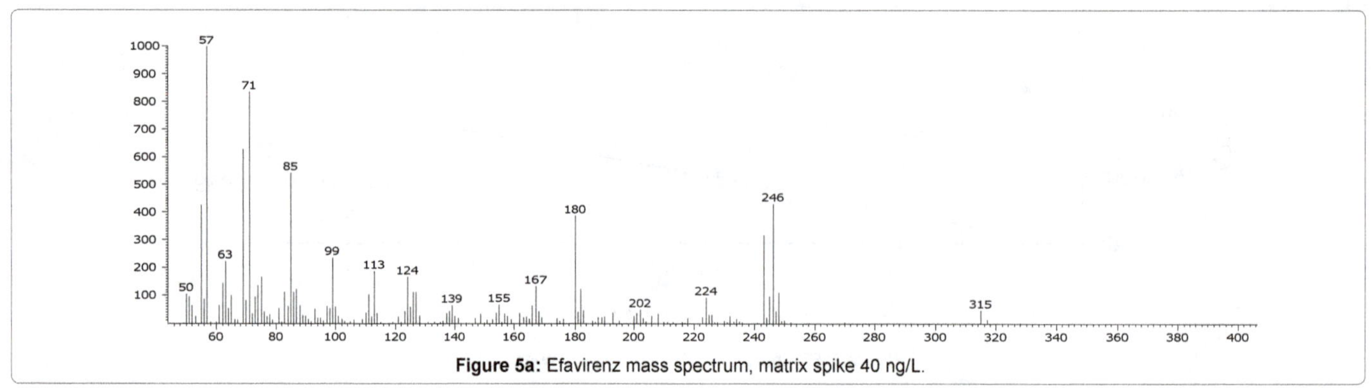

Figure 5a: Efavirenz mass spectrum, matrix spike 40 ng/L.

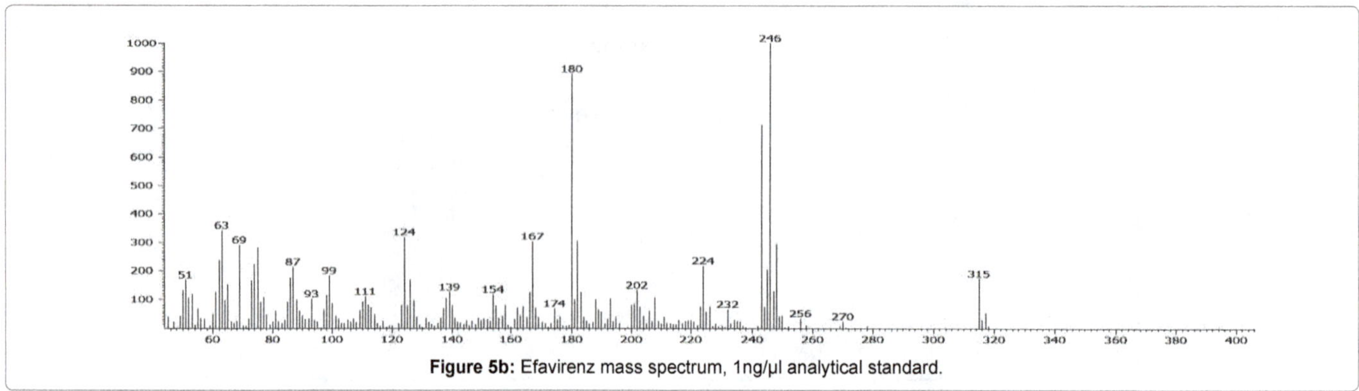

Figure 5b: Efavirenz mass spectrum, 1ng/μl analytical standard.

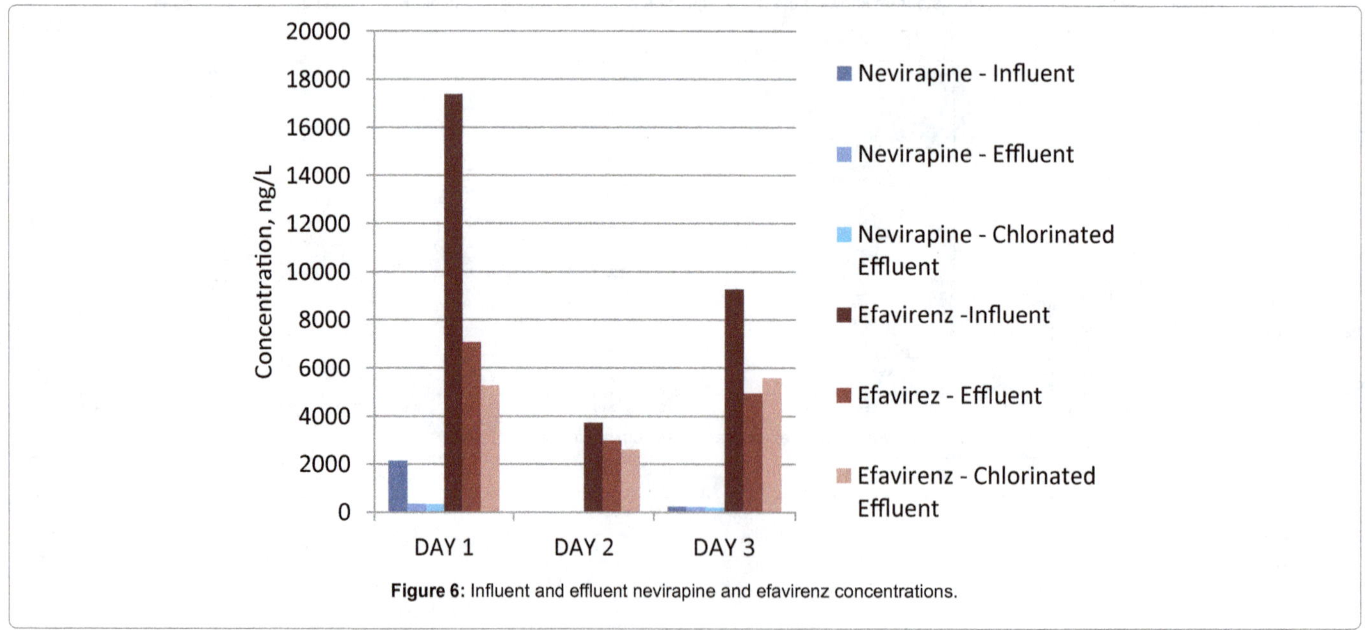

Figure 6: Influent and effluent nevirapine and efavirenz concentrations.

It has been shown that GC-TOFMS can be used to determine nevirapine and efavirenz in wastewater samples. The technique is reliable and robust and is a viable alternative to an LC-MS/MS. Further studies will be necessary to determine which of the processes in the wastewater treatment works were most effective for the removal of nevirapine and efavirenz from wastewater. To better evaluate the passage of these compounds though a wastewater treatment works the use of passive samplers would be advisable.

Acknowledgement

Rand Water Scientific Services for the provision of the technical environment and the Tshwane University of Technology for support.

References

1. World Health Organisation Report (2011) Pharmaceuticals in Drinking Water. WHO/HSE/WSH/11.05.

2. World Health Organisation Report (2013) Global update on HIV treatment, Results, impact and opportunities.

3. Swanepoel C,Henk Bouwman, Peters R and Bezuidenhoudt C (2014)

Presence, levels and potential implications of HIV antiretrivirals in drinking, treated and natural waters. Water Research Commission Report K5/2144.

4. Önal A (2006) Analysis of Antiretroviral Drugs in Biological Matrices for Therapeutic Drug Monitoring. JFDA 14: 99- 119.

5. Al-Rajab AJ, Sabourin L, Chapman R, Lapen DR, Topp E (2010) Fate of the antiretroviral drug tenofovir in agricultural soil. Science of the Total Environment 408: 5559- 5564.

6. Choi SO, Rezk NL, Kashuba AD (2007) High-performance liquid chromatography assay for the determination of the HIV-protease inhibitor Tipranavir in human plasma in combination with nine other antiretroviral medications. J Pharm Biomed Anal 43: 1562- 1567.

7. Shah SAB, Mullin R, Jones G, Shah I, Barker J, et al. (2013) Simultaneous analysis of antiretroviral drugs abacavir and tenofovir in human hair by liquid chromatography–tandem mass spectrometry. Journal of Pharmaceutical and Biomedical Analysis 74: 308- 313.

8. Vogel M, Bertram N, Wasmuth JC, Emmelkamp J, Rockstroh JK, et al. (2010) Determination of Nevirapine in Plasma by GC-MS. Journal of Chromatographic Science 48: 90- 94.

9. Prasse C, Schlüsener MP, Schulz R, Ternes TA (2010) Antiviral Drugs in Wastewater and Surface Waters: A New Pharmaceutical Class of Environmental Relevance?. Environmental Science and Technology 44 (5): 1728- 1735.

10. Wood TP, Duvenage CSJ, Rohwer E (2015) The occurrence of anti-retroviral compounds used for HIV treatment in South African water. Environmental Pollution 199: 235- 243.

11. Pedrouzo M, Borrull F, Marce´ RM, Pocurull E (2013) Analytical methods for personal-care products in environmental waters. Trends in Analytical Chemistry 30: 749- 760.

12. Langford KH, Thomas KV (2008) Inputs of chemicals from recreational activities into the Norwegian coastal zone. Journal of Environmental Monitoring 2008 10: 894- 898.

13. Matamoros V, Jover E, Bayona JM (2010) Part-per-trillion determination of pharmaceuticals, pesticides, and related organic contaminants in river water by solid-phase extraction followed by comprehensive two-dimensional gas chromatography time-of-flight mass spectrometry. Anal Chem 82: 699- 706.

14. Mackay D, Shui WY, Ma K, Lee SC (2006) Handbook of Physical-Chemical Properties and Environmenyal Fate for Organic Chemicals, 2nd Edition. Science pp: 4216.

15. Chan CC, Lee YC, Lam H, Zhang X (2004) Analytical Method Validation and Instrument Performance Verification. A John Wiley& Sons, Inc, Publication, 1st edition pp: 320.

16. Aebia B, Sturny-Jungo R, Bernhard W, Blanke R, Hirsch R (2002) Quantitation using GC–TOF-MS: example of bromazepam. Forensic Sci Int 128: 84- 89.

17. Bergknut M, Frech K, Andersson PL, Haglund P, Tysklind M (2006) Characterization and classification of complex PAH samples using GC–qMS and GC–TOFMS. Chemosphere 65: 2208- 2215.

18. Vanková M (2010) Biodegradability Analysis of Pharmaceuticals Used in Developing Countries; Screening with Oxi Top C-110.

Tributylamine Facilitated Separations of Fucosylated Chondroitin Sulfate (Fucs) by High Performance Liquid Chromatography (HPLC) into its Component Using 1-Phenyl-3-Methyl-5-Pyrazolone (PMP) Derivatization

Myron P[1], Siddiquee S[2]*, Azad SA[1] and Yong YS[2]

[1]*Borneo Marine Research Institute, University Malaysia Sabah, 88400 Kota Kinabalu, Sabah, Malaysia*
[2]*Biotechnology Research Institute, University Malaysia Sabah, 88400 Kota Kinabalu, Sabah, Malaysia*

Abstract

Monosaccharide characterization study can give valuable information on newly discovered intact mucopolysaccharide. The compound fucosylated chondroitin sulfate (FuCS) is a unique mucopolysaccharide having various bioactive properties reported only in targeted holothurians species. The monosaccharide composition of FuCS from a sea cucumber, *Holothuria arenicola* was studied by comparing a modified and conventional 1-phenyl-3-methyl-5-pyrazolone (PMP) derivatization method where tributylamine addition was found to improve the chromatographic conditions. The monosaccharide proportion of FuCS from *H. arenicola* was determined to be 1:1:0.44 for glucuronic acid: N-acetylgalactosamine: fucose, respectively. Thus, the FuCS monosaccharide ratio of *H. arenicola* was first reported and this method is more useful for structurally similar compound analysis.

Keywords: Fucosylated chondroitin sulfate; Monosaccharide; 1-phenyl-3-methyl-5-pyrazolone (PMP); Tributylamine; HPLC

Introduction

Fucosylated chondroitin sulfate is an unusual compound having various physicochemical activity found in sea cucumbers [1]. The mucopolysaccharide consists of a common chondroitin sulfate chain with the *β-D*-glucuronic acid moiety *O*-substituted with sulfated fucosyl at the carbon-3 position [2]. The hydrolysis sequence of this mucopolysccharide into its component starts by removal of sulfate esters at the vicinity of fucose branch followed by fucose release and sulfates in the chondroitin backbone. After that, the chondroitin backbone itself is hydrolyzed into glucuronic acid and acetylgalactosamine. Anticoagulation activity is one of most studied bioactivity expressed by this compound and the structural configuration is a determining factor for various physicochemical properties. Apart from that, the derived components are molecules having novelty pharmaceutical benefit. For example, glucuronic acid or its reduced form, glucuronolactone are natural antioxidant having hepatoprotective function [3], whereas, sulfated fucose are known anticancer agent [4].

Analyzing glycosaminoglycans is challenging due to their polydispersity, sequence heterogeneity and high negative charge density [5]. Monosaccharide composition is often characterized to understand the function of intact carbohydrate chain but saccharides are naturally low in UV absorbency. Therefore, derivatization is often required to label the compound with chromophore tag that allows ionic property changes to promote the desired separation conditions. 1-phenyl-3-methyl-5-pyrazolone (PMP) derivatization is particularly well documented for carbohydrates separation using HPLC since PMP derived carbohydrates are more hydrophobic and thus better retained in the reverse phase material. Separation techniques are involving the use of PMP for improving with the current knowledge that ion-pairing reagent facilitates carbohydrates derivatization. To the best of our knowledge, there is currently no reported on the use of ion-pairing techniques in the derivatization steps for determination of FuCS monosaccharide composition from sea cucumbers. Here, in this research, we were described the FuCS monosaccharide composition from a sea cucumber, *Holothuria arenicola* and the efficacy of using tributylamine facilitated derivatization strategy.

Materials and Methods

Materials

D-Glucuronic acid, *N*-Acetyl-*D*-Galactosamine, *L*-fucose, 1-phenyl-3-methyl-5-pyrazolone, trifluoroacetic acid were purchased from Sigma. Extraction solvents, methanol, *n*-hexane, acetonitrile, ethyl-acetate, chloroform and (HPLC grade) were purchased from Merck (USA). All other analytical grade reagents were available commercially.

Preparation of fucosylated chondroitin sulfate from sea cucumber

FuCS was extracted from the body wall of sea cucumber *Holothuria arenicola* harvested in Semporna, Sabah, Malaysia (mean weight 60 g) using a modified method from Mourão et al. [6]. Briefly, the sea cucumber was eviscerated and cleaned under running tap water to remove visible impurities. The body wall was carefully cut to separate from other tissue and homogenized. The homogenate was treated with acetone to dehydrate for 24 h. The dried residue was then digested with papain in a 0.1 M sodium acetate buffer solution (pH 6) containing 5 mM EDTA and 5 mM cysteine at 60°C for 24 h to remove proteins in the sample. The supernatant was collected, precipitated with 5% cetylpyridinium chloride solution and kept in room temperature. The clear precipitate was collected with centrifugation and the resultant

***Corresponding author:** Siddiquee S, Biotechnology Research Institute, University Malaysia Sabah, Jln UMS, 88400 Kota Kinabalu, Sabah, Malaysia
E-mail: shafiqpab@ums.edu.my, shafiqpab@hotmail.com

pellet was further precipitate first with 95% ethanol and washed twice with 80% isopropanol. The final precipitate was dialyzed against distilled water at room temperature using Spectra/Por membranes (MWCO 1000) for 3 buffer changes (2h; 6h; overnight) and finally lyophilize. The crude extract was fractionated on a FPLC systems (AkTAPRime) equipped with a DEAE Sephadex A-50 column with elution gradient from 0.2 to 2 M NaCl buffer (pH 7). Aliquots of each collected fractions were analyzed by phenol-sulfuric assay in microplate format adapted from Masuko et al. [7]. Positive fractions were lyophilized and afforded 5.7% FuCS extract.

Monosaccharide analysis

Purified FuCS (~1 mg) was hydrolyzed with 2 M TFA (1 ml) at 110°C for 4 h. The hydrolyzate was dried under a stream of dry nitrogen gas to remove TFA and was derivatized with PMP according to the following conventional and modified method.

Conventional PMP derivatization method

The dried hydrolyzate was derivatized with PMP based on Honda et al. [8]. Briefly, 1 mg of TFA hydrolyzed samples were added into a 2 ml reaction vial containing 100 µl of 0.3 mol/L methanolic PMP solution and subsequently added 50 µl of 0.3 mol/L NaOH. The mixture was left to react at 60°C for 60 min. After cooling to ambient temperature, the mixture was neutralized with 50 µl of 0.3 mol/L of HCl and diluted to a final volume of 1 ml. Excess PMP reagent was removed with three times 1 ml chloroform extraction. The final aqueous layer was diluted with deionized HPLC grade water prior to HPLC analysis.

Modified PMP derivatization method

Derivatization procedure was adapted from Zhang et al. [9] and replacement used NaOH with 50 µl tributylamine as the catalyst. Without NaOH, the neutralization step with HCl was omitted. The mixture (sample + tributylamine + methanolic PMP) was vortexed and left to react in a thermomixer for 30 min at 60°C with 300 rpm agitation. The mixture was then centrifuged at 6000 rpm for 10 min upon completely reaction and formed two-layers. The aqueous, lower portion was made to 1 ml. The upper organic layer consisted of tributylamine with excess PMP reagent removing by extraction with three times of equal volume of chloroform. The final aqueous layer was diluted for final concentration prior to HPLC analysis.

HPLC analysis

The PMP derivatized samples (n=3), individual sugar standards (n=3) and mix sugar standards (n=3) were carried out on Waters HPLC system equipped with a model 1515 quaternary pump, a model 2717W autosampler and a 2487 dual UV wavelength detector. The column (Agilent, Zorbax Eclipse XDB-C18, 250 × 4.5 mm) was optimized for PMP separations on Breeze IIHPLC system software. The flow rate was set constant at 1 ml/min. Mobile phase A and B were consisted of water and acetonitrile; used with a gradient run of 10 to 90% B in 18 min and hold for another 3 min during run. The injection volume was set to 20 µl. Single wavelength at 245 nm was used for data acquisition with a sampling rate of 2, 2.0 AUFS and 0.1 times constant. Each samples analysis was done in triplicates.

Sample preparation for GCMS analysis

A confirmation study was done using Gas Chromatography Mass Spectrometry (GCMS) for identification of monosaccharide composition using the *N,O*-Bis(trimethylsilyl)trifluoroacetamide (BSTFA) derivatization method. The hydrolyzed syrup was subjected to silica gel (60-200 µm mesh size) column chromatography purification using hexane (20 ml), methanol (20 ml) and acetonitrile:water (20 ml, 1:1, v/v) fractions. Fractions were collected and dried using reduced pressure and redissolve with 50 µl of BSTFA regent. The mixture was incubated at 60°C for 30 min in a heating block and subsequently purge with a constant nitrogen stream to remove excess reagent. The derivatives were made for the final dilutions using acetonitrile before injecting into GCMS.

GCMS analysis

The prepared polar fractions (methanol and ACN:water, n=3) were injected into GCMS system consisting of an Agilent 7890A gas chromatograph system tandem with an Agilent 5975C mass spectrometry detector. Separation was done using capillary column HP-5MS (30 m × 0.25 mm) of 0.25 µm coated film thickness in splitless mode. Injector temperature was adjusted at 250°C and the oven temperature ramp settings were as follow: initial temperature was held constant for 3 min at 156°C and ramp to 180°C in 25 min at a rate of 1°C/min. The flow rate used was a constant 1 ml/min of high purity helium gas (99.9% pure) as the carrier gas. Compound identification was done with matching scan spectral from National Institute of Standards and Technology (NIST) library and the compositions were computed with reference to the abundance of the compounds in chromatogram.

Results and Discussion

HPLC analysis of monosaccharide composition

The study objective was to investigate the influence of tributylamine on separation of FuCS monosaccharide composition and thus conventional and modified derivatization methods were compared with mix sugar standards. Figure 1a and b were the chromatograms for conventional derivatization method and modified method respectively on derived FuCS monosaccharide composition. The three main monosaccharide retention time were successfully separated among each other as expected. Each of the derivatization peaks were matched with standard monosaccharide derived using the same derivatization manner (Figure 2) and the order of separations were fucose (Fuc), glucuronic acid (GlcUA) and *N*-acetyl-galactosamine (GalNAc).

Conventional derivatization method gave ubiquitous spurious peaks probably due to the organic salt formed by addition of sodium hydroxide and during hydrochloric acid neutralization. Zhang et al. [9] demonstrated the addition of triethylamine can facilitate reaction to completeness in a mixture of sugar standards. In our conventional method, GalNAc-PMP derivative had lower peak intensity than GlcUA-PMP. We noticed that the concentration of NaOH was not sufficiently high to facilitate complete reaction and thus have the most obvious decrease in average peak area of 28.13% relative to GlcUA-PMP. The relative standard deviation for reproducibility was calculated to be 2.83% which was acceptable range while stability of derivatized samples when monitored within a 48 h frame at 8 h interval. Peak area was satisfactory with a decrease of 3.87 % and 11.14% at 24 h and 48 h, respectively.

The FuCS structure is common backbone of chondroitin sulfate reported by several authors [6,10-12]. In fact, previous studies on similar compound found almost same amounts of GlcUA and GalNAc with different molar range of Fuc and sulfate [13]. Hence, we were presented the monosaccharide proportion of *H. arenicola* FuCS to be 1.0:1.0: 0.44 (Table 1) based on the modified method.

Zhang et al. had successfully determined several types of

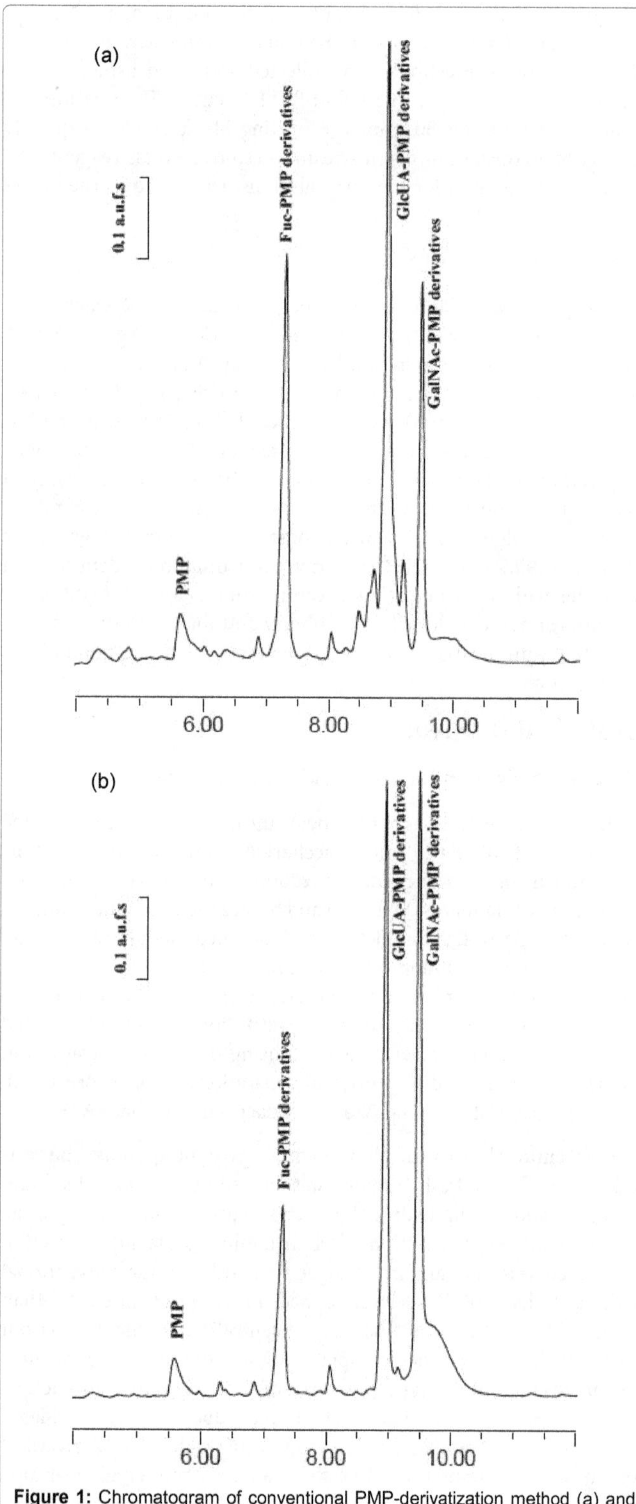

Figure 1: Chromatogram of conventional PMP-derivatization method (a) and modified method (b)

monosaccharides using triethylamine as a catalyst during PMP derivatization [9]. The basis of choosing tributylamine in this experiment is that longer alkyl chain length showed better compatibility with on-column retention of saccharides for a typical C18 type stationary phase [5]. Furthermore, longer alkyl chain was presumably better catalyst during PMP derivatization since longer alkyl chain length can give a higher pKa at 10.89 compared to 10.78 for triethylamine which eventually affects by charge delocalization. Therefore, tributylamine is more readily to accept the protons from the deprotonated monosaccharide species. The ability of the deprotonated hydronium ions was stabilized by tributylamine result in a rapid formation of ducing region which is essential for the PMP reagent to bind (Scheme 1).

To give a better confidence on the monosaccharide composition of purified FuCS, our research group determined the monosaccharide composition using GCMS. Methanol fraction afforded most of the sugar component while the ACN:water fraction has little to trace amount of other sugar species. The hexane fraction consisting of non-polar compounds in trace amount is presumably impurities and byproducts produced during TFA hydrolysis. The main monosaccharide component was identified from HPLC analysis and also presence in the methanol fraction, however, not clear separation (Figure 3) when the use of a semi-polar stationary phase column, HP-5MS. The column used is not specific to carbohydrate analysis and the separation is solely based on the oven temperature, where the volatility of the compounds is reflected in the retention time. Failing the distinct separation by using non-specific column is due to the compounds nature, as the interested

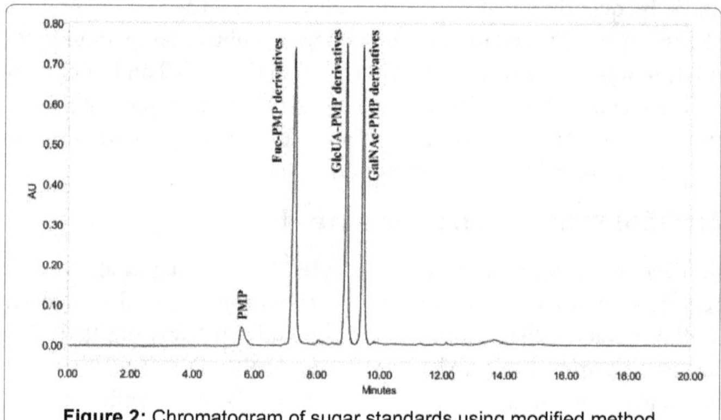

Figure 2: Chromatogram of sugar standards using modified method.

	GlcUA	GalNAc	Fuc
conventional	1.00	0.48	0.69
modified	1.00	1.00	0.44

Table 1: Molar ratio of FuCS monosaccharide component from conventional and modified derivatization method

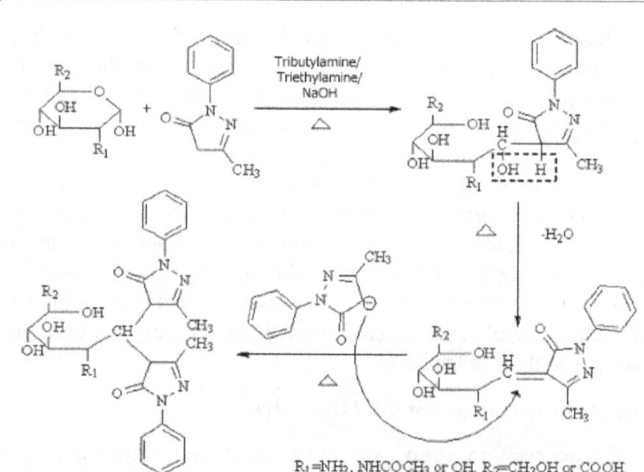

Scheme 1: Reaction pathway of monosaccharides with PMP reagents in presence of catalyst [9].

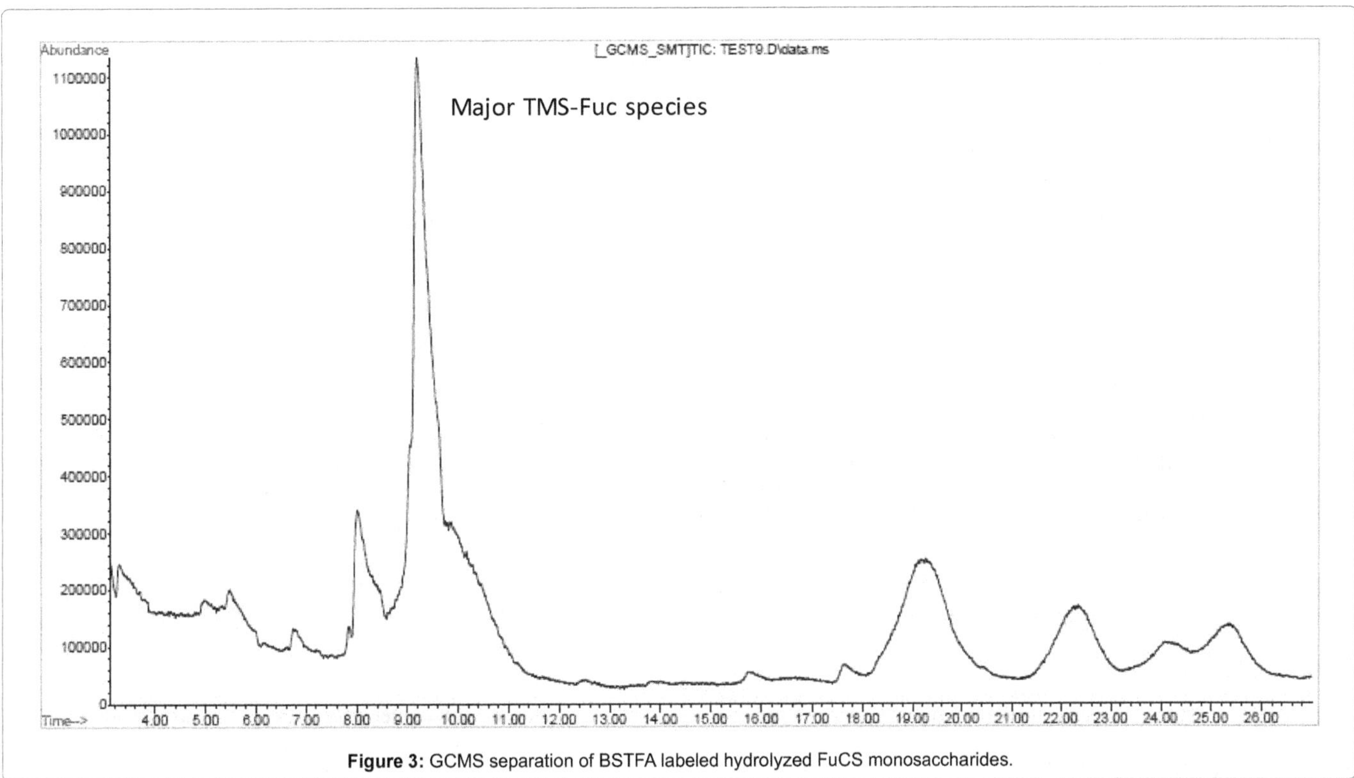

Figure 3: GCMS separation of BSTFA labeled hydrolyzed FuCS monosaccharides.

compounds (GlcUA, GalNAc, and Fuc) are homological similar. Thus, the obtained total ion chromatograms from GCMS were further tested by using deconvolution strategy for determination its separation integrity. Derivatized fucose with trimethylsilyl was seen clustering in unresolved peaks within retention time 7.846 to 9.314 while derivatized galactosamine signal appear at later retention time within 12.578 to 13.452. Derivatized glucuronic was not observed in methanol fraction but appear within a peak range of retention time 15.770 to 16.312.

Conclusions

The present study using a modified PMP-derivatization with addition of tributylamine is a robust method to determine hydrolyzed FuCS component. The improved chromatographic properties are allowed good separation on main monosaccharide composition of FuCS from *H. arenicola*. Nevertheless, it is tentatively proposed that non-commercial sea cucumber species can be served as a source of FuCS compound for potential pharmaceutical industry.

Acknowledgement

This work was supported by grants from the UMS Research Priority Scheme Area, Project Code SBK0084-SG-2013. The authors are thankful to the editor and two anonymous reviewers for their valuable comments that help to improve in this manuscript.

References

1. Yamada S, Sugahara K, Ozbek S (2011) Evolution of glycosaminoglycan. Communicative and Integrative Biology 4: 150–158.
2. Pomin V (2014) Holothurian fucosylated chondroitin sulfate. Marine Drugs 12: 232–254.
3. Vīna I, Linde R, Patetko A, Semjonovs P (2013) Glucuronic acid from fermented beverages: biochemical functions in humans and its role in health protection. IJRRAS 14: 217-230.
4. Hung LD, Sato Y, Hori K (2011) High-mannose N-gylcan-specific lectin from the red alga *Kappaphycus striatum* (Carrageenophyte). Phytochemistry 72: 855-861.
5. Volpi N, Linhardt R (2010) High-perforamnce liquid chromatography- mass spectrometry for mapping and sequencing glycosaminoglycan-derived oligosaccharides. Nat Protoc 5: 993-1004
6. Mourão P, Pereira M, Pavão M, Mulloy B, Tollefsen D, et al. (1996) Structure and anticoagulant activity of fucosylated chondroitin sulfate from echinoderm. Journal of Biological Chemistry 271: 23973–23984.
7. Masuko T, Minami A, Iwasaki N, Majima T, Nishimura S, et al. (2005) Carbohydrate analysis by phenol-sulphuric acid method in microplate format. Anal Biochem 339: 69-72.
8. Honda S, Akao E, Suzuki S, Okuda M, Kakehi K, et al. (1989) High-Performance Liquid Chromatography of Reducing Carbohydrates as Strongly Ultraviolet-Absorbing and Electrochemically Sensitive 1-phenyl-3-methyl5-pyrazolone Derivatives. Anal Biochem 180: 351–357.
9. Zhang Y, Zhang P, Wang Z, Huang L (2011) An innovative derivatization method for simultaneous determination of uronic acids and neutral and amino sugars in coexisting samples by HPLC-ESI-MS/MS[2]. Journal of Liquid Chromatography& Related Technologies 34: 1754-1771.
10. Kariya Y, Watabe S, Kyogashima M, Ishihara M, Ishii T (1997) Structure of fucose branches in the glycosaminoglycan from the body wall of the sea cucumber *Stichopusjaponicus*. Carbohydr Res 297: 273-279.
11. Chen S, Xue C, Yin L, Tang Q, Yu G et al. (2011) Comparison of structures and anticoagulant activities of fucosylated chondroitin sulfate from different sea cucumbers. Carbohydrate Polymers 83: 688-696.
12. Wu M, Huang R, Wen D, Gao N, He J et al. (2012) Structure and effect of sulfated fucose branches on anticoagulant activity of the fucosylated chondroitin sulfate from sea cucucmber *Thelenotaananas*. Carbohydrate Polymers 87: 862-868.
13. Myron P, Siddiquee S, Al-Azad S (2014) Fucosylated chondroitin sulfate diversity in sea cucumbers: a revew. Carbohydrate Polymers 112: 173-178.

Speciation of Cr (III) and Cr (VI) Ions via Fabric Phase Sorptive Extraction for their Quantification via HPLC with UV Detection

Heena[1], Gaurav[2], Susheela Rani[1], Ashok Kumar Malik[1]*, Abuzar Kabir[3]* and Kenneth G Furton[3]

[1]*Department of Chemistry, Punjabi University, Patiala, Punjab, India*
[2]*Punjabi University College of Engineering and Management, Rampura Phul, Punjab, India*
[3]*Department of Chemistry and Biochemistry, International Forensic Research Institute, Florida International University, Miami, FL 33193, USA*

Abstract

Simultaneous extraction of the morpholino dithiocarbamate (MDTC) chelates of Cr(III) and Cr(VI) ion from aqueous samples was accomplished by applying fabric phase sorptive extraction (FPSE) using a cellulose fabric modified with a sol-gel/polytetrahydrofuran composite. Following extraction, the chelates were separated by HPLC on a reverse phase C18 column with acetonitrile:water (65:35; v/v) as the mobile phase and UV detection at 320 nm. A single peak is found for the $Cr(MDTC)_3$ but two peaks are caused by Cr(VI) due to the formation of the complexes $Cr(MDTC)_2(OMDTC)$ and $Cr(MDTC)_3$ (where OMDTC stands for oxy-MDTC). The limits of detection for the Cr(III) and Cr(VI) complexes are 0.001 and 0.003 ng·mL^{-1}, respectively. The method can be used for the determination of chromium species in aqueous samples and displays high precision, sensitivity and reliability.

Keywords: Chromium; Speciation; HPLC; Morpholino dithiocarbamate; MDTC; Fabric phase sorptive extraction; FPSE

Introduction

Metal ions exist in variable oxidation states and have different impacts on health and environment. For an accurate assessment of their impact, it is essential to quantify the individual oxidation states of these metal ions within a sample rather than the total metallic concentration. Chromium is the heavy metal ions that exist in two different oxidation states i.e., trivalent chromium; Cr(III) and hexavalent chromium; Cr(VI) [1]. Cr(III) ion is an essential nutrient that maintains a glucose tolerance factor in human body [2], and involved in mechanism of action of pancreatic insulin [3]. Cr(VI) ion is toxic and does not occur naturally. It is introduced to the environment by various kinds of manufacturing units such as plastics, dyes, ink, metal casting and paint industries [4]. Unlike Cr(III), Cr(VI) ion may cross the cellular membranes by non-specific ion carriers and exert its noxious influence in the cell [5]. It affects the air quality through coal burning which eventually lead to water and soil contamination. After inhalation or oral exposure it effects on liver, kidney, immune system and even ulceration of skin [6]. Due to varied impact of different oxidation states, metal speciation has become an important research direction in past decade to obtain their respective toxicological assessment [7]. Therefore, interest and demand is continuously increasing for chromium speciation in biological and environmental samples.

Different methods used for the preconcentration of chromium complexes are liquid-liquid extraction [8], cloud point extraction [9], solid phase microextraction [10] and solid phase extraction [11]. For the analysis of chromium ions, various techniques are used frequently but high performance liquid chromatography (HPLC) meets most of the analytical requirements of metal determination. HPLC provides several advantages over other methods for separation and quantitation of Cr ions down to the trace level concentration [12]. A number of HPLC methods have been evaluated [13,14] and a selective determination of both species can be achieved successfully by using HPLC coupled to inductively coupled plasma mass spectrometry [15] ion chromatography-ICP-MS [16], flame atomic absorption spectrophotometry [17] and capillary electrophoresis [18].

Dithiocarbamate constitute the most exhaustively studied group of ligands used in the chromatographic speciation of Cr ions [19], diethyldithiocarbamate [20] and ammounium pyrrolidine dithiocarbamate [21] are the reported complexing reagents for HPLC-UV determination. But being highly hydrophilic, morpholino dithiocarbamate (MDTC) is used as a preferred chelating agent in the developed method.

Fabric phase sorptive extraction has already established itself as a viable alternative to the other conventional preconcentration techniques [22]. Fabric phase sorptive extraction used for the preconcentration of chromium metal ions followed by HPLC-UV detection. It ensures the faster preconcentration of target analytes, while at the same time minimizes the number of laborious and cumbersome sample preparation steps.

FPSE incorporates most of the advantageous features of other microextraction techniques such as SPME by utilizing flexible fabric surface as the substrate platform for sol-gel derived high-efficiency sorbent coating resulting into an efficient microextraction device [23]. High sample capacity of sol-gel fabric sorbent and the inherently porous formation of cellulose substrate synergistically facilitate the achievement of high preconcentration factors for the target analytes in a relatively short period of time from an original and unmodified sample matrix. The flexibility of FPSE media allows direct insertion into the sampling container holding the sample. FPSE media can face the harsh chemical environment (pH 1-12) for prolonged period of time without compromising the extraction performance. The strong chemical bonding between the extraction sorbent and the substrate allows exposing the FPSE media to any organic solvent of choice for elution process.

***Corresponding author:** Abuzar Kabir, International Forensic Research Institute, Department of Chemistry and Biochemistry, Florida International University, Miami, FL 33193, USA, E-mail: akabir@fiu.edu

The aim of this study is to evaluate the speciation of Cr(III) and Cr (IV) ions by chelating it with MDTC followed by preconcentration with FPSE media and HPLC-UV determination.

Experimental

Instrumentation

The Dionex HPLC unit consists of a P680 solvent delivery pump, a UVD 170 detector capable of detecting four wavelengths was interfaced to a computer loaded with Chromeleon software (version 6.70). A Supelco Ascentis Express reversed phase column of size 10 cm × 4.6 mm filled with C_{18} material (2.7 µm) was used for separation. The IR spectra were recorded on FTIR (PerkinElmer). Elico SL-164 double beam UV-visible spectrophotometer loaded with Spectra Treatz software was used to record the spectra with quartz cuvettes. A digital pH meter-101 (Delux, India) was used to adjust the pH of solutions. Gaussian 03 software was used for the optimization of the structure of metal complexes.

A digital vortex mixer (Fisher Scientific, USA) was employed for thoroughly mixing of sol solutions. An ultrasonic cleaner-2510 (Branson Inc., USA) was used to make sol solution free of trapped gas or bubbles. Centrifugation of sol solution, to obtain particle free solution, was carried out in an Eppendorf centrifuge model 5415 R. A Barnstead Nano Pure Diamond (Model D11911) deionized water system was used to obtain ultra-pure deionized water (18.2 MΩ cm) for sol-gel synthesis.

Materials chemicals and reagents

Cellulose substrates used to create FPSE media were purchased from Jo-Ann Fabric (Miami, FL, USA). Polytetrahydrofuran, acetone, dichloromethane, methyltrimethoxysilane (MTMS), trifluoroacetic acid (TFA) were purchased from Sigma-Aldrich (https://www.sigmaaldrich.com) (St. Louis, MO, USA). Sodium hydroxide and hydrochloric acid were purchased from Thermo Fisher Scientific (http://www.thermofisher.com) (Milwaukee, WI, USA). All the solvents used were of HPLC-grade and purchased from J.T. Baker Chemicals (Phillipsburg, NJ), and filtered by using Nylon-6,6 membrane filters (Rankem, India) in a filtration assembly (Perfit, India). Cr(III) nitrate nonahydrate, potassium dichromate, sodium acetate trihydrate, glacial acetic acid, and sodium hydroxide were obtained from Merck (http://www.merck.com/index.html). Prior to injection, all samples were filtered through 0.22 µm syringe filters (Rankem, India).

Synthesis of MDTC

MDTC was used as a chelating agent and prepared by method reported by Macrotrigiano [24].

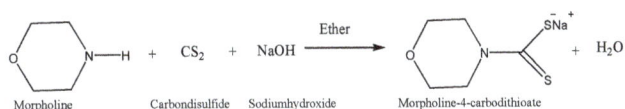

Morpholine + Carbondisulfide (CS_2) + Sodiumhydroxide (NaOH) $\xrightarrow{\text{Ether}}$ Morpholine-4-carbodithioate (S^-Na^+) + H_2O

In brief, morpholine with dry ether was taken into round bottom flask. To this, stoichiometric amount of CS_2 was added with constant stirring with magnetic bead and then reaction mixture was cooled. To this solution a stoichiometric quantity of NaOH was added with constant stirring for 3-4 h. Then the product formed was filtered, washed with ether and re-crystallized from alcohol. White needle shaped crystals were obtained with melting point 170°-174°C. In IR spectra, the presence of characteristic C=S stretching bands at 1083 cm^{-1} and two bands at 1437 and 1164 cm^{-1} due to C-N and C-O stretching, respectively confirmed the formation of MDTC (Figure 1).

A 10% w/v solution of sodium morpholine-4-carbodithioate was prepared in distilled water and it was standardized titrimetrically using mercuric acetate as titrant and diphenycarbazone as an internal indicator [25]. The sol solution for creating the sol-gel poly-THF coating was prepared by using a modified version of a previously described method [26].

Complexation procedure

10 mL solution containing Cr(III) and Cr(VI) ions (2 milli-moles and 7 milli-moles respectively) was transferred to a well stoppered bottle. To this, 1 mL (0.4 moles) of MDTC solution (10% w/v) was added. The total volume of solution was made up to 25 mL with water and acetate buffer solution (0.1 M) such that the pH of the resulting solution was 4.0. The prepared solution was heated on water bath at 55°C temperature. The complex obtained was then dissolved in 10 mL acetonitrile solvent to obtain the stock solution. The stock solution was then diluted further as per requirement.

Spectrometric analysis of chromium complexes

The different kinds of interaction of both chromium ions towards

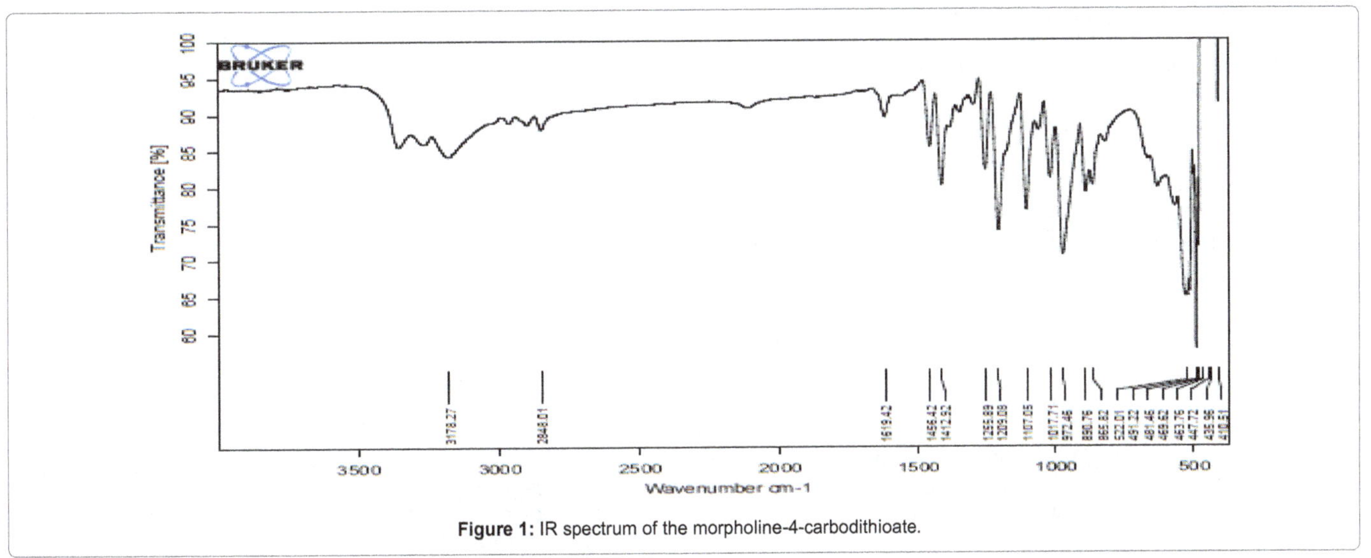

Figure 1: IR spectrum of the morpholine-4-carbodithioate.

dithiocarbamates make their determination easier. The IR spectra of 1:3 adduct of tris(morpholinedithiocarbamto)Cr(III) was recorded using KBr pellets. The shift of C-S band from 1083 cm^{-1} to 1050 cm^{-1} after complexation indicates that the ligand binds with metal through sulfur atoms. A band at 870 cm^{-1} may be attributed to S-O stretching vibration which confirmed the formation of oxy complex. The oxy complex formation of $Cr(MDTC)_2(OMDTC)$ was due to reduction of Cr (VI) with the reagent where Cr-O-S bond is formed in the complex [27]. The formation of complexes was also confirmed by thin layer chromatography. With solvent system methanol : ether : toluene (5:70:25), one spot for Cr(III) complex with Rf value 0.154 and two spots for Cr(VI) complex with Rf values 0.154 and 0.67 were obtained. Single absorption maximum (λ_{max}=298 nm) for Cr(III)-MDTC complex and two absorption maxima (λ_{max}=298, 350 nm) for Cr(VI)-MDTC complex were obtained in their respective UV-Vis spectra. It clearly indicates that Cr(III) ion form single product as $Cr(MDTC)_3$ and Cr(VI) ion form two distinct products as $Cr(MDTC)_2(OMDTC)$ and $Cr(MDTC)_3$. The complete optimization and frequency calculations for the complexes were carried out using Hartree Fock (HF) level of theory in combination with the STO-3G basis set via the Gaussian 03 software package. The energy minimization studies of chromium complexes interpreted that both complexes are energetically stable (Figure 2a and 2b).

Sample preparation

Ground water was obtained from the bore well located within Punjabi university campus, Patiala, India. The bore well water is pumped out at the depth of around 400 feet from ground level. Drinking water was obtained from water purification appliances installed in Punjabi university Patiala. The bore well water is fed to the appliance as raw water. Water is allowed to pass through an ultra-violet radiation treatment system and a set of activated carbon cartridges (particle size 1 and 5 μm) and is made ready for drinking. Industrial waste water was obtained from the effluent stream coming out of industrial area located in vicinity of Chandigarh city, India. Since industrial waste water contains high concentrations of particulate matter and suspended impurities, samples were first filtered with Whatman filter paper (grade no. 1) and then with 0.45 μm pore size Nylon-6, 6 membrane filters in a filtration assembly. Prior to the FPSE process, all water samples were degassed with ultrasonic bath.

Results and Discussion

Chemistry of sol-gel coated FPSE media

Sol-gel technology for creating microextraction sorbents was developed by Malik et al. [22] and based on this technology hundreds of sol-gel based sorbents with unique selectivity, extraction sensitivity and applications have emerged. It was prepared by dissolving 10 mL of sol-gel precursor methyltrimethoxysilane (MTMS), 10 g THF polymer, 20 mL methylene chloride: acetone (50:50 v/v) as the organic solvent, 4 mL 5% trifluoroacetic acid as sol-gel catalyst. The sol mixture was vortexed, centrifuged and sonicated for 2-3 minutes and finally clear supernatant sol solution part was transferred to a clear amber glass bottle. Due to its capability to extract both polar and non-polar analytes, polytetrahydrofuran, a medium polarity polymer containing tetramethylene oxide repeating units and terminal hydroxyl groups, was selected as the organic polymer. In addition to the organic polymer, methyltrimethoxysilane (MTMS), trifluoroacetic acid (containing 5% water v/v), and methylene chloride/acetone (50/50 v/v) were used as the inorganic precursor, sol-gel catalyst and solvent system, respectively.

During the condensation, the growing sol-gel poly-THF network reacts with available surface hydroxyl groups of cellulose microfibrils, resulting in a covalently bonded sol-gel poly-THF ultra-thin film uniformly distributed throughout the substrate with characteristic high solvent and chemical stability as well as highly accessible active sites for efficient and fast analyte extraction. The characterization of sol-gel coated poly-THF fabric media was performed with scanning electron microscopy and spectroscopic techniques in our previous paper [22].

FPSE procedure

To extract and preconcentrate the Cr-MDTC complex using FPSE, 10 mL aqueous solution containing the analytes (1 $ng.mL^{-1}$) was transferred to the glass vial. A clean Teflon coated magnetic stirring bar was inserted into it along with FPSE media. The solution was stirred for optimized duration of time. After this, the FPSE media was taken out of the solution, wiped gently to remove residual water and transferred into the vial containing 500 μL acetonitrile for desorption of analytes. The eluent was injected into injection loop of HPLC for chromatographic separation (Figure 3). The schematic representation of FPSE procedure is described in Figure 4.

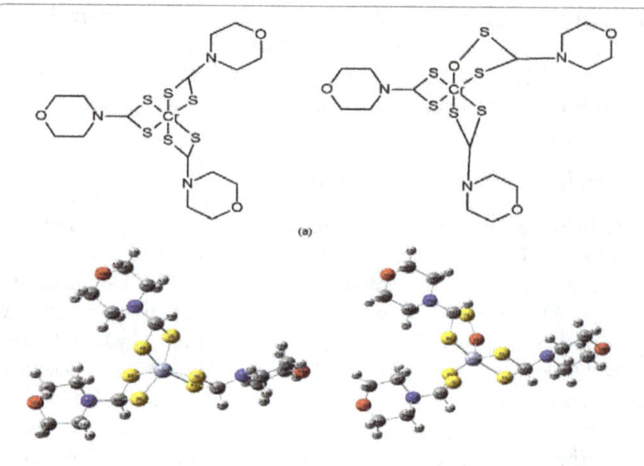

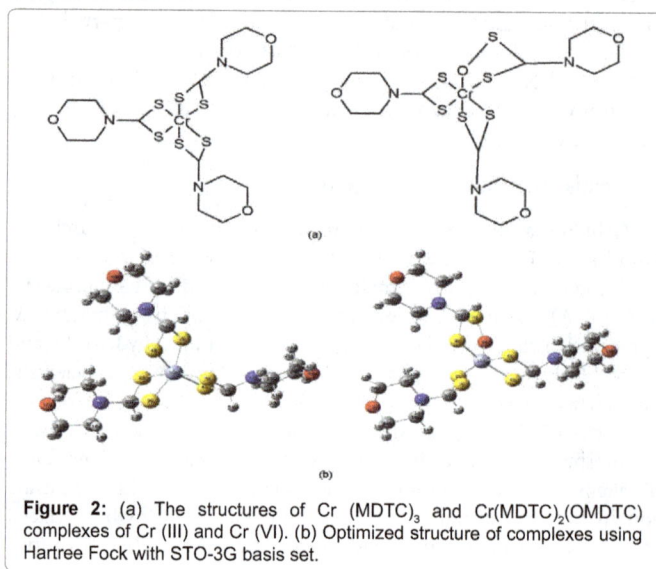

Figure 2: (a) The structures of Cr $(MDTC)_3$ and $Cr(MDTC)_2(OMDTC)$ complexes of Cr (III) and Cr (VI). (b) Optimized structure of complexes using Hartree Fock with STO-3G basis set.

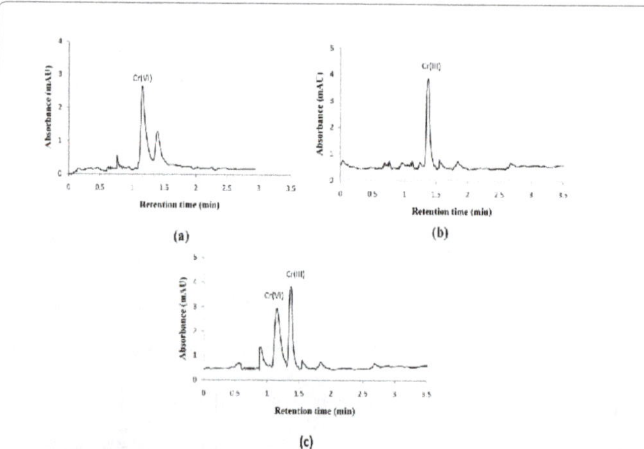

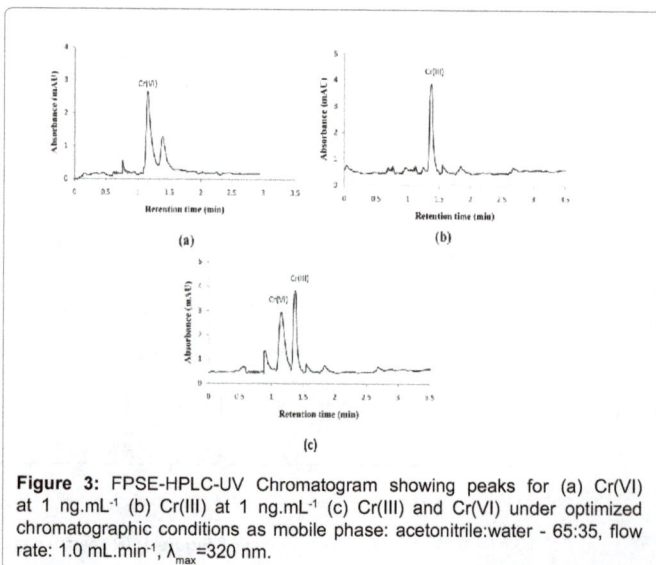

Figure 3: FPSE-HPLC-UV Chromatogram showing peaks for (a) Cr(VI) at 1 $ng.mL^{-1}$ (b) Cr(III) at 1 $ng.mL^{-1}$ (c) Cr(III) and Cr(VI) under optimized chromatographic conditions as mobile phase: acetonitrile:water - 65:35, flow rate: 1.0 $mL.min^{-1}$, λ_{max}=320 nm.

HPLC analysis

Chromatographic system was conditioned by passing the mobile phase through the column until a stable signal was obtained. The selection of mobile phase depends upon resolution of peak as well as signal response. Efficient separation of the chromium complex was achieved with acetonitrile and water. The optimized ratio of solvents in mobile phase was acetonitrile: water (65:35 v/v). The flow rate of mobile phase was maintained at 1 mL.min^{-1} with UV detection at 320 nm. The peaks were obtained at retention time of 1.2 and 1.4 min for $Cr(MDTC)_2(OMDTC)$ and $Cr(MDTC)_3$ complexes respectively.

Optimization of FPSE conditions

Different parameters affecting the performance of FPSE procedure towards extracting the metal complexes were optimized. Aqueous sample (10 mL) was taken for subsequent experimentations. Optimized conditions were used for validation and application to the environmental samples. The following parameters were optimized:

Optimization of sample extraction time: Equilibrium extraction time is one of the most important FPSE parameters. Once the extraction equilibrium is reached, FPSE media cannot extract the target analyte any further under the given conditions. The porous sol-gel sorbent network, high primary contact surface area and the permeable cellulose substrate synergistically reduces the extraction equilibrium time. The sample containing analytes (1 ng.mL^{-1}) was examined to study the effect of extraction time. Extraction times from 5 to 25 min were taken for observation. As the extraction time increases, extraction of target analytes onto the FPSE fiber increases and becomes almost constant at 15 min. So the extraction time of 15 min was optimized for further experimentation shown in Figure 5.

Effect of eluting/ back-extraction solvent: Once the target analytes are extracted into the FPSE media, a quantitative desorption into a suitable organic solvent is needed. The solvent breaks the analyte-sorbent interaction and dissolves the analyte into it. Different solvent mixtures were tried for the back-extraction of target analytes. Acetonitrile was the best elution/back-extraction solvent shown in Figure 6.

Effect of elution/back-extraction time: Elution/back-extraction time was optimized by performing the experimentation with back-extraction time ranging from 2 to 10 min. An elution/back-extraction time as 6 min was selected while keeping the other optimized parameters to their fixed values (Figure 7).

In addition to all these factors carry over effect was also studied for FPSE sorbent. A thorough clean up step was required in order to remove the matrix effect. Fiber was washed thoroughly with acetonitrile before the subsequent experimentation. The carry over effect was checked by injecting the eluent 3-4 times before analyzing the next sample. The matrix effect can also be minimized by using the new fiber for each extraction process, as it is inexpensive.

Method validation

The optimized FPSE-HPLC-UV conditions were used to prepare the calibration curves in the range 1-100 ng.mL^{-1} for all kinds of samples. Over the range, a linear response with the regression coefficient 0.997 and 0.995 was obtained for Cr(III) and Cr(VI) complexes, respectively. In the unknown samples, the concentrations of Cr(VI) can be determined by measuring the peak areas obtained due to $Cr(MDTC)_2(OMDTC)$ using the calibration curve. Proposed method is suitable for speciation of both Cr(III) and Cr(VI) in a sample

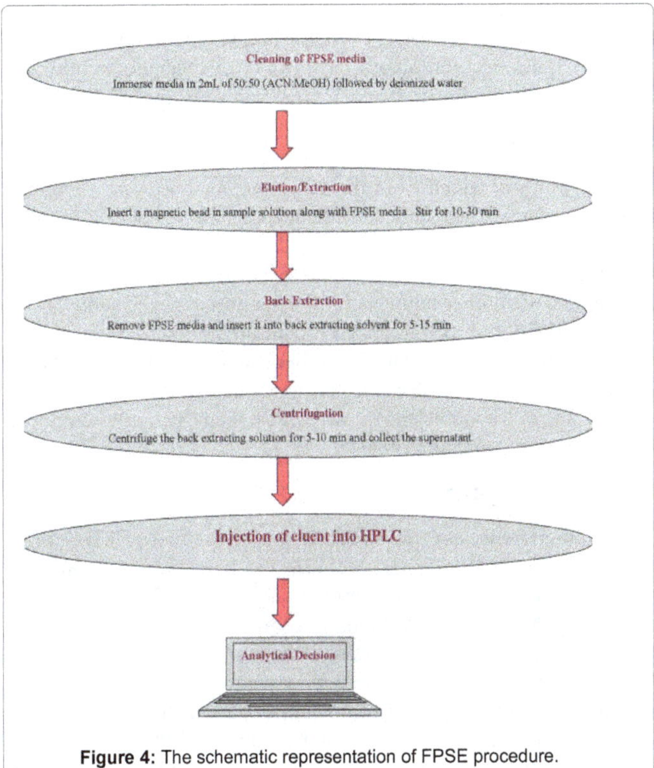

Figure 4: The schematic representation of FPSE procedure.

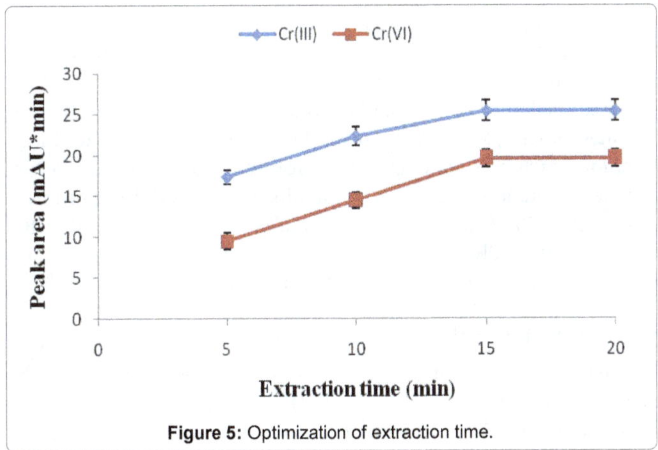

Figure 5: Optimization of extraction time.

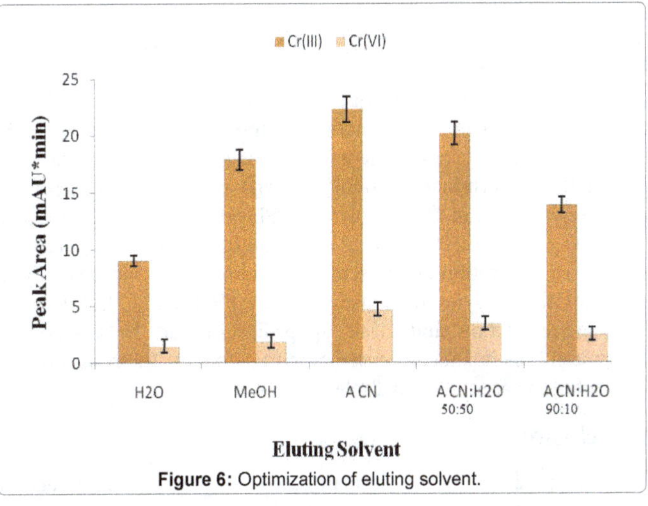

Figure 6: Optimization of eluting solvent.

using quantitative calculations based on the ratio of respective peak areas of absolute Cr(III)-MDTC complex and absolute Cr(VI)-MDTC complex. The concentration of Cr(III) can be calculated by subtracting the value corresponding to $Cr(MDTC)_3$ obtained from Cr(VI).

Comparison of method

The results obtained from the developed method were compared with the same from other techniques such as FAAS [27], ICP-MS [28] etc. and LOD values obtained were lower than the earlier published reports (Table 1). The values of accuracy and precision were comparable with other methods for the analysis of Cr(III) and Cr(VI) complex. The use of Ascentis Express column has reduced the analysis time to greater extent compared to earlier studies. In earlier reports, the chromatographic run time was from 5 to 6 min [8] but in the developed method 3 min was sufficient to carry out the analysis. The developed method is more rapid and robust.

Interference of other ions

The interference of various metal ions such as Co(II), Ni(II), Pd(II), Zn(II), Sn(II), Ag(I), Cu(II), Mo(VI), V(V) was studied. Different amount of these diverse metal ions were added individually to aliquots containing Cr(III) and Cr(VI) ions. At room temperature Co(II), Ni(II) and Pd(II) ions form complexes with morpholinedithiocarbamate (MDTC) and can be removed by pre-extraction in chloroform. As Cr ions do not form complex at room temperature with MDTC so these could be left in aqueous sample and be analyzed by the developed procedure. Vanadium and molybdenum ions require highly acidic conditions (pH<4) for complexation, so no interference was observed. Cr(III) or Cr(VI) ions have shown different tolerable concentration limits for different metal ions during complexation phenomenon. The tolerable concentration ratio added to 100 $ng.mL^{-1}$ Cr(III) or Cr(VI) ions was 1000 times for Na^+ and K^+, 100 times for Zn^{2+}, Hg^{2+}, Cd^{2+} and 10 times for Cu^{2+}, Mn^{2+}, Pb^{2+}, Co^{2+} and Fe^{3+} ions. Effect of the additions of anions such as sulfate, chloride, bromide, iodide, nitrite, acetate, oxalate, nitrate ions etc. was also examined. Anions did not interfere much up-to 50-60 $mg.L^{-1}$ concentration. This demonstrates that the common potentially interfering ions do not have a significant effect on the speciation of Cr (III) and Cr (VI) ions.

Analytical applications

The effectiveness of the developed method was verified by the determination of Cr (III) and Cr (VI) complex in various kinds of aqueous samples. The environmental aqueous samples were analysed under optimized extraction conditions to demonstrate the performance and practical applicability of the developed method. A good agreement was obtained between the added and recovered analytes concentration by the developed FPSE-HPLC-UV method (Table 2). The results obtained were satisfactory with recovery more than 95% for a broad concentration range. This proves that the developed procedure is suitable for genuine environmental and consuming water applications. During their determination no major interfering peaks were present at the retention time of these analytes. FPSE-HPLC-UV chromatogram of spiked ground water, drinking water and industrial waste water at 1 $ng.mL^{-1}$ concentration are presented in Figure 8a-8c. The welldefined peaks of the analytes demonstrated that FPSE-HPLC-UV was an adequate extraction and clean up procedure for the analysis of chromium ions in aqueous environmental samples. The results of the determination are shown in Table 3.

Conclusion

Compared to other sorbent-based sorptive microextraction

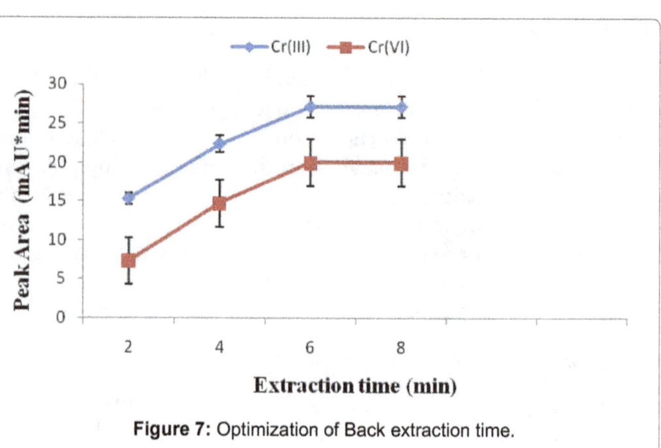

Figure 7: Optimization of Back extraction time.

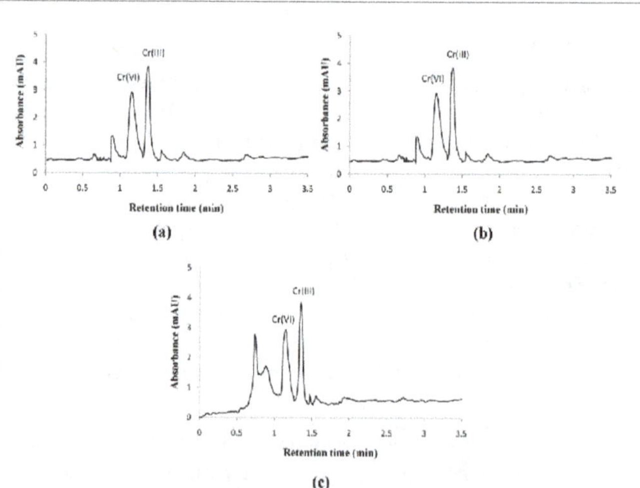

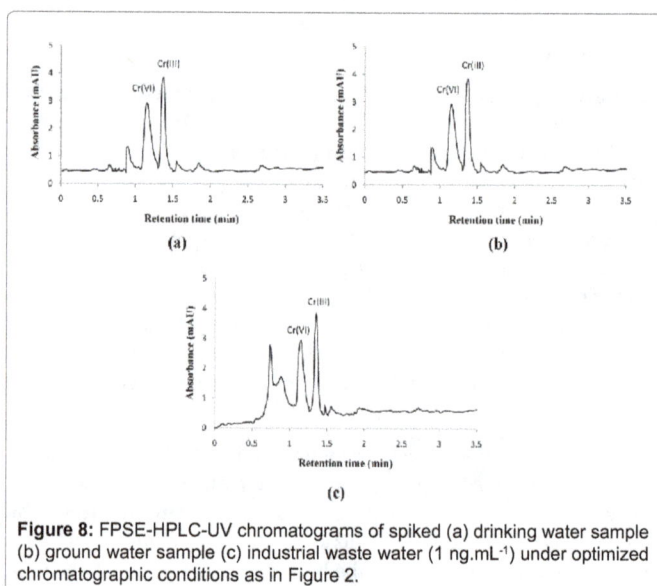

Figure 8: FPSE-HPLC-UV chromatograms of spiked (a) drinking water sample (b) ground water sample (c) industrial waste water (1 $ng.mL^{-1}$) under optimized chromatographic conditions as in Figure 2.

S No	Analytical technique	LOD [Cr(III)] ($ng.mL^{-1}$)	LOD [Cr(VI)] ($ng.mL^{-1}$)
1	*CPE-HPLC	3.4	5.2
2	IC-ICP-MS	>0.2	>0.2
3	HPLC-UV	0.005	0.007
4	*IC	0.02	0.02
5	HPLC-UV	400	1000
6	HPLC-ICP-MS	0.13	0.13
7	*LLE-GFAAS	N/A	0.2
8	FPSE-HPLC-UV	0.001	0.003

*CPE: Cloud Point Extraction; *IC: Ion Chromatography; *LLE: Liquid Liquid Extraction

Table 1: Comparison of detection limits of developed method with the previously reported methods.

Parameters	Sample preparation using FPSE	
	$Cr(MDTC)_3$ in Cr(III)	$Cr(MDTC)_2OMDTC$ in Cr(VI)
Wavelength (nm)	320	320
Correlation coefficient (R^2)	0.997	0.995
Retention time (min)	1.4	1.2
Regression equation	0.50x+4.8	0.275x+1.85
LOD ($ng.mL^{-1}$)	0.001	0.003
LOQ ($ng.mL^{-1}$)	0.003	0.009
RSD (%)	2.1-2.4	1.9-2.1

Table 2: Analytical characteristics of developed FPSE-HPC-UV method.

Analyte	Spiked Conc. (ppb)	Industrial Water		Bore well Water		Drinking Water	
		Obtained Conc. (RSD)	Recovery (%)	Obtained Conc. (RSD)	Recovery (%)	Obtained Conc. (RSD)	Recovery (%)
Cr(III) complex	1	0.89 (2.7)	89.60	0.98 (2.1)	98.70	0.97 (2.3)	97.60
	5	4.50 (2.5)	90.00	4.88 (2.0)	97.60	4.78 (2.3)	95.60
	10	9.16 (2.0)	91.60	9.75 (1.9)	97.59	9.65 (2.2)	96.70
Cr(VI) complex	1	0.87 (2.9)	87.00	0.97 (2.3)	97.50	0.98 (2.1)	98.50
	5	4.56 (2.6)	91.20	4.82 (2.2)	96.50	4.90 (2.0)	98.00
	10	9.13 (2.5)	91.30	9.86 (2.1)	98.60	9.68 (1.9)	96.80

Table 3: Recoveries of Cr (III) and Cr (VI) using FPSE-HPLC-UV.

techniques, FPSE possessed many advantages and has been proved to be a simple, economic, high extraction efficiency, faster extraction equilibrium and efficient sample preparation technique for the detection of chromium species using HPLC-UV. Trace level determination of Cr species is accomplished by using FPSE. Real samples were analysed after spiking with chromium complexes in order to confirm the validity of the developed method. This method shows potential as a rapid screening method to determine chromium species quantitatively in three kinds of water samples.

Acknowledgements

The authors are thankful to University Grant Commission (UGC), Delhi, India for providing the Project No. 42-304/2013 and financial support.

References

1. Cervantes C, Campos-Garcla J, Devars S, Gutierrez-Corona F, Loza-Tavera H, et al. (2001) Interactions of Chromium with microorganisms and plants. FEMS Microbiol Rev 25: 335–347.
2. Vincent JB (2014) Is chromium pharmacologically relevant?. J Trace Elem Med Biol 28: 397-405.
3. Tang AN, Jiang DQ, Jiang Y, Wang SW, Yan XP, et al. (2004) Surfactant sensitized calix arenes fluorescence quenching method for seperation of Cr(vI)/Cr(III) in Water samples. J Chromatogr A 1036: 183-188.
4. Rakhunde R, Deshpande L, Juneja HD (2012) Chemical separation of chromium in water: A review. Crit Rev Environ Sci Technol 42: 776-810.
5. Brien PO, Wang G (1989) Coordination chemistry and the carcinogenicity and mutagenicity of chromium. Environ Geochem Health 11: 77-79.
6. Unceta N, Séby F, Malherbe J, Donard OF (2010) Chromium speciation in solid matrices and regulation: a review. Anal Bioanal Chem 397: 1097-1111.
7. Barnowski C, Jakubowski N, Stuewer D, Broekaert JAC (1997) Speciation of chromium by direct coupling of ion exchange chromatography with inductively coupled plasma mass spectrometry. J Anal At Spectrom 12: 1155-1161.
8. Kaur V, Malik AK (2009) Speciation of chromium metal ions by RP-HPLC. J Chromatogr Sci 47: 238-242.
9. López-García I, Vicente-Martínez Y, Hernández-Córdoba M (2015) Non-Chromatographic speciation of chromium at sub-ppb levels using cloud point extraction in the presence of unmodified silver nanoparticles. Talanta 132: 23-28.
10. Chahal VK, Singh R, Malik AK, Matysik FM, Puri JK (2012) Preconcentration method on modified silica fiber for chromium speciation. J Chromatogr Sci 50: 26-32.
11. Peng H, Zhang N, He M, Chen B, Hu B (2015) Simultaneous speciation analysis of inorganic arsenic, chromium and selenium in environmental waters by 3-(2-aminoethylamino) propyltrimethoxysilane modified multi-wall carbon nanotubes packed microcolumn solid phase extraction and ICP-MS. Talanta 131: 266–272.
12. Ali I, Aboul-Enein HY (2002) Template-Free Hydrothermal Synthesis of β-FeOOH Nanorods and Their Catalytic Activity in the Degradation of Methyl Orange by a Photo-Fenton-Like Process. Chemosphere 48: 275-278.
13. Markiewicz B, Komorowicz I, Sajnag A, Belter M, Bara Akiewicz D (2015) Chromium and its speciation in water samples by HPLC/ICP-MS--technique establishing metrological traceability: a review since 2000. Talanta 132: 814-828.
14. Threeproem J, Purachaka S, Potipan L (2005) Application of molecularly imprinted polymers in food analysis: clean-up and chromatographic improvements. J Chromatogr A 1073: 291-295.
15. Laborda F, Gorriz MP, Bolea E, Castillo (2007) A Speciation Methodology To Study The Contributions Of Humic-Like And Fulvic-Like Acids To The Mobilization Of Metals From Compost Using Size Exclusion Chromatography-Ultraviolet Absorption-Inductively Coupled Plasma Mass Spectrometry And Deconvolution Analysis. Sci Total Environ. 373: 383-390.
16. Chen J, Meghraj M, Naidu R (2007) Micro-XRF Studies of Sediment Cores. Talanta 72: 394-400.
17. Anthemidis AN, Koussoroplis SV (2007) Phase formation in sodium dodecylsulfate solutions in the presence of salicylic acid for preconcentration purposes. Talanta 71: 1728-1733.
18. Capote FP, Castro MDL (2006) Chiral phosphines in nucleophilic organocatalysis. J Chromatogr A 1113: 244-250.
19. Wang JS, Chiu K (2004) Simultaneous extraction of Cr(III) and Cr(VI) with a dithiocarbamate reagent followed by HPLC separation for chromium speciation. Anal Sci 20: 841-846.
20. Tande T, Paettersen TE, Torgrimse T (1980) Modified thermoresponsive Poloxamer 407 and chitosan sol–gels as potential sustained-release vaccine delivery systems. Chromatographia 13: 607-610.
21. Hossain MA, Kumita M, Michigami Y, Islam TS, Mori S (2005) Rapid speciation analysis of Cr(VI) and Cr(III) by reversed-phase high-performance liquid chromatography with UV detection. J Chromatogr Sci 43: 98-103.
22. Kumar R, Gaurav, Heena, Malik AK, Kabir A, et al. (2014) Efficient analysis of selected estrogens using fabric phase sorptive extraction and high performance liquid chromatography-fluorescence detection. J Chromatogr A 1359: 16-25.
23. Kabir A, Furton KG, Malik A (2013) Innovations in sol-gel microextraction phases for solvent-free sample preparation in analytical chemistry Trends. Anal Chem 45: 197-218.
24. Macrotrigiano G, Pellacani GC, Preti C, Tosi G (1975) Dithiocarbamate Complexes of Rhodium(III), Iridium(III), Palladium(II) and Platinum(II). Bull Chem Soc Jpn 48: 1018-1020.
25. Kumar A, Paul Y, Rao ALJ (1992) Calculation of MAS Spectra Influenced by Slow Molecular Tumbling. J Indian Acad Forensic Sci 31: 19-24.
26. Kabir A, Hamlet C, Malik A (2004) Dynamics of protein hydration water. J Chromatogr A 1047: 1-13.
27. Beni A, Karosi R, Posta J (2007) Organotin Compounds in Sediments of Northern Lakes. J Microchem 85: 103-108.
28. Martinez-Bravo Y, Roig-Navarro AF, Lopez FJ, Hernandez F (2001) Microfluidic extraction, stretching and analysis of human chromosomal DNA from single cells. J Chromatogr A 926: 265-274.

Validated HPLC and Thin Layer-Densitometric Methods for Determination of Quetiapine Fumarate in Presence of its Related Compounds

Sawsan MA[1], Hesham S[2], Marianne N[1,3] and El-Maraghy MC[4*]

[1]*Analytical Chemistry department, Faculty of Pharmacy, Cairo University, Kasr-El Aini Street, 11562Cairo, Egypt*
[2]*Pharmaceutical Analytical Chemistry department, Faculty of Pharmacy, Deraya University, Minia, Egypt*
[3]*Pharmaceutical Analytical Chemistry department, Faculty of Pharmacy & Drug Technology, Heliopolis University, 3 Cairo Belbeis desert road, 2834El- Horria, Cairo, Egypt*
[4*]*Analytical Chemistry department, Faculty of Pharmacy, October University for Modern Sciences and Arts (MSA), 11787 6th October city, Egypt*

Abstract

Two chromatographic methods were developed for determination of quetiapine fumarate in presence of three related compounds; namely quetiapine N-oxide, des-ethanol quetiapine and quetiapine lactam, in pure form and pharmaceutical preparation. The first method depended on densitometric thin layer chromatography where the separation was achieved using silica gel $60F_{(254)}$ plates as stationary phase and toluene:1,4-dioxane:dimethylamine (5:8:2, v/v/v) as a mobile phase. The second method utilized the reverse phase high performance liquid chromatographic technique, using C18 column and methanol: acetonitrile: phosphate buffer (pH 5.3) in a ratio (19:40:41, v/v/v) as a mobile phase. The flow rate was1 mL min^{-1} and UV-detection was at wavelength 220 nm. The validation parameters of the developed methods were calculated and the results obtained were statistically compared with those of the HPLC manufacturer method.

Keywords: HPLC; Quetiapine fumarate related compounds; Thin layer-densitometry

Introduction

Quetiapine Fumarate (QTF) is a psychotropic agent belonging to a chemical class of dibenzothiazepine derivatives. The chemical designation is 2-(2-(4-dibenzo (b,f)(1,4)thiazepin-11-yl-1-piperazinyl) ethoxy)-ethanol fumarate (2:1) (salt). It is present in tablets as the fumarate salt. It is a white or almost white powder, moderately soluble in water and soluble in methanol and 0.1N HCl. It is used to treat psychosis associated with parkinson's disease and chronic schizophrenia [1], synthesized originally by Warawa and Migler [2] and can be used alone or in combination with other medications to treat schizophrenia and bipolar disorder [3,4]. Several methods have been reported for the determination of QTF in bulk powder, pharmaceutical preparations and biological samples. These included UV-Visible spectrophotometric methods [5-9], HPTLC methods [10-12], capillary zone electrophoretic method [13], voltammetry [14] and HPLC methods [15-17].

The evaluation of QTF related compounds has been an important issue that was recommended by regulatory agencies. Few papers were published for estimation of QTF in presence of its related compounds including; a GC method for determination of QTF in presence of 2-(2-chloroethoxy)ethanol and n-methyl-2-pyrrolidinon [18] and four HPLC methods, where it was determined in presence of piperazine (PI), quetiapine lactam (QL) and dibenzothiazepine piperazinyl ethanol hemifumarate compounds in the first method [19], and in presence of two related compounds quetiapine N-oxide (QO) and (QL) in the second one [20], also in presence of (QL),(PI) and Des-ethanol quetiapine (DQ) in the third one with long run time (18 mins) [21] , finally with (PI), (DQ) and quetiapine dimer using binary gradient mode in the fourth one [22]. And a RP-UPLC method for determination of QTF in presence of (QO), (PI), (DQ), S-oxide and quetiapine dimer using complicated binary gradient elution with linearity range (62.5-187.5 μg mL^{-1}) [23]. No method of the mentioned HPLC methods determined QTF in presence of this combination of related compounds; QO, DQ and QL in a simple isocratic isocratic elution mode for the mobile phase with good resolution between the four proposed components compared.

So far to our knowledge, only one stability-indicating HPTLC method for the determination of QTF in presence of its degradation products has been reported [24] but no TLC-densitometric method was reported for the determination of QTF in the presence of its related compounds.

The aim of our work was to develop more sensitive HPLC method with higher throughput using a mobile phase in a simple isocratic mode of elution and also to develop a selective, accurate, reproducible HPTLC-densitometric method which has the advantage of being of low cost and a faster technique when compared to HPLC method for determination of QTF in presence of the three related compounds; QO, DQ and QL (Figure 1) in raw material and in dosage form.

Materials and Methods

Instruments

Camag TLC scanner (Camag, Muttenz, Switzerland) operated with winCATS software version 3.15, Linomat IV autosampler (Camag, Muttenz, Switzerland). 100-μL Camag microsyringe (Hamilton, Bonaduz, Switzerland), Precoated silica gel aluminium Plates 60 F254 (20 cm × 20 cm) 250 μm thicknesses (E. Merck, Darmstadt, Germany).

Agilent 1200 series chromatographic system equipped with quaternary pump, microvacuum degasser, thermostatic column compartment and variable wavelength UV–VIS detector was used. Sample injections were made through an Agilent 1200 series

***Corresponding author:** Christine M. El-Maraghy, Analytical Chemistry Department, Faculty of Pharmacy, October University for Modern Sciences and Arts (MSA), 11787 6th October City, Egypt
E-mail: christine_elmaraghy@hotmail.com

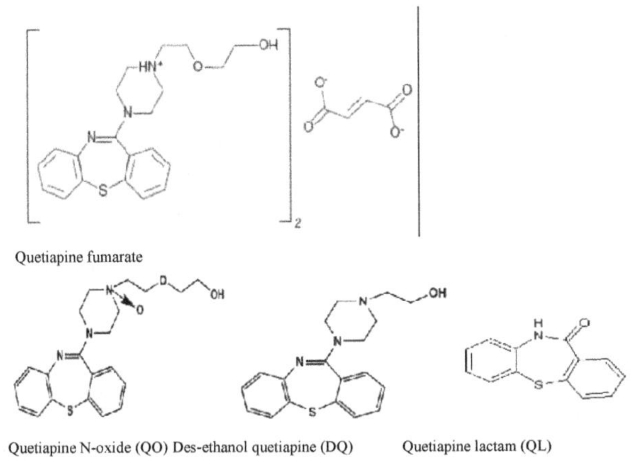

Quetiapine fumarate

Quetiapine N-oxide (QO) Des-ethanol quetiapine (DQ) Quetiapine lactam (QL)

Figure 1: Structures of quetiapine fumarate and its related compounds.

autosampler. Data collection and processing were performed using Agilent ChemStation software, version A.10.01. Column SN (USUXB06014) Agilent-C_{18} column (150 mm × 4.6 mm, 5 μm particle size i.d.) was from (Agilent Technologies, Polo Alto, CA, USA).

pH-meter (Jenway3505, UK), equipped with combined glass electrode was used for pH adjustment.

Ultrasound sonicator (Crest Ultrasonics, New York).

Chemicals and solvents

All chemicals used throughout the work were of analytical grade and solvents were of HPLC grade; Acetonitrile, methanol and orthophosphoric acid (Riedel-dehaen, Sigma-Aldrich, Germany), potassium dihydrogen orthophosphate and dimethyamine (ADWIC, Egypt), toluene (Euromedex, France), 1,4-dioxane (Alpha Chemika, India).

Samples

Quetiapine fumarate, (99.4%) and its related compounds, (99.8%) were kindly supplied by National Organization of Drug Control and Research (NODCAR) institute, Cairo, Egypt.

Seroquel® tablets were manufactured by Astra Zeneca, Egypt. Each tablet was labeled to contain 25 mg of quetiapine fumarate.

Chromatographic conditions

HPTLC-densitometry: The mobile phase was selected as mixture of toluene, 1,4-dioxane and dimethylamine in a ratio of (5:8:2, v/v/v). The densitometric scanning was performed at 225 nm. Analysis was performed on precoated 20 × 20 cm silica gel 60 F_{254} aluminium sheets (E. Merck, Darmstadt, Germany). Samples were applied to the plates using Camag Linomat IV applicator along with 100 μL Camag microsyringe. Spots were applied 1.5 cm apart from each other and 2 cm from the bottom edge. The chromatographic chamber was pre-saturated with the mobile phase for 15 min and the developing distance on TLC-plate was 180 mm.

HPLC: The mobile phase used was mixture of methanol, acetonitrile and phosphate buffer (pH adjusted to 5.3 using orthophosphoric acid) in a ratio of (19:40:41, v/v/v). The mobile phase was freshly prepared and filtered by vacuum filtration through 0.45 μm filter and degassed by ultrasound sonicator for 50 minutes just prior to use. The analysis was done under isocratic condition at a flow rate 1 mL min^{-1} and at room temperature using UV detector at 220 nm.

Standard solutions

Stock standard solutions (1.0 mg mL^{-1}) each of QTF, QO, DQ and QL were prepared in methanol. The working standard solution (0.1 mg mL^{-1}) of each one was prepared by further dilution of each stock solution with methanol.

Construction of calibration curve

HPTLC-densitometry: The linearity was evaluated by analyzing a series of different concentrations of the drug in the range of 1.0-11.0 μg $spot^{-1}$. Each concentration was repeated three times, in order to provide information on the variation in peak areas values among samples of the same concentration. The plates were developed using the specified mobile phase. The spots were scanned at 225 nm. The average peak area was calculated for each concentration of QTF and was plotted versus their concentration to obtain the calibration graph and the regression equation was then computed.

HPLC: Aliquots (0.1-3.0 mL) from QTF standard working solution were transferred into a series of 10 mL volumetric flasks and then diluted with methanol to obtain a concentration range of (1.0-30.0 μg mL^{-1}). Twenty microliters were injected for each concentration in triplicate and chromatographed using the HPLC conditions described above. The average peak area was calculated for each concentration of QTF and was plotted versus their concentration to obtain the calibration graph and the regression equation was then computed.

Assay of laboratory prepared mixtures

Solutions containing different ratios of QTF, QO, DQ and QL were prepared from their respective working standard solutions and diluted with methanol. The average peak areas of the laboratory prepared mixtures were calculated and processed as described above for the two proposed methods. The concentration of QTF was calculated using the computed regression equations.

Application to pharmaceutical preparation (Seroquel® tablet)

Seroquel® tablet was individually weighed to get the average weight of the tablet. A sample of the powdered tablets, claimed to contain 100.0 mg of drug was transferred separately to 100 mL volumetric flask, sonicated for 20 minutes with 50 mL methanol, then the volume was brought to 100 mL with the same solvent and filtered to prepare stock solution, having a concentration of (1.0 mg mL^{-1}). A working standard solution (0.1 mg mL^{-1}) was prepared by further dilution of the stock solution with methanol.

In HPTLC-densitometric method, aliquot of 5.0 mL was transferred from the prepared stock solution into 10 mL volumetric flask, completed to the volume with methanol. Then, 10 μL were applied onto HPTLC plates (n=5). While, in HPLC method, aliquot of 1.0 mL was transferred from the prepared working solution into10-mL volumetric flask and the volume was completed with methanol. Then, 20 μL from this dilution were injected (n=5).

The general procedure described above for each method was followed. Then, the concentration of QTF in its pharmaceutical preparation was calculated.

Application of standard addition technique

To check the validity of the proposed chromatographic methods, standard addition technique was applied by adding known amounts of the pure drug to the previously analyzed tablets. In HPTLC-densitometry, three aliquots (5.0 mL) of the previously prepared stock solution of tablet (1.0 mg mL^{-1}) were mixed with aliquots (1.0, 1.5, 2.0 mL) of pure stock standard solution of QTF, separately 10 mL volumetric flasks and the volume was completed with methanol. Then, 10 μL from each dilution were applied onto HPTLC plates in triplicate. In HPLC method, three aliquots (1.0 mL) of the working standard solution of tablet (0.1 mg mL^{-1}) were mixed with aliquots (0.5, 1.0, 1.5 mL) of pure working standard solution of QTF, separately into 10 mL volumetric flasks and the volume was completed with methanol. Then, 20 μL from each final dilution were injected in triplicate.

The general procedure described above for each method was followed and the concentration of the added pure QTF standard was calculated from the specified regression equation.

Results and Discussion

Development of analytical methods for the determination of pharmaceuticals in the presence of related compounds without previous chemical separation is always a matter of interest. The aim of this work was to establish sensitive, accurate and precise HPTLC-densitometric and HPLC methods for determination of QTF in presence of its related compounds namely; quetiapine N-oxide (QO), des-ethanol quetiapine (DQ) and lactam (QL), in its bulk powder and commercial tablets.

Related compounds of quetiapine fumarate

Seven potential impurities, including by-products, starting materials and intermediates were identified in pharmaceutical substance QTF and characterized by spectroscopic methods (MS, IR, and NMR) [25].

During the stability studies of QTF in the laboratory, several batches have been analyzed for purity by HPLC. QO (oxidation product) at level 0.1% was detected by ion-pair reversed-phase high performance liquid chromatography (HPLC) [26]. As per the stringent regulatory requirements recommended by ICH the impurities ≥0.1% must be identified and characterized and determined [27].

DQ is a byproduct in the process of QTF preparation, Based on the spectral data, the impurity was characterized as2-(4-dibenzo(b,f)(1,4) thiazepine-11-yl-1-piperazinyl)1-2-ethanol [28].

According to the Scheme (Warawa and Migler) for synthesis of QTF; QL is considered the starting material [28].

Method optimization

HPTLC-densitometry: The proposed HPTLC-densitometric method is based on the difference between the R_f values of QTF and its three related compounds due to the difference in their polarities and their migration rates on TLC plates. The chromatographic conditions were optimized by spotting the drug with its related compounds on TLC plates and developing in different solvent systems to achieve best separation. Different solvent systems were tried. Initially a system of toluene and methanol in a ratio of (8:2, v/v) was tried, but only the drug moved from the baseline as the three related compounds were highly polar. So, a mixture of toluene, dioxane and dimethylamine was tried in different ratios to obtain a good separation between QTF and its related compounds. The optimum mobile phase used was toluene:1,4-dioxane:dimethylamine in a ratio of (5:8:2, v/v/v). The chromatographic system described in this work allowed complete separation of QTF from its related compounds, as shown in Figure 2. Calibration graph was obtained by plotting the average peak area against the concentration of QTF. Linearity range was found to be 1.0–11.0 μg $spot^{-1}$ using the following regression equation:

A= 684.53 C – 278.13 r=0.9997

Where A represents the average peak area, C is the concentration in μg $spot^{-1}$ and r is the correlation coefficient.

HPLC: The chromatographic conditions, especially the composition of mobile phase, were optimized to achieve a good resolution and symmetric peak shapes for the drug and its related compounds, as well as a short analytical time. Initially a mixture of methanol, acetonitrile and phosphate buffer (pH adjusted to 6.4) in a ratio of (20:50:30, v/v/v) was used as a mobile. The peaks of QTF and QL were overlapped. Increasing the ratio of buffer to (20:40:40, v/v/v) resulted in increasing the retention time of QTF and QL peaks to more than 10 minutes and there was still some overlapping between QTF and QL peaks. By decreasing pH to 5.3, the separation between QTF and QL overlapped peaks was resolved. The optimum mobile phase used for the simultaneous determination of QTF and its related compounds was a mixture of methanol, acetonitrile and phosphate buffer (pH adjusted to 5.3) in a ratio of (19:40:41, v/v/v). The average retention time (R_t)± SD, for 6 replicate injections for QTF, QO, DQ and QL were found to be 3.61 ± 0.04, 2.53 ± 0.01, 3.20 ± 0.02 and 4.24 ± 0.04; respectively. A typical chromatogram of bank injection and QTF and its related compounds are shown in Figures 3 and 4, respectively.

Calibration graph was obtained by plotting the average peak area against concentration of QTF (μg mL^{-1}). Linearity range was found to be 1.0-30.0 μg mL^{-1} using the following regression equation:

A =68.611 C - 2.2421 r=0.9998

Where A represents the average peak area, C is the concentration in μg mL^{-1} and r is the correlation coefficient.

Method validation

International conference on Harmonization (ICH) guidelines [29] for method validation was followed for validation of the suggested methods.

Linearity: The linearity of the proposed methods for determination of QTF was evaluated by analyzing a series of different concentrations of the drug in the range of 1.0-11.0 μg $spot^{-1}$ for HPTLC-densitometric method and 1.0-30.0 μg mL^{-1} for HPLC method. Each concentration was repeated three times, in order to provide information on the variation in peak areas values among samples of the same concentration. Linear relationships were obtained by plotting the drug concentrations against the average peak area obtained for each concentration of QTF. The validation parameters for the regression equation of the adopted

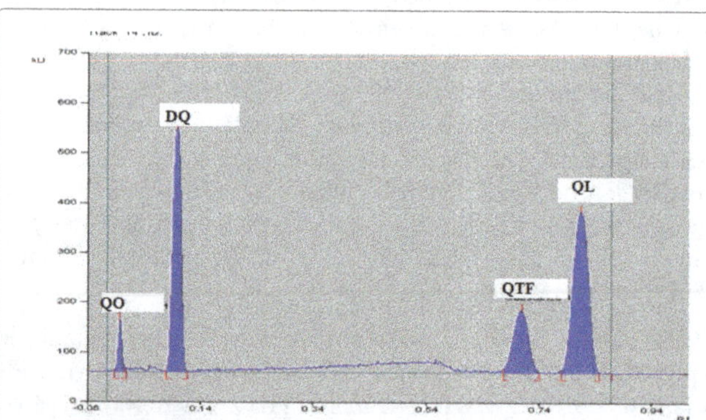

Figure 2: HPTLC chromatogram of QO, DQ, QTF and QL mixture of concentration (10.0 μg $spot^{-1}$) each, using toluene:dioxane:dimethylamine (5:8:2, by volume) at 225 nm.

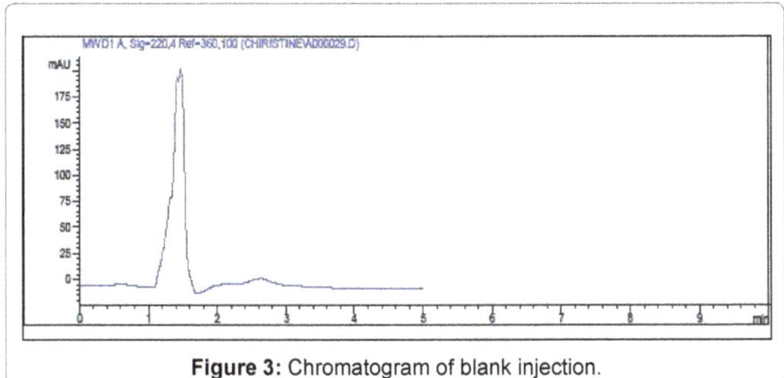

Figure 3: Chromatogram of blank injection.

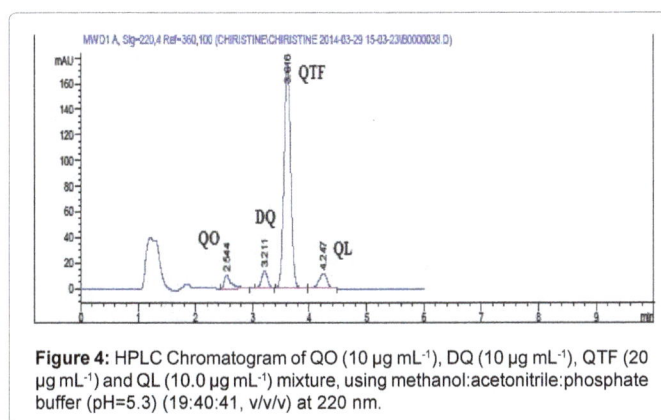

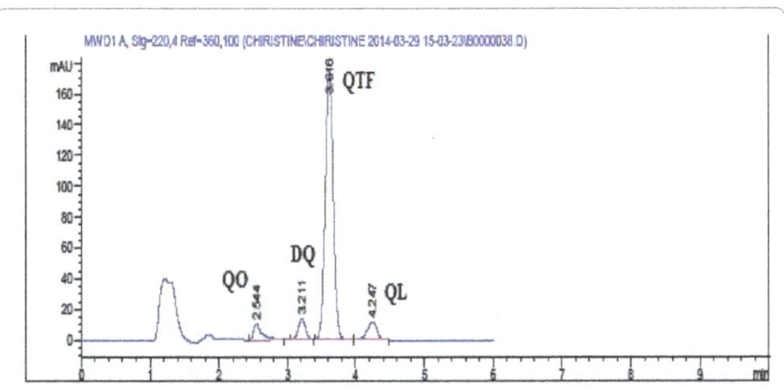

Figure 4: HPLC Chromatogram of QO (10 μg mL^{-1}), DQ (10 μg mL^{-1}), QTF (20 μg mL^{-1}) and QL (10.0 μg mL^{-1}) mixture, using methanol:acetonitrile:phosphate buffer (pH=5.3) (19:40:41, v/v/v) at 220 nm.

chromatographic methods are given in Table 1.

Accuracy: The accuracy of the proposed method was checked by the analysis of five different concentrations of authentic samples in triplicate. The concentrations of QTF were calculated using the regression equation for each of the two proposed chromatographic methods and then the mean percentage recovery and the relative standard deviation (RSD%) were calculated, as shown in Table 1 indicating the satisfactory accuracy of the proposed methods.

Precision: Three replicates of each concentration were analyzed on the same day to determine the intra-day precision of the methods. To confirm the inter-day precision, three replicates of each concentration were analyzed at three separate days using the developed chromatographic methods and calculating RSD%. Results in Table 1 indicate satisfactory precision of the proposed methods.

Detection and quantitation limits: According to (ICH) recommendation [29] the limits of detection (LOD) and limits of quantitation (LOQ) were determined using the formula: LOD or LOQ=k SD/slope of the response, where k=3.3 for LOD and 10 for LOQ and SD is the residual standard deviation of the regression line and results are presented in Table 1.

Specificity: The specificity of a method is the extent to which it can be used for analysis of a particular analyte in a mixture or matrix without interference from other components. The specificity of the proposed methods was tested by the analysis of six laboratory prepared mixtures containing different ratios of QTF and its three related compounds; QO, DQ and QL. The laboratory prepared mixtures were analyzed according to the previous procedures described under each of the proposed methods. The specificity was assessed to show that QTF could be measured without interference from its related compounds. Well resolved peaks for QTF from its related compounds indicated that there was no interference observed at the retention times of the drug. Satisfactory results were obtained (Table 2) indicating the high specificity of the proposed methods.

System suitability: System suitability test parameters must be checked to ensure that the system was working correctly during the analysis. Method performance data including capacity factor, selectivity, resolution, and tailing factor are listed in Table 3. All data was satisfactory and indicative of the good specificity of the method for determination of QTF in presence of its related compounds.

Robustness: The robustness of the chromatographic methods was investigated by the analysis of samples under a variety of experimental conditions such as small changes in TLC mobile phase ratio; toluene:dioxane:dimethylamine (6:8:2 and 5:9:2, v/v/v) and deliberate variations in HPLC mobile phase pH (5.5 and 5.6). Results presented in Table 4, indicate that the capacity of the utilized methods remain unaffected by these small deliberate variations, providing an indication for the reliability of the proposed chromatographic methods during routine work.

Analysis of pharmaceutical preparation

The proposed methods were applied for the determination of QTF in Seroquel® tablets, Figures 5 and 6. The results, shown in Table 5, are satisfactory and with good agreement with the labeled amount.

Standard addition technique

The interference of excipients in the pharmaceutical formulations was studied using standard addition method. The mean percentage recoveries and the standard deviation were calculated (Table 5). According to the obtained results a good accuracy and precision was observed. Consequently, the excipients in pharmaceutical formulations did not interfere in the analysis of QTF in its pharmaceutical formulation.

Statistical analysis

The results obtained by applying the proposed chromatographic methods were statistically compared to the reference HPLC method used for QTF analysis. The calculated t and F values were less than the theoretical ones indicating that there was no significant difference between the proposed and the manufacturer method with respect to accuracy and precision, as presented in Table 6.

Validation parameters		HPTLC-densitometry	HPLC
Linearity range		1.0-11.0 μg spot^{-1}	1.0-30.0 μg ml^{-1}
Correlation coefficient (r)		0.9997	0.9998
Slope		684.53	68.61
Intercept		-278.13	-2.24
LOD		0.29 μg spot^{-1}	0.20 μg ml^{-1}
LOQ		0.90 μg spot^{-1}	0.56 μg ml^{-1}
Accuracy (Recovery% ± RSD)		100.36 ± 1.41	100.21 ± 0.41
Precision (RSD%, n=9)	Intra-day	1.65	0.36
	Inter-day	1.93	1.93

Table 1: Characteristic parameters for the regression equations of the proposed methods, for determination of QTF.

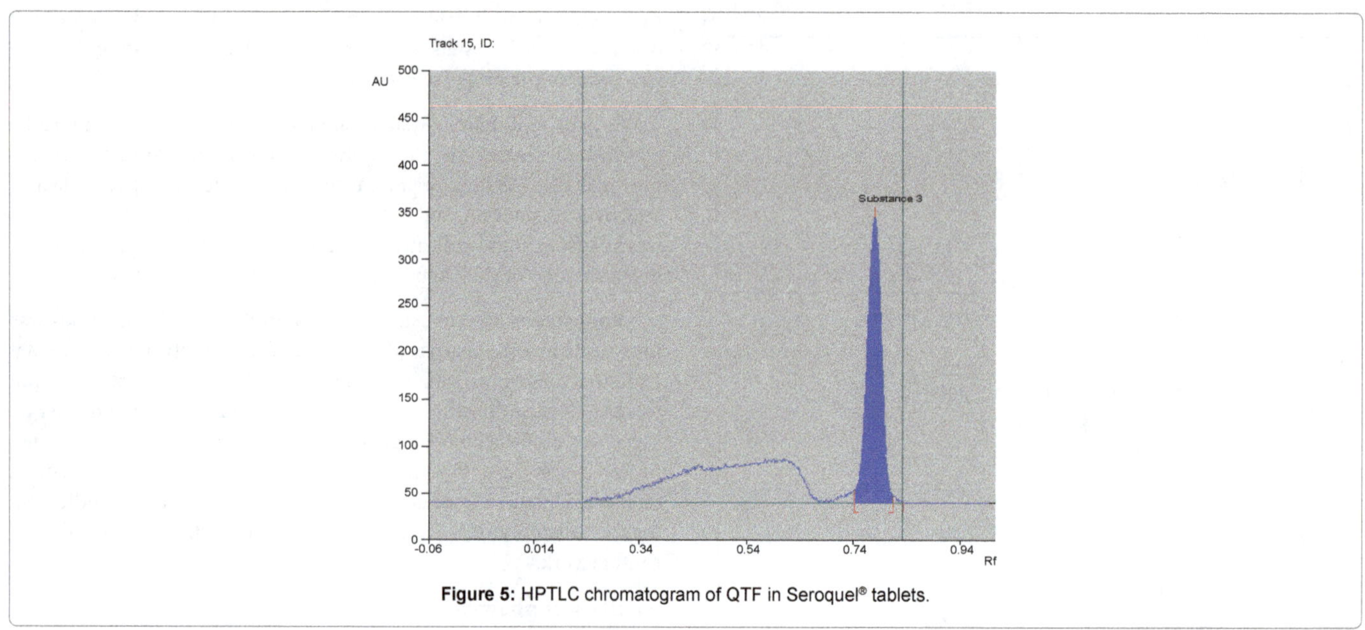

Figure 5: HPTLC chromatogram of QTF in Seroquel® tablets.

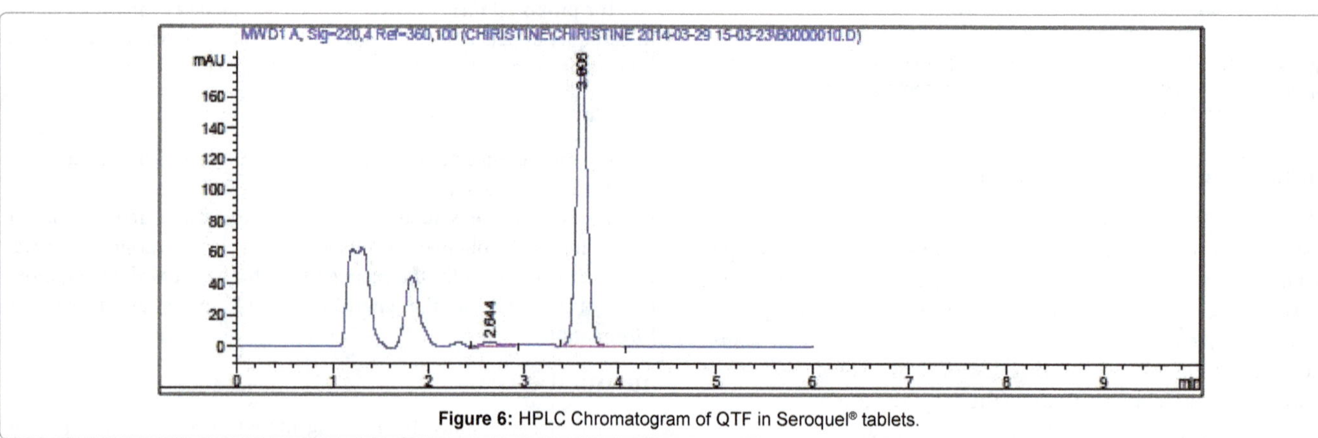

Figure 6: HPLC Chromatogram of QTF in Seroquel® tablets.

Ratio(QTF:QO:DQ:QL)	HPTLC (%Recovery[a] of QTF)	HPLC (%Recovery[a] of QTF)
10:0.01:0.01:0.01	99.76	100.16
10:0.02:0.02:0.02	99.34	100.18
10:0.03:0.03:0.03	101.20	101.22
10:0.06:0.06:0.06	100.65	100.17
10:0.1:0.1:0.1	101.12	100.79
10:0.2:0.2:0.2	101.75	101.15
Mean % ± SD	100.63 ± 0.91	100.61 ± 0.50

[a]Average of three determinations.

Table 2: Results obtained for the analysis of laboratory prepared mixtures containing different ratios of the intact QTF with its related compounds, by the proposed chromatographic methods.

Parameters	HPTLC-densitometry				HPLC				Reference values
	QO	DQ	QTF	QL	QO	DQ	QTF	QL	
t_R, min (HPLC)/R_f (HPTLC)	0.01	0.08	0.70	0.81	2.53	3.20	3.61	4.24	
Tailing factor (T)	0.90	0.76	0.93	1.02	0.73	0.96	1.01	1.18	T≤ 2
Plates number (N)					3040	4608	4688	3991	N>2000
Height equivalent to theoretical plate (HETP; cm plate^{-1})					0.049	0.032	0.031	0.037	The smaller the value, the higher the column efficiency
Resolution (R_s)	16	10.47	-	1.6	2.37	2.03	-	2.64	$R_s \geq 2$

Table 3: System suitability test results of the developed chromatographic methods for determination of QTF.

Parameters	HPTLC		HPLC	
	6:8:2, v/v/v	5:9:2, v/v/v	pH=5.5	pH=5.6
R_f (HPTLC) / t_R,min (HPLC) of QTF	1.015	0.97	3.56	3.58
T	0.97	1.04	1.02	1.07
N			4753	4634
R_s [b]	1.37	1.24	2.00	1.96

Table 4: Results [a] of robustness testing of the proposed chromatographic methods for determination of QTF.

Preparation	Claimed		%Recovery [a] ± SD		Standard addition technique			
					HPTLC		HPLC	
	HPTLC µg spot^{-1}	HPLC µg ml^{-1}	HPTLC	HPLC	Pure added µg spot^{-1}	%Recovery [b]	Pure added µg ml^{-1}	%Recovery [b]
Seroquel® tablets labeled to contain 25.0mg QTF	5.0	10.0	99.16 ± 1.27	100.67 ± 0.16	1.0	100.43	5.0	100.21
					1.5	99.52	10.0	100.35
					2.0	100.82	15.0	100.63
					Mean	100.27	**Mean**	100.39
					± SD	0.68	**± SD**	0.21

[a] Average of five determinations.
[b] Average of three determinations

Table 5: Quantitative determination of QTF in the pharmaceutical preparation and application of standard addition technique, using the proposed chromatographic methods.

Items	HPTLC- densitometry	HPLC	HPLC method [a]
Mean	99.16	100.67	99.21
RSD	1.17	0.16	0.97
variance	1.36	0.025	0.94
n	5	5	5
Student's t-test (2.44)	2.36	1.05	-
F-test (6.338)	2.45	1.18	-

The values between parenthesis are the theoretical values of t- and F-at P=0.05.
[a] HPLC method supplied by Astra Zeneca Company through personal communication using, the stationary phase; C18 column and mobile phase; methanol: acetonitrile: 0.02M dibasic ammonium phosphate (54:7:39, v/v/v) , UV detection at 254 nm and retention time 6mins.

Table 6: Statistical comparison between the results obtained, by applying the proposed methods and the manufacturer method for determination of QTF in pharmaceutical preparation.

Conclusion

The quality of pharmaceutical products is of vital importance for patient safety. The presence of related compounds or impurities may affect the efficacy and safety of pharmaceuticals. In this work, sensitive, accurate, precise and reproducible HPTLC-densitometric and HPLC methods were developed for the determination of QTF in the presence of three related compounds (QO, DQ and QL) which may be present in the pharmaceutical products in bulk powder and pharmaceutical preparation. The HPTLC-densitometric method has the advantage of being of low cost and is a faster technique when compared to HPLC. The proposed HPLC method offers high sensitivity, short run time and the use of isocratic elution mode for the mobile phase with good resolution between the four proposed components compared with the reported HPLC methods. Thus both methods can be used for routine analysis and quality control labs.

References

1. Cutler AJ, Goldstein JM, Tumas JA (2002) Dosing and switching strategies for quetiapine fumarate. Clin Ther 24: 209-222.
2. Warawa EJ, Migler BM (1988) Novel dibenzothiazepine antipsychotic. U.S. Patent 4879288 A.
3. Sweetman S (2006) Matrindale, in the complete Drug Reference. Electronic Version. Pharmaceutical Press, London.
4. Cheer SM, Wagstaff AJ (2004) Quetiapine. A review of its use in the management of schizophrenia. CNS Drugs 18: 173-199.
5. Rajendraprasad N, Basavaiah K, Vinay KB (2010) Sensitive and selective extraction-free spectrophotometric determination of quetiapine fumarate in pharmaceuticals using two sulphonthalein dyes. Journal of Pre-Clinical and Clinical Research 4: 24-31.
6. Basavaiah K, Rajendraprasad N, Ramesh PJ, Vinay KB (2010) Sensitive ultraviolet spectrophotometric determination of quetiapine fumarate in pharmaceuticals. Thai J Pharm Sci 34: 146-154.
7. Rajendraprasad N, Basavaiah K, Vinay KB (2012) Extractive spectrophotometric determination of quetiapine fumarate in pharmaceuticals and spiked human urine. Croat Chem Acta 85: 9-17.
8. Bagade SB, Narkhede SP, Nikam DS, Sachde CK (2009) Development and validation of UV-Spectrophotometric method for determination of Quetiapine fumarate in two different dose tablets. Int J ChemTech Res 1: 898-904.
9. Hiraman BB, Sandip V, Lohiya TR, Umekar JM (2009) Spectrophotometric determination of an atypicalantipsychotic compound in pharmaceutical formulation. Int J ChemTech Res 1: 1153-1161.
10. Mahadik MV, Patre NG, Dhaneshwar SR (2009) Stability-indicating HPTLC method for quantitation of quetiapine fumarate in the pharmaceutical dosage form. Acta Chromatogr 21: 83-93.
11. Sathiya R, Krishnaraj K, Muralidharan S, Muruganantham N (2010) A simple and validated HPTLC method of evaluation for quetiapine fumarate in oral solid dosage form. Eurasian J Anal Chem 5: 246-253.
12. Dhandapani B, Somasundaram A, Raseed SH, Raja M, Dhanabal K (2009) Development and validation of HPTLC method for estimation of quetiapine in bulk drug and in tablet dosage form. Int J PharmTech Res 1: 139-141.

13. Hillaert S, Snoeck L, Van den Bossche W (2004) Optimization and validation of a capillary zone electrophoretic method for the simultaneous analysis of four atypical antipsychotics. J Chromatogr A 1033: 357-362.

14. Ozkan SA, Dogan B, Uslu B (2006) Voltammetric Analysis of the Novel Atypical Antipsychotic Drug Quetiapine in Human Serum and Urine. Microchim Acta 153: 27-35.

15. Suneetha D, Rao A L (2010) A validated RP-HPLC method for the estimation of quetiapine in bulk and pharmaceutical formulations. E J Chem 7: 61-66.

16. Venkata BK, Battula SR, Dubey S (2013) Validation of quetiapine fumarate in pharmaceutical dosage by reverse-phase HPLC with internal standard method. J Chem 2013: 1-8.

17. Ingale PL, Dalvi SD, Gudi SV, Patil LD, Jadav DD, et al. (2013) Development of analytical method for determination of quetiapine fumarate in bulk & tablet dosage form. Der Pharma Chemica 5: 26-30.

18. Stolarczyk EU, Groman A, Kaczmarek S, Goebiewski P (2007) GC method for quantitative determination of residual 2-(2-chloroethoxy) ethanol (CEE) and N-methyl-2-pyrrolidinone (NMP) in quetiapine. Acta Pol Pharm 64: 187-189.

19. Raju IVS, Raghuram P, Sriramulu J (2009) Development and Validation of a New Analytical Method for the Determination of Related Components in Quetiapine Hemifumarate. Chromatographia 70: 545-550.

20. Belal F, Elbrashy A, Eid M, Nasr JJ (2008) Stability-indicating HPLC method for the determination of quetiapine: Application to tablets and human plasma. J Liq Chromatogr Related Technol 31: 1283-1298.

21. Rosa PCP, Pires IFR, Markman BEO, Perazzo FF (2013) Development and Validation of RP-HPLC Method for the Determination of Related Compounds in Quetiapine Hemifumarate Raw Material and Tablets. Journal of Applied Pharmaceutical Science 3: 6-15.

22. Krishna SR, Rao BM, Rao NS (2008) A validated stability indicating HPLC method for the determination of related substances in quetiapine fumarate. Rasayan jchem 1: 466-474.

23. Trivedi RK, Patel MC (2011) Development and validation of a stability indicating RP-UPLC method for determination of quetiapine in pharmaceutical dosage form. Sci Pharm 79: 97-111.

24. Dedania ZR, Sheth NR, Dedania RR (2013) Stability indicating high-performance thin-layer chromatographic determination of Quetiapine Fumarate. IJPSR 4: 2406-2414.

25. Stolarczyk EU, Kaczmarek L, Eksanow K, Kubiszewski M, Glice M, et al. (2009) Identification and characterization of potential impurities of quetiapine fumarate. Pharm Dev Technol 14: 27-37.

26. Mittapelli V, Vadali L, Sivalakshmi AD, Suryanarayana MV (2010) Identification, isolation, synthesis and characterization of principal oxidation impurities in Quetiapine. Rasayan Jchem 3: 677-680.

27. Food and Drug Administration, HHS (2003) International Conference on Harmonisation; revised guidance on Q3B(R) Impurities in New Drug Products; Availability. Notice. Fed Regist 68: 64628-64629.

28. Bharathi Ch, Prabahar KJ, Prasad ChS, Srinivasa Rao M, Trinadhachary GN, et al. (2008) Identification, isolation, synthesis and characterization of impurities of quetiapine fumarate. Pharmazie 63: 14-19.

29. ICH-International Conference on Harmonization, of Technical Requirements for registration of Pharmaceuticals for Human Use Topic Q2 (R1), (2005) Validation of Analytical Procedures: Text and Methodology, Geneva, Switzerland.

26

Retention Mechanism in Hydrophilic Interaction Liquid Chromatography New Insights Revealed From the Combination of Chromatographic and Molecular Dynamics Data

Fabrice Gritti*

Waters Corporations, 34 Maple St., Milford, MA, USA

Abstract

The retention mechanism of polar/charged analytes in hydrophilic interaction liquid chromatography (HILIC) remains ambiguous from the sole measurement of their retention factors. It is because thermodynamic properties only relate to the equilibrium relationship between the concentration in the mobile phase and that in the stationary phase. They do not provide any insights regarding their microscopic distribution across the mesopore volume. Chromatographers cannot unambiguously conclude whether analytes are adsorbed onto the surface of HILIC adsorbents, partitioned between the bulk and the water-rich interfacial layer or if both adsorption and partitioning mechanisms participate to the retention mechanism.

In order to solve this ambiguity, it is proposed to combine chromatographic data (retention factor, inverse-size exclusion, sample diffusivity along the bed) with molecular dynamics (MD) data. The latter provide microscopic information regarding the structure of the eluent and the average mobility of the analyte across the mesopore. This enables 1) the clear delimitation between three pore regions: the rigid water layer adsorbed onto the solid surface, the interfacial diffuse water layer, and the bulk region, 2) the measurement of the adsorption and partitioning equilibrium constants of the analyte between these three pore regions, and 3) the concentration distribution of the analyte in the pore volume.

The benefits of this new approach are demonstrated for a weakly retentive HILIC adsorbent (3.5 µm hybrid organic/inorganic silica particles) in contact with ternary eluent mixtures (acetonitrile/water pH 5/third solvent, 90/5/5, v/v/v). The third solvent has various polarities from water to n-hexane. The results show that, despite having nearly identical retention factors, the retention process of nortriptyline is essentially controlled by a partitioning mechanism while that of cytosine is governed by an adsorption mechanism. On the application side, it is shown how to significantly increase sample retention in HILIC by adding a third solvent in the mobile phase.

Keywords: Retention mechanism; Molecular dynamics; Hydrophilic interaction liquid chromatography; Adsorption-partitioning mechanism; Ternary eluent mixtures; Small molecules

Introduction

The physico-chemical description of the retention mechanism in either reversed-phase liquid chromatography (RPLC) [1-5], supercritical fluid chromatography (SFC) [6-8], or HILIC [9-17] has been the source of intense researches, discussions, controversies, and speculations over the past decades. These fundamental riddles have been debated since the early ages of liquid chromatography to the modern era of chromatography. From a fundamental viewpoint, these on-going enigmas found their origin in 1) the complexity of the eluent structure (at thermodynamic equilibrium) between the solid phase (bare or grafted solid adsorbents) and the bulk eluent and 2) in the subsequent and unknown distribution of the local sample concentration across the whole diameter of mesopores in HPLC particles. For instance, in HILIC, questions are often raised whether the retention process is controlled by the adsorption of the analyte onto the solid surface or by its partitioning between the bulk eluent (far from the surface) and the interfacial eluent layer that still remains under the influence of the adsorbent chemistry.

In order to elucidate the retention mechanisms in LC, MD simulations have been performed during the last two decades by taking advantage of the rapid increase of computational resources. They provide a wealth of microscopic information that can be combined to the measurement of retention factors and adsorption isotherms by dynamic chromatographic methods [5]. After calibration, these MD simulations provide unprecedented results for the concentration profiles of solvent and analyte molecules across the pores of RPLC silica-C_{18} [18-26] or HILIC silica-based particles [27-32]. Most interestingly, MD results show that the analyte molecules do not solely concentrate at the solid surface by adsorption but, also, they can accumulate in a pore region far from the wall of the mesopores (interfacial region) according to a partitioning process. This non-uniformity of the distribution of the sample mass in the mesopores inevitably makes the interpretation of chromatographic data speculative, ambiguous, or even impossible.

In the case of HILIC, MD simulations have confirmed the existence of a water-rich layer with rigid and diffuse parts at the interface between the solid silica surface and the bulk liquid phase [27-31]. Typically, the thickness of the rigid water layer is close to 4 Å and the mobility of the solvent molecules is zero in it. The diffuse water layer is thicker (ca. 11 Å wide), the solvent molecules have a finite degree of mobility and their concentration rapidly changes across this layer to reach their bulk concentration far enough from the adsorbent surface. Similar

***Corresponding author:** Fabrice Gritti, Waters Corporations, 34 Maple St., Milford, MA, USA; E-mail: Fabrice_Gritti@waters.com

heterogeneity of the eluent structure can be observed by MD in RPLC with the addition of tethered C_{18} chains amidst the solvent molecules [25,26].

As a result, it is established that the pore volume can be segmented into three distinct regions in which the local sample concentration can be drastically different. Analyte molecules can partition between the bulk and the diffuse water regions but, also, they can be adsorbed from the latter to the rigid water region. By definition, the sole observation of the chromatographic retention factors in HILIC does not allow the experimenter to distinguish between the number of sample molecules present in the rigid and diffuse parts of the water-rich layer: the chromatographer is inevitably facing indetermination. A new method combining chromatographic and MD data is then proposed to differentiate between the contributions of adsorption and partitioning to the overall retention of small, polar, and ionizable analytes in HILIC.

In this work, both the retention factor (elution time, information on the total sample mass present in the pores) and the effective intra-particle sample diffusivity (peak parking data [27-29], information on the average mobility of the analyte across the pores) are measured. They are combined with MD data which provides the local mobility of the analyte across the pores of silica HILIC particles and the delimitation between the rigid water, diffuse water, and bulk regions inside mesopores. This method eventually enables to determine the relative contributions of the adsorption and partitioning processes to the overall retention in HILIC. For the sake of illustration, it is applied to the retention behavior of a few ionizable analytes onto a 3.5 μm **hybrid organic/inorganic** HILIC stationary phase using ternary mixtures of acetonitrile (90% in volume), water (5%), and of a third solvent (water, ethanol, THF, ACN, or n-hexane, 5%).

Theory

Definitions

The inter-particle and particle porosities of the HILIC chromatographic column are noted ε_e and ε_p, respectively. The column hold-up volume is V_0. The eluent inside the mesopores of bare silica HILIC particles consists in three layers : an adsorbed or rigid water layer (volume $V_{ads.}$), a diffuse water layer (volume $V_{part.}$), and the bulk eluent (volume V_{bulk}) far from the surface [27,28]. The volume fractions occupied by the bulk phase and the diffuse water layer in the pores are noted f_b and f_d, respectively. The volume fraction occupied by the rigid water layer inside the pores is then f_a=1- f_b-f_d. The total pore volume is V_p.

The mole number of the analyte in the rigid water layer, in the diffuse water layer, in the bulk region, in the inter-particle volume are n_a, n_d, n_b, and n_e. The number of mole present in a hypothetically inert (no preferential adsorption of the solvent and analyte molecules) pore is n_p.

The reference sample concentration in the bulk eluent is c_b. Its average concentration in the diffuse and rigid water layers are c_d and c_a, respectively. These three concentrations define two independent equilibrium constants K_d and K_a, relative to the partitionning between the bulk and diffuse water layer and to the adsorption between the diffuse water layer and the rigid water layer. Accordingly:

$$K_d = \frac{c_d}{c_b} \tag{1}$$

and

$$K_a = \frac{c_a}{c_d} \tag{2}$$

The local diffusion coefficients (or mobilities) of the analyte in the bulk phase and in the diffuse water layer are D_m and δD_m, respectively, whereby the value of δ depends on the acetonitrile volume fraction in the bulk mobile phase [27-29]. The local mobility in the rigid water layer is by definition equal to zero (due to the physisorption of the analyte).

The effective sample diffusivity along the chromatographic bed is D_{bed}. The coefficient Ω is defined as the ratio of the sample diffusivity across each single particle, D_p, to the bulk diffusion coefficient (note that D_p is defined with the convention that the reference concentration gradient is defined with respect to the bulk concentration c_b):

$$\Omega = \frac{D_p}{D_m} \tag{3}$$

Finally, the effective sample diffusivity across the mesopores of the particles is $D_{mesopore}$.

Retention factor

The retention factor also called capacity factor k' is defined as the amount of sample present in the column minus the amount of sample present in the column for an inert adsorbent and divided by the latter amount. From the definitions given in the previous section and after simplification, k' is written:

$$k' = \frac{f_b(1-K_aK_d) + f_dK_d(1-K_a) + K_aK_d - 1}{1 + \frac{\varepsilon_e}{\varepsilon_p(1-\varepsilon_e)}} \tag{4}$$

It is noteworthy that the observed retention factor depends on the column (ε_e) and particle (ε_p) porosities, on the volume fractions occupied by two of the three regions inside the mesopores (f_b and f_d), and by the equilibrium constants K_d and K_a. Equation (3) describes quantitatively the inherent ambiguity that the chromatographer is facing when measuring only retention factors: no information is received regarding the distribution of the analyte across the mesopore volume in the three pore regions. For this reason, the average mobility of the analyte across the mesopore is needed. It is given from diffusivity data as explained in the next section.

Effective sample diffusivity through the bed, the porous particles and the mesopores

The only sample diffusivity that the chromatographer can directly observe is the effective diffusivity D_{bed} along the packed bed immersed in the mobile phase. This can be achieved by the peak parking method [33-35].

For columns packed with fully porous particles, the most relevant model of effective diffusion in heterogeneous binary (1-particles filled with eluent, 2-external eluent) media is the Torquato model [36]. Accordingly, D_{bed} is written:

$$D_{bed} = \frac{1}{\varepsilon_e(1-k_1)}\left[\frac{1+2(1-\varepsilon_e)\beta - 2\varepsilon_e\xi_2\beta^2}{1-(1-\varepsilon_e)\beta - 2\varepsilon_e\xi_2\beta^2}\right] \tag{5}$$

Where k_1 is the zone retention factor defined by:

$$\varepsilon_e(1+k_1) = [\varepsilon_e + (1-\varepsilon_e)\varepsilon_p](1+k') \tag{6}$$

β is a function of the ratio Ω:

$$\beta = \frac{\Omega - 1}{\Omega + 2} \tag{7}$$

and $_2$ is a constant (=0.627) given by the external porosity (=0.39) and the external obstruction factor ξ_e (=0.57) of the bed:

$$\gamma_e = \frac{2\left(1-\frac{\xi_2}{2}\right)}{3-\varepsilon_e(1+\xi_2)} \tag{8}$$

The experimental protocol applied to extract the coefficient Ω from the measurement of the longitudinal diffusion coefficient $B=2(1+k_1)D_{bed}$ has already been described elsewhere [37-39]. For porous particles, the effective sample diffusivity in a heterogeneous binary (1-solid silica, 2-mesopore eluent) medium is well described by a classical Landauer model [40]. According to this model, the effective sample diffusivity across the mesopores is directly related to the above-defined diffusivity coefficient D_p [39]:

$$\Omega = \frac{3\varepsilon_p - 1}{2}(f_b + f_d K_d \delta) \tag{9}$$

A time-averaged model of effective diffusion was considered for the effective diffusion of the analyte in the mesopores because these three regions can be considered as parallel to each other in space. Therefore, a new relationship is written between Ω, ε_p, f_b, f_d, and K_d [39]:

$$\Omega = \frac{3\varepsilon_p - 1}{2}(f_b + f_d K_d \delta) \tag{10}$$

Figure 1 schematizes this succession of calculations that lead to the measurement of the effective sample diffusivities along the bed, across the particles, and through the mesopores.

Chromatographic data

Standard elution and inverse size-exclusion chromatography are applied for the measurement of the retention factor (*k'*), the total mesopore volume (V_p), and the hold-up volume (V_0). Accordingly, these data provide two independent relationships. One for the total pore volume:

$$V_{ads.} + V_{part.} + V_{bulk} = V_p \tag{11}$$

The second for the retention factor:

$$K_{ads.} K_{part.} V_{ads.} + K_{part.} V_{part.} + V_{bulk} = k' V_0 + V_p \tag{12}$$

Since the retention behavior (combining adsorption and partitioning) is fully determined from the knowledge of the six independent variables (V_0, $V_{ads.}$, $V_{part.}$, V_{bulk}, $K_{ads.}$, and $K_{part.}$) and V_0 is known, three additional relationships are missing. They will be given by molecular dynamics data.

Molecular dynamics data

All the details regarding the MD simulation of the equilibrium process between the solvent molecules in a ternary mixtures and the silica surface are given in reference [27,28]. To summarize: an open cylindrical silica-based pore is generated by computer (β-cristolabillite structure), it is in equilibrium with a solvent reservoir flanked at each side of the cylinder, the force fields for the different solvent components are selected so that the simulated results match the experimental properties of the bulk mobile phase mixtures, the total number of solvent molecules is around 15 000 in this single pore, and the recorded simulation time last ~ 6 ns after the sytem was conditionned during nearly 150 ns. The equations of motion were integrated with a 1 fs time step so that 6 millions molecular configurations were recorded before calculating the density and mobility profiles of the solvent molecules across the pore diameter.

Eventually, MD allows to unambigously decouple the roles of adsorption and partitioning on the overall retention process in HILIC once k' (Equation 4) and Ω (Equation 10) are known from chromatographic data. Indeed, three independent and additional information are directly derived from the results of the MD calculations: the volume fractions of the rigid and diffuse water layer (f_a and f_d from the calculated density profiles) as well as the local mobility, δD_m, of the solvent molecules in the diffuse water region (from the calculated mobility profiles). Figures 2 and 3 show the MD results for the density and mobility profiles of the solvent molecules across the cylindrical pore.

Decoupling adsorption from partitioning retention mechanisms

Chromatographic and MD data are combined to solve the equilibrium problem set by Equations 11 and 12. First, the measurements of Ω (chromatography) and δ (MD) provides the partitioning constant $K_{part.}$ from Equation 10. Secondly, the measurements of *k'*, V_p, V_0 (chromatography), f_a, and f_d (MD) enables the measurement of the adsorption constant K_a from Equation 4. Therefore [39]:

$$K_d = \frac{(2\Omega/3\varepsilon_p - 1) - f_b}{f_d \delta} \tag{13}$$

and

$$K_a = \frac{k'\left(1 + \frac{\varepsilon_e}{(1-\varepsilon_e)\varepsilon_p}\right) + 1 - f_b - f_d K_d}{K_d(1 - f_d - f_b)} \tag{14}$$

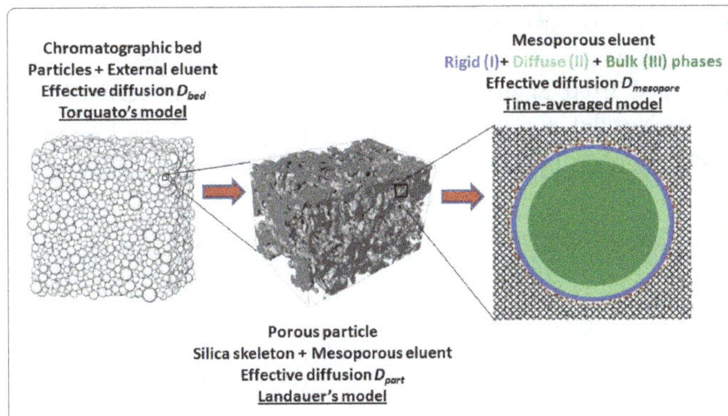

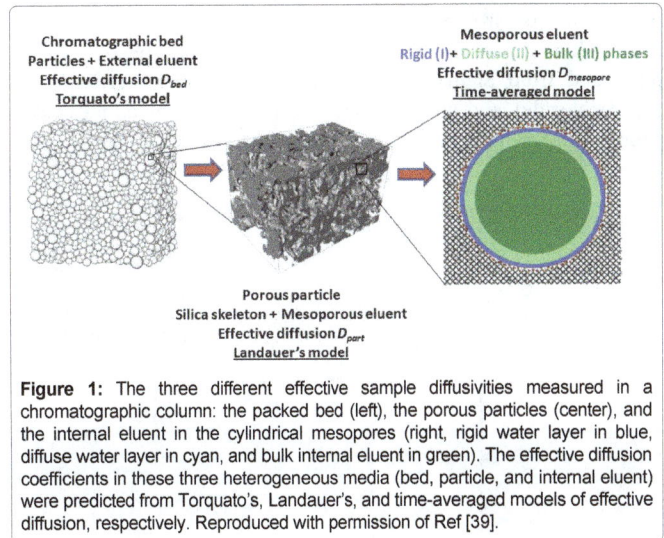

Figure 1: The three different effective sample diffusivities measured in a chromatographic column: the packed bed (left), the porous particles (center), and the internal eluent in the cylindrical mesopores (right, rigid water layer in blue, diffuse water layer in cyan, and bulk internal eluent in green). The effective diffusion coefficients in these three heterogeneous media (bed, particle, and internal eluent) were predicted from Torquato's, Landauer's, and time-averaged models of effective diffusion, respectively. Reproduced with permission of Ref [39].

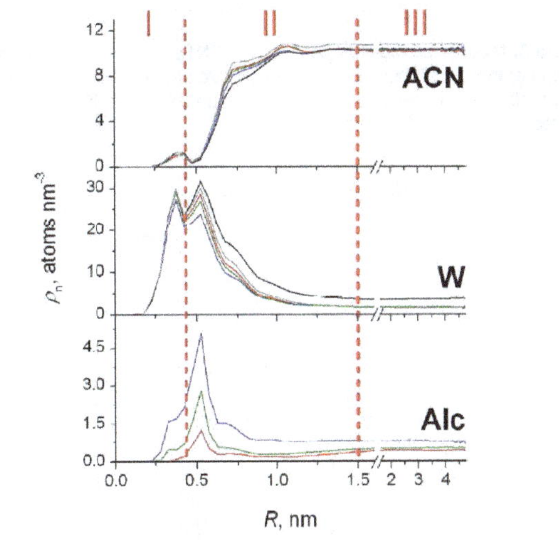

Figure 2: Density profiles of acetonitrile (ACN), water (W), and of the third solvent (Alc) across the silica-based cylindrical pore. The volume fractions of ACN, W, and Alc in the bulk ternary eluent are 90%, 5%, and 5%, respectively. Alc=water (black solid line), methanol (blue solid line), ethanol (green solid line), isopropanol (red solid line), or acetonitrile (gray solid line). Note the delimitation between the pore regions I (rigid water layer), II (diffuse water layer), and III (bulk region). Reproduced with permission of Ref [30].

Equation 13 and 14 can then be used to determine the molar fractions of the sample molecules present in the rigid water layer, in the diffuse water layer, and in the bulk region of the mesopore.

Experiments

Chemicals

The mobile phases were prepared by mixing 225 mL of acetonitrile, 12.5 mL of a buffer stock solution (pH 5) prepared from 200mM ammonium acetate and glacial acetic acid, and 12.5 mL of a third solvent (water, ethanol, tetrahydrofuran, acetonitrile, or *n*-hexane). The buffer concentration in the mobile phase is then 10mM. All solvents were HPLC grade from Fisher Scientific (Fair Lawn, NJ, USA). Acetonitrile was filtered before use on a surfactant-free cellulose acetate filter membrane, 0.20 μm pore size purchased from Sigma-Aldrich (Suwannee, GA, USA). Eleven polystyrene standards (MW=590, 1100, 3680, 6400, 13200, 31600, 90000, 171000, 560900, 900000 and 1870000) were purchased from Phenomenex (Torrance, CA, USA) and used to perform inverse size-exclusion chromatography (ISEC) experiments. Ammonium acetate and glacial acetic acid for buffer preparationwere purchased from Sigma–Aldrich. Toluene, cystosine, nortrriptyline (hydrochloride), and nicotinic acid (see Figure 4) were all purchased from Fisher Scientific, with a minimumpurity of 99%.

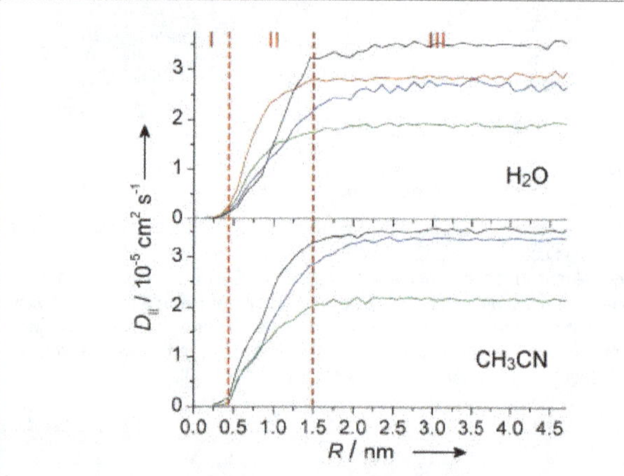

Figure 3: Mobility profiles of acetonitrile (ACN) and water (W) across the three regions of the silica-based cylindrical pore. Volume fractions of ACN in the bulk eluent: 70% (solid green line), 95% (solid blue line), and 99% (solid black line). Reproduced with permission of Ref [29].

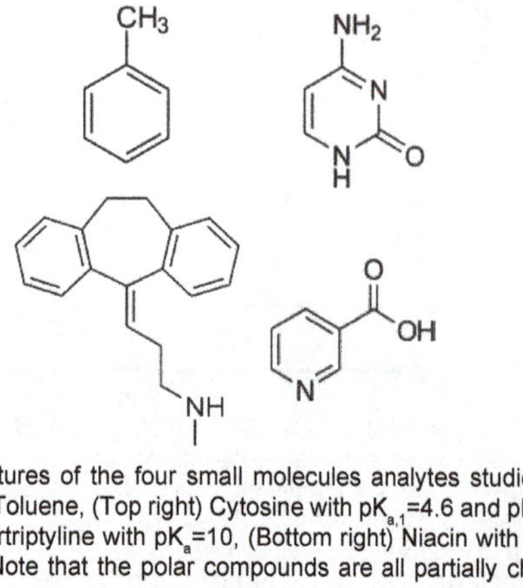

Figure 4: Structures of the four small molecules analytes studied in this work. (Top left) Toluene, (Top right) Cytosine with $pK_{a,1}$=4.6 and $pK_{a,2}$=12.2, (Bottom left) Nortriptyline with pK_a=10, (Bottom right) Niacin with $pK_{a,1}$=2.1 and $pK_{a,2}$=4.8. Note that the polar compounds are all partially charged at pH 5.

Apparatus

Chromatography measurements were all performed on a 1290 Infinity HPLC system (Agilent Technologies, Waldbroon, Germany) liquid chromatograph. This system includes a 1290 Infinity Binary Pump with solvent selection valves and a programmable auto-sampler. The injection volume is drawn into one end of the 20 μL injection loop. The instrument includes a two-compartment oven and a multi-diode array UV-Vis detection system. The system is controlled by the Chemstation software. The sample trajectory in the equipment involves the successive passage through the 20 μL injection loop attached to the injection needle, a small volume needle seat capillary (~ 1 μL), a small volume injection valve (~ 1.2 μL), two 130 μm × 250 mm long Viper connecting capillaries (3.3 μL each) and a standard volume detection cell (2.4 μL). The extra-column volume is then close to 10 μL.

Column

The HILIC column used was a 4.6 mm × 100 mm column packed with 3.5 μm hybrid organic/inorganic fully porous (130 Å) silica particles. This silica material designed to enhance pH stability of the column from pH 2 to pH 10 was provided by the manufacturer (Waters, Milford, MA, USA). The external, total, and particles porosities measured from ISEC were 0.39, 0.72 and 0.54 respectively.

Measurement of the retention factor k'

By convention, all the retention factors were measured with reference to the elution time of toluene in pure THF for the hold-up time. All the retention times were corrected for the extra-column time. The volumetric flow rate was fixed at 1 mL/min. Even though there are as many t_0 as tracers used to measure it, the hold-up time should always be measured from a pure solvent in order to avoid the possible exclusion of toluene from the water-rich layers when using an acetonitrile-rich mobile phase. The temperature was set by the lab air-conditioning system (297 ± 1 K). The measurement of k' was performed after one to two hours equilibration time. The relative standard deviations were smaller than 0.5% for water, ethanol, and THF as the third solvent and smaller than 3% for acetonitrile and *n*-hexane.

Measurement of the bed diffusivity D_{bed}

The effective sample diffusivity along the chromatographic bed was measured from the classical peak parking method as previously described in references [33-35]. It is directly proportional to the slope of the time variance, σ_t^2, of the eluted peak *versus* the parking time, t_p, and to the square of the linear migration velocity u_R. Accordingly,

$$D_{bed} = \frac{1}{2}\frac{\Delta\sigma_t^2}{\Delta t_p}u_R^2 \quad (15)$$

Measurement of the intra-particle diffusivity Ω

The coefficient Ω is measured by combining Equations 5-8 (Torquato's model) with Equation 15 (peak parking data). The mathematical solution of this problem (a second order algebraic equation has to be solved) is given in reference [39].

Results and Discussion

Chromatographic data

Retention factor: Figure 5 shows the eluted peak profiles of toluene, cytosine, nortriptyline, and niacin in mobile phases of 90/5/5 (v/v/v) acetonitrile/aqueous acetate buffer (pH 5)/third solvent for water, ethanol, tetrahydrofuran, acetonitrile and *n*-hexane as the third solvent. The polarity of the third solvent decreases in that order. Its

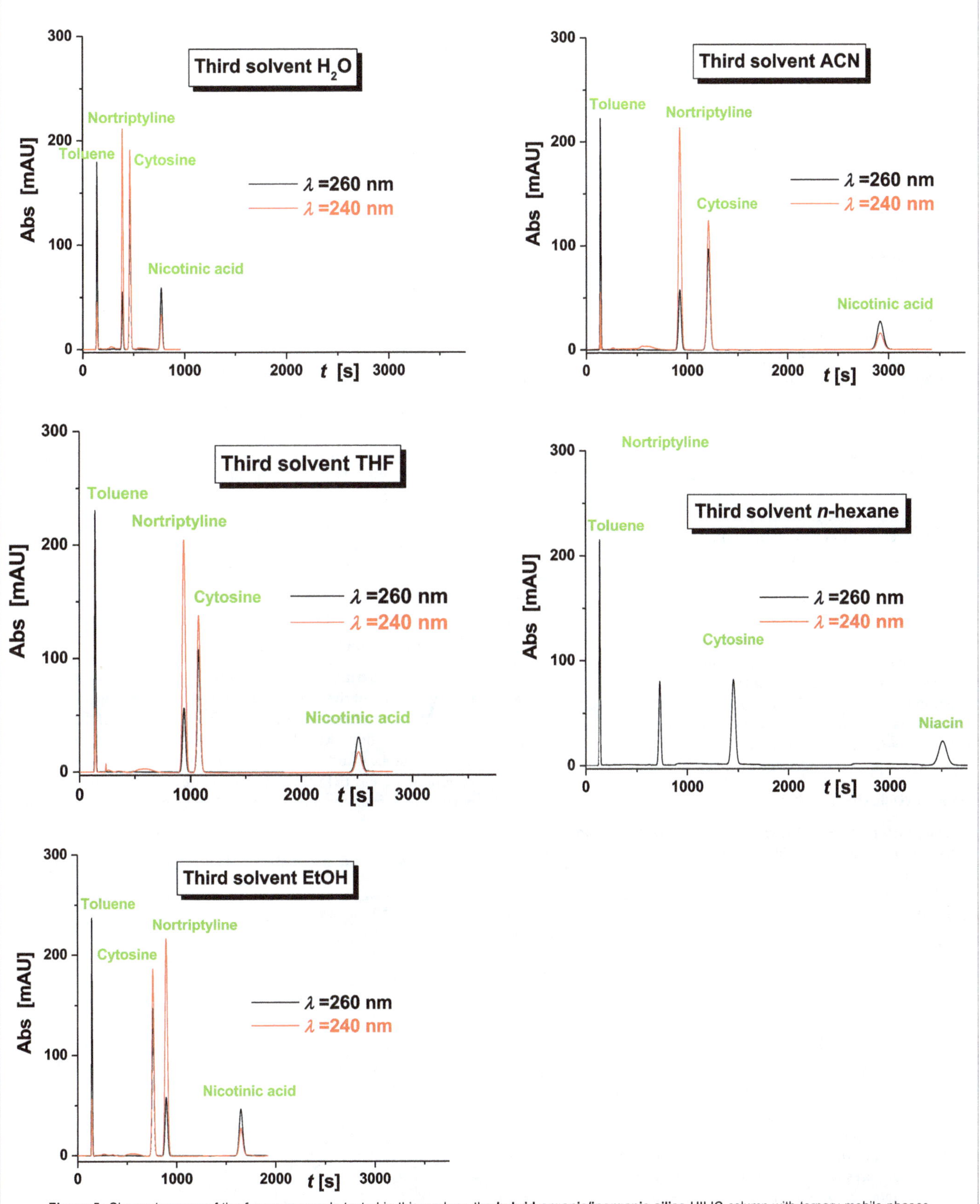

Figure 5: Chromatograms of the four compounds tested in this work on the **hybrid organic/inorganic silica** HILIC column with ternary mobile phases of 90/5/5 (v/v/v) acetonitrile/200 mM acetate buffer pH 5/third solvent. See the experiments section for the experimental conditions selected. Reproduced with permission of Ref [39].

role is to enhance the retention factor of polar analytes, which are too poorly retained onto the hybrid organic/inorganic silica HILIC column [41]. Expectedly, toluene, the apolar compound, is not retained irrespective of the nature of the thrid solvent. Niacin, which carries a partial negative charge at pH 5, is the most retained compound. The partially positively charged compounds, cytosine and nortriptyline, show intermediate retention. It is striking to observe that, unlike niacin, the retention time of the positively charged compounds do not increase monotously with decreasing the polarity of third solvent. In particular (see the red and black colored chromatograms corresponding to two different absorption wavelengths), the elution order of cytosine and nortriptyline changes from water to ethanol and from ethanol to tetrahydrofuran. Additionally, the retention time of nortriptyline goes through a maximum for tetrahydrofuran as the third solvent while the retention time of cytosine increases monotoneously from the most to the least polar third solvent. One hypothesized explanation for the increase of retention with decreasing the polarity of the thrid solvent is the relative increase of the water content in the diffuse water region with respect to the bulk phase [27] and the increasing importance of a partitioning mechanism.

Figure 6 summarizes these retention data by plotting the experimental retention factor of the four compounds as a function of the nature of the third eluent. Whereas the retention behavior of toluene (non-retained compound) is clear, those of cytosine, nortriptyline, and niacin remain puzzling. They illustrate why chromatography alone and the mere measurement of retention factor cannot provide relevant clues regarding the observed differences in retention behavior. At this point, it is not possible to tell whether their retention is mostly controlled by an adsorption, a partitioning, or by both mechanisms and how the intensity of these two contributions is changing when decreasing the polarity of the third solvent. Additional information is needed to conclude where the sample molecules spend much of their time inside the mesopores : do they accumulate at the surface (in the rigid water layer) or in the interfacial region (in the diffuse water layer)? This enigma can only be solved by measuring complimentary data. Next, it is shown how the measurement of the intra-particle porosity can solve this problem from a qualitative viewpoint.

Intra-particle diffusivity

The intra-particle diffusivity coefficient Ω (Equation 10) is a measure of the average mobility of the analyte through the particle. It is obviously hindered by the presence of the solid silica walls. In the case of hybrid organic/inorganic silica particles, the volume fraction occupied by the internal eluent is ε_p=54%. The sample molecules diffuses accross the particles when present in either the bulk region III (local mobility D_m) or in the diffuse water region II (local mobility δD_m) of the mesopores (see Figure 2). The sample molecules present in the rigid water layer I do not contribute to the mobility across the particle because they are physisorbed (zero local mobility). The coefficient Ω is expressed quantitatively by Equation 10.

In order to validate the selection of the Landauer's and time-averaged models for the effective sample diffusivity across each individual HILIC particle (see Equation 9), the coefficient Ω was roughly estimated for a non-retained apolar compound that is fully excluded from the water-rich regions I and II. The volume fraction occupied by the bulk phase in cylindrical pores is f_b=62% (see delimitations between regions I, II, and III in Figure 2). The corrected particle porosity is then ε_p=0.54 × 0.62 ~ 0.33. Considering a physically relevant pore obstruction factor (combining internal tortuosity and constriction) γ_i~0.60 [42] and a diffusion hindrance parameter F()=0.81 for small molecules (Renkin correlation, pore size 140 Å and sample size 7 Å) [43], the expected coefficient Ω would take a value of $\varepsilon_p \times \gamma_i \times F(\lambda)=0.33 \times 0.60 \times 0.81=0.16$. According to the Landauer's and time averaged models (Equation 10), Ω would be ½ (3 × 0.54 -1) × 0.62=0.19. The combination of these two models is then justified. Finally, the experimental values of Ω for toluene were measured at 0.22 [39]. This slighly larger value could be consistent with the fact that toluene is actually not fully excluded from the diffuse water layer.

Therefore, for the three retained compounds and according to Equation 10, the measurement of the coefficients Ω reveals how important the contribution of the partionning mechanism to the overall retention can be with respect to toluene. Figure 7 shows the plot of Ω as a function of the nature of the third eluent. As expected, Ω values are all larger than 0.22. This confirms that these polar analytes are all present in the diffuse water layer at equilibrium. Most remarkable is the difference between the W values of nortriptyline (large) and those of cytosine for ethanol and tetrahydrofuran as the third eluent. Whereas their retention factors are quasi-identical (see Figure 6), their mobility inside the mesopores of the **hybrid organic/inorganic silica particles** are much different. This demonstrate that nortriptyline molecules accumulates more than cytosine in the diffuse water layer. A partionning mechanism is then clearly relevant for nortriptyline. This conclusion cannot be drawn from simple measurement of the retention factor (Figure 6).

Finally, Figure 7 shows that no general conclusion can be drawn about the effect of the third solvent on the retention mechanism of the analytes. It inevitably depends on the nature of the analyte and on its distribution between the bulk, the diffuse water region, and the rigid water region.

Molecular dynamics data

Eluent structure in the mesopore: Figure 2 shows the calculated density profiles of the solvent molecules (acetonitrile, water, and third solvent) as a function of the distance z from the surface of the adsorbent. All the calculations were based on the MD simulations as presented in the theoretical section 5. The MD data enables to distinguish three regions (I, II, and III) within the mesopores of the HILIC particles: 1) beyond 15 Å from the pore wall, the eluent composition is equal to the

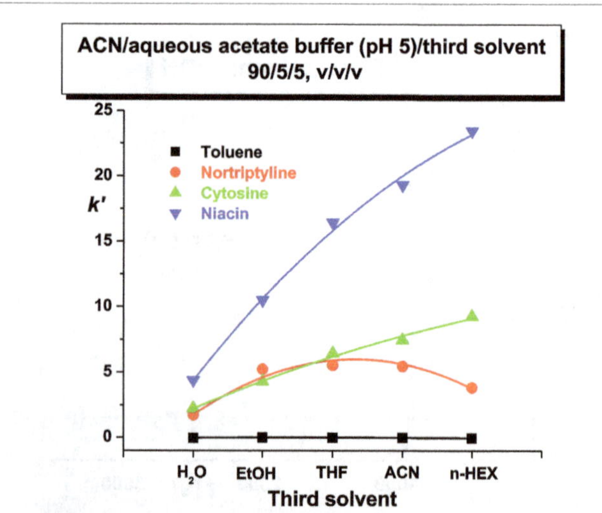

Figure 6: Plots of the experimental retention factors k' of the four analytes as a function of the nature of the third solvent in a mobile phase of acetonitrile/ aqueous acetate buffer (pH 5)/third solvent, 90/5/5, v/v/v. Reproduced with permission of Ref [39].

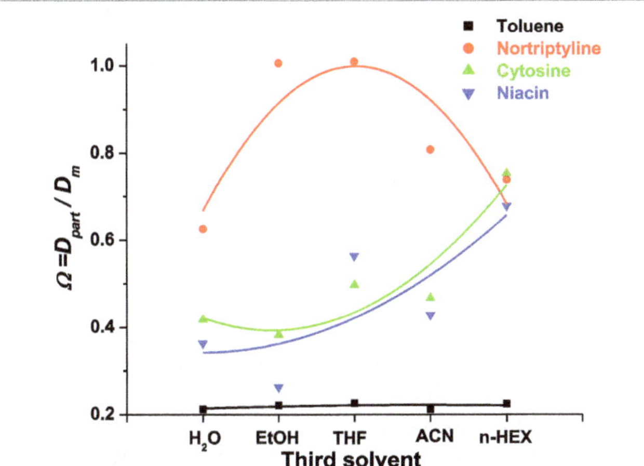

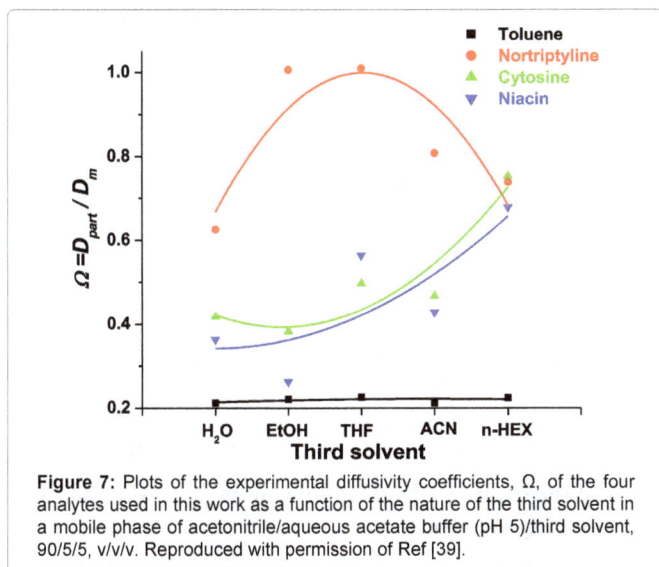

Figure 7: Plots of the experimental diffusivity coefficients, Ω, of the four analytes used in this work as a function of the nature of the third solvent in a mobile phase of acetonitrile/aqueous acetate buffer (pH 5)/third solvent, 90/5/5, v/v/v. Reproduced with permission of Ref [39].

bulk eluent composition. 2) from 0 to 4 Å, all the solvent molecules are physisorbed and the region II is called the rigid water layer. Water is in large excess in this region relative to its bulk concentration. Acetontrile is mostly excluded from the surface of the adsorbent while the amount of third solvent molecules adsorbed increases from isopropanol to ethanol and to methanol. This is consistent with a reduction of the steric hindrance from large to small molecules easing the way to the surface. Finally, in between z=4 Å and z=15 Å, the concentration of all the solvent molecules is changing dramatically which defines the pore region II or the diffuse water layer. Alltogether, the pore regions I and II form the interface volume between the the solid adsorbent and the bulk eluent. Its physico-chemical properties (polarity, mobility, relative permittivity, etc.) contrast with those of the bulk : for instance, its polarity is much larger than that of the acetonitrile-rich eluent, the local molecular mobility is severely reduced in the water-rich interface region (higher viscosity and lesser degree of freedom under the action of the solid adsorbent), and its relative permittivity is close to that of pure water (ε_r=80.1 versus ε_r=37.5 for pure acetonitrile). In conclusion, the distribution of the solvent molecules in the mesoporous volume of HILIC particles is highly non-uniform. The same heterogeneity in the concentration of the analyte molecules can be expected based on their distribution between the three delimited pore regions.

These MD results for the density profiles of the eluent molecules enable the calculation of the volume fractions f_b (region III) and f_d (region II). Assuming cylindrical pores and a unimodal pore size of 140 Å, f_b =62%, f_d=27%, and, so, f_a=12% (region I). These numerical data are indispensable for the measurement of the equilibrium constant K_d (Equation 13) and K_d (Equation 14).

Solvent mobility in the mesopore

Figure 3 plots the mobilities (in cm^2/s) of the solvent molecules along the direction normal to the HILIC surface. They were calculated from MD. Irrespective of the solvent molecule (water, acetonitrile, and third solvent), the local diffusion coefficient monotoneously increases with increasing the distance from the adsorbent surface. The molecular mobility is rigorously zero in region I (due to physisorption) and it is equal to the bulk diffusion coefficient in region III (far from the surface field). Inbetween, it is striking to observe that the local diffusion coefficient of all solvent molecules vary nearly linearly with the distance *R* from the silica wall. Accordingly, in the rest of this work, it is assumed that the average mobility of the solvent and analyte molecules in the interfacial region II is 0.5 Dm. so the unknow coefficient d is equal to 0.5. This data was required to measure the equilibrium constant K_d in Equation 14.

Combining chromatographic and MD data to solve the retention mechanism in HILIC

The combination of the chromatographic (k', ε_e, ε_p, and Ω) and MD (f_b, f_d and δ) data enable the unambiguous determination of the equilibrium constants K_d and K_a. In the next two sections, we are representing the distribution of the analyte molecules in the three pore regions and the relative importance of the adsorption and partitioning mechanism on the overal retention factor of the four analytes used in this work. Five different third eluents are considered in the ternary mobile phase mixture (acetonitrile/water pH 5/third solvent): water, ethanol, tetrahydrofuran, acetonitrile, and *n*-hexane.

Analyte distribution in the mesopore of HILIC particles

The molar fractions, x_a, x_d, and x_b, of analytes present in the rigid water layer, diffuse water layer, and in the bulk region of the mesopore are given as a function of the previously measured parameters f_b, f_d, K_d, and K_a. Accordingly, by definition and after some basic algebra:

$$x_a = \frac{n_a}{n_a + n_d + n_b} = \frac{K_a K_d (1 - f_a - f_d)}{K_a K_d (1 - f_a - f_d) + K_d f_d + f_b} \quad (16)$$

$$x_d = \frac{n_d}{n_a + n_d + n_b} = \frac{K_d f_d}{K_a K_d (1 - f_a - f_d) + K_d f_d + f_b} \quad (17)$$

$$x_b = \frac{n_b}{n_a + n_d + n_b} = \frac{f_b}{K_a K_d (1 - f_a - f_d) + K_d f_d + f_b} \quad (18)$$

Figures 8-12 show these results for water, ethanol, tetrahydrofuran, acetonitrile, and *n*-hexane as the third solvent in the ternary mobile phase mixture (acetonitrile/water pH 5/ third eluent, 90/5/5, v/v/v). For the sake of reference, the mole fractions of a fully inert analyte uniformly distributed in the whole mesopore volume are 12% (in the rigid water layer), 29% (in the diffuse water layer), and 59% (in the bulk solvent), respectively. Irrespective of the nature of the thrid solvent, it is noteworthy that toluene is not severely excluded from the whole water-rich layer: the sum of x_a and x_d is nearly constant around 35% instead of 43% for the fully inert adsorption system. Overall, toluene is only

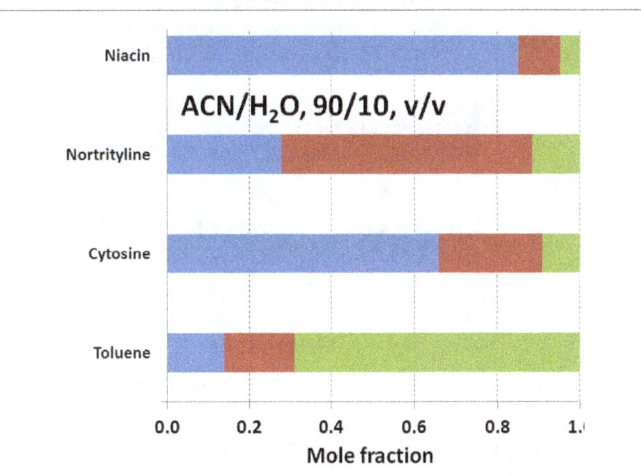

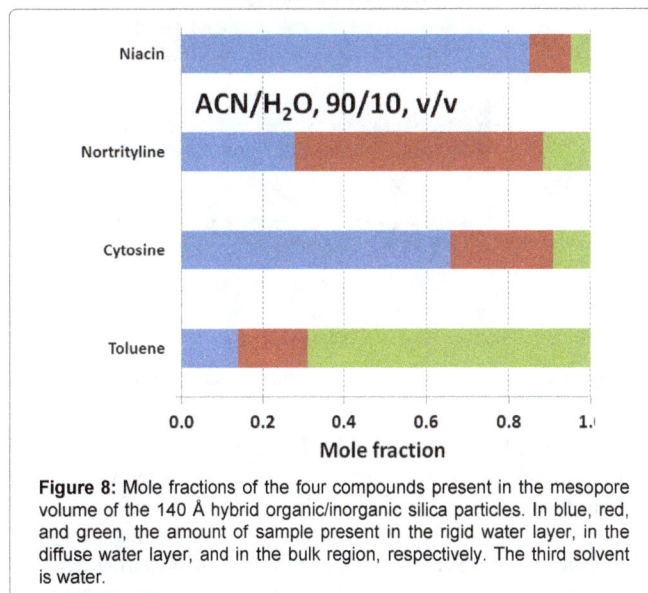

Figure 8: Mole fractions of the four compounds present in the mesopore volume of the 140 Å hybrid organic/inorganic silica particles. In blue, red, and green, the amount of sample present in the rigid water layer, in the diffuse water layer, and in the bulk region, respectively. The third solvent is water.

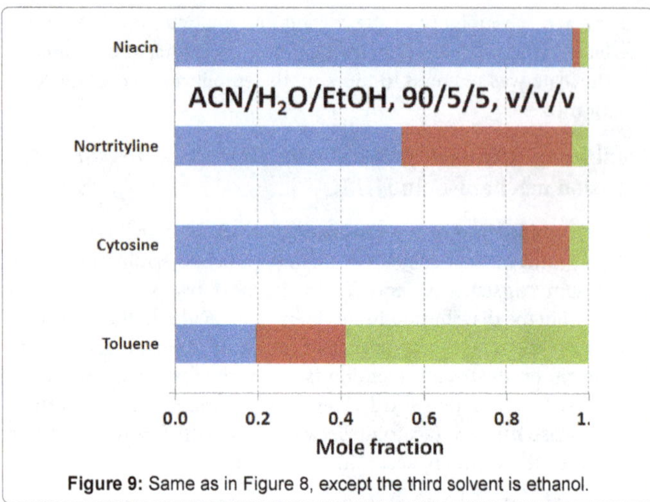

Figure 9: Same as in Figure 8, except the third solvent is ethanol.

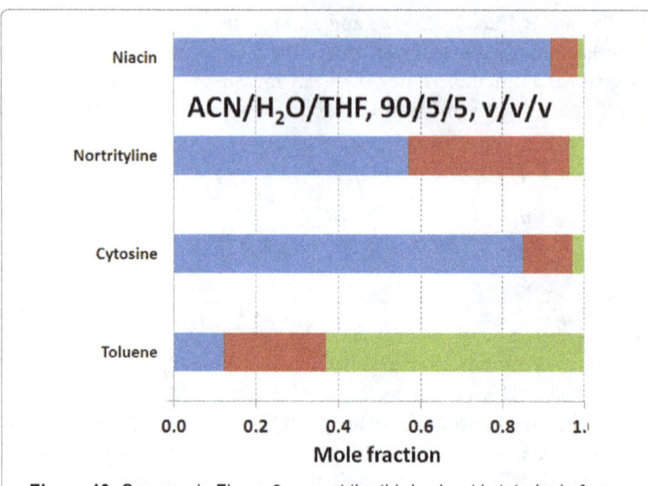

Figure 10: Same as in Figure 8, except the third solvent is tetrahydrofuran.

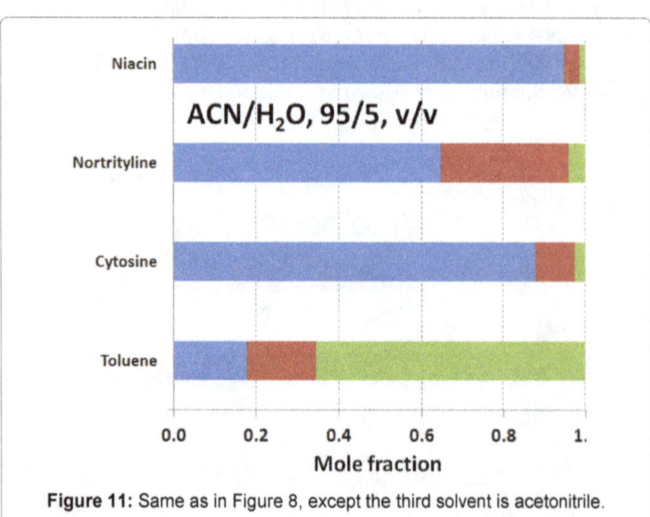

Figure 11: Same as in Figure 8, except the third solvent is acetonitrile.

slightly excluded from the water-rich layer. It is mostly repelled from the diffuse water layer since the average molar fractions x_d are close to 20% *versus* 29% for the reference inert system. Therefore, toluene is also present at the silica surface where it is weakly adsorbed onto the hydrophobic siloxane bridges. Regarding niacin (partial negative charge, -0.6 at pH 5), it is mostly present in the rigid water layer (from 85% for water to 97% for ethanol), e.g., it is strongly adsorbed onto specific sites such as silanols and/or metal catalyst residuals. Remarkably, niacin is partially excluded from the diffuse water layer: x_d varies from 2% for ethanol to no more than 11% for water as the third solvent. The most surprising results are for the two positively charged compound cytosine (+0.3 at pH 5) and nortriptyline (+1): the retention factors of these two compounds are nearly the same, yet, the former analyte accumulates preferentially in the rigid water layer through adsorption (x_a varies from 65% for water to 87% for ethanol as the third solvent, $10<x_d<20\%$) while the latter is both dissolved into the diffuse water layer (x_d varies from 35% for acetonitrile to 60% for water) and adsorbed ($25\%<x_d<65\%$) onto the surface of the hybrid organic/inorganic silica particles. To summarize, it is very difficult to predict *a priori* the retention mechansim of ionizable compounds in HILIC: it is not general and it depends on the specific affinities of each anlyte for the solid adsorbent surface and a water-rich environment.

Effect of the nature of the third solvent on the adsorption and partitionning mechanisms

Figures 13-16 represents quantitative results for the relative importance of the adsorption and partitionning retention mechanisms on the overall retention factor of toluene, cytosine, nortriptyline, and niacin. By definition, the overall retention factor is written:

$$k' = \frac{n_a + n_d + n_b - n_p}{n_e + n_p}$$

It can also be written as the sum of three different terms: two positive terms for the presence of the analyte in the rigid and diffuse water layer and a negative term for its presence in the bulk region of the pore. Accordingly,

$$k' = \frac{K_a K_d f_a \varepsilon_p (1-\varepsilon_e)}{\varepsilon_e + \varepsilon_p(1-\varepsilon_e)} + \frac{K_d f_d \varepsilon_p (1-\varepsilon_e)}{\varepsilon_e + \varepsilon_p(1-\varepsilon_e)} - \frac{\varepsilon_p (1-\varepsilon_e)(f_a + f_d)}{\varepsilon_e + \varepsilon_p(1-\varepsilon_e)}$$

These three contributions to the total experimental retention factor are shown in Figures 13-16 by the vertical blue, red, and green 3D bars, respectively.

Figure 13 reveals that the overall retention factor of toluene is slightly negative (from -0.05 to 0) since it is slightly excluded from the water-rich layer present at the surface of the hybrid organic/inorganic silica particles. The nature of the third solvent has no measurable impact on the distribution of toluene between the rigid and the diffuse water layers. The retention increase of cytosine (Figure 14) from

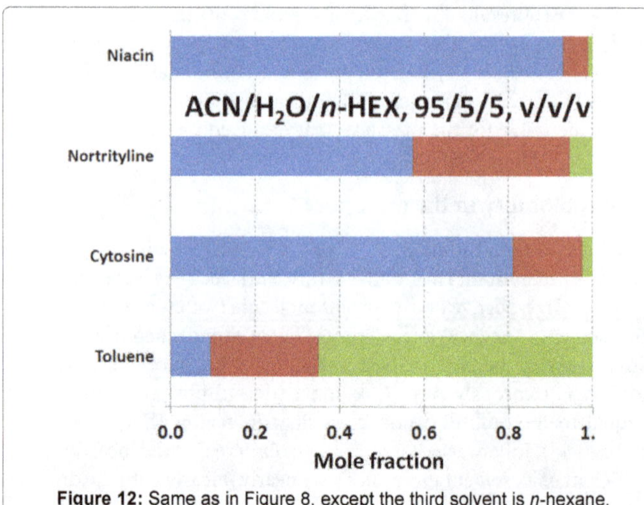

Figure 12: Same as in Figure 8, except the third solvent is *n*-hexane.

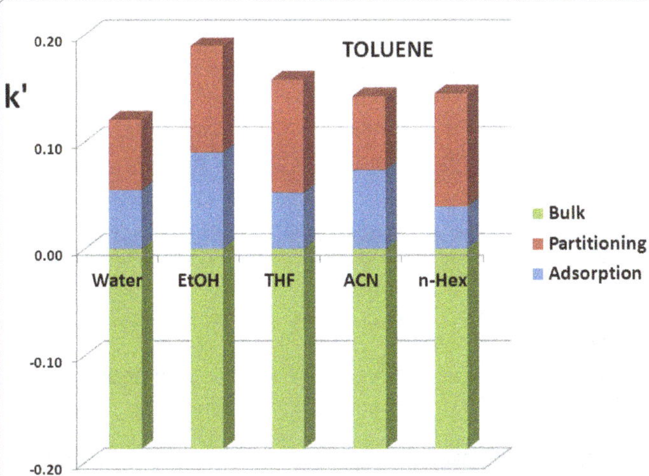

Figure 13: Contributions of the adsorption (blue color) and partitionning (red color) processes on the intensity of the observed retention factor, *k'*, of toluene as a function of the nature of the third solvent. The constant height of the green 3D bars represent the expected retention factor (k'=-0.17) if the sample molecules were fully excluded from the water-rich layer. The polarity of the third solvent is decreasing from water to ethanol (EtOH), tetrahydrofuran (THF), acetonitrile (ACN), and to *n*-hexane (*n*-Hex).

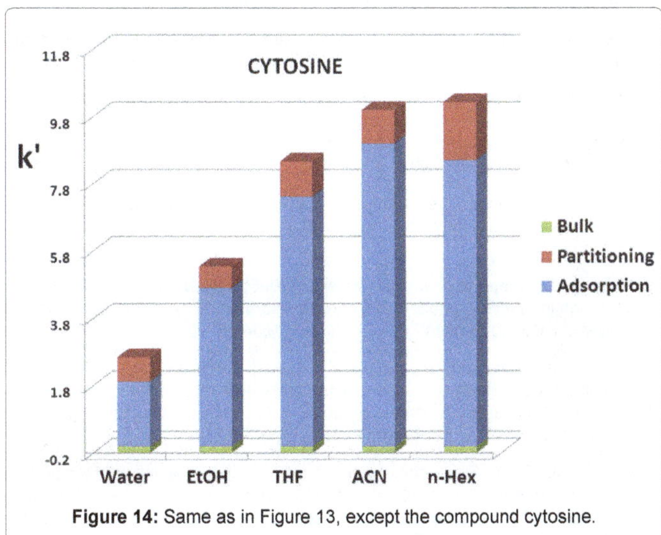

Figure 14: Same as in Figure 13, except the compound cytosine.

water to acetonitrile used as the third eluent is fully explained from the increasing amount of cytosine in the rigid water layer relatively to its amount in the diffuse water layer. However, the slight increase of retention from acetonitrile to n-hexane used as the third eluent is justified by a relative increase of the importance of the partitioning mechasnim, the adsorption process remaining dominant. The apparent riddle regarding the reversed U-shaped retention behavior of nortriptyline can now be explained unambigously: retention increases from water to tetrahydrofuran as third solvent because the importance of both adsorption over partitioning processes increases. However, from tetrahydrofuran to acetonitrile and to *n*-hexane, partitionning and adsorption mechanisms become less intense causing a diminution of the overall retention factor of nortriptyline. Finally, the trend observed for niacin in Figure 15 is clear: the retention factor monotoneously increases from water to *n*-hexane used as the third eluent because the adsorption mechanism is continuously reinforced. Overall, among the three retained ionizable compounds, none has shown a retention behavior strictly identical to the other when decreasing the polarity of third solvent. Yet, a tendency can be seen: the use of a weekly polar third solvent usually generates an increase of the retention because adsorption is reinforced. Consistent with the MD data given in [27], the molar fractions of water in the water-rich layer is continuously increasing when decreasing the polarity of the third solvent: as a result, the transfer of the partially charged analyte from the bulk to the diffuse and to the rigid water layers becomes more favourable.

Conclusions

This work demonstrates that the measurement of the retention factor k' in liquid chromatography is not sufficient to draw definitive and unambiguous conclusions regarding the retention mechanism of any solid-liquid equilibrium systems. The fundamental explanation for such ambiguity is the non-uniformity of the composition and structure of the eluent mixture across the whole diameter of the mesopores. Solids are never inert to the solvent molecules, so, preferential adsorption occurs for one particular solvent component and the analyte molecules are not homogeneously distributed in the mesopore volume. They are retained because they can be adsorbed at the very surface of the adsorbent (adsorption mechanism) or they can partition between the bulk and the wide diffuse water layer (partitionning mechanism). The existence of this large interfacial layer was revealed in both HILIC [44]

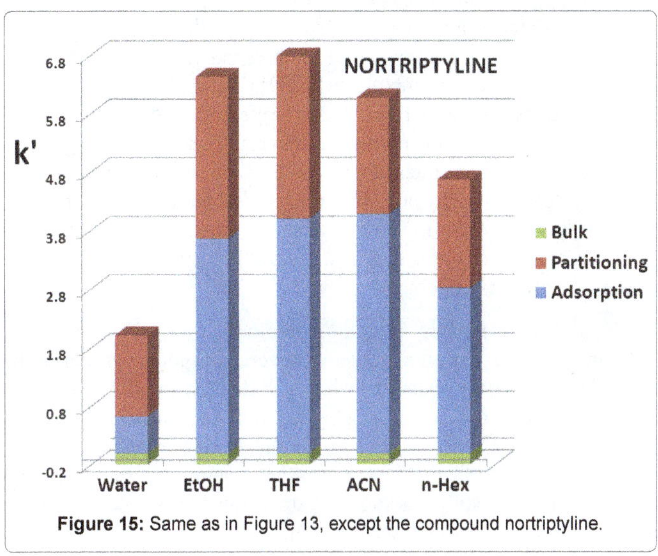

Figure 15: Same as in Figure 13, except the compound nortriptyline.

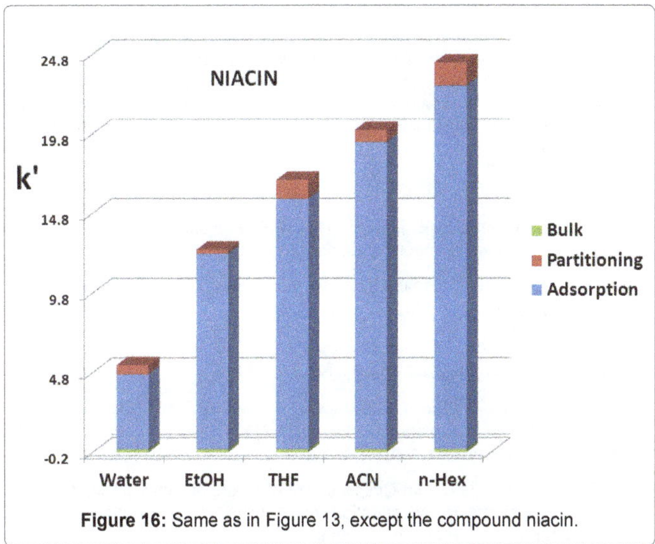

Figure 16: Same as in Figure 13, except the compound niacin.

and RPLC [45] retention modes. This riddle was illustrated in HILIC with a **hybrid organic/inorganic silica adsorbent** in contact with a ternary mixture of acetonitrile, water (pH 5, acetate buffer, 200mM), and third solvent (water, ethanol, tetrahydrofuran, acetonitrile, and *n*-hexane).

It was shown that the relative importance of adsorption and partitionning could be quantitatively determined by combining HPLC (retention factor and column porosities) and MD (microscopic information pertained to the structure and dynamics of the internal eluent mixture) data. Unexpected interpretations were then provided for the overall retention behavior of positively and negatively charged compounds when continuously decreasing the polarity of the third solvent. For instance, two compounds with comparable net charges can have very close retention factors but well different retention mechanisms in terms of adsoprption and partitioning retention mechanism.

The implications of this work go way beyond the case of HILIC separations with unmodified silica surface. It appllies to any HILIC stationary phases such as monomeric and polymeric grafted silica phases. The very same ambiguities apply to the determination of the retention mechanisms in RPLC-C_{18} or in supercritical fluid chromatography (SFC). They can be alleviated by performing MD with silica-C_{18} bonded phases in contact with aqueous mixtures of methanol or acetonitrile and with silica-ethylpyridine bonded phases in equilibrium with CO_2/organic modifier eluent mixtures. In RPLC and SFC, it was shown by minor disturbance methods that a preferential adsorption of the organic modifier onto the derivatized silica adsorbents is taking place. Similarly to HILIC, the internal composition and structure of the RPLC and SFC eluents is highy non-uniform across the mesopores. This will necessarily affect the sample distribution in the pores depending on its molecular descriptors (polarity, hydrophobicity, hydrogen-bond donor/acceptor, π-π interactions, etc.).

It is also noteworthy that the combination of HPLC and MD data in RPLC is expected to solve several on-going mysteries in chromatography: for instance, what is hidden from a microscopic viepoint behind the so-called phenomenon of "surface diffusion" observed in RPLC with neutral apolar (alkybenzene [46]) or weakly polar compounds (alkanophenones [47])? A sound physico-chemical interpretation is still missing and this topic has been recently approached from chromatographic [45] and MD investigations.

Acknowledgements

This work was supported in part by the cooperative agreement between the University of Tennessee and the Oak Ridge National Laboratory. We thank Prof. Ulrich Tallarek and Dr. Alexandra Höltzel (Philipps-Universität Marburg, Marburg, Germany) for providing us with the relevant MD data. We also thank Martin Gilar (Waters, Milford, MA, USA) for the generous gift of the **hybrid organic/inorganic silica** HILIC column.

References

1. Horvath C, Melander WR, Molnar I (1976) Solvophobic interactions in liquid chromatography with nonpolar stationary phases. J Chromatogr 125: 129-156.
2. Vailaya A, Horváth C (1998) Retention in reversed-phase chromatography: partition or adsorption? J Chromatogr A 829: 1-27.
3. Dorsey JG, Dill KA (1989) The molecular mechanism of retention in reversed-phase liquid chromatography. Chem Rev 89: 331-346.
4. Martire D, Boehm R (1983) Unified theory of retention and selectivity in liquid chromatography. 2. Reversed-phase liquid chromatography with chemically bonded phases. J Phys Chem 87: 1045-1062.
5. Gritti F, Guiochon G (2005) Critical contribution of nonlinear chromatography to the understanding of retention mechanism in reversed-phase liquid chromatography. J Chromatogr A 1099: 1-42.
6. Lesellier E, West C (2015) The many faces of packed column supercritical fluid chromatography--a critical review. J Chromatogr A 1382: 2-46.
7. Berger TA (2005) Supercritical Fluid Chromatography: Overview. Encyclopedia of Analytical Science, 2nd edn.
8. Fornstedt T (2015) Introduction to "Fundamental challenges and opportunities for preparative supercritical fluid chromatography by G. Guiochon, A. Tarafder [J. Chromatogr. A 1218 (2011) 1037-1114]". J Chromatogr A S0021-9673:01617-01619.
9. Alpert AJ (1990) Hydrophilic-interaction chromatography for the separation of peptides, nucleic acids and other polar compounds. J Chromatogr 499: 177-196.
10. Hemström P, Irgum K (2006) Hydrophilic interaction chromatography. J Sep Sci 29: 1784-1821.
11. Ikegami T, Tomomatsu K, Takubo H, Horie K, Tanaka N (2008) Separation efficiencies in hydrophilic interaction chromatography. J Chromatogr A 1184: 474-503.
12. Jandera P (2011) Stationary and mobile phases in hydrophilic interaction chromatography: a review. Anal Chim Acta 692: 1-25.
13. Jandera P (2008) Stationary phases for hydrophilic interaction chromatography, their characterization and implementation into multidimensional chromatography concepts. J Sep Sci 31: 1421-1437.
14. Yoshida T (2004) Peptide separation by Hydrophilic-Interaction Chromatography: a review. J Biochem Biophys Methods 60: 265-280.
15. Buszewski B, Noga S (2012) Hydrophilic interaction liquid chromatography (HILIC)--a powerful separation technique. Anal Bioanal Chem 402: 231-247.
16. Greco G, Letzel T (2013) Main interactions and influences of the chromatographic parameters in HILIC separations. J Chromatogr Sci 51: 684-693.
17. Zhang L, Rafferty JL, Siepmann JI, Chen B, Schure MR (2006) Chain conformation and solvent partitioning in reversed-phase liquid chromatography: Monte Carlo simulations for various water/methanol concentrations. J Chromatogr A 1126: 219-231.
18. Rafferty JL, Zhang L, Siepmann JI, Schure MR (2007) Retention mechanism in reversed-phase liquid chromatography: a molecular perspective. Anal Chem 79: 6551-6558.
19. Rafferty JL, Siepmann JI, Schure MR (2008) Influence of bonded-phase coverage in reversed-phase liquid chromatography via molecular simulation II. Effects on solute retention. J Chromatogr A 1204: 20-27.
20. Rafferty JL, Siepmann JI, Schure MR (2009) The effects of chain length, embedded polar groups, pressure, and pore shape on structure and retention in reversed-phase liquid chromatography: molecular-level insights from Monte Carlo simulations. J Chromatogr A 1216: 2320-2331.
21. Rafferty JL, Sun L, Siepmann JI, Schure MR (2010) Investigation of the driving forces for retention in reversed-phase liquid chromatography: Monte Carlo simulations of solute partitioning between n-hexadecane and various aqueous-organic mixtures. Fluid Phase Equilibr 290: 25-35.
22. Rafferty JL, Siepmann JI, Schure MR (2012) A molecular simulation study of the effects of stationary phase and solute chain length in reversed-phase liquid chromatography. J Chromatogr A 1223: 24-34.
23. Rafferty JL, Siepmann JI, Schure MR (2011) Mobile phase effects in reversed-phase liquid chromatography: a comparison of acetonitrile/water and methanol/water solvents as studied by molecular simulation. J Chromatogr A 1218: 2203-2213.
24. Lindsey RK, Rafferty JL, Eggimann BL, Siepmann JI, Schure MR (2013) Molecular simulation studies of reversed-phase liquid chromatography. J Chromatogr A 1287: 60-82.
25. Rybka J, Höltzel A, Melnikov SM, Seidel-Morgenstern A, Tallarek U (2016) A new view on surface diffusion from molecular dynamics simulations of solute mobility at chromatographic interfaces. Fluid Phase Equilibr 407: 177-187.
26. Melnikov SM, Höltzel A, Seidel-Morgenstern A, Tallarek U (2013) How ternary mobile phases allow tuning of analyte retention in hydrophilic interaction liquid chromatography. Anal Chem 85: 8850-8856.
27. Melnikov SM, Höltzel A, Seidel-Morgenstern A, Tallarek U (2013) Adsorption of Water-Acetonitrile Mixtures to Model Silica Surfaces. J Phys Chem C 117: 6620-6631.

28. Melnikov SM, Höltzel A, Seidel-Morgenstern A, Tallarek U (2012) A molecular dynamics study on the partitioning mechanism in hydrophilic interaction chromatography. Angew Chem Int Ed Engl 51: 6251-6254.

29. Melnikov SM, Höltzel A, Seidel-Morgenstern A, Tallarek U (2011) Composition, structure, and mobility of water-acetonitrile mixtures in a silica nanopore studied by molecular dynamics simulations. Anal Chem 83: 2569-2575.

30. Melnikov SM, Höltzel A, Seidel-Morgenstern A, Tallarek U (2009) Influence of Residual Silanol Groups on Solvent and Ion Distribution at a Chemically Modified Silica Surface. J Phys Chem C 113: 9230-9238.

31. Braun J, Fouqueau A, Bemish RJ, Meuwly M (2008) Solvent structures of mixed water/acetonitrile mixtures at chromatographic interfaces from computer simulations. Phys Chem Chem Phys 10: 4765-4777.

32. Knox J, McLaren L (1964) A New Gas Chromatographic Method for Measuring Gaseous Diffusion Coefficients and Obstructive Factors. Anal Chem 36: 1477-1482.

33. Knox J, Scott H (1983) B and C terms in the Van Deemter equation for liquid chromatography. J Chromatogr 282: 297-313.

34. Gritti F, Guiochon G (2006) Effect of the surface coverage of C18-bonded silica particles on the obstructive factor and intraparticle diffusion mechanism. Chem Eng Sci 61: 7636-7650.

35. Torquato S (1985) Effective electrical conductivity of two-phase disordered composite media. J Appl Phys 58: 3790-3797.

36. Torquato S (2002) Random Heterogeneous Materials. Microstructure and Macro-scopic Properties. Springer, New York, USA.

37. Gritti F (2014) LC GC North America 32: 928-940.

38. Gritti F, Höltzel A, Tallarek U, Guiochon G (2015) The relative importance of the adsorption and partitioning mechanisms in hydrophilic interaction liquid chromatography. J Chromatogr A 1376: 112-125.

39. Landauer R (1952) The Electrical Resistance of Binary Metallic Mixtures. J Appl Phys 23: 779-784.

40. Grumbach ES, Diehl DM, Neue UD (2008) The application of novel 1.7 microm ethylene bridged hybrid particles for hydrophilic interaction chromatography. J Sep Sci 31: 1511-1518.

41. Mitzithras FMCA, Strange JH (1992) J Mol Liq 260: 273-281.

42. Renkin EM (1954) Filtration, diffusion, and molecular sieving through porous cellulose membranes. J Gen Physiol 38: 225-243.

43. Dinh NP, Jonsson T, Irgum K (2013) Water uptake on polar stationary phases under conditions for hydrophilic interaction chromatography and its relation to solute retention. J Chromatogr A 1320: 33-47.

44. Gritti F, Guiochon G (2005) Adsorption mechanism in RPLC. Effect of the nature of the organic modifier. Anal Chem 77: 4257-4272.

45. Miyabe K, Guiochon G (2010) Surface diffusion in reversed-phase liquid chromatography. J Chromatogr A 1217: 1713-1734.

46. Gritti F, Guiochon G (2014) The rationale for the optimum efficiency of columns packed with new 1.9 µm fully porous Titan-C18 particles-a detailed investigation of the intra-particle diffusivity. J Chromatogr A 1355: 164-178.

47. Gritti F (2015) Determination of the solvent density profiles across mesopores of silica-C18 bonded phases in contact with acetonitrile/water mixtures: A semi-empirical approach. J Chromatogr A 1410: 90-98.

Physical, Thermal and Spectroscopical Characterization of Biofield Treated Triphenylmethane: An Impact of Biofield Treatment

Trivedi MK[1], Branton A[1], Trivedi D[1], Nayak G[1], Bairwa K[2] and Jana S[2]*

[1]*Trivedi Global Inc., 10624 S Eastern Avenue Suite A-969, Henderson, NV 89052, USA*
[2]*Trivedi Science Research Laboratory Pvt. Ltd., Hall-A, Chinar Mega Mall, Chinar Fortune City, Hoshangabad Rd., Bhopal, Madhya Pradesh, India*

Abstract

Triphenylmethane is a synthetic dye used as antimicrobial agent and for the chemical visualization in thin layer chromatography of higher fatty acids, fatty alcohols, and aliphatic amines. The present study was an attempt to investigate the impact of biofield treatment on physical, thermal and spectroscopical charecteristics of triphenylmethane. The study was performed in two groups i.e., control and treatment. The treatment group subjected to Mr. Trivedi's biofield treatment. The control and treated groups of triphenylmethane samples were characterized using X-ray diffraction (XRD), surface area analyzer, differential scanning calorimetry (DSC), thermogravimetric analysis (TGA), Fourier transform infrared (FT-IR), ultraviolet-visible (UV-Vis) spectroscopy, and gas chromatography-mass spectrometry (GC-MS). XRD study revealed decreases in average crystallite size (14.22%) of treated triphenylmethane as compared to control sample. Surface area analysis showed a slight increase (0.42%) in surface area of treated sample with respect to control. DSC thermogram of treated triphenylmethane showed the slight increase in melting point and latent heat of fusion with respect to control. TGA analysis of control triphenylmethane showed weight loss by 45.99% and treated sample showed weight loss by 64.40%. The T_{max} was also decreased by 7.17% in treated sample as compared to control. The FT-IR and UV spectroscopic result showed the similar pattern of spectra. The GC-MS analysis suggested a significant decrease in carbon isotopic abundance (expressed in $\delta^{13}C$, ‰) in treated sample (about 380 to 524‰) as compared to control. Based on these results, it is found that biofield treatment has the impact on physical, thermal and carbon isotopic abundance of treated triphenylmethane with respect to control.

Keywords: Triphenylmethane; Biofield treatment; X-ray diffraction; Differential scanning calorimetry; Thermogravimetric analysis; Gas chromatography-Mass Spectrometry (GC-MS)

Abbreviations

XRD: X-ray diffraction; DSC: Differential scanning calorimetry; TGA: Thermogravimetric analysis; DTA: Differential thermal analysis; FT-IR: Fourier transform infrared; UV-Vis: Ultraviolet-visible; GC-MS: Gas chromatography-mass spectrometry; PM: Primary molecule

Introduction

Triphenylmethane is a hydrocarbon with molecular formula $(C_6H_5)_3CH$. It builds the basic skeleton of many synthetic dyes such as bromocresol green, malachite green etc., and used as pH indicator and fluorescence agent [1]. Boulos RA has reported its antimicrobial property [2]. Triphenylmethane reported to inhibit 3-methyl-cholanthrene-induced neoplastic transformation of 10T1/2 cells in a dose-dependent manner and as a novel chemo preventive agent [3]. This has been used as visualizing agentin thin-layer chromatography of higher chain fatty acids, fatty alcohols, and aliphatic amines [4]. Recently, triphenylmethane was reported as an alternative for mediated electronic transfer systems in glucose oxidase biofuel cells (enzymatic biofuel cell). The enzymatic biofuel cell is a type of fuel cell wherein the enzymes are used as a catalyst to oxidize its fuel, instead of costly metals [5]. Hugle et al. used triphenylmethane as a possible moderator material to reduces the speed of neutrons in nuclear chain reactions, especially promising as cold neutrons moderator. It has a unique structure i.e., three aromatic phenyl groups surrounding one central carbon atom that is able to generate a stable radical ion [6]. Diverse applications of triphenylmethane especially as florescent indicator, mediator in biofuel and as a moderator had been suggested the importance of physicochemical property of triphenylmethane. It was previously reported that physical and thermal properties of molecule also affect its reactivity [7,8]. Hence, it is beneficial to find out an alternate approach that can improved the physicochemical properties of compounds like triphenylmethane, which can enhance its usability. Recently, biofield treatment reported to alter the spectral properties of various pharmaceutical drugs like chloramphenicol and tetracycline, and physicochemical properties of metals, beef extract and meat infusion powder [9-11].

Relation between mass and energy ($E=mc^2$) is well reported in literature [12]. The mass (solid matter) is consist of energy and when this energy vibrates at a certain frequency, it provides physical, atomic and structural properties like size, shape, texture, crystal structure, and atomic weight to the matter [13]. Similarly, human body also consists with vibratory energy particles like protons, neutrons, and electrons [14]. Due to vibrations in these particles, an electrical impulse is generated that cumulatively forms electromagnetic field, which is known as biofield [15]. The human has the ability to harness the energy from the environment or Universe and transmit this energy into any object (living or nonliving) on the Globe. The object(s) receive the energy and respond into useful way, this process is known as biofield treatment.

***Corresponding author:** Snehasis Jana, Trivedi Science Research Laboratory Pvt. Ltd., Hall-A, Chinar Mega Mall, Chinar Fortune City, Hoshangabad Rd., Bhopal-462 026, Madhya Pradesh, India, E-mail: publication@trivedisrl.com

Mr. Trivedi's unique biofield energy is also called as The Trivedi Effect®, and reported to change various physicochemical, thermal and structural properties of several metals [10,16,17] and ceramics [18]. In addition, biofield treatment has been extensively studied in different fields such as agricultural science [19,20], biotechnology research [21], and microbiology research [22-24].

Conceiving the impact of biofield treatment on various living and nonliving things, the study aimed to evaluate the impact of biofield treatment on spectral and physicochemical properties of triphenylmethane using different analytical techniques.

Materials and Methods

Study design

Triphenylmethane was procured from Sisco Research Laboratories, India. The study was performed in two groups i.e., control and treatment. The control sample was remained as untreated; and treatment sample was handed over in sealed pack to Mr. Trivedi for biofield treatment under laboratory conditions. Mr. Trivedi provided the biofield treatment through his energy transmission process to the treatment group without touching the sample [9]. The control and treated samples of triphenylmethane were evaluated using various analytical techniques like X-ray diffraction (XRD), surface area analyzer, differential scanning calorimetry (DSC), thermogravimetric analysis (TGA), Fourier transform infrared (FT-IR), ultraviolet-visible (UV-Vis) spectroscopy, and gas chromatography-mass spectrometry (GC-MS).

X-ray diffraction (XRD) study

XRD analysis of triphenylmethane was performed on Phillips, Holland PW 1710 X-ray diffractometer with copper anode and nickel filter. Wavelength of X-ray used in XRD system was 1.54056 Å with scanning rate of 0.05° 2/s, and a chart speed of 10 mm/2. Data obtained from XRD system were in the form of a chart of 2θ (10-100°) *vs.* intensity. The average crystallite size (G) of triphenylmethane was calculated using the following equation [25].

$$G = k\lambda/(b\cos\theta)$$

Percent change in average crystallite size = $[(G_t-G_c)/G_c] \times 100$

Where, G_c and G_t are average crystallite size of control and treated powder samples respectively.

Surface area analysis

Surface area of control and treated triphenylmethane was measured using the Brunauer–Emmett–Teller (BET) surface area analyzer, Smart SORB 90. Percent changes in surface area were calculated using following equation:

$$\%\,\text{change in surface area} = \frac{[S_{Treated} - S_{Control}]}{S_{Control}} \times 100$$

Where, $S_{Control}$ and $S_{Treated}$ are the surface area of control and treated samples, respectively.

Differential scanning calorimetry (DSC) study

The control and treatment samples of triphenylmethane were analyzed using a Pyris-6 Perkin Elmer differential scanning calorimeter (DSC) on a heating rate of 10°C/min under air atmosphere with air flow rate of 5 mL/min. An empty pan sealed with covered aluminum pan was used as a reference. The melting temperature (T_m) and latent heat of fusion (ΔH) were obtained from the DSC curve.

Thermogravimetric analysis-differential thermal analysis (TGA-DTA)

Thermal stability of control and treated triphenylmethane were analyzed using Mettler Toledo simultaneous TGA and differential thermal analyzer (DTA). The samples were heated from room temperature to 400°C with a heating rate of 5°C/min under air atmosphere. Percent change in temperature at which maximum weight loss occurs in sample was calculated.

Spectroscopic studies

For determination of FT-IR and UV-Vis spectroscopic characters, the treated sample was divided into two groups i.e., T1 and T2. Both treated groups were analyzed for their spectral characteristics using FT-IR and UV-Vis spectroscopy as compared to control triphenylmethane sample. While, for GC-MS analysis, the treated sample was divided into four groups i.e., T1, T2, T3, and T4 and all treated groups were analyzed along with control sample for isotopic abundance of carbon-13.

FT-IR spectroscopic characterization

FT-IR spectra of control and treated samples of triphenylmethane were recorded on Shimadzu's Fourier transform infrared spectrometer (Japan) with frequency range of 4000-500 cm^{-1}. The analysis was accomplished to evaluate the effect of biofield treatment at atomic level like dipole moment, force constant and bond strength in chemical structure [26].

UV-Vis spectroscopic analysis

UV spectra of control and treated samples of triphenylmethane were obtained from Shimadzu UV-2400 PC series spectrophotometer. A quartz cell with 1 cm and a slit width of 2.0 nm was used for analysis. The study was carried out using wavelength in the range of 200-400 nm. The UV spectra were analyzed to determine the effect of biofield treatment on the energy gap of bonding and nonbonding transition of electrons [26].

Gas chromatography-mass spectroscopy (GC-MS) analysis

The GC-MS analysis of control and treatment samples (T1, T2, T3, and T4) of triphenylmethane were performed on Perkin Elmer/auto system XL with Turbo mass and electron ionization mode, USA. Detection limit was set to 1 Pico gram, and mass range was set to 10-650 amu. The isotopic ratio $^{13}C/^{12}C$ was expressed by its deviation in treated triphenylmethane sample with respect to control. The isotopic abundance of ^{13}C was computed on a delta scale per thousand. The values of $\delta^{13}C$ of treated samples were calculated using following equation [27].

$$\delta^{13}C\,(‰) = \frac{R_{Treated} - R_{Control}}{R_{Control}} \times 1000 \quad (1)$$

Where, $R_{Treated}$ and $R_{Control}$ are the ratio of intensity at *m/z*=245/ *m/z*=244 for $\delta^{13}C$ in mass spectra of treated and control samples respectively.

Results and Discussion

XRD analysis

XRD of control and treated triphenylmethane are presented in Figure 1. The control triphenylmethane showed the XRD peaks at 2θ equals to 11.70°, 11.91°, 15.00°, 18.24°, 19.67°, 22.52°, 22.66°, 23.99°, 24.96°, 26.19°, and 28.71°. However, the XRD diffractogram of treated triphenylmethane showed the decrease in intensity of the peaks. XRD peaks in treated sample were appeared at 2θ equals to 11.92°, 12.33°, 14.87°, 15.03°, 18.25°, 19.66°, 20.76°, 22.52°, 23.99°, 24.24°, and 26.19°.

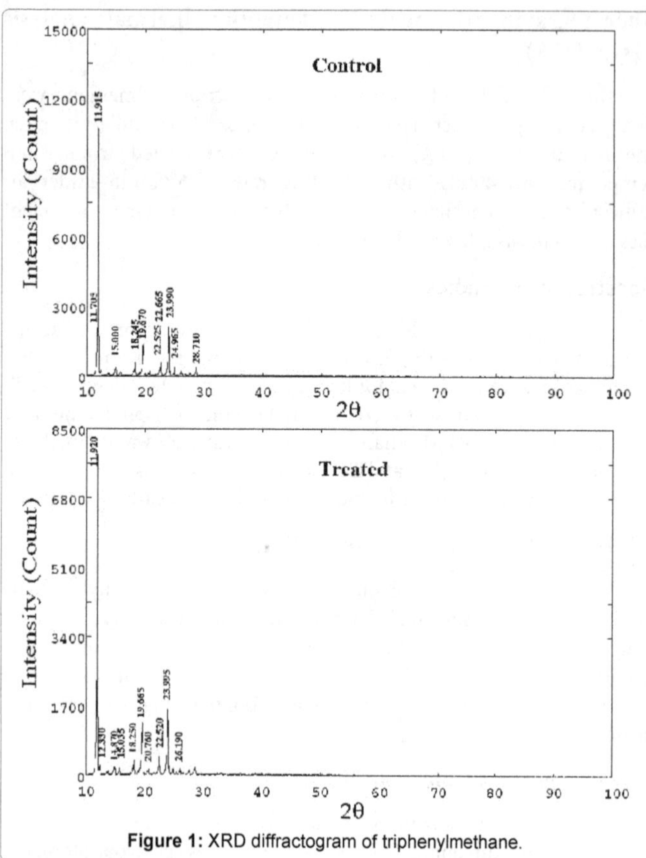

Figure 1: XRD diffractogram of triphenylmethane.

The sharp and intense peak in XRD diffractogram of control and treated samples suggested the crystalline nature of triphenylmethane in both samples. The result showed that the XRD peaks were shifted after biofield treatment as 1191°→1192°; 1500°→15.03°; 18.24°→18.25°; 19.67°→19.66°; 24.96°→24.24° etc. moreover, few peaks like 11.70° and 28.7° in control sample are disappeared or their intensity is decreased after biofield treatment. The decrease in intensity of XRD peaks in biofield treated triphenylmethane might be attributed to decrease in long-range order of the molecules. The average crystallite size was calculated using Scherrer formula and the result are shown in Figure 2. The average crystallite size of control triphenylmethane was found as 117.17 nm that was decreased to 100.51 nm in treated sample. The result showed about 14.22% decrease in average crystallite size in treated sample as compared to control. It was reported that increase in internal micro strain leads to decrease the corresponding crystallite size of the material [27]. Moreover, Zhang et al. showed that presence of strain and increase atomic displacement from their ideal lattice positions causes reduction in crystallite size [28]. Hence, it is assumed that biofield treatment may induce the internal strain in triphenylmethane. This might be the responsible for decrease in average crystallite size of the treated triphenylmethane as compared to control.

Surface area analysis

The surface area of control and treated samples of triphenylmethane was determined using BET surface area analyzer and data are presented in Figure 3. The control sample showed a surface area of 0.8243 m^2/g; however, the treated sample of triphenylmethane showed a surface area of 0.8278 m^2/g. The result showed a slight increase in surface area (0.42%) in the treated triphenylmethane sample as compared to control. The increase in surface area might be correlated to particle size reduction due to high internal strain produced by biofield treatment [29]. The increase in surface area may lead to increase in solubility [30] and reactivity of triphenylmethane as compared to control.

DSC analysis

DSC was used to analyze the melting temperature and latent heat of fusion (ΔH) of control and treated triphenylmethane. In solid materials, substantial amount of interacting forces exist in atomic level that hold the atoms at their positions. ΔH is defined as the energy needed to overcome the interaction force to change the solid phase into liquid phase. Hence, the energy provided during phase change i.e., ΔH is stored as potential energy of atoms. However, melting point is related with kinetic energy of the atoms [31]. DSC thermogram showed the melting temperature at 94.57°C in control and 95.11°C in treated sample (Table 1), which revealed about 0.57% increase in melting temperature in treated sample of triphenylmethane with respect to control. The melting temperature of triphenylmethane was well supported by literature data [32]. Likewise, the DSC thermogram exhibited the latent heat of fusion i.e., 85.05 J/g in control and 85.27 J/g in treated sample of triphenylmethane. The result depicted about 0.26% change in latent heat of fusion of treated sample as compared to control.

Thermogravimetric analysis (TGA)/derivative thermogravimetry (DTG) analysis

The TGA and DTG analysis of control and treated samples of triphenylmethane are shown in Table 1. TGA thermogram of control triphenylmethane exhibited the onset temperature around 216.00°C that was end-set around 257.00°C with 45.99% weight loss. However, the treated triphenylmethane started losing weight around 193.00°C and end-set around at 248.00°C with 64.40% weight loss (Figure 4). The result showed decrease in onset temperature of treated triphenylmethane by 10.65% as compared to control. Moreover, DTG thermogram exhibited the maximum thermal decomposition

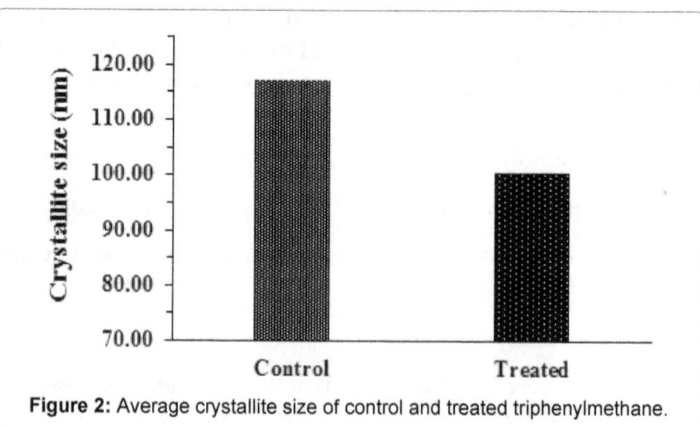

Figure 2: Average crystallite size of control and treated triphenylmethane.

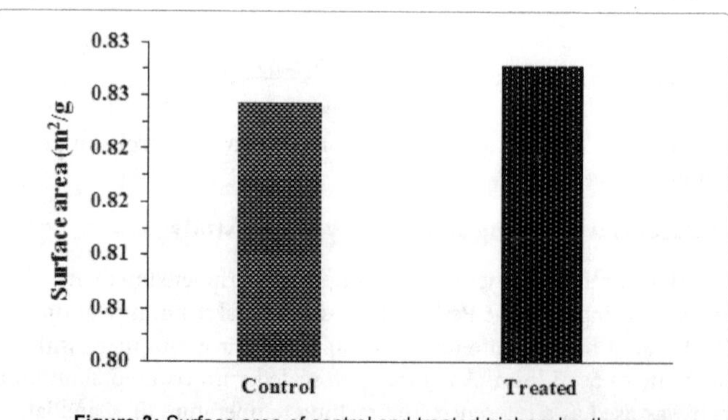

Figure 3: Surface area of control and treated triphenylmethane.

temperature (T_{max}) at 232.77°C in control sample and at 216.07°C in treated sample of triphenylmethane. The result suggested about 7.17% decrease in T_{max} of treated sample with respect to control. The decrease in T_{max} of treated sample might be correlated with increase in vaporization or volatilization of treated triphenylmethane after biofield treatment. It might be due to alteration in internal energy that results into earlier vaporization of treated triphenylmethane sample as compared to control.

S No	Parameter	Control	Treated
1	Latent heat of fusion (J/g)	85.05	85.27
2	Melting point (°C)	94.57	95.11
3	Onset temperature (°C)	216.00	193.00
4	T_{max} (°C)	232.77	216.07

Table 1: Thermal analysis of control and treated samples of triphenylmethane. T_{max}: Temperature at maximum weight loss occurs.

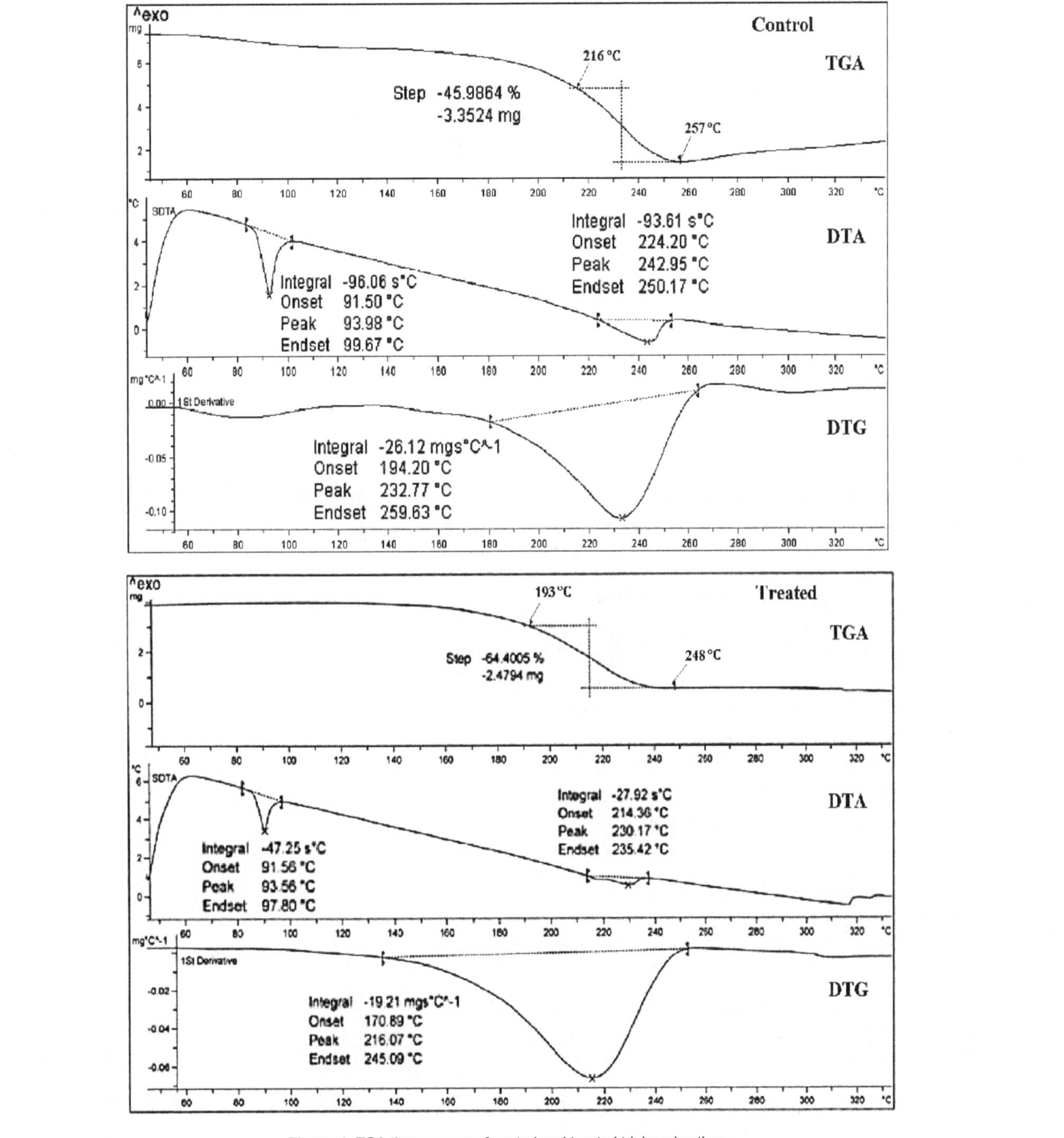

Figure 4: TGA thermogram of control and treated triphenylmethane.

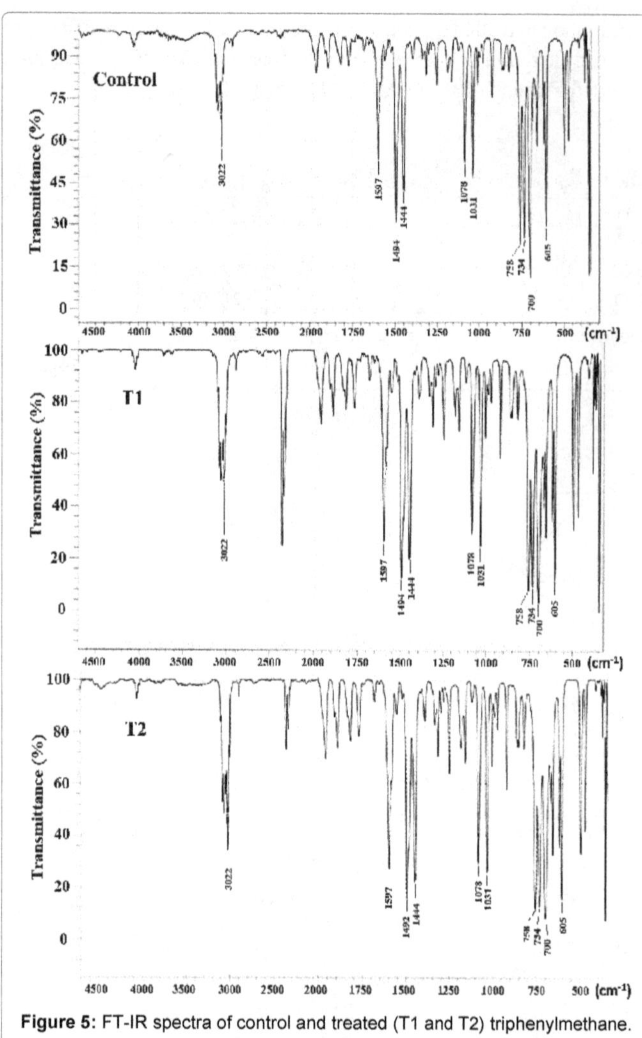

Figure 5: FT-IR spectra of control and treated (T1 and T2) triphenylmethane.

Wave number (cm^{-1})			Frequency assigned
Control	T1	T2	
3022	3022	3022	C-H stretching
1444-1597	1444-1597	1444-1597	C=C stretching
1031-1078	1031-1078	1031-1078	C-H in-plane deformation
605-758	605-758	605-758	C-H out of plane deformation

Table 2: FT-IR Spectral analysis of triphenylmethane.

FT-IR spectroscopic analysis

The recorded FT-IR spectra (Figure 5) were analyzed based on theoretically predicted wavenumber and presented in Table 2. The chemical structure of triphenylmethane consists with three phenyl groups attached with a methyl carbon. Therefore, The FT-IR spectra of triphenylmethane should contains the stretching and bending peaks mainly due to C-H and C=C groups. The C-H stretching was assigned to peak appeared at 3022 cm^{-1} in control and treated (T1 and T2) samples. The C=C stretchings of phenyl ring carbon were assigned to vibrational peak observed at 1444-1597 cm^{-1} in all three samples i.e., control, T1, and T2. The C-H in-plane deformation peaks were attributed to vibrational peaks observed at 1031-1078 cm^{-1} in all three samples i.e., control, T1, and T2. In addition, the C-H out of plane deformation peaks were assigned to IR peak appeared at 605-758 cm^{-1} in control, T1, and T2 samples. The FT-IR spectrum of triphenylmethane was well supported with the literature data [33]. The FT-IR result suggested that the biofield treatment did not induce any structural changes in the triphenylmethane sample with respect to control.

UV-Vis spectroscopy

UV spectrum of control triphenylmethane showed absorbance maxima (λ_{max}) at 207.0, 262.0 and 269.4 nm. Similar pattern of λ_{max} was observed for both the treated samples (T1 and T2) i.e., at 213.0, 261.8, and 269.2 nm in T1 and 206.5, 261.5, and 269.0 nm in T2. The result exhibited a slight bathochromic shift of peak at 213.0 nm in T1 as compared to control (207.0 nm). Apart from this, other peaks were observed at the similar wavelength in all three samples i.e., control, T1 and T2. Overall, the UV spectral analysis suggests that biofield treatment did not cause any significant alterations in the λ_{max} of treated molecules as compared to control. The findings of UV analysis were also supported by the FT-IR results.

GC-MS analysis

The GC-MS spectra of control and treated (T1, T2, T3, and T4) samples of triphenylmethane are shown in Figures 6a and 6b, and the peak intensity of molecular ion and most probable isotopes are illustrated in Table 3. The GC-MS spectra of control and treated samples showed the primary molecule (PM, triphenylmethane) peak at *m/z* 244, which suggested the molecular weight of triphenylmethane. In addition, the peak at *m/z* 245 was assigned to isotopic abundance peaks due to PM^{+1} ($^{13}C/^{12}C$). It is assumed that isotopic abundance ratio of PM^{+1} was mainly due to ^{13}C and ^{2}H isotopes in triphenylmethane. Based on this, it is speculated that the peak at *m/z* 244 is may be due to

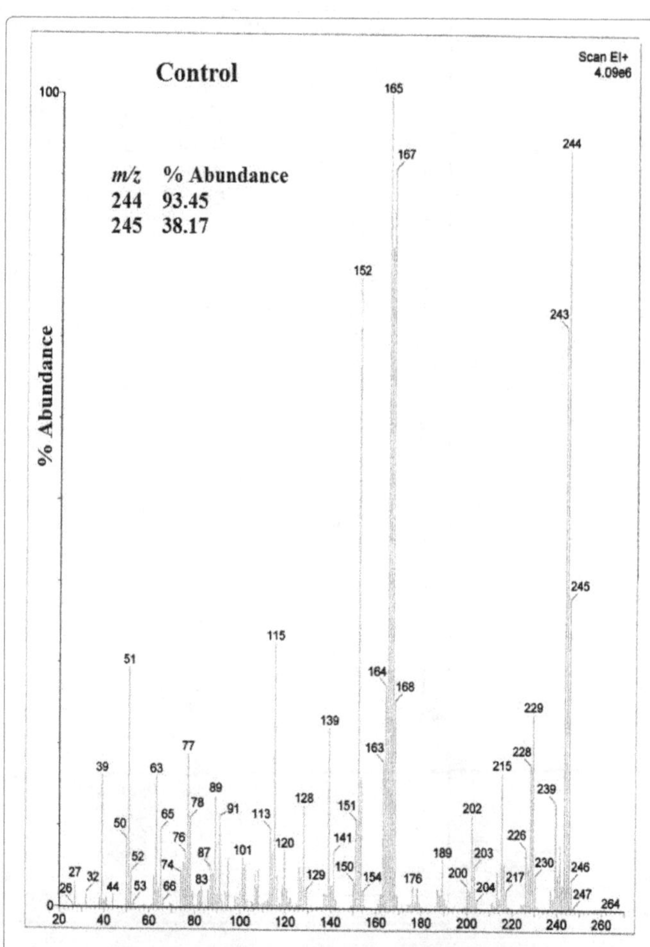

Figure 6a: GC-MS spectra of triphenylmethane control.

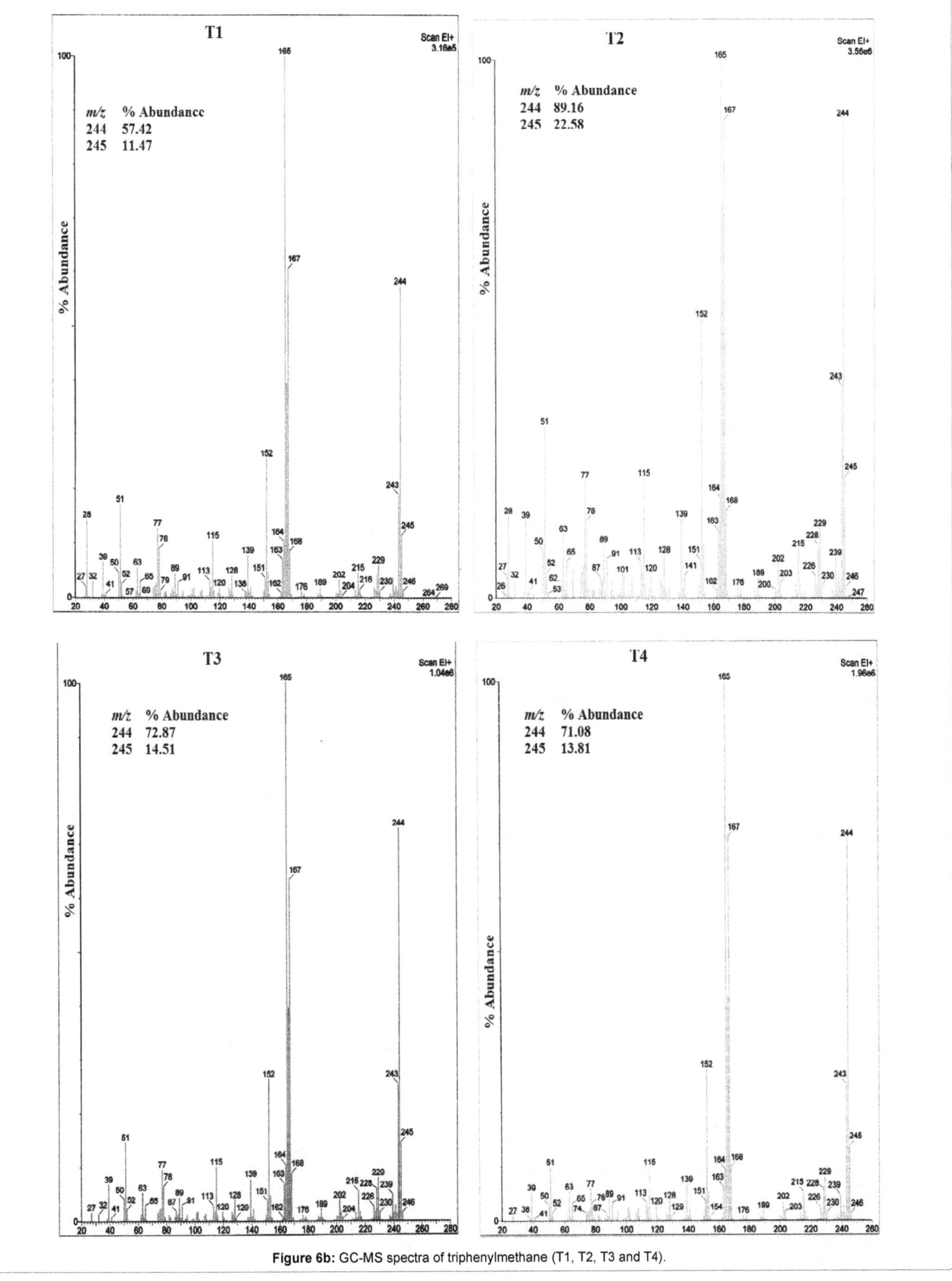

Figure 6b: GC-MS spectra of triphenylmethane (T1, T2, T3 and T4).

Parameter	Control	Treated			
		T1	T2	T3	T4
Peak intensity at m/z=244 (PM)	93.45	57.42	89.16	72.87	71.08
Peak intensity at m/z=245 (PM^{+1})	38.17	11.47	22.58	14.51	13.81
Ration of peak intensity [(100 × (PM^{+1}/ PM)]	40.845	19.976	25.325	19.912	19.429
Carbon isotopic abundance ($\delta^{13}C$ ‰) with respect to control		-510.9	-380.0	-512.5	-524.3

Table 3: GC-MS isotopic abundance analysis of triphenylmethane. PM: Primary molecule (triphenylmethane).

$^{12}C_{19}{}^{1}H_{16}$ and *m/z* 245 is due to $^{13}C_{1}{}^{12}C_{18}{}^{1}H_{16}$ and $^{12}C_{19}{}^{2}H_{1}{}^{1}H_{15}$. The GC-MS analysis result (Table 3) showed that carbon isotopic abundance (expressed in $\delta^{13}C$, ‰) was changed as -510.9, -380.0, -512.5, and -524.3‰ in T1, T2, T3, and T4, respectively with respect to control. The result depicted that in the entire treated samples (T1, T2, T3, and T4) the ^{13}C and ^{2}H were transformed into ^{12}C and ^{1}H by releasing a neutron from their higher isotopes. This conversion of $^{13}C \rightarrow {}^{12}C$ and $^{2}H \rightarrow {}^{1}H$ might be possible if a nuclear level reaction occurred due to influence of biofield treatment. Based on this, it is postulated that biofield treatment possibly induced the nuclear level reactions in triphenylmethane, which may lead to convert the ^{13}C into ^{12}C and ^{2}H into ^{1}H in treated sample as compared to control.

Conclusions

XRD diffractogram of biofield treated triphenylmethane showed the alteration in intensity of XRD peaks and average crystallite size (14.22%) as compared to control. The surface area analysis showed the slight increase in surface area of treated triphenylmethane with respect to control. The thermal analysis (DSC, TGA/DTG) showed a slight change in melting temperature and latent heat of fusion in treated triphenylmethane as compared to control. The T_{max}, was also decreased by 7.17% in treated sample as compared to control. The spectroscopic analysis (FT-IR and UV-Vis) showed that biofield treatment did not affect the dipole moment, bond force constant and the absorbance maxima (λ_{max}) of treated sample as compared to control. GC-MS analysis showed the alteration in carbon isotopic abundance ($\delta^{13}C$) as -510.9, -380.0, -512.5, and -524.3‰ in T1, T2, T3, and T4, respectively as compared to control.

Overall, the physical, thermal and spectroscopical study suggests the impact of biofield treatment on physicochemical properties of treated triphenylmethane with respect to control. Based on this it is assumed that treated triphenylmethane could be more useful as compared to control.

Acknowledgements

The authors would like to acknowledge the whole team of MGV pharmacy college, Nashik for providing the instrumental facility. We would also like to thank Trivedi Science™, Trivedi Master Wellness™ and Trivedi Testimonials for their consistent support during the work.

References

1. Kolling OW, Smith ML (1959) Selected triphenylmethane dyes as acid-base indicators in glacial acetic acid. Anal Chem 31: 1876-1879.
2. Boulos RA (2013) Antimicrobial dyes and mechanosensitive channels. Antonie Van Leeuwenhoek 104: 155-167.
3. Cooney RV, Pung A, Harwood PJ, Boynton AL, Zhang LX, et al. (1992) Inhibition of cellular transformation by triphenylmethane: a novel chemopreventive agent. Carcinogenesis 13: 1107-1112.
4. Kwapniewski Z, Cichon R (1979) The application of triphenylmethane dyes to visualization of selected aliphatic compounds in thin-layer chromatography. Microchem J 24: 298-299.
5. La Rotta H CE, Ciniciato GP, González ER (2011) Triphenylmethane dyes, an alternative for mediated electronic transfer systems in glucose oxidase biofuel cells. Enzyme Microb Technol 48: 487-497.
6. Hugle T, Mocko M, Hartl MA, Daemen LL, Muhrer G (2014) Triphenylmethane, a possible moderator material. Nucl Instrum Methods Phys Res A 738: 1-5.
7. Carballo LM, Wolf EE (1978) Crystallite size effects during the catalytic oxidation of propylene on $^{Pt}/\gamma$-Al_2O_3. J Catal 53: 366-373.
8. Chaudhary AL, Sheppard DA, Paskevicius M, Pistidda C, Dornheim M, et al. (2015) Reaction kinetic behaviour with relation to crystallite/grain size dependency in the Mg-Si-H system. Acta Mater 95: 244-253.
9. Trivedi MK, Patil S, Shettigar H, Bairwa K, Jana S (2015) Spectroscopic characterization of chloramphenicol and tetracycline: An impact of biofield. Pharm Anal Acta 6: 1-5.
10. Trivedi MK, Patil S, Tallapragada RM (2013) Effect of bio field treatment on the physical and thermal characteristics of silicon, tin and lead powders. J Material Sci Eng 2: 1-7.
11. Trivedi MK, Nayak G, Patil S, Tallapragada RM, Jana S, et al. (2015) Bio-field treatment: An effective strategy to improve the quality of beef extract and meat infusion powder. J Nutr Food Sci 5: 389.
12. Einstein A (1905) Does the inertia of a body depend upon its energy-content? Ann Phys 18: 639-641.
13. Becker RO, Selden G (1985) The body electric: Electromagnetism and the foundation of life. New York City, William Morrow and Company.
14. BARNES RB (1963) Thermography of the human body. Science 140: 870-877.
15. Rubik B (2002) The biofield hypothesis: its biophysical basis and role in medicine. J Altern Complement Med 8: 703-717.
16. Trivedi MK, Tallapragada RR (2008) A transcendental to changing metal powder characteristics. Met Powder Rep 63: 22-28.
17. Trivedi MK, Patil S, Tallapragada RM (2012) Thought intervention through biofield changing metal powder characteristics experiments on powder characterisation at a PM Plant. Future Control and Automation LNEE 173: 247-252.
18. Trivedi MK, Patil S, Tallapragada RM (2013) Effect of biofield treatment on the physical and thermal characteristics of vanadium pentoxide powders. J Material Sci Eng S11: 001.
19. Sances F, Flora E, Patil S, Spence A, Shinde V (2013) Impact of biofield treatment on ginseng and organic blueberry yield. Agrivita J Agric Sci 35.
20. Lenssen AW (2013) Biofield and fungicide seed treatment influences on soybean productivity, seed quality and weed community. Agricultural Journal 8: 138-143.
21. Patil SA, Nayak GB, Barve SS, Tembe RP, Khan RR (2012) Impact of biofield treatment on growth and anatomical characteristics of Pogostemon cablin (Benth.). Biotechnology 11: 154-162.
22. Trivedi MK, Patil S (2008) Impact of an external energy on Staphylococcus epidermis [ATCC-13518] in relation to antibiotic susceptibility and biochemical reactions-an experimental study. J Accord Integr Med 4: 230-235.
23. Trivedi MK, Patil S (2008) Impact of an external energy on Yersinia enterocolitica [ATCC-23715] in relation to antibiotic susceptibility and biochemical reactions: an experimental study. Internet J Alternat Med 6: 1-6.
24. Trivedi MK, Patil S, Shettigar H, Gangwar M, Jana S (2015) Antimicrobial sensitivity pattern of Pseudomonas fluorescens after biofield treatment. J Infect Dis Ther 3: 222.
25. Patterson AL (1939) The Scherrer formula for X-Ray particle size determination. Phys Rev 56: 978-982.
26. Pavia DL, Lampman GM, Kriz GS (2001) Introduction to spectroscopy. 3rd edn, Thomson Learning, Singapore.
27. Paiva-Santos CO, Gouveia H, Las WC, Varela JA (1999) Gauss-Lorentz size-strain broadening and cell parameters analysis of Mn doped SnO_2 prepared by organic route. Mat Structure 6: 111-115.
28. Zhang K, Alexandrov IV, Kilmametov AR, Valiev RZ, Lu K (1997) The crystallite-size dependence of structural parameters in pure ultrafine-grained copper. J Phys D Appl Phys 30: 3008-3015.
29. Trivedi MK, Nayak G, Tallapragada RM, Patil S, Latiyal O, et al. (2015) Effect of biofield treatment on structural and morphological properties of silicon carbide. J Powder Metall Min 4: 1-4.

30. Hansen CM (2007) Hansen Solubility Parameters: A User's Handbook. 2nd edn, CRC press.

31. Jamin E, Martin F, Martin GG (2004) Determination of the 13C/12C ratio of ethanol derived from fruit juices and maple syrup by isotope ratio mass spectrometry: collaborative study. J AOAC Int 87: 621-631.

32. Cornish hh, Zamora E, Bahor RE (1964) Metabolism of Triphenylmethane and Triphenylcarbinol. Arch Biochem Biophys 107: 319-324.

33. Cheriaa J, Khaireddine M, Rouabhia M, Bakhrouf A (2012) Removal of triphenylmethane dyes by bacterial consortium. ScientificWorld Journal 2012: 512454.

Validated HPLC Method for Determining Related Substances in Compatibility Studies and Novel Extended Release Formulation for Ranolazine

Suresh Babu VV[1]*, Sudhakar V[1] and Murthy TEGK[2]

[1]Natco Pharma Limited, Hyderabad, Andhra Pradesh, India
[2]Department of Pharmaceutical Analysis & Quality Assurance, Bapatla College of Pharmacy, Bapatla, Andhra Pradesh, India

Abstract

HPLC method has been developed for determination of Ranolazine together with its related substances in a laboratory mixture for drug and excipient compatibility studies as well as in a novel extended release tablet developed in-house. Efficient chromatographic separation was achieved on a Supelcosil C 18, (250×4.6 mm, 5 μm) column with mobile phase containing combination of Phosphate buffer pH 7.0 and Methanol in ratio of 350:650 at flow rate of 1.0 ml/minute and eluent was monitored at 220 nm. Linearity, regression value, recovey, % RSD of method precision, LOD and LOQ values were found with in the limits. In this method impurities were well seperated from the main peak. This method was found to be satisfactory. Limits for reporting threshold and total impurities were 0.1% and 2.0%, respectively, as per Q3B(R) Impurities in New drug Products. Compatibility studies are essential for preformulation studies of formulation development. In the present study, the possible interactions between Ranolazine and some excipients (Hypromellose phthalate grade HP-55, Ethocel 7 FP premium, Natrosol type 250 HHX, Klucel HF pharm, Avicel PH 101 and Magnesium Stearate) were evaluated by examining the pure drug or drug-excipient powder mixtures which were stored under different conditions (55°C after 15 days and 40°C/75% RH after 30 days) using High Performance Liquid Chromatography (HPLC). The results demonstrate the suitability of Ranolazine with Hypromellose phthalate grade HP55, Ethocel7FP premium, Natrosol type 250 HHX, Klucel HF pharm, Avicel PH101 and Magnesium Stearate. Same method also used in novel extended release tablet for determination of related substances. Based upon obtained results shows developed formulation was stable at 40°C/75% RH for 3 months.

Keywords: HPLC method; Ranolazine; Chromatographic separation

Introduction

Ranolazine is used for the treatment of Cardiac ischemia it affects sodium dependent calcium channels during myocardial ischemia [1]. The International Conference on Harmonization (ICH). Recommends Regulatory requirements for the identification, qualification and control of impurities in drug substances and their formulated products. As recommended by ICH all routine impurities at or above 0.1% level, should be identified through suitable analytical methods. Ranolazine is not cited in the any pharmacopoeia and contamination by impurities as per supplier specifications [2]. Compatibility studies thus allow in systematic selection of excipients, for formulation development. Excipients may contribute to incompatibility by altering the moisture content, altering the micro-environment pH, and acting as a catalyst for degradation or contributing an impurity that causes degradation [3-7]. After through literature search, it was observed that no HPLC method was found for estimation of impurities in Ranolazine. Therefore, it was thought worth determining the impurities of Ranolazine to ensure the quality, efficacy and safety of the final pharmaceutical formulation. The purpose of the present study was to validate the HPLC method for determining the related substances to evaluate the physical & chemical stability of Ranolazine when mixed with Excipients and in a novel extended release tablet at different storage conditions in accordance with the ICH guidance document.

Experimental

Reagents and chemicals

The following materials where used in the present investigation Ranolazine,Hypromellose phthalate grade HP 55, Ethocel standard 7 FP premium, Natrosol type 250 HHX, Klucel HF pharm, Avicel pH101 and Magnesium Stearate, HPLC grade chemicals were gifted by Natco Pharma Limited (Hyderabad, India). Novel extended release tablets of Ranolazine and a placebo were formulated in Formulation research and development, Natco Pharma Limited, Kothur, India.

Chromatography apparatus and conditions

The chromatograph consisted of an Water alliance HPLC system with 2489 detector.

Column	: Supelcosil C18, (250×4.6 mm), 5 μm
Flow rate	: 1.0 mL/minute
Wavelength	: UV-220 nm
Column temperature	: 40°c
Sample temperature	: 5°c
Injection Volume	: 20 μL
Runtime	: 60 minutes

Preparation of solutions

Mobile phase preparation: Weighed and transferred about 0.5 g of Disodium hydrogen ortho phosphate (Na_2HPO_4) into 1000 mL beaker

***Corresponding author:** Suresh Babu VV, Natco Pharma Limited, Hyderabad, Andhra Pradesh, India, E-mail: vvsbabu@gmail.com

and 350 mL of Purified water was added and mixed well. pH of the solution was adjusted to 7.0 with diluted orthophosphoric acid. This solution and methanol were mixed in the ratio of 350:650 respectively. Above solution was filtered through 0.45 μ membrane filter and degasify.

Standard preparation: Accurately weighed and transferred about 20 mg of Ranolazine working standard into a 100 mL volumetric flask. Then added about 30 mL of mobile phase and sonicate to dissolve then cooled toroom temperature and diluted to volume with mobile phase. 1.0 mL of the solution was transferred into a 200 mL volumetric flask, and make up the volume with mobile phase.

Placebo Preparation: Accurately weighed placebo powder (without drug but having same composition) equivalent to formulation sample containing 500 mg Ranolazine was transferred into 100 mL volumetric flask, 50 mL of mobile phase was added and sonicate for 20 min. Cooled to room temperature and diluted to volume with mobile phase and mixed well. The solution was filtered through a 0.45 μm membrane filter. 5.0 mL of the solution was transferred into a 25 mL volumetric flask, and made up the volume with mobile phase.

Sample preparation: Not fewer than 5 tablets were grounded to fine powder and transferred an accurately weighed portion of the powder, equivalent to 500 mg of Ranolazine, to a 100 mL volumetric flask. Then added about 50 mL of mobile phase, and sonicate for 20 minutes with occasional shakings. Cooled to room temperature and diluted to volume with mobile phase and mixed well. The solution was filtered through a 0.45 μm membrane filter.

5.0 mL of the solution was transferred into a 25 mL volumetric flask, and made up the volume with mobile phase.

System suitability: Chromatograph the standard preparation (six replicate injections), measured the peak area responses for the analyte peak and evaluated the system suitability parameters as directed.

Acceptance criteria:

- %RSD for replicate injections of peak area response for Ranolazine peak from the standard preparation should be not more than 2.0.
- The Tailing factor for Ranolazine peak should be not more than 2.0.
- Theoretical plate for Ranolazine peak should be not less than 2000.

Procedure: Separately injected equal volumes (about 20 μL) of mobile phase as blank, placebo and sample preparations into the chromatograph and recorded the peak area response for the analyte peaks and calculated by using formula as mentioned in below formula.

UASW110025PAvg.wt

-----x------- x ------- x ------- --------x---------x------------- x 100

SA100200TW5100LA

Total Impurities: Sum of Unknown impurities

Where,

SA : Peak area response due to Ranolazine from sample preparation

UA : Peak area response due to Un-known impurity from sample preparation

SW : Weight of Ranolazine working standard taken, in mg

TW : Weight of sample taken in, mg

P : Purity of Ranolazine working standard taken on, as is basis

Preparation of physical powder mixture: In order to evaluate Ranolazine drug excipient interactions, The excipients and drug were taken in different ratios as reported in Table 1, required to prepare extended release tablets [Ranolazine with Hypromellose phthalate grade HP 55, Ethocel standard 7 FP premium, Natrosol type 250 HHX, Klucel HF pharm, Avicel pH 101 in (1:0.5) and Magnesium Stearate in (1:0.2)] and homogeneously mixed with a mortar and pestle for 10 min, then powder mixture was placed in glass vials with a rubber stoppers. These vials were stored in at 55°C for 14 days and 40°C/75% RH for 28 days and evaluated periodically using HPLC.

Preparation of extended release tablets: The tablets were prepared by wet granulation technique. Drug and other excipients were accurately weighed, mixed and sifted through ASTM (American society of Testing and Materials) 40 mesh. The wet mass was passed through ASTM 12 mesh and granules were dried. Dried granules were further passed through ASTM 18 mesh. The granules were lubricated and compressed into oblong shaped (16.5×8.0 mm for 500 mg) tablets using 12-station rotary compression machine (Rimekminipress–II MT).

Analytical method validation

Specificity: Specificity is to validate the ability to assess unequivocally the analyte in the presence of components, which may be expected to be present such as matrix components, impurities and degradation products.

Acceptance criteria: The interference of the placebo and diluent is considered insignificant, if the chromatogram of the diluent and placebo shows no peak at the retention time of Ranolazine and impurity peaks.

Limit of detection and Limit of Quantification of Ranolazine (un-known impurity): For calculating the LOD and LOQ values, solutions with known decreased concentrations of analyte were injected into the HPLC system. The limit of detection (LOD) and quantification (LOQ) were then measured by calculating the minimum level at which the analyte can be readily detected (signal to noise ratio of 3:1) and quantified (signal to noise ratio of 10:1).

S.No	Ingredients	Sample ID	Ratio
1	Ranolazine	RZ01	1:0
2	Hypromellose phthalate grade HP-55	RZ02	1:0
3	Ethocel standard 7FP Premium	RZ03	1:0
4	Natrosol Type 250 HHX	RZ04	1:0
5	Klucel HF pharm	RZ05	1:0
6	Avicel pH101	RZ06	1:0
7	Magnesium Stearate	RZ07	1:0
8	Hypromellose phthalate grade HP-55+ Ethocel standard 7FP Premium+ Natrosol Type 250 HHX+ Klucel HF pharm+ Avicel pH101+ Magnesium Stearate	RZ08	0.5:0.5:0.5: 0.5:0.5:0.2
9	Ranolazine+Hypromellose phthalate grade HP-55	RZ09	1:0.5
10	Ranolazine+Ethocel standard 7FP Premium	RZ10	1:0.5
11	Ranolazine+Natrosol Type 250HHX	RZ11	1:0.5
12	Ranolazine+Klucel HF pharm	RZ12	1:0.5
13	Ranolazine+Avicel pH101	RZ13	1:0.5
14	Ranolazine+Magnesium Stearate	RZ14	1:0.2
15	Ranolazine+Hypromellose phthalate grade HP-55+ Ethocel standard 7FP Premium+ Natrosol Type 250HHX+ Klucel HF pharm+ Avicel pH101+ Magnesium Stearate	RZ15	1:0.5:0.5:0.5: 0.5:0.5:0.2

Table 1: Ranolazine and Excipients Ratio Used In Compatibility Study.

Linearity: Linearity of detector response was established by plotting a graph to concentration versus area of Ranolazine (un-known impurity). Determined the correlation coefficient and Y-intercept/ response at 100% of working concentration. A series of solutions Ranolazine (un-known impurity) solutions in the concentration ranging from about LOQ% level to about 150% (LOQ, 25%, 50%, 100%, 125%, & 150%) of the target concentration were prepared and injected into the HPLC system.

Acceptance Criteria:

• The correlation coefficient (r) is NLT 0.995

• Regression coefficient (r^2), Y-intercept, slope of regression line should be reported.

• %Y-intercept at 100% target concentration should be NMT ± 5.0%

Precision: The method precision of the method is ascertained by injecting 6 replicates of test sample % impurity and % RSD was calculated.

Accuracy: Accuracy of the proposed method was determined by recovery studies using standard addition method. The percentage recovery studies of Ranolazine was carried out in triplicate 3 different levels 50%, 100%, 150% by spikingstandard drug solution to the placebo [8,9].

Acceptance criteria: The mean recovery of the impurities at each level should be between 85.0% and 115.0%.

Results and Discussion

The purpose of the present study was to validate the HPLC method for determining the related substances to evaluate the physical & chemical stability of Ranolazine when mixed with excipients and in a novel extended release tablet at different storage conditions in accordance with the ICH guidance document. The HPLC method used in this investigation was Accurate, Precise, and linear it is validated by using Water Alliance HPLC system with 2489 detector and Supelcosil C 18, (250×4.6 mm, 5 μm particle size).Limits for reporting threshold and total impurities were 0.1% and 2.0%, respectively, as per Q3B(R) Impurities in New drug Products.

Analytical method validation

Specificity: A study was conducted for the interference of diluent and placebo with Ranolazineand impurities. Samples were prepared in triplicate by taking the placebo equivalent to about the weight in portion of test preparation as per the test method and injected into the HPLC system. The results were summarized in Table 2. No peaks are eluted at the retention time of Ranolazine and impurities.

LOD and LOQ: Determined the limit of detection (LOD) of Ranolazine (un-known impurity) by the analysis of samples with known concentrations of analyte by establishing the minimum level at which the analyte can be reliably detected and results tabulated in Table 3. S/N Ratio of 3.10:1 was observed for LOD.

S No.	Sample name	Retention time (minutes)	% Interference (Respective to Ranolazineand Impurities)
1	Diluent	NA	NO
2	Placebo 500 mg	NA	NO
3	Impurity 1	5.565	NO
4	Impurity 2	8.331	NO
5	Unknown impurity	3.113	NO
6	Ranolazine	10.464	NA

Table 2: Specificity results.

Component	Limit of Detection (LOD)		
	Concentration (mcg/mL)	Concentration (%)	S/N Ratio
Ranolazine	0.0273	0.003	3.10 : 1

Table 3: Limit of detection Results.

Component	Limit of Detection (LOD)		
	Concentration (mcg/mL)	Concentration (%)	S/N Ratio
Ranolazine	0.0818	0.008	10.19 : 1

Table 4: Limit of Quantification Results.

Injection	Limit of Quantification (Ranolazine Peak areas)
1	2810
2	2816
3	2754
4	2796
5	2856
6	2896
Average	2821.0
% RSD	1.7

Table 5: Ranolazine Precision results at LOQ.

% Level	Concentration (mcg/mL)	Peak Area (Average)
LOQ	0.0818	2800
25	0.2557	8812
50	0.5115	18623
100	1.0229	37551
125	1.2786	47091
150	1.5344	56320
Correlation coefficient (r)		**1.0000**
Regression coefficient (r^2)		**0.9999**
Slope		37034.797
Y-intercept		-384.55307
Y-Intercept/response at 100% of Concentration×100		-1.0

Table 6: Linearity results.

Determined the limit of quantification (LOQ) of Ranolazine (un-known impurity) by the analysis of samples with known concentrations of analyte by establishing the concentration, which will give a concentration of about 3.3 times of LOD with an RSD of 10% for peak area, with acceptable precision and accuracy at LOQ level and results tabulated in Table 4. S/N Ratio of 10.19:1 was observed for LOD.% RSD for six replicate injections of lower (LOQ level) was 1.7 for Ranolazine precision at LOQ and these results tabulated in Table 5. From the results, it can be concluded that the proposed method can quantify the small quantity of impurities in Ranolazine samples. The Limit of Detection and Limit of Quantification test results met with the acceptance criteria.

Linearity: Linearity of the related substances test method covers from LOQ Level to 150% of target concentration for Ranolazine and results tabulated in Table 6. Hence the test method was found linear for Ranolazine.

Precision: Precision of the test method was determined by injecting test preparation and tested through the complete analytical procedure from sample preparation to final result. Repeatability assessed using a minimum of 6 determinations and calculated% relative standard

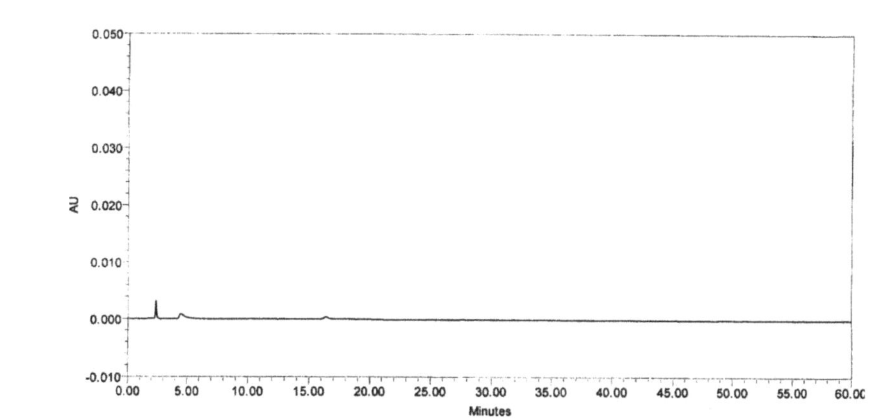

Figure 1: Related Substance Chromatogram for Mobile phase.

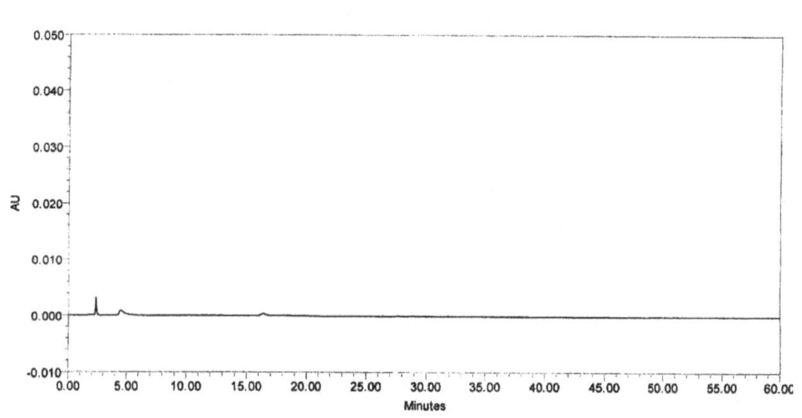

Figure 2: Related Substance Chromatogram for RZ08 sample at 40°C/75%RH for 28 days.

Sample No	% Of Related substances		
	Un-known imp-1	Un-known imp-2	Total impurities
1	0.051	0.042	0.093
2	0.049	0.043	0.092
3	0.052	0.045	0.097
4	0.051	0.044	0.095
5	0.049	0.042	0.091
6	0.051	0.043	0.094
Average	0.051	0.043	0.094
%RSD	**2.43**	**2.71**	**2.31**

Table 7: Precision results for Ranolazine.

deviation of impurities. The results were given in Table 7. Related substances results meet the specification limits.

Accuracy: The Accuracy of the related substances test procedure was determined by Spiking of Ranolazine on placebo and injecting samples in triplicate at LOQ, 50%, 100%, and 150% of the target concentration 0.1% v/v of Ranolazine (Un-known imp). Calculated the% recovery of impurities and the results are given in Table 8. The accuracy test results were met with the acceptance criteria.

Drug-Excipient Compatibility studies

The sample was subjected to above mentioned conditions, chromatograms were shown in Figure 1 for mobile phase, (Figure 2) for RZ08 sample and (Figure 3) for RZ15 sample respectively. System suitability test results were within the USP limits.Same peaks observed in mobile phase and RZ08 sample indicates only blank peaks appeared on chromatogram and no excipient peaks. Chromatograms of Hypromellose phthalate HP55, Ethocel 7FP premium, Natrosol type 250 HHX, Klucel HF pharm, Avicel PH 101, Magnesium Stearate and mixture of these excipients does not shown the peaks at all storage conditions. In the sample chromatogram for RZ15, drug was eluted at a retention time of 10.491 min and the unknown impurities were observed at a retention time of 5.588, 8.349 min respectively as shown in Figure 3. The percentage of impurities was calculated by dilute standardization method. No significant variation in unknown impurity maximum and total impurities for Pure drug Ranolazine (RZ01), Combination of Ranolazine and excipient (RZ09, RZ10, RZ11, RZ12, RZ13 and RZ14) and combination of Ranolazine and all excipients (RZ15) at different storage conditions. Single unknown impurity were found to be not more than 0.03% and the total impurities were found to be not more than 0.06% in all combinations at different storage conditions shown in Table 9, i.e., impurities are within the limit as per ICH guidelines.

Pure drug Ranolazine or Ranolazine in binary mixtures was stable for 14 days at 55°C and 28 days at 40 ± 2°C/75 ± 5.0% RH. Since there is no change in peak area or change in Rt and the percentage of impurities was within limits as per ICH guidelines It can be deduced that the drug was stable in pure form or in the presence of excipients tested under these elevated temperature and Humidity conditions. Therefore these

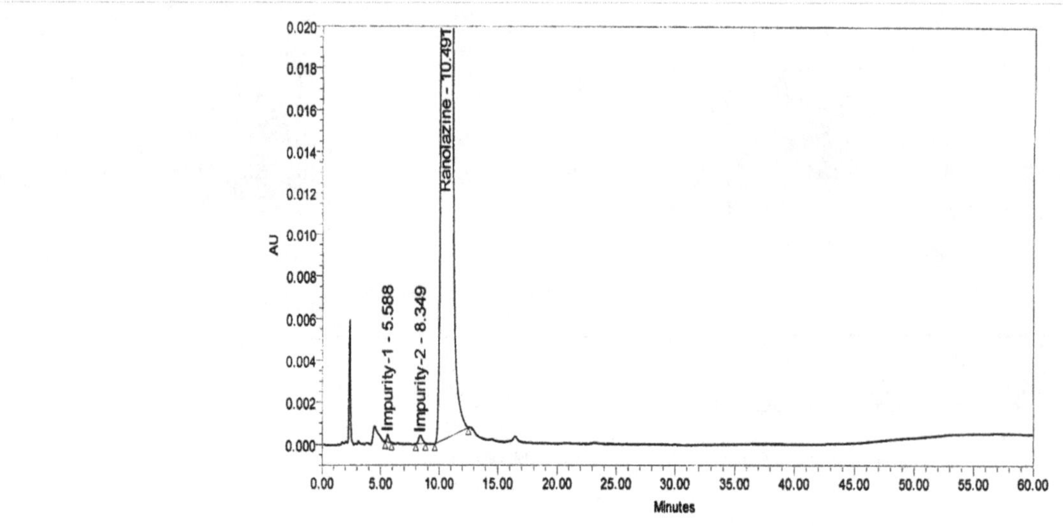

Figure 3: Related Substances Chromatogram of Ranolazine +All Excipients (RZ15) stored at 40°C/75%RH (after 28 days).

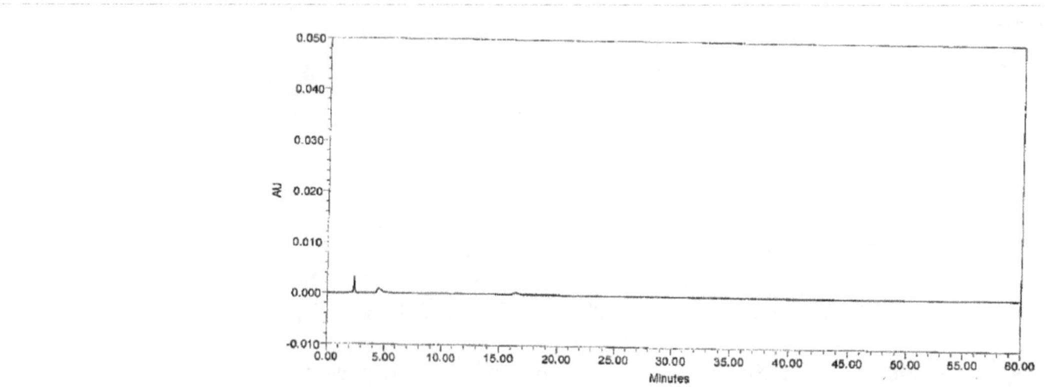

Figure 4: Related Substance Chromatogram for Mobile phase.

S. No.	% Level	mcg/ml Added	mcg/ml Found	%Recovery	Average Recovery (%)
1	LOQ	0.0812	0.0828	101.97	101.9
2			0.0834	102.71	
3			0.0821	101.11	
4	50	0.5121	0.5183	101.20	101.4
5			0.5165	100.90	
6			0.5236	102.20	
7	100	1.0326	1.0395	100.67	101.2
8			1.0496	101.65	
9			1.0456	101.26	
10	150	1.5456	1.5359	99.37	100.4
11			1.5665	101.35	
12			1.5552	100.62	

Table 8: Accuracy results for Ranolazine.

excipients were found to be compatible with Ranolazine.

Extended release tablets

Same method also used in novel extended release tablet for determination of related substances and results were tabulated in Table 4. Same peaks observed in the mobile phase and placebo sample indicates only blank peaks appeared on chromatogram and no excipient peaks as shown in Figures 4 and 5. In the sample chromatogram drug was eluted at a retention time of 10.464 min and the unknown impurities were observed at a retention time of 3.113, 5.565, 8.331 min, respectively as shown in Figure 6. The percentage of impurities was calculated by dilute standardization method. No significant variation was found in unknown impurity maximum and total impurities for novel extended release tablets at different storage conditions. With respect to drug excipient mixtures in compatibility studies, one additional impurity was observed in novel extended release formulation. Single unknown impurity were found to be not more than 0.05% and the total impurities were found to be not more than 0.09% in extended release tablets at different storage conditions shown in Table 10, i.e., impurities are within the limit as per ICH guidelines. Slight trending was observed in tablets when compared with compatibility samples which may be due to additional storage time period. Based on reporting threshold and total impurities results, developed formulation was stable at 40°C/75% RH for 3 months.

Conclusion

The proposed HPLC method for estimation of related substances for Ranolazine is analyzed in bulk drug and Ranolazine extended release tablets as per ICH guidelines. This method is found to be specific for estimation of unknown impurities. Knowledge of drug-excipient interactions is a necessary prerequisite to the development of dosage forms that are stable and of good quality. The results of HPLC

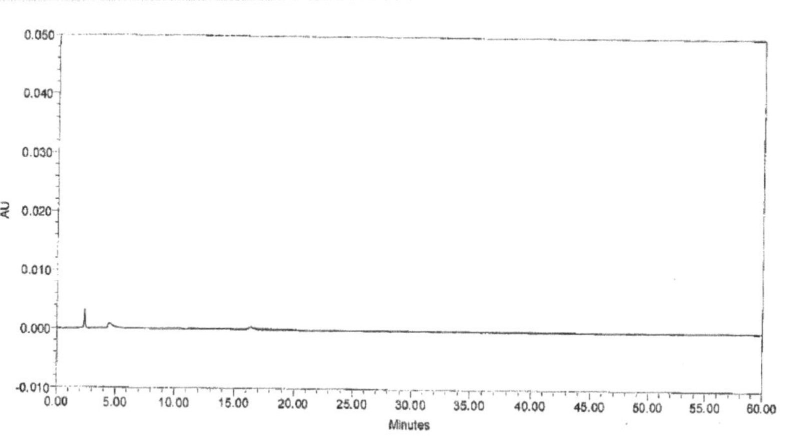

Figure 5: Related Substance Chromatogram for placebo sample.

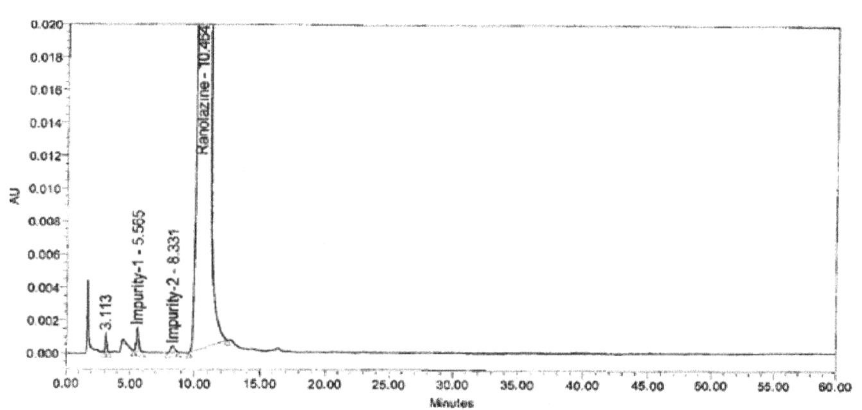

Figure 6: Related Substances Chromatogram of Ranolazine extended release tablets Stored at 40°C/75% RH (after 3 months).

Sample	Impurities (%) (Initial)		Impurities (%) (550C) 14 days		Impurities (%) (400C/75%RH) 28 Days	
	U_{max}*	Total Impurities	U_{max}*	Total Impurities	U_{max}*	Total Impurities
RZ01	0.02	0.04	0.03	0.05	0.02	0.04
RZ09	0.02	0.04	0.02	0.04	0.02	0.05
RZ10	0.03	0.03	0.03	0.04	0.03	0.04
RZ11	0.02	0.03	0.02	0.04	0.02	0.04
RZ12	0.03	0.04	0.03	0.05	0.02	0.04
RZ13	0.03	0.04	0.02	0.04	0.03	0.04
RZ14	0.03	0.04	0.03	0.04	0.03	0.03
RZ15	0.03	0.04	0.02	0.04	0.03	0.06

Note: Umax* means Unknown Impurity Maximum

Table 9: Related impurities results for Ranolazine with different excipients in compatibility studies.

Impurities (%)							
Initial		400C/75%RH					
		1st Month		2nd Month		3rd Month	
U_{max}*	Total Impurities	U_{max}*	Total Impurities	U_{max}*	Total Impurities	Umax*	Total Impurities
0.03	0.07	0.03	0.07	0.05	0.09	0.05	0.09

Note: U_{max}* means Unknown Impurity Maximum

Table 10: Related impurities results for Ranolazine extended release tablets 500 mg stored at accelerated condition for different time intervals.

studies showed the method is suitable for the assay of Ranolazine in the given Extended release formulation and that there were no possible interactions between drug and selected excipients like Hypromellose phthalate HP55, Ethocel7FP premium, Natrosol type 250 HHX, Klucel HF pharm, Avicel PH 101 and Magnesium Stearate during 14 days at 55°C and 28 days at 40 ± 2°C/75 ± 5.0% RH storage conditions.

Acknowledgments

The author grateful to Natco Pharma Limited and Principal, Bapatla college of Pharmacy Bapatla for encouragement and providing the necessary facilities during the course of investigation.

References

1. LuizB, JohnCS (2006) The mechanism of Ranolazine action to reduce ischemia-induced diastolic dysfunction. Eur Heart J Suppliments 8: A10-A13.
2. Vivekanad AC, Pawan KP, Neeraj U (2012) Validated gradient stability indicating HPLC method for determining Diltiazem Hydrochloride and related substances in bulk drug and novel tablet formulation. J Pharmaceutical Analysis: 226-237.
3. Fathima N, Mamatha T, HusnaKQ, Anitha N, Venkateswara Rao J (2011) Drug-excipient interaction and its importance in dosage form development. J Appl Pharmaceutical Sci 6: 66-71.
4. Compatibility Studies; Dr.Arvind Bansal Module 9: Pharmaceutical Preformulation: Basics and Industrial Applications
5. Crowley P, Luigi GM (2001) Drug excipient interactions.Advanstar Publications, USA.

6. Bozdağ-Pehlivan S, Subaşi B, Vural I, Unlü N, Capan Y (2011) Evaluation of Drug-Excipient Interaction in the Formulation of Celecoxib Tablets. Acta Poloniae Pharmaceutica 68: 423-433.

7. Murthy TEGK., BalaVishnu priya M, Suresh Babu VV (2012) Compatibility studies of Acetazolamide with excipients by using High performance liquid chromatography. Indian Drugs 49: 39-45.

8. International conference on Harmonization (ICH) (1995) Guidelines on validation of analytical procedure definition & terminology. Federal Registar 60: 11260.

9. Praveen Kumar SN, BhadreGowda DG (2012) Development and Validation of HPLC Method for the Determination of Simvastatin in Bulk and Pharmaceutical Formulation. J Chem Pharmaceutical Res 4: 2404-2408.

Retention Profile and Selective Separation of Trace Concentrations of Phenols from Water onto Iron(III) Physically Loaded Polyurethane Foam Solid Sorbent: Kinetics and Thermodynamic Study

El-Shahawi MS *, Hamza A, Alwael H, Bashammakh AS, Al-Sibaai AA and Saigl ZM

Department of Chemistry, Faculty of Science, Damietta University, Damietta, Egypt

Abstract

Phenols are included in the list of potential toxins for motor neurons. Thus, fast and selective method for removal of phenols has been developed. The method was based upon the extraction of iron (III) phenolate complex anions $[Fe(phenolate)_6]^{3-}$ by tri-*n*-butylphosphine (TBP)-plasticized iron(III) immobilized polyurethane foam (PUF) from the test aqueous solution adjusted to pH≈1 followed by subsequent formation of ternary ion association complex of these species with the protonated urethane and/or ether linkages on the PUF solid phase extractor (SPE). The rate of removal of phenols from the aqueous solution by TBP plasticized iron (III) immobilized PUF were studied in batch conditions employing a series of kinetic models. Initially, phenols uptake onto plasticized iron (III) treated PUF was fast followed by kinetically pseudo-second order rate equation with an overall rate constant (k) of 0.014 and 0.018 g $(mg\ min)^{-1}$ for phenol and 2-chlorophenol (2-CP), respectively. Endothermic nature of phenols sorption is governed by the positive value of ΔH where as the negative values of ΔG dictate that the uptake of the analyte onto the sorbent is spontaneous phenomena. The positive values of ΔS for both phenols may be indicative of moderate sorption step of the complex anion $[Fe(C_6H_5O)_6]^{3-}$ and ordering of ionic charges with a compensatory disordering of the sorbed species onto the sorbents. PUF sorbent provides efficient removal of phenols traces from various water samples.

Keywords: Phenols; Retention; Pseudo-second order model; Thermodynamic; Polyurethane foam

Introduction

Phenolic compounds present in the aquatic environment as a result of their industrial applications such as a production of plastics, dyes, drugs, pesticides, anti-oxidants, paper and petrochemical [1]. The toxicity and unpleasant organoleptic properties of phenols at trace and ultra-trace concentrations in water and fish have been reported by the US Environmental Protection Agency (EPA) [2,3]. Relatively high concentrations have been measured in coastal seawater (up to 1500 ng L^{-1}) for both 2,4-dichlorophenol (DCP), and penta chlorophenol (PCP) (PCP) [3,4]. Gao et al. [5] have reported the median levels of 5, 2, and 50 ng $L^{-1}$for 2,4-DCP, 2,4,6-TCP and PCP, respectively, in China's Rivers.

Numerous analytical techniques have been reported for determination of different phenolic compounds [6-10,11]. Simultaneous determination of 14 chlorophenols (CPs) and chloroanisoles (CAs) in wine samples have been carried out using stir bar sorptive extraction (SBSE) followed by gas chromatography-mass spectrometry (GC-MS) [11]. Chlorophenols have been isolated from aquatic media and analyzed by high performance liquid chromatography with ultraviolet detection (HPLC-UV) [12-14]. However, none of these combinations can achieve the quantification limits required for direct determination of phenols in drinking water, resulting in a necessity for a preconcentration step. Separation and enrichment techniques such as liquid-liquid extraction (LLE) [15], solid phase extraction (SPE) [16-23] and cloud point extraction (CPE) [24] have been widely used as a sample pretreatment step. Among these techniques, SPE has been used most frequently due to the availability, easy recovery of the solid phase and the high preconcentration factors that can be achieved [25].

PUF is a good sorbent material in SPE and it has been the subject of several review articles [15]. Recent years have seen an upsurge of interest for developing low cost, rapid and sensitive analytical methods for preconcentration and removal and subsequent determination of phenols in large sample volumes [7-11,26,27]. Thus, the work presented in this article is focused on studying the kinetics and thermodynamic characteristics and the most probable retention mechanism of the separation of some selected phenols onto PUF immobilized with iron (III).

Experimental

Reagents and materials

All chemicals used were of analytical reagent grade and were used as received. All glassware's were rinsed with doubly deionized water prior to use. Commercial white sheets of open cell PUF (polyether type) were purchased from a local market in Jeddah City, Saudi Arabia. Foam cubes of approximately 1.0 cm^3 were cut from the PUF sheets and were washed and dried as reported [26]. Phenol and 2-CP were purchased from BDH (Poole, England). A standard stock solution (1000 $\mu g\ mL^{-1}$) of each phenol was prepared in ultra-pure water in presence of few drops of ethanol. More diluted solutions (50-100 $\mu g\ mL^{-1}$) of each phenol were prepared by appropriate dilution of the stock solutions with water. BDH Tri-n-butylphosphine (Poole, England) (0.01-0.05% v/v) was prepared in water in the presence of drops of ethanol. A series of Britton-Rrobinson (B-R) buffer solutions was prepared as reported [28].

***Corresponding author:** El-Shahawi MS, Department of Chemistry, Faculty of Science, King Abdulaziz University, Jeddah, Saudi Arabia
E-mail: malsaeed@kau.edu.sa, mohammad_el_shahawi@yahoo.co.uk

Apparatus

A Perkin-Elmer UV-Vis spectrophotometer (Lambda 25, USA) with 10 mm (path width) was used for recording the electronic spectra of phenolic compounds. A Corporation Precision Scientific mechanical shaker (Chicago, CH, USA) was used. Deionized water was obtained from Milli-Q Waters Plus system (Milford, MA, USA). A Thermo Fisher Scientific Orion model 720 pH Meter (Milford, MA, USA) were also used.

Preparation of iron (III)-immobilized PUF

Accurate weights (0.2 g) of PUF cubes were shaken with 200 mL of different concentrations (0.01-0.05% m/v) of $FeCl_3.6H_2O$ with efficient stirring for 30 min. The cubes were separated out, pressed between two sheets of filter papers and finally dried [27,28]. The amount of iron (III) retained onto the PUF was caculated from the difference between iron (III) before (C_o) and after (C) shaking employing the following equation:

$$a = (C_0 - C)V / W \quad (1)$$

Where, V is the volume of solution in (mL) and W is the mass (g) of PUF sorbent. Plasticized iron (III)-immobilized PUF cubes were prepared by shaking iron (III) treated PUF with different concentrations of TBP (0.01-0.1% v/v) with efficient stirring for 30 min. The foam cubes were then treated as reportd [27].

Recommended extraction procedures

In a series of conical flasks (250 mL), an accurate weight of the unplasticized and plasticized iron (III)-immobilized PUF cubes (0.2 ± 0.002g) were equilibrated with 100-200 mL of an aqueous solution containing 50-100 μg mL^{-1} of the phenolic compounds. The solution pH was adjusted at pH ≈1. The solutions were shaken for 2 hours at 25 ± 1°C on a mechanical shaker. After phase separation, the aqueous phase was separated out by decantation and the amount of the phenols remained in the aqueous phase was determined by measuring the absorbance of the solution before and after extraction at λ_{max} against a reagent blank [27,29]. Phenols retained on the PUF cubes were then calculated from the difference between the absorbance of the aqueous phase before (A_b) and after extraction (A_f). The sorption percentage (%E), the amount of phenol retained at equilibrium (q_e) and the distribution ratio (D) of the phenol uptake were calculated as reported [29]. Following these procedures, the influence of shaking time and temperature on retention of the target analytes by the PUF sorbents was fully studied. All experiments were carried out at least in triplicate and the results are the average of three independent measurements and the precision in most cases was ± 2%.

Results and Discussion

Solid PUF concentrates inorganic and organic substances from different media by phase distribution mechanism rather than adsorption [29,30]. The membrane-like structure of PUF, together with efficient aero- and hydrodynamic characteristics offer high concentrating ability and high flow rate compared to other solid supports [30]. The non-selective sorption characteristic of the PUF has been rendered more selective by controlling the experimental conditions. Preliminary study has shown that, on shaking plasticized TBP iron (III)-immobilized PUF individually with the aqueous solutions of phenols considerable amount of the analyte retained in the PUF in a short time. Hence, a detailed study was performed to assign the retention profile, kinetics, thermodynamics and the most probable sorption mechanism of phenols from the aqueous solutions by the proposed solid sorbent.

Retention profile of phenols from aqueous solution onto iron (III)-treated PUF

Iron (III) forms orange-red colored complexes with phenols in aqueous acidic solutions [31]. Thus, the sorption profile of aqueous solutions of phenols at different pH by iron (III)-loaded PUF was studied after 1h shaking time. After equilibrium, the amount of phenolic species in the aqueous phase was determined photometrically [31]. The %E and the D phenols sorption onto the PUF decreased markedly on increasing pH and reached maximum at pH ≈1 (Figure 1). At pH >1, the sorption of iron (III)-treated PUF towards phenols decreased markedly. This behavior is most likely attributed to the deprotonation of the phenolic species i.e. formation of polar phenolate species, absence of protonated ether oxygen ($-CH_2-O-CH_2-$) and/or urethane nitrogen (-NH-COO-) of treated iron (III) PUF, instability and/ or hydrolysis or incomplete extraction of the produced iron (III) phenolate complex and PUF solid sorbent.

The retention of phenols in strong acidic media is most likely attributed to the protonation of ether and/or urethane linkages available in PUF sorbent. This effect enhanced the retention of iron (III) phenolate complex anions on PUF via formation of a ternary ion associate with the protonated form of the PUF sorbent. Thus, in the extraction media of pH <1, the interaction of the complexed iron (III) phenolate complex speciesis easily proceeded with the protonated ether group of the PUF. Based on these data and the results reported on the retention of $AuCl_4^-$ by PUF [29], a sorption mechanism involving a weak base anion ion exchange and/or solvent extraction of iron (III) phenolate complex $[Fe(phenolate)_6]^{3-}$ by the protonated ether ($-CH_2-HO^+-CH_2-$) oxygen linkage of the PUF as a ternary complex ion associate is most likely proceeded [26-30] as follows:

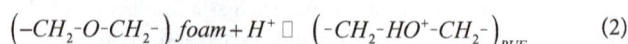

$$(-CH_2\text{-}O\text{-}CH_2\text{-})\ foam + H^+ \rightleftharpoons (-CH_2\text{-}HO^+\text{-}CH_2\text{-})_{PUF} \quad (2)$$

$$(-CH_2\text{-}HO^+\text{-}CH_2\text{-})_{PUF} + [Fe\ (phenolate)_6]^{3-}_{PUF} \rightleftharpoons [-CH_2\text{-}HO^+\text{-}CH_2-]_3 \cdot [Fe\ (phenolate)_6]^{3-}_{PUF} \quad (3)$$

The influence of the plasticizer TBP (0.01-0.05% v/v) on phenols uptake from the aqueous solutions onto the iron (III)-loaded PUF was studied. Phenols sorption onto the sorbent dramatically increased (D $=1.6 \times 10^4$mL g^{-1}) in the presence of TBP (0.01% v/v). The fact, TBP enhanced the swollen character of the sorbent and decreases the glass transition temperature of the sorbent [30]. This effect enhanced the diffusion of the species within the PUF membranes [30]. These results suggested the possible use of the TBP plasticized iron (III)-loaded PUF in packed column for enrichment and subsequent analysis of phenols

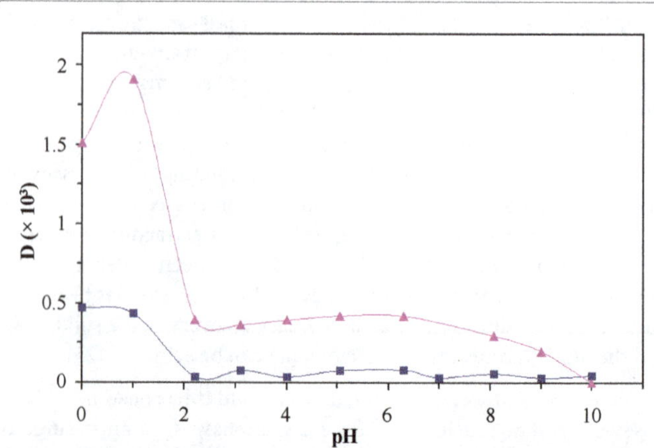

Figure 1: Influence of pH on the extraction of phenol (■) and 2-CP (▲) from aqueous solution by TBP-iron(III) immobilized PUF after 1 h shaking at 25°C.

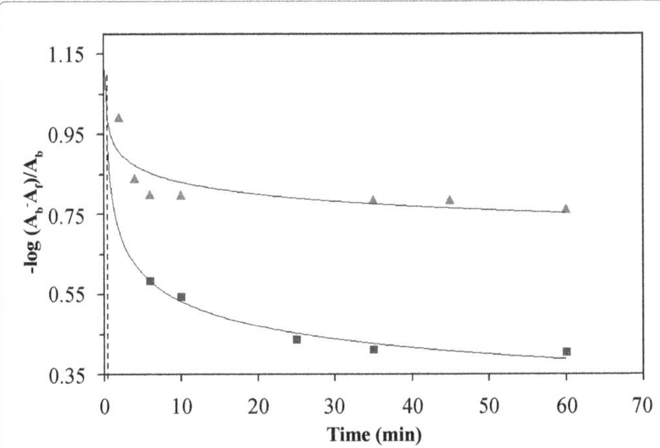

Figure 2: Rate of phenols sorption from aqueous solution onto iron (III)-loaded PUF at pH ≤ 2, phenol (■) and 2-CP (▲).

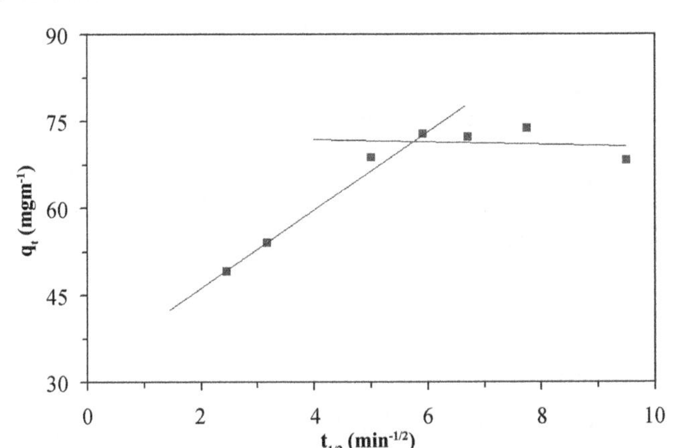

Figure 3: Weber-Morris plot for phenol sorption from aqueous solution onto the iron (III)-immobilized PUF.

from large volumes of aqueous solutions. The retained phenols species were successfully recovered with NaOH (10 ml, 0.01 mol dm^{-3}) and finally analyzed [31].

Kinetic behaviour of phenols

The influence of shaking time (0-120 min) on the uptake of the phenols from the aqueous acidic media at pH≤1 was investigated. The sorption of the phenols onto the iron (III)-immobilized PUF sorbent was fast and reached equilibrium after 60 min shaking time. This conclusion was supported by calculating the half-life time ($t_{1/2}$) of phenols sorption from the aqueous solutions onto the PUF sorbents (Figure 2). The values of $t_{1/2}$ were calculated from the plot of -log A_b-A_f/A_b *versus* time as shown in Figure 2. The values of $t_{1/2}$ were found to be1.2 ± 0.04 min for phenol and 1.05 ± 0.008 min for 2-CP. Thus, gel diffusion is not only the rate-controlling step for iron (III)-loaded PUF as in the case of common ion exchange resins [29,30] and the kinetic of phenols sorption onto the sorbent depends on film and intraparticle diffusion. Thus, the results were subjected to a number of kinetic models.

Weber-Morris model [32] can be expressed as follows:

$$q_t = R_d (t)^{1/2} + C \quad (4)$$

Where R_d is the rate constant of intraparticle transport (mg g^{-1} min$^{-1/2}$), q_t is the sorbed phenol concentration (mg g^{-1}) at time *t* and *C* (mg g^{-1}) is the intercept. The plots of q_t *versus* time were slightly linear with R^2=0.9438 and 0.9652 for Phenol and 2-CP, respectively, in the initial stage (Figure 3) up to 16 ± 1 min and 2-CP deviate on increasing the shaking time in the initial stage. The change in the slope is most likely due to the existence of different pore size [29]. The plot does not pass through the origin, thus it can be assumed that a boundary layer effect occurs at a given degree and intraparticle diffusion process is not the unique rate controlling step [33].

The results were further subjected to Lagergren model for pseudo-first order [34]:

$$log\ (q_e - q_t) = log\ q_e - (K_{Lager} / 2.303)t \quad (5)$$

Where, q_e is the amount of phenol sorbed at equilibrium per unit mass of sorbent (mg g^{-1}) and K_{Lager} is the first order overall rate constant of pseudo-first order kinetic for the retention process per min and *t* is the time in min. The values of Lagergren parameters K_{Lager} and q_e calculated from the linear plot of log (q_e-q_t) *vs.* time (Figure 4) are summarized in Table 1. Because of the large difference between the calculated and experimental values of q_e [35], the order of the sorption process changed after elapsed time and the experimental data could not be adjusted to pseudo-first order equation (4). This trend could indicate that, the first stage of the retention process (external transport of phenol from the bulk solution to sorbate surface) follows pseudo-first order kinetics while the posterior processes (film diffusion and intraparticle diffusion) do not follow the same order. Thus, the results of phenols uptake were subjected to the pseudo-second order kinetic model [35,36]. This model can be expressed by the following equation:

$$t / q_t = 1/\ (k_2 q^2_e) + (1/q_e)t \quad (6)$$

Wherek_2 (g mg^{-1} min^{-1}) is the pseudo second-order rate constant, The plots of t/q_t *versus* t for both phenols were linear (Figure 4). Kinetic data obtained gave very good fit with the pseudo-second order model as shown by the excellent correlation coefficients R^2=0.996-0.999

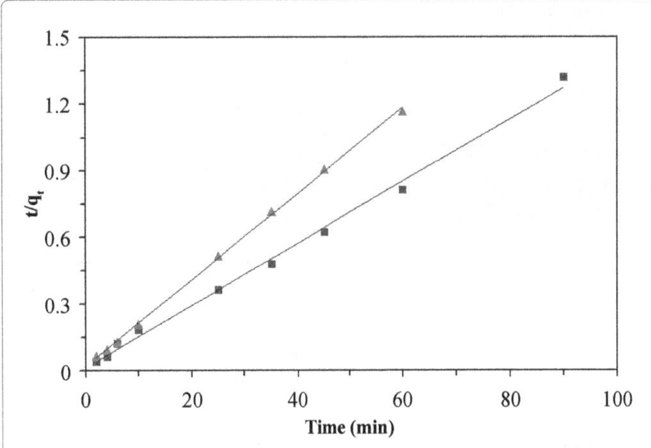

Figure 4: Second-order plot for the retention of phenol (■) and 2-CP (▲) onto the iron (III)-immobilized PUF.

Compound	q_e (mg g^{-1}) experimental	First-order kinetic model (Lagergren model)			Second-order kinetic model		
		K_1 (min^{-1})	q_e	R^2	K_2×10^{-2}g (mg.min)$^{-1}$	q_e	R^2
Phenol	73.916	0.108	55.39	0.9727	1.40	71.43	0.9955
2-CP	51.45	0.021	4.85	0.9759	1.80	51.54	0.9992

Table 1: Kinetic parameters for pseudo first-order and second-order kinetic models for sorption of phenol and 2-CP onto TBP plasticized iron (III)-immobilized PUF at 25 ± 1°C.

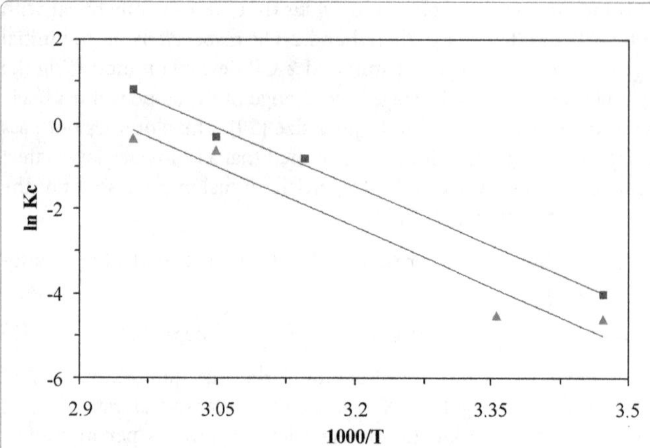

Figure 5: Plot of ln K_C *versus* 1000/T for phenol sorption from the aqueous solution onto unplasticized (■) and plasticized (▲)iron(III)-immobilized PUF.

Compound	ΔH**, Jmol^{-1}	ΔS, J mol^{-1} K^{-1}	ΔG, KJmol^{-1}	R^2
Phenol	76.75 (78.58)	233.3 (231.3)	-69.45 (-68.85)	0.9938 (0.9568)
2-CP	38.64 (47.8)	104.51 (120.8)	-31.11 (-35.95)	0.9162 (0.9682)

*The values of thermodynamic parameters using plasticized iron(Ш)-immobilized PUF are given in parentheses.

**The calculated values of ΔH from Vant·hoff equation for phenol and 2-CP retention onto unplasticized- and plasticized iron (III) immobilized PUF were found 75.55 (78.72) and 38.7 (48.33) Jmol^{-1}, respectively.

Table 2: Numerical thermodynamic parameters (ΔH, ΔS and ΔG) for the tested phenols onto unplasticized and TBP plasticized iron (III)-immobilized PUF*.

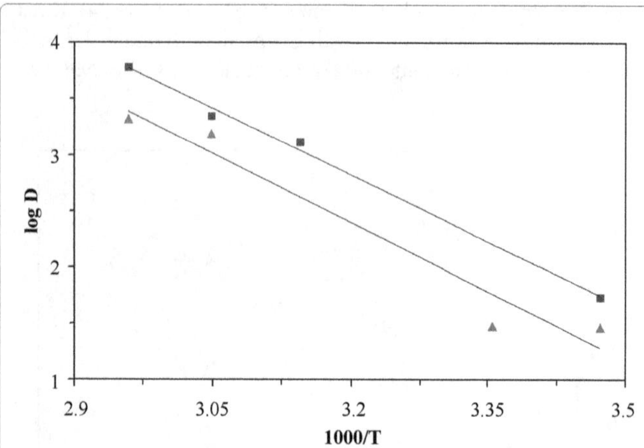

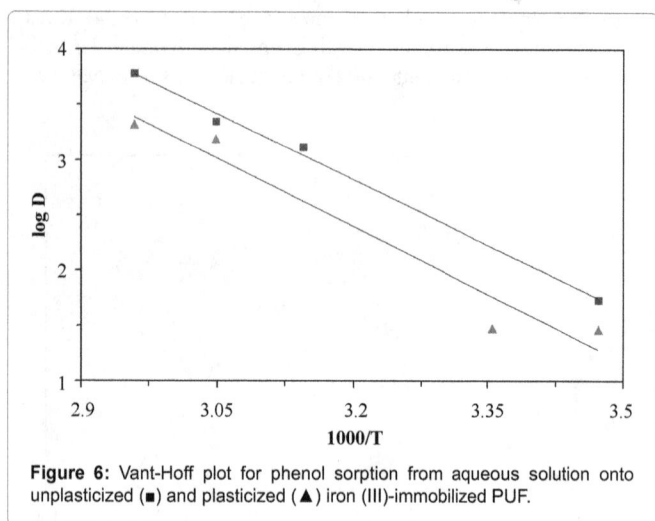

Figure 6: Vant-Hoff plot for phenol sorption from aqueous solution onto unplasticized (■) and plasticized (▲) iron (III)-immobilized PUF.

compared to Lagergren model 0.973-0.976 (Table 1). The second-order rate constant (k_2), equilibrium capacity (q_e), and the correlation coefficient (R^2) for the tested phenols are summarized in Table 1. The pseudo-second order constants (k_2) for phenol and 2-CP were 1.40 and 4.85 (g mg^{-1}.min^{-1}), respectively. Based on analyzing the kinetic data, it can be concluded that, the first step of the process (mass transport from solution until sorbent) followed pseudo-first order kinetics and at the early stage of extraction and the whole sorption process is governed by a pseudo-second-order kinetics [35,36].Thus, a chemi-sorption reaction is most likely predominate in the rate-controlling step [37,38]. The adsorption capacity calculated for phenol and 2-CP were 55.39 and 4.85 mg g^{-1}, respectively.

Thermodynamic characteristics

A thermodynamic characteristics of sorption of phenols from the aqueous solution by plasticized TBP- and un plasticized iron (III)-treated PUF was critically performed over a wide range of temperature (288-338 K). The thermodynamic parameters e.g. ΔH, ΔS and ΔG) were evaluated using the following equations [39]:

$$\log K_C = -\Delta H / 2.303RT + \Delta S / 2.303R \quad (7)$$

$$\Delta G = \Delta H - T\Delta S \quad (8)$$

$$\Delta G = -RT \ln K_c \quad (9)$$

Where ΔH, ΔS, ΔG, T and R are the enthalpy, entropy, Gibbs free energy changes, temperature in Kelvin, and the gas constant (8.314 J K^{-1} mol^{-1}), respectively. K_C is the equilibrium constant depending on the fractional attainment (F_e) of the sorption process. The K_C for retention of phenols at equilibrium onto the sorbent was calculated for each temperature employing the following equation:

$$K_C = F_e / 1 - F_e \quad (10)$$

The plot of ln K_C *versus* 1000/T for phenol retention onto plasticized and un plasticized iron (III)-immobilized PUF were linear (Figure 5) over the temperatures range (288-338 K). Similar trend was also achieved for 2-CP. The equilibrium constant increased on decreasing temperature revealing that the retention process of $[Fe(C_6H_5O)_6]^{3-}$ species onto the PUF sorbent is an endothermic process for both phenols. The numerical values of ΔH, ΔS and ΔG for phenol and 2-CP retention calculated from the slopes and intercepts of the linear plots of ln K_C against 1000/T are summarized in Table 2.

According to Van`t Hoff equation [38], the distribution coefficient (D) of the phenol retention is correlated with temperature according to the following expression:

$$\log D = -\Delta H / 2.30\ RT + C \quad (11)$$

Where, C is constant. The plots of log D *versus* 1000/T for phenol retention onto plasticized and unplasticized iron (III)-immobilized PUF were linear (Figure 6). Similar trend was also achieved from 2-CP. The computed values of ΔH for the sorption of both phenols from the slopes of the plots are summarized in Table 2. These results are quite close to the values computed from equations 7 and 8.

The positive values of ΔH and the values of D and K_C reflect the endothermic behavior of phenols retention and the non-electrostatics bonding formation between the sorbent and the sorbate. The negative values of ΔG (Table 2) imply the spontaneous and physical sorption nature of the phenols. The positive values of ΔS for both phenols may be indicative of moderate sorption step of the complex anion $[Fe(C_6H_5O)_6]^{3-}$and ordering of ionic charges with a compensatory disordering of the sorbed species onto the sorbents. Thus, the freedom of $[Fe(C_6H_5O)_6]^{3-}$motion is less restricted in the PUF membrane than in solution and the sorption process involves a decrease in the free energy. Thus, the physical structure of the PUF membrane may be changing and affecting the strength of the intermolecular interactions between PUF membrane and the $[Fe(C_6H_5O)_6]^{3-}$species. The high temperature may also make the membrane matrix become more structured and does not affect the ability of the polar segments to engage in unstable hydrogen bonding with $[Fe(C_6H_5O)_6]^{3-}$species resulting in higher extraction. The energy of urethane nitrogen and/or ether-oxygen sites of the PUF provided by raising the temperature on the phenol and 2-CP uptake by the PUF sorbent is most likely maximizes the possible interaction between the active sites of the PUF and the bulky complex anion $[Fe(C_6H_5O)_6]^{3-}$ resulting in a higher sorption related to "Solvent extraction" and an added component for "surface adsorption". Thus, the dual mode sorption mechanism of these analytes by the used PUF sorbent seems a more probable model for tested compounds.

Conclusion

Iron (III)-immobilized PUF can be used as an effective sorbent for the removal of phenols from water samples in strong acidic conditions. The formation of non-ionized species of phenols and the formation of ternary complex ion associate between the complex anion of iron (III) with phenols i.e. $[Fe(phenolate)_6]^{3-}$and the protonated PUF results in a better extraction. The retention kinetics followed a pseudo-second order model and according to Weber-Morris model, an intraparticle diffusion process participates in the adsorption rate. The amount of retained phenols increased with the increase of temperature, indicating that, the adsorption is an endothermic process. Work is still continuing for increasing the number of phenols that can be retained, separated out and simultaneously determined by the proposed method. Further studies will involve an investigation of these extractions for a wider range of organic compounds in order to obtain a clearer view of the extraction mechanism.

References

1. Martínez D, Pocurull E, Marcé RM, Borrull F, Calull M (1996) Separation of eleven priority phenols by capillary zone electrophoresis with ultraviolet detection. J Chromatogr A 734: 367-373.
2. Pizarro C, Pérez-del-Notario N, González-Sáiz JM (2007) Optimisation of a microwave-assisted extraction method for the simultaneous determination of haloanisoles and halophenols in cork stoppers. J Chromatogr A 1149: 138-144.
3. Environmental Protection Agency (1995) Phenols by Gas Chromatography: Capillary Column Technique. EPA Method 8041 pp: 1-26.
4. Basheer C, Lee HK (2004) Analysis of endocrine disrupting alkylphenols, chlorophenols and bisphenol-A using hollow fiber-protected liquid-phase microextraction coupled with injection port-derivatization gas chromatography-mass spectrometry. J Chromatogr A 1057: 163-169.
5. Gao J, Liu L, Liu X, Zhou H, Huang S, et al. (2008) Levels and spatial distribution of chlorophenols - 2,4-dichlorophenol, 2,4,6-trichlorophenol, and pentachlorophenol in surface water of China. Chemosphere 71: 1181-1187.
6. Elci L, Kolbe N, Elci SG, Anderson JT (2011) Solid phase extractive preconcentration coupled to gas chromatography-atomic emission detection for the determination of chlorophenols in water samples. Talanta 85: 551-555.
7. Padilla-Sánchez JA, Plaza-Bolaños P, Romero-González R, Barco-Bonilla N, Martínez-Vidal JL, et al. (2011) Simultaneous analysis of chlorophenols, alkylphenols, nitrophenols and cresols in wastewater effluents, using solid phase extraction and further determination by gas chromatography-tandem mass spectrometry. Talanta 85: 2397-2404.
8. Guo L, Lee HK (2012) Electro membrane extraction followed by low-density solvent based ultrasound-assisted emulsification microextraction combined with derivatization for determining chlorophenols and analysis by gas chromatography-mass spectrometry. J Chromatogr A 1243: 14-22.
9. Kovács Á, Mörtl M, Kende A (2011) Development and optimization of a method for the analysis of phenols and chlorophenols from aqueous samples by gas chromatography-mass spectrometry, after solid-phase extraction and trimethylsilylation. Microchem J 99: 125-131.
10. Ortiz-Arzate Z, Jiménez-Bravo TS, Sánchez-Flores NA, Pacheco-Malagón G, Bulbulian S, et al. (2013) Adsorption of phenol and chlorophenol mixtures on silicalite-1 determined by GC-MS method. J MexChemSoc 57: 111-117.
11. Cacho JI, Campillo N, Viñas P, Hernández-Córdoba M (2014) Stir bar sorptive extraction polar coatings for the determination of chlorophenols and chloroanisoles in wines using gas chromatography and mass spectrometry. Talanta 118: 30-36.
12. Lee J, Khalilian F, Bagheri H, Lee HK (2009) Optimization of some experimental parameters in the electro membrane extraction of chlorophenols from seawater. J Chromatogr A 1216: 7687-7693.
13. Chao YY, TuYM, Jian ZX, Wang, HW, Huang YL (2013) Direct determination of chlorophenols in water samples through ultrasound-assisted hollow fiber liquid-liquid-liquid microextraction on-line coupled with high-performance liquid chromatography. J Chromatogr A 1271: 41-49.
14. Cheng Q, Qu F, Li NB, Luo HQ (2012) Mixed hemimicelles solid-phase extraction of chlorophenols in environmental water samples with 1-hexadecyl-3-methylimidazolium bromide-coated Fe3O4 magnetic nanoparticles with high-performance liquid chromatographic analysis. Anal Chim Acta 715: 113-119.
15. Nishihama S, Hirai T, Komasawa I (2001) Review of advanced liquid-liquid extraction systems for the separation of metal ions by a combination of conversion of the metal species with chemical reaction. IndEngChem Res 40: 3085-3091.
16. Camel V (2003) Solid phase extraction of trace elements. Spectrochimica Acta Part B: Atomic Spectroscopy 58: 1177-1233.
17. Pereira MG, Arruda MAZ (2003) Trends in Preconcentration Procedures for Metal Determination Using Atomic Spectrometry Techniques. Microchim Acta 141: 115-131.
18. Mester Z, Sturgeon R (2005) Trace element speciation using solid phase microextraction. Spectrochimica Acta Part B: Atomic Spectroscopy 60: 1243-1269.
19. Davis AC, Wu P, Zhang X, Hou X, Jones BT (2006) Determination of cadmium in biological samples. ApplSpectrosc Rev 41: 35-75.
20. El-Sheikh AH, Alzawahreh AM, Sweileh JA (2011) Preparation of an efficient sorbent by washing then pyrolysis of olive wood for simultaneous solid phase extraction of chloro-phenols and nitro-phenols from water. Talanta 85: 1034-1042.
21. Kamaraj M, Salam HA, Sivaraj R Venckatesh R (2013) Detection of bisphenol-a in various environment samples collected from tamilnadu, india by solid-phase extraction and gc analysis. Advances in Bioresearch 4: 59-64.
22. Guo F, Liu Q, Shi JB, Wei FS, Jiang GB (2014) Direct analysis of eight chlorophenols in urine by large volume injection online turbulent flow solid-phase extraction liquid chromatography with multiple wavelength ultraviolet detection. Talanta 119: 396-400.
23. Bezerra MA, Arruda MAZ, Ferreira SLC (2005) Cloud point extraction as a procedure of separation and pre-concentration for metal determination using spectroanalytical techniques: A review. ApplSpectrosc Rev 40: 269-299.
24. Cerutti S, Wuilloud RG, Martinez LD (2005) Knotted reactors and their role in flow-injection on-line preconcentration systems coupled to atomic spectrometry-based detectors. ApplSpectrosc Rev 40: 71-101.
25. Thurman EM, Mills MS (1998) Solid-Phase Extraction: Principles and Practice. John Wiley and Sons, New York.
26. El-Shahawi MS, Farag AB, Mostafa MR (1994) Pre-concentration and Separation of Phenols from Water by Polyurethane Foams Separation. SciTechnol 29: 289-299.
27. El-Shahawi MS, Nassif HA (2003) Kinetics and retention characteristics of some nitrophenols onto polyurethane foams. Anal Chim Acta 487: 249-259.
28. Bashammakh AS,Bahaffi SO, Al-Shareef FM, El-Shahawi MS (2009) Development of an analytical method for trace gold in aqueous solution using polyurethane foam sorbents: kinetic and thermodynamic characteristic of gold(III) sorption. Anal Sci 25: 413-418.
29. Dabrowski A (2001) Adsorption--from theory to practice. Adv Colloid Interface Sci 93: 135-224.
30. Palágyi S, Braun T (1992) Unloaded polyether type polyurethane foams as solid extractants for trace elements. J RadioanalNucl Chemistry 163: 69-79.
31. Marczenko Z (1986) Spectrophotometric Determination of Elements. (3rd edn), EllisHorwoodChichester, UK.
32. Weber WJ, Morris JC (1964) J SanitEngDiv 90: 70-77.
33. Crini G, Peindy HN, Gimbert F (2007) Removal of C.I. basic green 4 (Malchite Green) from aqueous solutions by adsorption using cyclodextrin based adsorbent: Kinetic and equilibrium studies. Sep PurifTechnol 53: 97-110.
34. Lagergren SY (1898) On the theory of so-called adsorption of solutes. The Royal Swedish Academy of Sciences Stockholm Sweden 1-39.
35. Ho YS, McKay G (1999) Pseudo-second order model for sorption processes. Process Biochem 34: 451-465.

36. Kuleyin A (2007) Removal of phenol and 4-chlorophenol by surfactant-modified natural zeolite. J Hazard Mater 144: 307-315.

37. Ekpete OA, Jnr MH, Tarawou T (2011) Sorption kinetic study on the removal of phenol using fluted pumpkin and commercial activated carbon. International Journal of Biological and Chemical Sciences 5: 1143-1152.

38. Senturk HB,Ozdes D, Gundogdu A, Duran C, Soylak M (2009) Removal of phenol from aqueous solutions by adsorption onto organomodifiedTirebolubentonite: equilibrium, kinetic and thermodynamic study. J Hazard Mater 172: 353-362.

39. Somorjai GA (1994) Introduction to Surface Chemistry and Catalysis. John Wiley& Sons, Inc., New York.

Quantitative Determination of Amlodipine Besylate, Losartan Potassium, Valsartan and Atorvastatin Calcium by HPLC in their Pharmaceutical Formulations

Hafez HM[1]*, Abdullah AE[2], Abdelaziz LM[2] and Kamal MM[3]

[1]*Department of Pharmaceutical Science, Zagazig University, Zagazig, Egypt*

[2]*Department of Medicinal Chemistry, Faculty of Pharmacy, Zagazig University, Zagazig, Egypt*

[3]*Department of Chemistry, Nancy University, EIPICO, Egypt*

Abstract

Amlodipine besylate is a calcium channel blocker which is used in treatment of hypertension alone or in combination with other antihypertensive drugs like angiotensin-II-receptor antagonists (ARA II) group (Losartan potassium and Valsartan) or in combination with anti hyperlipidemic agent like Atorvastatin calcium. RP- HPLC method was developed for the assay of these drugs. The method was performed by reversed phase high performance liquid chromatography using a mobile phase 0.01 M ammonium acetate buffer (pH 5.5): acetonitrile with detection at 240 nm on a spherical monomeric C18 column (250 mm × 4.6 mm, 5 µm) at flow rate of 1.5 ml/min. The proposed method was validated in terms of linearity ranged between [(2-12, 10-60, 16-96, 4-24 µg/ml) corresponding levels of 20-120% w/w of the nominal analytical concentration] with linear regression equations were [{y=64.627x – 3.6383 (r= 0.9998), y=75.385x – 8.3856 (r= 0.9997), y=64.492x – 25.981 (r= 0.9998), y=70.964x – 28.505 (r= 0.9998}], accuracy [100.18 ± 1.38, 100.79 ± 0.59, 100.45 ± 0.58 and 100.8 ± 1.69%], precision [99.29, 99.33, 99.30 and 99.30%], limits of detection [0.03, 0.18, 0.15, 0.007 µg/ml] and limits of quantitation [0.1, 0.54, 0.45, 0.024 µg/ml] for Amlodipine besylate, Losartan potassium, Valsartan and Atorvastatin calcium respectively. Method validation was developed following the recommendations for analytical method validation of International Conference on Harmonization (ICH) and Food and Drug Administration (FDA) organizations.

Keywords: HPLC; Amlodipine Besylate; Losartan potassium; Valsartan; Atorvastatin calcium

Introduction

Hypertension is the "silent killer" of humans because this disease is usually asymptomatic until the damaging effects of hypertension such as coronary heart disease and stroke. *Amlodipine besylate* is chemically described as (3-Ethyl 5-methyl (4RS)-2-[(2-aminoethoxy) methyl]-4-(2-chlorophenyl)-6-methyl-1, 4-dihydropyridine-3, 5-dicarboxylate benzene sulphonate). *Amlodipine besylate* is a calcium-channel blocking agent; a dihydropyridine derivative with an intrinsically long duration of action. *Amlodipine besylate* is an anti-hypertensive and an antianginal agent in the form of the besylate salt [1-3]. *Losartan Potassium* is chemically described as (1H-Imidazole-5-methanol, 2-butyl-4-chloro-1-[[2'-(lH-tetrazol-5-yl)[1,1'-biphenyl]-4-yl]methyl]monopotassiumsalt). *Losartan Potassium* is an angiotensin II receptor (type AT1) antagonist. It is indicated for the treatment of hypertension. It is indicated to reduce the risk of stroke in patients with hypertension and left ventricular hypertrophy [1-3]. Valsartan is chemically described as N-(1-oxopentyl)-N-[[2'-(1H-tetrazol-5-yl) [1,1'-biphenyl]-4-yl] methyl]-L-valine. Valsartan is an angiotensin II receptor antagonist with actions similar to those of Losartan. It is used in the management of hypertension to reduce cardiovascular mortality in patients with left ventricular dysfunction after myocardial infarction and in the management of heart failure [1-3]. *Atorvastatin calcium* is [R-(βR,δR)]-2-(4-fluorophenyl)-β, δ-dihydroxy-5-(1-methylethyl)-3-phenyl-4-[(phenylamino)carbonyl]-1Hpyrrole-1-heptanoic acid, calcium salt (2:1) trihydrate. *Atorvastatin calcium* is a synthetic lipid-lowering agent. Atorvastatin calcium is an inhibitor of 3-hydroxy-3-methylglutaryl-coenzyme A (HMG-CoA) reductase. This enzyme catalyzes the conversion of HMG-CoA to mevalonate, an early and rate-limiting step in cholesterol biosynthesis [1-3] (Figure 1).

The preparation of new combinations of drugs in pharmaceuticals for pharmacological activity development, as well as the requirements of modern industrial-scale pharmaceutical analysis, encourages researchers to develop new and efficient methods for multi-quantification with separation procedures. High performance liquid chromatography is a dominant separation technique, especially in pharmaceutical analysis [4]. So, it is necessary to develop a validated analytical method for assay of these drugs in combination with each other in its pharmaceutical preparations. Literature review revealed that USP described RP-HPLC methods for assay of Atorvastatin calcium, Losartan potassium and Valsartan individually and ion pair HPLC for *Amlodipine besylate* [2]. BP described a RP-HPLC method for assay of *Amlodipine besylate* and potentiometric titrations for assay of Atorvastatin calcium, Losartan potassium and Valsartan [3]. Several methods have been published for simultaneous determination of studied drugs in their combinations with each other, these methods depends on different analytical technique like Spectrophotometry [4-11], Spectrofluorimetry [12-14], Capillary Electrophoresis [15-18], HPTLC and TLC [19-21] and HPLC coupled with UV detector [21-29], fluorescence detector [30] and mass spectrometer detector [31-34]. Our scope is development of a validated analytical method for assay

***Corresponding author:** Hafez HM, Bachelor degree of Pharmaceutical Science, Zagazig University, Zagazig, Egypt
E-mail: hanyhaf_1982@yahoo.com

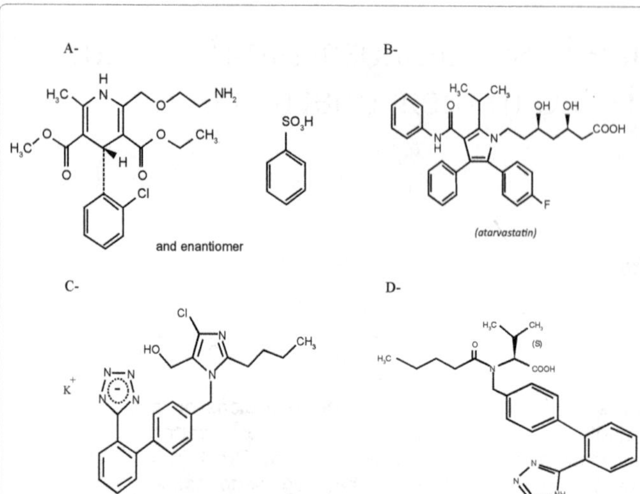

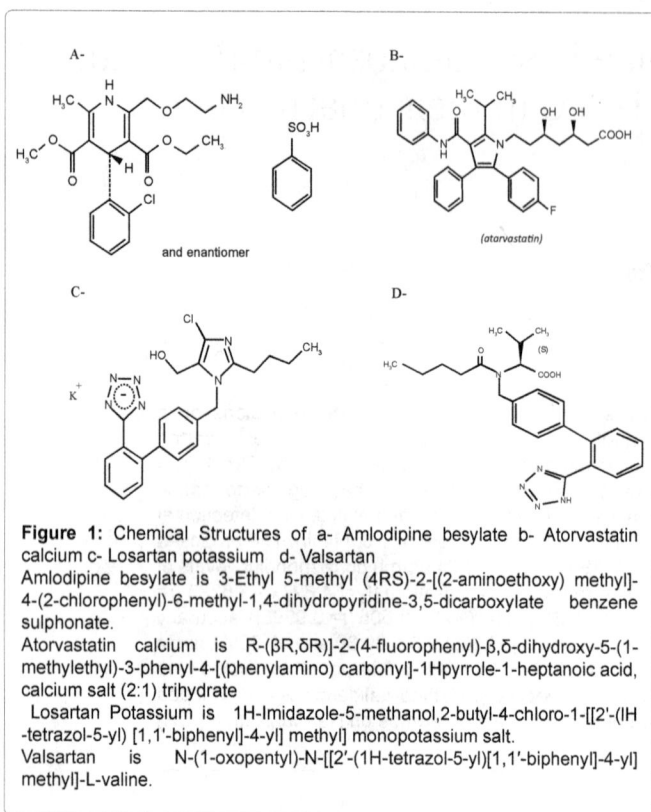

Figure 1: Chemical Structures of a- Amlodipine besylate b- Atorvastatin calcium c- Losartan potassium d- Valsartan
Amlodipine besylate is 3-Ethyl 5-methyl (4RS)-2-[(2-aminoethoxy) methyl]-4-(2-chlorophenyl)-6-methyl-1,4-dihydropyridine-3,5-dicarboxylate benzene sulphonate.
Atorvastatin calcium is R-(βR,δR)]-2-(4-fluorophenyl)-β,δ-dihydroxy-5-(1-methylethyl)-3-phenyl-4-[(phenylamino) carbonyl]-1Hpyrrole-1-heptanoic acid, calcium salt (2:1) trihydrate
Losartan Potassium is 1H-Imidazole-5-methanol,2-butyl-4-chloro-1-[[2'-(lH-tetrazol-5-yl) [1,1'-biphenyl]-4-yl] methyl] monopotassium salt.
Valsartan is N-(1-oxopentyl)-N-[[2'-(1H-tetrazol-5-yl)[1,1'-biphenyl]-4-yl] methyl]-L-valine.

of these drugs in combination with each other in its pharmaceutical preparations and it should be characterized by a simplicity, accuracy, preciseness and sensitivity.

Experimental

Instrumentation and chromatographic condition

Analysis was performed on a chromatographic system of AGILENT 1200 separation module connected to AGILENT 1200 Diode Array detector (DAD) detector. The system equipped by Agilent chemistation PC program. The chromatographic separation was achieved by injection of 50 μl of drugs sample solutions on Spherical monomeric C18 column (250×4.6 mm, 5μ) and the mobile phase is consisting of Ammonium acetate (pH 5.5, 0.01M) - acetonitrile (45:55, V/V) which pumped at a flow rate equals to 1.5 ml/min at 40°C and monitored at lambda=240 nm.

Ammonium acetate (0.01 M) was prepared by dissolving 0.77 g Ammonium acetate in approximately 950 ml distilled water. The pH was adjusted to 5.5 with glacial acetic acid. Water was added to 1000 ml. Mobile phase was filtered through a 0.45 μl Nylon membrane filter (Millipore, Milford, MA, USA) under vacuum and degassed by ultrasonication (Cole Palmer, Vernon Hills, USA) before usage.

Chemicals and reagents

All reagents used were of analytical grade or HPLC grade. Ammonium acetate, Glacial acetic acid and Sodium hydroxide (NaOH) were supplied by (Merck, Darmstadt, Germany), Acetonitrile and Methanol HPLC grade were supplied by (Fischer scientific, U.K.) and Distilled water. (Note: The water used in all the experiments was obtained from Milli-RO and Milli-Q systems (Millipore, Bedford, MA). *Amlodipine besylate* (99.8% as anhydrous), Losartan potassium (99.8% as is), Valsartan (99.3% as anhydrous) and Atorvastatin calcium (99.5% as anhydrous) working standard powders and all used excipients were kindly supplied by Egyptian international pharmaceutical industries company (EIPICO) (10th Ramadan, Egypt), and were used without further purification.

Pharmaceutical preparation

Exforge 10/160 tablets Novartis (Egypt) contain (10 mg Amlodipine as *Amlodipine besylate* and 160 mg Valsartan) per tablet B.NO: 10028/S0098. Caduet 5/10 tablets Pfizer Company (Egypt) contains (5 mg Amlodipine as *Amlodipine besylate* and 10 mg Atorvastatin as Atorvastatin calcium) per tablet B.NO: 0996099. Ator 10 tablets EIPICO (Egypt) contain (10 mg Atorvastatin as Atorvastatin calcium) per tablet B.NO: 1102999. Losarmepha tablets Mepha (Switzerland) contain 50 mg Losartan potassium per tablet B.NO:02862.

Preparation of stock standard solutions

Stock standard solutions containing (0.1, 0.5, 0.8, 0.2 mg/ml) of Amlodipine besylate, Losartan potassium, Valsartan and Atorvastatin calcium respectively were prepared by dissolving (5, 25, 40, 10 mg) of each in methanol in 50 ml volumetric flask respectively. It was then sonicated for 15 minutes and the final volume of solutions was made up to 50 ml with methanol to get stock standard solutions.

Preparation of calibration plot (working standard solutions)

To construct calibration plots, The stock standard solutions were diluted with the mobile phase to prepare working solutions in the concentration ranges (2-12, 10-60, 16-96, 4-24 μg/ml) for Amlodipine besylate, Losartan potassium, Valsartan and Atorvastatin calcium respectively. Each solution (n=5) was injected in triplicate and chromatographed under the mentioned conditions above. Linear relationships were obtained when average drug standard peak area were plotted against the corresponding concentrations for each drug. Regression equation was computed.

Sample preparation

Ten units of Exforge tablets, Caduet tablets, LosarMepha tablets and Ator tablets were prepared by grinding them to a fine, uniform size powder, triturated using mortar and pestle. After calculating the average tablet weight, amounts of powder equivalent to (10, 80, 20 and 50 mg) for *Amlodipine besylate* , Valsartan, Atorvastatin calcium and Losartan potassium of tablets were accurately weighed and transferred separately to 100 ml volumetric flasks respectively. Complete with methanol up to 100 ml. Solutions were sonicated for 15 min and the solutions were then filtered through 0.45 lm Nylon membrane filters (Millipore, Milford, MA, USA). Aliquots of appropriate volume (10 ml) were transferred to 100 ml calibrated flasks and diluted to volume with mobile phase to furnish the mentioned concentration above. The diluted solutions were analyzed under optimized chromatographic conditions and chromatogram is depicted in (Figure 2).

Results

Method validation

Selectivity: Specificity of the method was evaluated by assessing whether excipients present in the pharmaceutical formulations interfered with the analysis or not [35]. Inactive ingredients of studied tablets are Calcium Carbonate, Croscarmellose Sodium,

Microcrystalline Cellulose, Pregelatinized Starch, Polysorbate 80, Hydroxypropyl Cellulose, Colloidal Silicon Dioxide (anhydrous), Magnesium Stearate, titanium dioxide and talc. A placebo was prepared by mixing the respective excipients and solutions were prepared by

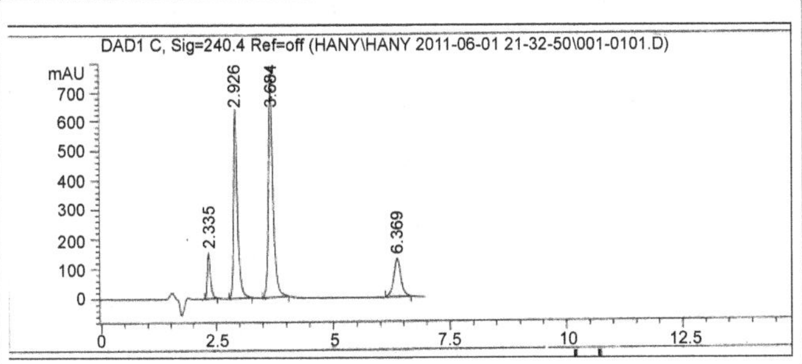

Figure 2: HPLC chromatogram of Amlodipine besylate (2.33 min), Losartan potassium(2.92 min), Valsartan (3.68 min) and Atorvastatin calcium (6.36 min) respectively on spherical monomeric C18 Column and mobile phase consisted of (A) acetonitrile and (B) acetate buffer pH=3.5 in ratio 55:45% at flow rate =1.5 ml/min by an isocratic technique.

Drug name	Average µg/ml	Average %	RSD
Amlodipine besylate	5.04	100.8	0.30%
Losartan potassium	49.60	99.20	0.70%
Valsartan	78.9	98.63	0.79%
Atorvastatin calcium	9.91	99.18	0.33%

Table 1: Repeatability of Amlodipine besylate, Losartan potassium, Valsartan and Atorvastatin calcium respectively.

Drug name	1st day µg/ml	2nd day µg/ml	3rd day µg/ml	pooled average	pooled average %	RSD
Amlodipine besylate	5.03	4.91	5.00	4.98	99.29	1.33%
Losartan potassium	49.60	49.21	50.18	49.66	99.33	0.98%
Valsartan	78.90	79.10	80.4	79.48	99.30	1.03%
Atorvastatin calcium	9.91	101.0	9.77	9.93	99.30	1.68%

Table 2: Intermediate precision Amlodipine besylate, Losartan potassium, Valsartan and Atorvastatin calcium respectively.

Drug name	Recovery at 80% conc. (%)	Recovery at 100% conc. (%)	Recovery at 120% conc. (%)	Average Recovery (%)	RSD
Amlodipine besylate	101.57	98.80	100.17	100.18	1.38%
Losartan potassium	99.91	99.68	99.87	100.79	0.59%
Valsartan	100.25	100.00	101.10	100.45	0.58%
Atorvastatin calcium	98.37	101.61	99.13	100.8	1.69%

Table 3: Recovery results for standard solution plus excipients for Amlodipine besylate, Losartan potassium, Valsartan and Atorvastatin calcium respectively.

following the procedure described in the section of sample preparation. The commonly used tablet excipients did not interfere with the method. The diluent chromatogram shows that the tablet diluent has negligible contribution after the void volume at the method detection wavelength of 240 nm.

Linearity and range: The linearity of the method was evaluated by analyzing different concentration of the drugs. According to ICH recommendations [35] at least five concentrations must be used. In this study five Concentrations were chosen, in the ranges (2-12, 10-60, 16-96, 4-24 µg/ml) corresponding levels of 20-120% w/w of the nominal analytical concentration for Amlodipine besylate, Losartan potassium, Valsartan and Atorvastatin calcium respectively. The linearity of peak area responses versus concentrations was demonstrated by linear least square regression analysis. The linear regression equations were {y=64.627x – 3.6383 (r=0.9998), y=75.385x – 8.3856 (r=0.9997), y=64.492x – 25.981 (r=0.9998), y=70.964x – 28.505(r=0.9998} for Amlodipine besylate, Losartan potassium, Valsartan and Atorvastatin calcium respectively. Where Y is the peak area of standard solution and X is the drug concentration.

Precision: According to ICH recommendations [35], The precision of the assay was investigated by measurement of both repeatability and Intermediate precision.

Repeatability: Repeatability was investigated by injecting 6 determinations at 100% of the test concentration percentage RSD were calculated in Table 1.

Intermediate precision: In the inter-day studies, standard and sample solutions prepared as described above, were analyzed in triplicate on three consecutive days at 100% of the test concentration and percentage RSD were calculated (Table 2).

Accuracy: According to ICH recommendations [35], Accuracy was assessed using 9 determinations over 3 concentration levels covering the specified range (80,100 and 120%). Accuracy was reported as percent recovery by the assay of known added amount of analyte in the sample (as standard addition method) (Table 3).

Limits of detection and Limits of quantitation: According to the ICH recommendations [35], Determination of limits of detection and quantitation was based on the standard deviation of the y-intercepts of regression lines and the slope of the calibration plots (Table 4).

Robustness: Robustness of an analytical procedure is a measure of its capacity to remain unaffected by small variations in method parameters and provides an indication of its reliability during normal usage [35]. Robustness was tested by studying the effect of changing mobile phase pH by ± 0.1, the amount of acetonitrile in the mobile phase by ± 2%, temperature ± 2°C, different column and flow rate ± 0.05 ml/min had no significant effect on the chromatographic resolution of the method.

Stability of analytical solution: Also as part of evaluation of robustness, solution stability was evaluated by monitoring the peak area response. Standard stock solutions in methanol were analyzed right after its preparation 1, 2 and 3 days after at 5°C and for a day at room temperature. The change in standard solution peak area response over 3 days was (1.67, 1.03, 1.10 and 1.56%) for Amlodipine Besylate, Losartan potassium, Valsartan and Atorvastatin calcium respectively. Their solutions were found to be stable for 3 days at 5°C and for a day at room temperature at least.

Item	Amlodipine besylate	Losartan potassium	Valsartan	Atorvastatin calcium
Linear range (µg/ml)	2-12	10-60	16-96	4-24
Detection limit (µg/ml)	0.03	0.18	0.15	0.007
Quantitation limit (µg/ml)	0.1	0.54	0.45	0.024
Regression data				
N	5	5	5	5
Slope (b)	64.62	75.39	64.49	70.96
Standard deviation of the slope	0.71	0.1	0.065	0.69
Intercept (a)	6.0	8.39	30.19	33.99
Standard deviation of the intercept	0.62	4.09	2.95	0.17
Correlation coefficient®	0.9998	0.9997	0.9998	0.9998
Standard error of regression	0.07	0.40	0.47	0.17

(Y = a + bC, where C is the concentration of the compound (µg/ml) and Y is the drug peak area

Table 4: Calibration data was resulted from method validation of Amlodipine besylate, Losartan potassium, Valsartan and Atorvastatin calcium respectively.

Drug name	Resolution	HETP	Capacity Factor	Tailing Factor
Amlodipine besylate	-	6070	1.12	0.65
Losartan potassium	4.61	7184	1.66	0.75
Valsartan	4.92	7534	2.35	0.8
Atorvastatin calcium	12.45	9564	4.79	0.95

Table 5: Chromatographic parameters for Amlodipine besylate, Losartan potassium, Vasartan and Atorvastatin calcium respectively.

Drug name		Recovery ± SD		Calculated t- values	Calculated F- values
		Proposed methods	Reference method		
Exforge®	Aml. (%)	95.81 ± 0.83	95.79 ± 0.65 [26]	0.06	1.64
	Val. (%)	97.46 ± 0.49	97.09 ± 0.51[26]	1.04	0.92
LosarMepha® Ator®	Los. (%)	101.15 ± 0.95	101.10 ± 0.45 [21]	0.11	4.54
	Ator. (%)	99.98 ± 1.35	100.65 ± 1.34 [21]	0.91	1.02
Caduet®	Aml. (%)	96.70 ± 0.61	96.68 ± 0.63 [22]	0.05	0.92
	Ator. (%)	96.41 ± 0.26	96.93 ± 0.36 [22]	2.26	0.54

(Where the Tabulated t-values and F-ratios at p = 0.05 are 2.57 and 5.05)
[21,22,26] are the reference numbers of reported method used in the comparison

Table 6: Statistical comparison of the proposed and published methods for determination of Amlodipine besylate, Losartan potassium, Valsartan and Atorvastatin calcium respectively in their dosage forms by reported method (T-student test) and (F–test for variance).

Discussion

Optimization of chromatographic condition

To establish and validate an accurate method for analysis of these drugs in pharmaceutical formulations, preliminary tests were performed with the objective of selecting optimum conditions. To reach our goal, BDS Hypersil column (25 cm) and Spherical monomeric C18 column (250 mm × 4.6 mm, 5 µm) were tried for simultaneous determination of the drugs. Spherical monomeric C18 column (25 cm) is less hydrophobic stationary phase so, it gave good separation between these drugs but drugs eluted very slowly especially Atorvastatin.

Acetonitrile was better than methanol in separation between all drugs, at higher percent (75% and 65%), interference between drugs and bad resolution obtained. At lower percent of acetonitrile (45% and 35%), atorvastatin eluted very lately. 55% of organic modifier is the most appropriate one. The effect of pH of aqueous mobile phase composition was also studied. Where at lower pH=3.5, all drugs except amlodipine eluted at higher retention time. After pH had been raised to 5.5 [greater than pka of ionizable drugs like Atorvastatin (4.5) and Valsartan (4.7)], they eluted faster due to containing of carboxylic acid group and Losartan (pKa=3.15) eluted faster due to presence of ionizable tetrazole group. Good separation between these drugs was achieved at pH=5.5. The optimum wavelength for detection was 240 nm at which much better detector responses for four drugs were obtained. The best resolution with reasonable retention time was obtained at 45% ammonium acetate and 55% acetonitrile as organic modifier. A major reason for using a concentration of 10 mM was achieving maximum sensitivity of UV detection at low wavelengths. After all previous trial had done, the most appropriated chromatographic condition was consisted of a mobile phase 0.01 M ammonium acetate buffer (pH 5.5): acetonitrile with detection at 240 nm on a spherical monomeric C18 column (250 mm × 4.6 mm, 5 µm) at flow rate of 1.5 ml/min. According to USP [2], system suitability tests are an integral part of an LC method. System suitability tests are used to verify that the Capacity Factor, Selectivity Factor (Resolution), Tailing Factor and Reproducibility of the chromatographic system are adequate for the analysis. System suitability tests were carried out on freshly prepared standard stock solutions of all drugs. The system was found to be suitable as shown in Table 5.

Application on pharmaceutical Preparation

The proposed methods were successfully used to determine Amlodipine besylate, Losartan potassium, Valsartan and Atorvastatin calcium respectively in their dosage forms e.g. Exforge tablets, LosarMepha tablets, Ator tablet and Caduet tablets respectively. Five replicate determinations were performed. Satisfactory results were obtained for each compound in good agreement with label claims. The results obtained were compared statistically with those from reported methods [21,22,26] by using Student's t-test and the variance ratio F-test. The results showed that the t and F values were smaller than the critical values. So, there were no significant differences between the results obtained from this method and published methods (Table 6).

Conclusion

A simple, accurate, precise, robust and reliable LC method has been established for simultaneous determination for Amlodipine besylate, Losartan potassium, Valsartan and Atorvastatin calcium in their formulations. The method has several advantages:

The first is using HPLC-UV which is the most available instrument in pharmaceutical analysis in companies with using of simple reversed phase mobile phase to lengthen column lifetime. High sensitive method has LOD range (0.007-0.18) µg/ml and LOQ range (0.024-0.54) µg/ml. Fast analysis of the four drugs had obtained, run time is less than 7 minutes, in addition to simplicity of sample preparation and extraction.

It is suitable for analysis of antihypertensive agents in their formulations in a single run, in contrast with previous methods. This makes the method suitable for routine analysis in quality-control laboratories.

References

1. Ahmed-Jushuf IH, Ah See KW, Allison SP, Badminton MN, Baglin TP et al. (2009) British National Formulary (57thedn) BMJ Group and RPS Publishing , London, UK.
2. http://www.healthwise.org/
3. http://www.pharmacopoeia.co.uk/
4. Alnajjar AO (2012) Simultaneous Determination of Amlodipine and Atorvastatin in Tablet Formulations and Plasma using Capillary Electrophoresis. LCGC Europe 25.
5. Bada PK and Sahu PK (2011) Simple Spectrophotometric Methods for Simultaneous Determination of Losartan Potassium and Atorvastatin Calcium in Combined Dosage Forms, J Chem Pharma sci 4:127-131.
6. Mohamed NG (2011) Simultaneous determination of amlodipine and valsartan. Anal Chem Insights 6: 53-59.
7. Mohamed NG (2011) Simultaneous determination of amlodipine and valsartan. Anal Chem Insights 6: 53-59.
8. Rakesh SU, Patil PR, Dhabale PN and Burade KB (2009) New spectrophotometric method applied to the simultaneous estimation of Losartan potassium and amlodipine besylate in tablet dosage form. J Pharm Res 2: 1252-1255.
9. Gupta KR, Mahapatra AD, Wadodkar AR, Wadodkar SG (2010) Simultaneous UV Spectrophotometric Determination of Valsartan and Amlodipine in Tablet. Int J Chem Res 2: 551-556.
10. Muthu AK, Gupta TR, Sharma S, Smith AA, Manavalan R and Kannappan N (2008) Simultaneous Estimation of Amlodipine and Atorvastatin in Tablets Using Orthogonal Function Ratio Spectrometry. Int J Chem Sci 6: 2233-2241.

11. Nagavalli D, Vaidhyalingam V, Santha A, Sankar AS, Divya O (2010) Simultaneous spectrophotometric determination of losartan potassium, amlodipine besilate and hydrochlorothiazide in pharmaceuticals by chemometric methods. Acta Pharm 60: 141-152.

12. Ramesh D, Ramakrishna S (2010) New Spectrophotometric Methods for Simultaneous Determination of Amlodipine Besylate and Atorvastatin Calcium In Tablet Dosage Forms. Int J Pharm Pharmal Sci 2: 215-219.

13. Cagigal E, González L, Alonso RM, Jiménez RM (2001) Experimental design methodologies to optimise the spectrofluorimetric determination of Losartan and Valsartan in human urine. Talanta 54: 1121-1133.

14. Shaalan RA, Belal TS (2010) Simultaneous spectrofluorimetric determination of amlodipine besylate and valsartan in their combined tablets. Drug Test Anal 2: 489-493.

15. Moussa BA, El-Zaher AA, Mahrouse MA, Ahmed MS (2013) Simultaneous determination of amlodipine besylate and atorvastatin calcium in binary mixture by spectrofluorimetry and HPLC coupled with fluorescence detection. Anal Chem Insights 8: 107-115.

16. Hillaert S, Van den Bossche W (2002) Optimization and validation of a capillary zone electrophoretic method for the analysis of several angiotensin-II-receptor antagonists. J Chromatogr A 979: 323-333.

17. Alnajjar AO (2011) Validation of a capillary electrophoresis method for the simultaneous determination of amlodipine besylate and valsartan in pharmaceuticals and human plasma. J AOAC Int 94: 498-502.

18. Hefnawy MM, Sultan M and Al-Johar H (2009) Development of Capillary Electrophoresis Technique for Simultaneous Measurement of Amlodipine and Atorvastatin from Their Combination Drug Formulations, J Liq Chromatogr. Rel Tech 32: 2923-2942.

19. Hillaert S, De Beer TR, De Beer JO, Van den Bossche W (2003) Optimization and validation of a micellar electrokinetic chromatographic method for the analysis of several angiotensin-II-receptor antagonists. J Chromatogr A 984: 135-146.

20. Santhana Lakshmi K1, Lakshmi S (2012) Simultaneous Analysis of Losartan Potassium, Amlodipine Besylate, and Hydrochlorothiazide in Bulk and in Tablets by High-Performance Thin Layer Chromatography with UV-Absorption Densitometry. J Anal Methods Chem 2012: 108281.

21. Ramadan A, Al-Akraa H, Maktabi M (2014) TLC Simultaneous Determination of Amlodipine, Atorvastatin, Rosuvastatin and Valsartan in Pure Form and In Tablets Using Phenyl-Modified Aleppo Bentonite, Int J Pharm Pharm Sci 6: 180-188

22. Panchal J and Suhagia BN (2010) Simultaneous analysis of Atorvastatin calcium and Losartan potassium in tablet dosage forms by RP-HPLC and HPTLC. Acta Chromatogr Chem 22: 173-187.

23. Abdallaha OM, Badawey AM (2011) Derivative- Ratio Spectrophotometric, Chemo metric and HPLC Validated methods for Simultaneous Determination of Amlodipine and Atorvastatin in Combined Dosage Form, Int J Ind Chem 2: 78-85.

24. Shankar GG, Madhusudana K, Sai Shalini N and Sistla R (2012) Simultaneous Determination of Losartan and Atorvastatin in Rat Plasma and Its Application to Pharmacokinetic Study. Int J Pharm 2: 260-270.

25. Bhatia NM, Gurav SB, Jadhav SD, Bhatia MS (2012) RP-HPLC Method for Simultaneous Estimation of Atorvastatin Calcium, Losartan Potassium, Atenolol, and Aspirin from Tablet Dosage Form and Plasma. J Liq Chrom Rel Tech 35: 428-443.

26. Patil PR, Rakesh SU, Dhabale PN and Burade KB (2009) RP- HPLC Method for Simultaneous Estimation of Losartan potassium and Amlodipine besylate in Tablet Formulation. Int J Chem Res 1: 464-469.

27. Chitlange SS, Bagri K and Sakarkar DM (2008) Stability Indicating RP- HPLC Method for Simultaneous Estimation of Valsartan and Amlodipine in Capsule Formulation, Asian J Res Chem 1: 15-18.

28. Patel SB, Chaudhari BG, Buch MK, Patel AB (2009) Stability indicating RP-HPLC method for simultaneous determination of valsartan and amlodipine from their combination drug product. Int J Chem Res 1: 1257-1267.

29. Brunetto MR, Contreras Y, Clavijo S, Torres D, Delgado Y et al. (2009) Determination of losartan, telmisartan, and valsartan by direct injection of human urine into a column-switching liquid chromatographic system with fluorescence detection. J Pharm Biomed Anal 50: 194-199.

30. Çelebier M, Kaynak M S, Altinöz S and Sahin S (2010) HPLC method development for the simultaneous analysis of Amlodipine and Valsartan in combined dosage forms and in vitro dissolution studies , Brazilian J Pharm Sci l: 761-768.

31. Phani RS, Prasad KR, Mallu UR (2013) High Resolution RP-HPLC Method for the Determination of Hypertensive Drug Products. Int J Pharm Bio Sci 4: 440-454.

32. Pilli NR, Inamadugu JK, Mullangi R, Karra VK, Vaidya JR, et al. (2011) Simultaneous determination of atorvastatin, amlodipine, ramipril and benazepril in human plasma by LC-MS/MS and its application to a human pharmacokinetic study. Biomed Chromatogr 25: 439-449.

33. Ramani AV, Sengupta P, Mullangi R (2009) Development and validation of a highly sensitive and robust LC-ESI-MS/MS method for simultaneous quantitation of simvastatin acid, amlodipine and valsartan in human plasma: application to a clinical pharmacokinetic study. Biomed Chromatogr 23: 615-622.

34. Gadepalli SG, Deme P, Kuncha M, Sistla R (2014) Simultaneous determination of amlodipine, Valsartan and hydrochlorothiazide by LC–ESI-MS/MS and its application to pharmacokinetics in rats, J pharm anal.

35. Yacoub M, Awwad AA, Alawi M, Arafat T (2013) Simultaneous determination of amlodipine and atorvastatin with its metabolites; ortho and para hydroxy atorvastatin; in human plasma by LC-MS/MS. J Chromatogr B Analyt Technol Biomed Life Sci 917-918: 36-47.

36. ICH Harmonized Tripartite Guideline, Validation of Analytical Procedures: Text and Methodology.

The Application of Portable GC-MS on the Petrochemical Wharf

Zhao Yun*

China Waterborne Transport Research Institute, China

Abstract

It is very important to quickly identify the liquid chemicals in the petrochemical wharf when they volatile in the air. Since most of the liquid chemicals transported in the petrochemical wharf are harmful to people, and they spread very fast in the air. If the exact kind of liquid chemicals is undefined or is wrongly defined, it will lead to a big mistake in detection result. The mass spectrum is very effective in determining the gas kinds, but its big size and complex sampling mode are not convenient for quick detection. To resolve this problem, a method based on portable GC_MS for VOCs determination is developed. The VOCs are directly extracted by solid phase micro extraction head for 120 seconds. After that, they are analysed on the portable GC-MS, determined by retention time and MS peaks, quantified by external standard method. It only takes 6.4 min from sample extracting to get the result. And its linear correlation is larger than 0.98, and the minimum detectable concentration is lower than half of their occupational exposure limits. In the simulation test, the relative error is lower than 4.4%. This method can directly sample in the air, rarely depends on external conditions. At the same time, it can determine and quantify multiple kinds of VOCs at a time. So, it will be very suitable for the VOCs emergency detection in the leakage scene.

Keywords: Portable GC-MS; Volatile organic compounds; Quick detection; Direct sample; Solid phase micro extraction; Petrochemical wharf

Introduction

There are many kinds of dangerous chemicals in the petrochemical wharf area. And most of them are volatile organic compounds(VOCs). The pollution accidents have occurred many times in the petrochemical wharf area [1,2]. The VOCs diffuses very fast in the air, and would pollute water and soil on a large scale. Also, there are many kinds of VOCs, and their toxicities are very complicated. Once the pollution accidents occurred, the people nearby would be in danger [3,4]. The identification of VOCs is not difficult, but there are few detecting method to quickly quantify the concentration of VOCs when they are identified. So, it is necessary to develop a quick detecting method to identify and quantify the concentration of VOCs in the petrochemical wharf area. The portable gas chromatograph-mass spectrometer (portable GC_MS) is quite small, and convenient to carry out. So, it is very suitable for quick detection on the scene. When it is used on the scene, the sampling method is very important. Solid phase micro extraction (SPME) is a sampling method without solvents. It can combine the sampling, concentration and injection [5-7]. Especially, the SPME could direct sample in the air, so it will save much time.

Experimental

Instruments and reagents

Portable GC_MS, gas distributing device (MF-3C), bladder tank (5 L), SPME sampling head, micro syringe (10 μL,100 μL), standard substance (benzene, toluene, paraxylene, aniline, trichloroethylene, tetrachloroethylene, cyclohexanone).

Analysis conditions

chromatographic condition: *Injection temperature:* 270°C temperature progress: hold 40°C for 10 s, then heat up at a speed of 2°C/s to 280°C, then hold for 20 s, flow speed: 0.15 mL/min.

Ionization source: EI, energy level: 70 eV, scan mode: Full scan, scan range: 45-500 m/z.

Sampling

The air sample is directly taken by the SPME sampling head for 120 s. Then it is injected into the portable GC_MS without further treatment.

Results and Discussion

The concentrations of standard series of VOCs

The most important purpose of quick detection is to evaluate whether the pollutant concentration will cause health damage and to determine the pollution scope. Here we could refer to the occupational exposure limits of each VOCs, set by national standard. When the pollutant concentrations in the field are higher than their occupational exposure limits, it should be identified as contaminated area. This requires that the minimum detectable concentration of each chemical hazardous agent must be lower than its occupational exposure limit. According to the <Occupational exposure limits for hazardous agents in the work place: Chemical hazardous agents> (GBZ 2.1-2007) [8], the occupational exposure limits of benzene, toluene, paraxylene, aniline, trichloroethylene, tetrachloro ethylene and cyclohexanone are 6, 50, 50, 3, 30, 200 and 50 mg/m^3, respectively. Based on this, we set the concentrations of standard series gases as 0.5, 1, 2, 4, and 8 times of their occupational exposure limits. The specific concentrations are shown in Table 1.

The sampling time

The more gases the SPME sample hand collected, the higher precision will be obtained. correspondingly, the sampling time needed will be longer. and if the sampling time is too long, it will not match the demand of the quick detection. So, a balance between the sampling time and detecting precision will be needed, at which, the analysis result could be obtained quickly and the analysis precision is accuracy enough. The analysis precisions at different sampling time are compared

***Corresponding author:** Zhao Yun, China Waterborne Transport Research Institute, China, E-mail: zhaoyun@wti.ac.cn

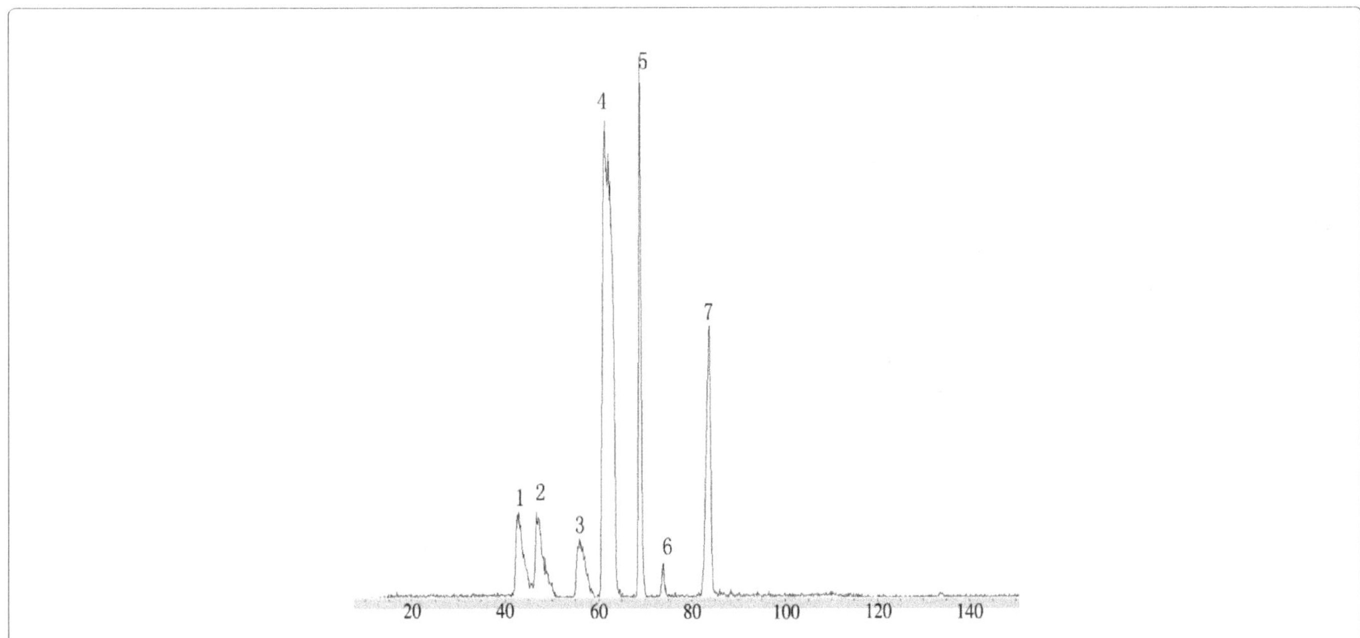

Figure 1: The separation effect of 7 kinds of VOCs (1. Benzene; 2. trichloroethylene; 3. Toluene; 4. Tetrachloroethylene; 5. Paraxylene; 6. Cyclohexanone; 7. Aniline).

Standard series	Benzene (mg/m^3)	Toluene (mg/m^3)	Paraxylene (mg/m^3)	Aniline (mg/m^3)	Trichloroethylene (mg/m^3)	Tetrachloroethylene (mg/m^3)	Cyclohexanone (mg/m^3)
Series 1	3	25	25	1.5	15	100	25
Series 2	6	50	50	3	30	200	50
Series 3	12	100	100	6	60	400	100
Series 4	24	200	200	12	120	800	200
Series 5	48	400	400	24	240	1600	400

Table 1: The concentration of standard series gases.

Compounds	Retention Time (s)	Mass Peak (m/z)	Linear Equation	Correlation Coefficient	Minimum Detectable Concentration (mg/m^3)
Benzene	43.0	77	y=4.369x-10.66	0.994	1.1
Toluene	61.4	91	y=0.891x-6.833	0.995	21
Paraxylene	68.9	104	y=2.931x-7.541	0.993	23
Aniline	83.4	91	y=4.513x-1.375	0.986	0.6
Trichloroethylene	46.1	129,131	y=10.86x-10.62	0.996	12
Tetrachloroethylene	61.2	160,162,164	y=0.099x-6.041	0.996	26
Cyclohexanone	74.0	98	y=0.262x+1.375	0.994	20

Table 2: The linear correlations and minimum detection concentrations of 7 kinds of VOCs by portable GC-MS.

using the detection of minimum concentration in the standard series gases. When the sampling time is 120 s, the analysis precision of every gas is good, and 120 s is short enough. So, the sampling time is set as 120 s. The separating effect is shown in Figure 1.

The standard curve

The standard mix gases of benzene, toluene, paraxylene, aniline, trichloroethylene, tetrachloroethylene and cyclohexanone are prepared with pure air. The concentrations are 48, 400, 400, 24, 240, 1600, 400 mg/m^3 respectively. And it is diluted into the series concentrations shown in Table 1. Then the series gases are analysed on portable GC_MS. All the gases are determined by the mass spectrum peak and retention time, and the standard curve are plotted through the peak area. Also, the equations of linear regression and correlation coefficients are obtained. The minimum detectable concentrations are calculated through 3 times noise signals. The equations of linear regression, correlation coefficients and the minimum detectable concentrations are shown in Table 2.

Discussion

A new quick detection method for VOCs based on portable GC_MS and SPME head was developed. And a mixture of 7 kinds of VOCs was analysed by this method. This detection method took only 6.4 mins to get the analysis result. And the SPME head was used in the sampling process, and it provided convenient sampling procedure. This combination of portable GC_MS and SPME head would make the detection of VOCs more convenient and fast. So, it would be very suitable for the VOCs quick detection in the petrochemical wharf area. And it is helpful to determine the pollution scope and improve the worker's occupational health for reduce the damage from VOCs [9-13].

References

1. Xia Q, Liu ZH, Qi WQ (2013) Application of on-site monitoring instruments in environmental emergency monitoring. Modern Scientific Instruments, pp: 30-32.
2. Tian-Feng LV, Xu XY, Liang X (2009) Application of portable GC- MS for determination of volatile organic compounds in emergency monitoring. Journal of Instrumental Analysis 28: 116-119.

3. Zou YD (2007) The Superiority on portable GC- MS in organic air emergency monitoring. Jiangsu Environmental Science and Technology 20: 44-46.

4. Zou YD (2008) The application of portable GC-MS in the environmental pollution accident emergency response. Science & Technology Information, pp: 369-370.

5. Ding H (2012) Portable GC-MS in traffic ambient air VOCs monitoring application. Chemical Engineering & Equipment, pp: 206-208.

6. Shen LJ (2013) Introduction to environmental emergency monitoring method. Environmental Study and Monitoring 26: 16-18.

7. Guan S (2012) Instruction of portable gas chromatograph and on the progress of its application to emergency monitoring in environmental pollution accidents. Physical Testing and Chemical Analysis Part B (Chemical Analysis) 48: 995-999.

8. Yao CH, Jia LM, Wei QB (2013) Determination of volatile organic compounds in soil by liquid leachinig and headspace and portable gas chromatograph-mass spectrometer. Chemical Engineer 216: 19-21.

9. Yu JF, Zhou YY, Gu QB (2011) Determination of BTEX in Soil by Head Space Portable Gas Chromatography. Environmental Monitoring in China 27: 28-31.

10. Yang CP, Lian JJ, Tan PH (2006) Rapid quantification of volatile organic compounds in environmental pollution accidents. Chemical Analysis and Meterage 15: 35-37.

11. Hu WL, Liu WY (1999) A novel solid phase extraction method-Solid phase microextraction. Progress in Pharmaceutical Sciences 23: 257-260.

12. Yuan HL, Gao ST, Han SK (2002) The progress in indoor air sampling method for volatile organic compounds. Environmental Pollution & Control 24: 297-299.

13. Occupational exposure limits for hazardous agents in the workplace. Chemical hazardous agents.

Rapid Peptide in-Solution Isoelectric Focusing Fractionation for Deep Proteome Analysis

Mastrobuoni G, Zasada C, Bindel F, Aeberhard L and Kempa S*

Integrative proteomics and metabolomics, Berlin Institute for Medical Systems Biology at the Max Delbrück Center for Molecular Medicine, Robert-Roessle-Strasse 10, 13125 Berlin, Germany

Abstract

The interplay between resolution, accuracy, sensitivity and speed of the mass spectrometer, as well as the complexity of the peptide mixture in relation to chromatographic separation and the analysis time finally determines the number of identified proteins within a 'shotgun' proteomic study. The improvement of one of these parameters can enhance the quality of the proteome analysis. Here we evaluated the technique of in-solution isoelectric focusing (IEF) for pre-fractionation of tryptic peptides prior to LC-MS/MS-based proteomics analysis. In-solution IEF turned out to be a simple and fast method for peptide fractionation prior to LC-MS analysis. By adapting the experimental procedures, this approach enabled identification of more than 44,000 peptides belonging to 5,800 proteins in less than 48 working hours, from protein extraction until the end of LC-MS analysis.

This technique was applied successfully to analyze the proteomes of mammalian cells and different model organisms, without additional efforts or special technical equipment. The in-solution IEF of peptides is very robust and can be applied in combination with different extraction procedures. The high number of identified peptides using a standard LC-MS system led to average protein coverage of 25%. Such a high average number of identified peptides per protein improved the discrimination of protein species as isoforms or splice variants. Thus, in solution IEF is a fast and robust alternative to gel-based proteomics or other gel free fractionation techniques upstream to LC-MS/MS analysis. The reduction of processing time and the high performance of this technique can speed up deep proteomics analyses significantly.

Keywords: LC-MS/MS; isoelectric focusing; peptide fractionation; shotgun proteomics; high resolution mass spectrometry; LTQ-Velos Orbitrap; SILAC; Microrotofor

Introduction

Technical advancements of mass spectrometry instrumentation are continuously improving the sensitivity, resolution, accuracy and the dynamic range of MS-based proteomics analyses. Thus, the number of identified peptides, and subsequently proteins, determined by the untargeted "shotgun" proteomics approach constantly increases [1,2].

The identification of tens of thousands of peptides and thousands of proteins can be routinely achieved with an Orbitrap mass spectrometer coupled with nano High Performance Liquid Chromatograpy (nHPLC) within a few hours of analysis time. Recently, using an exceptional chromatographic setup, including ultra-high performance liquid chromatography and ultra-low nano flow rates, Thakur et al. reported the identification of more than 4,500 protein and 30,000 peptides from human cell line in a single 8h chromatographic run [3].

However, the complexity and dynamic range of the protein sample, especially of higher organisms, is a challenging task and thus only a fraction of the proteome can be detected and identified [4,5]. The most abundant proteins are always identified in shotgun proteomic experiments; lower abundance proteins are often not reproducibly identified and quantified from experiment to experiment.

One possibility to tackle this issue and to increase protein coverage is fractionating the sample before protein digestion. This can be achieved by organelle fractionation and other protein fractionation methods. A common approach is the so called GeLC-MS/MS approach [6,7], where the protein sample is loaded on a Sodium Dodecyl Suphate-Polyacrilamide Gel Electrophoresis (SDS-PAGE), the lane is cut in several slices, subsequently proteins are *in gel* digested [8] and the resulting peptide mixtures are analyzed on the Liquid Chromatography-Mass Spectrometry (LC-MS) system.

The advantage of the GeLC-MS/MS method is that the information about the relative protein molecular weight for each identified protein is retained; the drawbacks are that only limited amount of protein can be separated, the low yield of the digestion process and the susceptibility to contaminations, beside the laboriousness of the protocol.

An alternative solution is fractionating the peptide mixture after protein digestion, in order to reduce the complexity of the sample prior to LC-MS/MS analysis. A possible approach makes use of the multidimensional protein identification technology (MudPIT) [9,10]; with this approach the peptide sample is first loaded on a strong cation exchange column, from where it is stepwise eluted and directly separated on a reverse-phase column before mass spectrometer detector. The entire process is completely automated but it requires a dedicated LC system and a time-consuming setup. Alternatively, the peptide sample can be fractionated by isoelectric-focusing (IEF); available systems make use of immobilized pH gradients (IPG) and allow the separation of the sample in up to 24 fractions before the LC-MS analysis [11,12]. In-solution IEF allows instead a faster fractionation; although it has been since long time used for protein fractionation [13-15] its application for

***Corresponding author:** Kempa S. Integrative proteomics and metabolomics, Berlin Institute for Medical Systems Biology at the Max Delbrück Center for Molecular Medicine, Robert-Roessle-Strasse 10, 13125 Berlin, Germany
E-mail: Stefan.kempa@mdc-berlin.de

peptide fraction was limited to simple proteomes [16] or was used for phospho-peptide enrichment [17].

In our laboratory, we were interested in a fractionation technique which allows a simple and fast fractionation of peptides without limitation for the amount of input material and which enables high identification rates. Here we applied for the first time the in-solution isoelectric focusing to highly complex peptide mixtures and we systematically compared the performances and the results of this technique with those obtained with other published methods. We applied this technique to separate peptide mixtures from various organisms of different complexity and performed deep proteome analyses. The applied technique worked fast and robustly and the amounts of input material could be adjusted without negative impact on the overall performances. Furthermore, our established workflow allowed deep proteome analyses within 48 hours from protein extraction till end of LC-MS/MS analyses. The analyses resulted in high numbers of identified proteins with remarkable sequence coverage. The workflow was also already used to validate the *de novo* assembled transcriptome of planarian [18] or for high proteome coverage in *C. elegans* [19].

Material and Methods

HEK cells and yeast culture and protein extraction

HEK293 cells were cultivated in Dulbecco's Modified Eagle Medium (DMEM [Invitrogen]) supplemented with 10% fetal bovine serum with 2 mM Glutamine and 0.9 g/L glucose. Isotopically labeled HEK293 cells were grown in DMEM supplemented with dialyzed fetal bovine serum and L-$^{13}C_6{}^{15}N_4$-arginine (Arg10) and L-$^{13}C_6{}^{15}N_2$-lysine (Lys8) replacing the natural amino acids. Heavy cells were cultured in the Stable Isotope Labeling by Amino acids in Cell culture (SILAC) medium approximately for seven generation to reach complete labeling. Confluent cultures were trypsinized and 1×10^6 cells were pelleted by centrifugation.

Heavy and light cells pellets were mixed and lysed in urea buffer (8 M Urea, 100 mM Tris·HCl, pH 8.3) and briefly sonicated. Cell debris was removed by centrifugation (14000g, 5 min). Protein concentration was then measured by Bradford colorimetric assay [20].

Yeast strain FY-4 was cultivated in standard yeast Synthetic Defined (SD) medium (Sigma), cells were filtered on nylon membrane (1µm pore size, 3M) and cells were disrupted using a CryoMill device (Retsch). Proteins were resolved using urea containing buffer.

Protein concentration was then measured by Bradford colorimetric assay [20] before enzymatic digestion.

In-solution digestion

Disulfide bridges were then reduced in dithiothreitol (DTT) 2mM for 30 minutes at 25°C and successively free cysteines were alkylated in 11 mM iodoacetamide for 20 minutes at room temperature in the darkness. LysC digestion was performed by adding LysC (Wako) in a ratio 1:40 (w/w) to the sample and incubating it for 18 hours under gentle shaking at 30°C. After LysC digestion, the samples were diluted 3 times with 50 mM ammonium bicarbonate solution, 7 µL of immobilized trypsin (Applied Biosystems) were added and samples were incubated 4 hours under rotation at 30°C. Digestion was stopped by acidification with 10 µL of trifluoroacetic acid and removal of trypsin beads by centrifugation.

The resulting peptide mixtures were loaded on Empore cartridges (3M) following the instructions of the manufacturer and eluted with 70% acetonitrile.

In-solution isoelectric focusing IEF

After partial removal by evaporation of the acetonitrile used for the elution from the Empore cartridges, the peptide mixture was diluted up to 2.5 mL with MilliQ water and 150 µL of ampholyte solution (pH range 3-10, 40% w/w, Bio-Rad) were added. The sample was then loaded into the focusing chamber of the Microrotofor device. Isoelectric focusing was performed following the manufacturer instruction. Briefly, constant 1 W power current was applied, while the maximum allowed voltage and current were set to 500V and 10 mA, respectively. After reaching stable voltage (~2.5 h time), the current was applied for further 30 minutes before collecting the fractionated peptides from the focusing chamber.

The resulting fractions were desalted on STAGE Tips (max 15 µg per StageTip), dried and reconstituted to 25 µL of 0.5 % acetic acid in water [21].

LC-MS analysis

5 microliters of the desalted peptides were injected on a LC-MS/MS system (Agilent 1200 [Agilent Technologies] and LTQ-Orbitrap Velos [Thermo]). Each analysis was run in duplicate. For the chromatographic separation, a linear binary gradient, ranging from 5% to 40% of organic buffer (80% acetonitrile, 20% water, 0.1% formic acid), in 155 or 240 minutes was used. As aqueous solvent, 5% acetonitrile in water with 0.1% formic acid was used. 20 cm long capillary (75 µm inner diameter), packed in-house with 3 µm C18 beads (ReprosilPur C18 AQ, Dr. Maisch), was used as chromatographic column. At one end of the capillary a nanospray tip was generated using a laser puller (P-2000 Laser Based Micropipette Puller, Sutter Instruments), allowing fretless packing.

The nanospray source of the mass spectrometer was operated with a spay voltage of 1.9 kV and with an ion transfer tube temperature of 260°C. Data were acquired in data dependent mode, with one survey MS scan in the Orbitrap mass analyzer (resolution 60000 at m\z 400), followed by up to 20 MS/MS scans in the ion trap on the most intense ions (intensity threshold=500 counts). Once selected for fragmentation, ions were excluded from further selection for the next 30 seconds, in order to increase the number of new sequencing events.

Data processing and analysis

Raw data were analyzed using the MaxQuant proteomics pipeline v1.1.1.36 and the built-in Andromeda search engine [22,23] with the International Protein Index Human version 3.71 database or the Saccharomyces Genome Database, version from 5 Jan 2010. Carbamidomethylation of cysteines was chosen as fixed modification, oxidation of methionine and acetylation of N-terminus were chosen as variable modifications. 2 missed cleavage site were allowed and peptide tolerance was set to 7 ppm. The search engine peptide assignments were filtered at 1% false discovery rate at both the peptide and protein level. The 'match between runs' feature was not enabled, 'second peptide' feature was enabled, while other parameters were left as default. For SILAC samples, two ratio counts were set as threshold for quantification.

Data analysis was performed using custom tools in Microsoft Excel and R!. Gene Ontology (GO) analysis was performed using David tool [24]. Enrichment of specific GO terms in the urea digestion/in-solution IEF dataset was calculated using the entire human genome as background; enrichment GO term enrichment was calculated using the entire human genome as background.

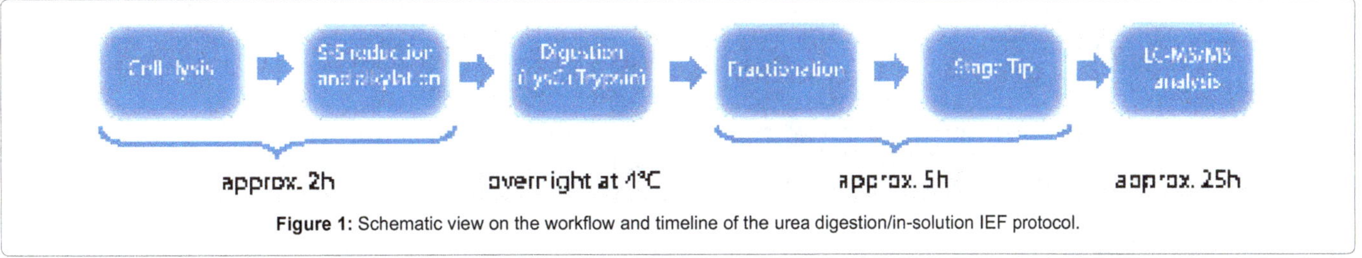

Figure 1: Schematic view on the workflow and timeline of the urea digestion/in-solution IEF protocol.

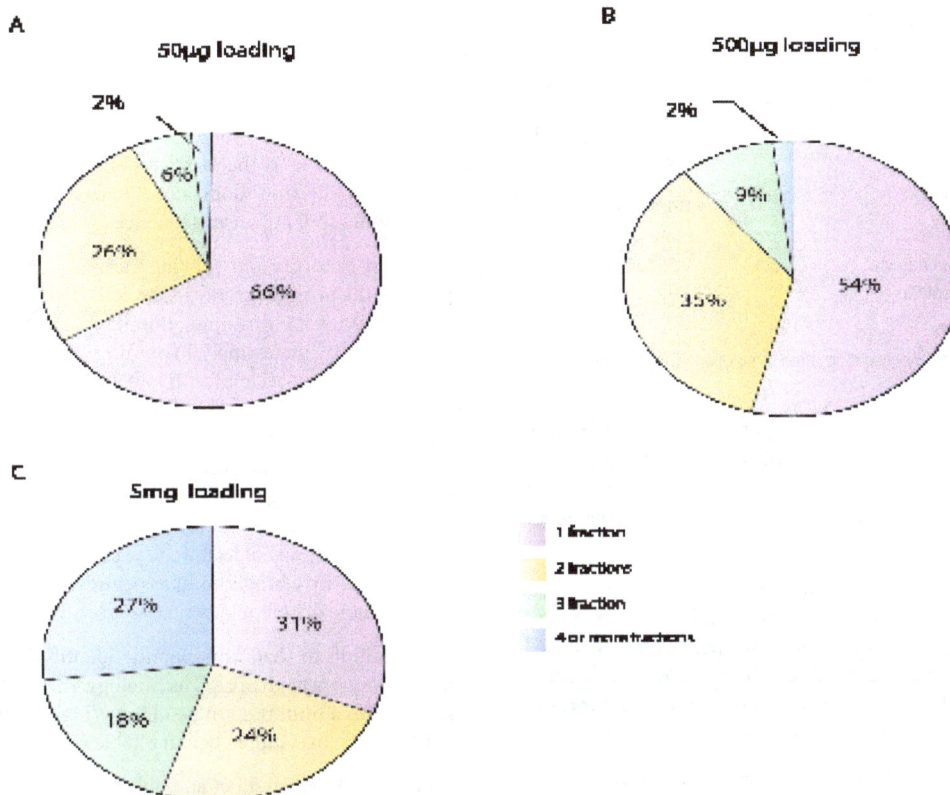

Figure 2: Influence of loaded sample amount on the focusing performance. 50µg (A) and 500µg (B) of HEK cells or 5 mg of yeast protein digests (C) were fractionated on the in-solution IEF device. Percentages represent the fraction of peptides identified in only one well or in two, three or four and more fractions.

Theoretical isoeletric point of identified peptides was calculated by a Perl script using the available tools in the BioPerl package (http://search.cpan.org/dist/BioPerl-1.6.1/). Default pK values for the charged aminoacids were used.

Results and discussion

In contrast to classical IPG isoelectric focusing techniques, which are time consuming and require higher applied voltages (500-4000V and 24hours separation time), in-solution IEF proceeds with lower voltages and finishes within 3 hours. Furthermore, the amount of input material can vary from few micrograms to several milligrams, offering a wide range of applications. The time required for the entire workflow we describe here is considerably reduced compared to GeLC-MS because the starting material is directly digested in solution and then loaded on the device for fractionation; the resulting fractions are purified using Stage Tips [21] and are ready for LC-MS/MS analysis (Figure 1).

To investigate the performance of the in-solution IEF technique we analyzed SILAC-labeled HEK293 cells and yeast proteomes and evaluated the number of identified proteins and peptides, sequence coverage and focusing efficiency using different amounts of peptides.

Evaluation of fractionation efficiency

To investigate the influence of sample amount on the focusing efficiency, we performed three separations using different amount of input spanning on two orders of magnitude. In particular we fractionated 50 and 500 µg of HEK cell digest or 5 mg of yeast digest.

As expected and already observed for other isoelectric focusing techniques, increasing the loading corresponded to a decreased peptide focusing [12].

Using 50 µg of sample, 66% of peptides were 'perfectly' focused, being identified in only one of the ten fractions, while additional 26% were found in two fractions (Figure 2A and supplementary Table 1). A ten-fold increase of the loaded sample did not significantly compromise the focusing efficiency, with 54% of peptides perfectly focused and other 35% found in two adjacent fractions (Figure 2B and supplementary table 2). Only in the case of an extreme sample loading (5 mg) the focusing performance dropped down, with only 31% of

	Replicate 1	Replicate 2	Combined
Peptides	44742	43931	50672
Proteins (1 unique peptide)	5884	5846	6150
Proteins (2 peptides*)	5072	5036	5834

In the combined column are reported the numbers of protein and peptides identified at least in one of the two runs. (*at least one unique peptide)

Table 1: Protein and peptide identification after in solution IEF of 500µg of HEK cells peptides.

	155 min gradient	240 min gradient	Combined
Peptides	17335	19115	22353
Proteins (1 unique peptide)	3276	3579	3782
Proteins (2 peptides*)	2525	2778	3384

In the combined column are reported the numbers of protein and peptides identified at least in one of the two runs. (*at least one unique peptide)

Table 2: Protein and peptide identification after in solution IEF of 50µg of HEK cells peptides.

	Replicate 1
Peptides	22102
Proteins (1 unique peptide)	3226
Proteins (2 peptides)	2769

(*at least one unique peptide)

Table 3: Protein and peptide identification after in solution IEF of 5 mg of yeast peptides.

perfectly focused peptides (Figure 2C and supplementary table 3). An *in silico* calculation of the isoelectric point of the identified peptides showed a good correlation (R^2= 0.86), with the expected pH of the fractions in which they were identified, supporting the effectiveness of the fractionation (supplementary Figure1).

Since the in-solution pH gradient cannot be as accurate as an immobilized pH gradient and is susceptible of diffusion phenomena, the resolving power of in-solution IEF is consequently lower than IPG-IEF, but the resulting fractionation still can be considered satisfactory for the downstream LC-MS/MS analysis. We compared then our results with those obtained with IPG-IEF reported by Hubner et al. [12].

In their work peptides were separated into 24 fractions; for small sample loadings (50 µg) the focusing efficiency of the IPG system is superior, with 99% of peptides focused in one or two fractions. However, the in-solution IEF is less sensitive to increase of sample loadings. The focusing results are better than those observed with the immobilized gradient (89% versus 54% focused in one or two fractions) when 500 µg of peptides were separated. Only the separation of 5 mg of peptides led to a clear decrease of the focusing efficiency, but resulting still comparable with those of the IPG-IEF system with 500 µg loaded (55% of peptides focused in one or two fractions for both methods).

Evaluation of HEK cells proteome coverage

The fractions from the in-solution IEF of 500 µg of peptides were analyzed within one day of measurement time (10 fractions, 155 minutes LC-MS/MS analysis for each fraction). This analysis resulted in the identification of 44,742 distinct peptides mapping to 5,884 proteins with at least one unique peptide. In average 7.6 peptides per protein were identified resulting in 24.9% sequence coverage.

Repetition of the analysis of each fraction led to a modest increase in the number of the identified peptides and proteins, suggesting that already a single analysis covers most of the detectable proteins with this approach (Table 1 and supplementary Table 4).

The use of reduced amounts of peptide material resulted in lower proteome coverage, even if a better focusing performance could be achieved. Using 50 µg of peptides led to identification of 3,276 proteins and 17,335 peptides, with an average of 5.2 peptides per protein and sequence coverage of 19%.

In order to test whether the use of a longer LC gradient may lead to a significant increase of protein identification, we repeated the analysis using a 240 minutes gradient. This analysis identified only 10% more proteins and the combination with the 155 minutes gradient analyses yielded only another 5% increase (Table 2 and supplementary Table 5). Instead, considering peptide identification the increase was 28%, with direct impact on the number of proteins identified with 2 or more peptides. In fact, in the combined dataset this number increased by 34%.

These results thus suggest that the single analysis of the in-solution IEF fractions on a 155 minutes gradient can already cover most of peptides detectable by the mass spectrometer; use of a longer gradient or replicate injection does not increase significantly the protein identifications, but enhances sequence coverage.

Protein isoforms and splicing variants can play a crucial role in regulating the normal activity of the cell. High peptide coverage is then fundamental for the discrimination of proteins that share a large part of the sequence. Interestingly, in our dataset we could detect an isoform of pyruvate kinase (Uniprot ID Q504U3) that currently has been observed only at transcript level. This finding is intriguing because this isoform constitute a shortened version of the isoform PKM1 (Uniprot ID P1618-2), which normally is expressed in organs that are strongly dependent upon a high rate of energy regeneration, such as muscle and brain. PKM1 presence was not reported in HEK cells and actually we could not identify any of its unique peptides. Since the protein Q504U3 was identified by a single unique peptide, the MS/MS spectrum was validated manually (supplementary Figure 2).

In addition to that, high peptide identification improves also the quantification through SILAC technology. In our experiments we could quantify with a minimum of two Heavy/Light counts up to 92% of the identified proteins (supplementary Table 6).

Recently, Wiśniewski et al. published the largest human proteome dataset obtained within a single experiment, using a detergent-based filter-aided sample preparation (FASP) and IPG-IEF fractionation [25]. Although the reported workflow differs from ours and we analyzed a SILAC labeled sample, resulting in doubled complexity, the results constitute an excellent term of comparison to further evaluate the urea extraction/in-solution IEF procedure and the quality of the results.

Considering the proteins identified in the combined dataset of the 500 µg HEK cell sample, we could find that the FASP dataset and in-solution IEF dataset overlap substantially (Figure 3B), whereas a substantial large proportion of identified peptides (45% of the FASP dataset and 56% of the in-solution IEF one) is specific for one of the two workflows (Figure 3A). Examination of the peptide sequences identified by only one of the two methods does not reveal a significant difference in the amino acid composition. On the contrary, the average and median lengths of peptides identified only in the FASP dataset are longer than those found only in the in solution IEF (14.6 versus 12.3 amino acids), while the peptides in common have an intermediate length of 13.4 amino acids. This may be due to the higher hydrophobicity of longer peptides, which are better covered by the FASP protocol.

Gene Ontology analysis of the identified proteins in the FASP dataset shows a significant enrichment of membrane proteins [25], while the proteins only identified by urea extraction/in-solution IEF

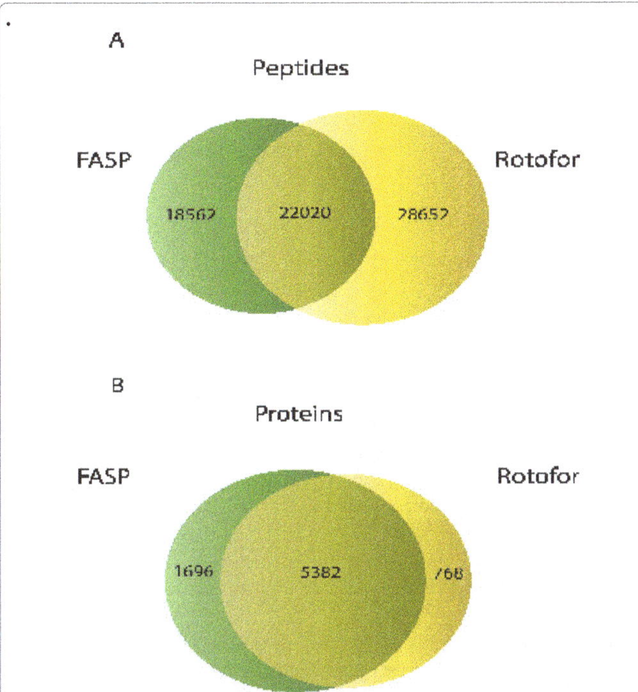

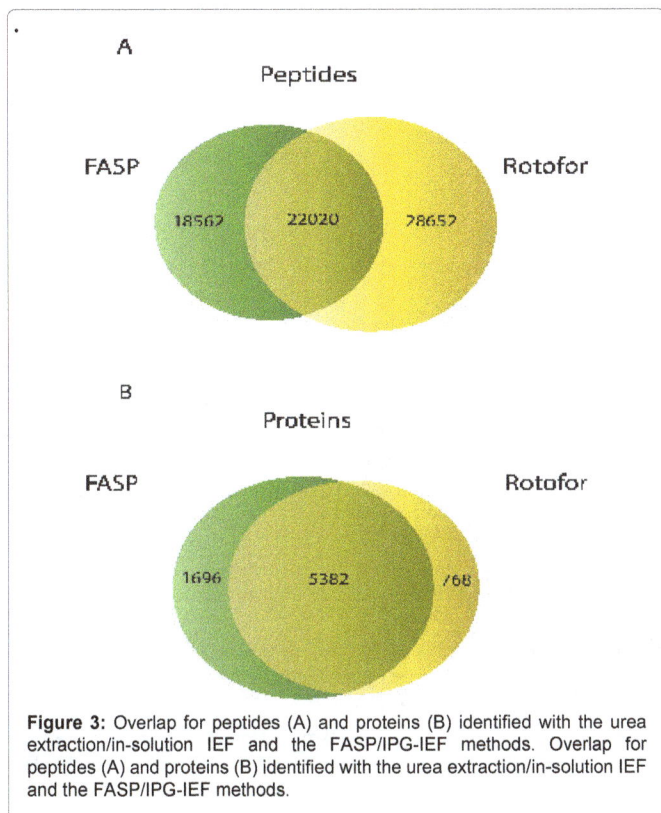

Figure 3: Overlap for peptides (A) and proteins (B) identified with the urea extraction/in-solution IEF and the FASP/IPG-IEF methods. Overlap for peptides (A) and proteins (B) identified with the urea extraction/in-solution IEF and the FASP/IPG-IEF methods.

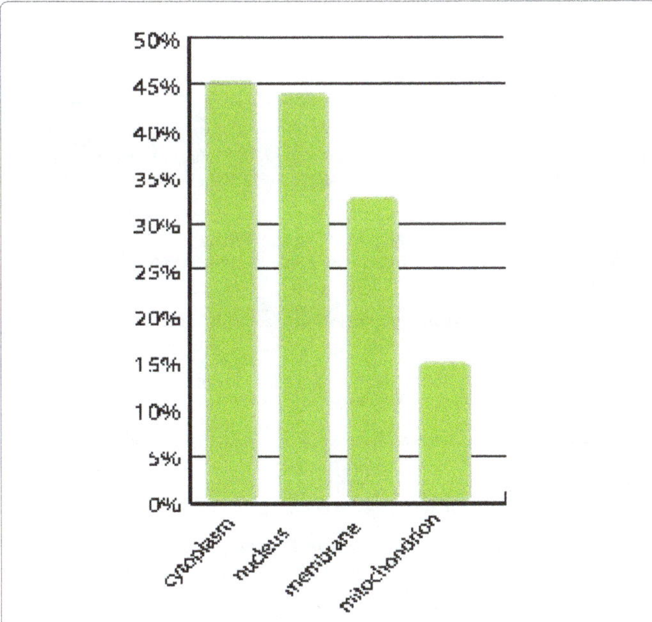

Figure 4: Percentages of proteins with a Gene Ontology cell component (GOCC) annotation, classified by the main cellular compartments.

show an enrichment in intracellular proteins . However, a significant presence of integral membrane proteins could be detected proving that the method, even without detergents, can still be used to detect hydrophobic proteins (Figure 4 and supplementary Table7).

Finally, in solution IEF can be applied with other than urea extraction protocols, as for example the FASP protein extraction technique.

Evaluation of yeast proteome coverage

To evaluate the robustness of the in solution IEF protocol, we fractionated 5 mg of yeast peptides and analyzed the resulting fractions on 240 minutes LC-gradient.

Using such a large sample amount the fractionation was completed after 3 hours as for lower sample loadings. Analyzing the ten fractions we could identify 22,102 different peptides mapping to 3,226 distinct proteins (Table 3 and supplementary Table 8).

In frame of a recent comprehensive yeast proteome analysis, that involved extensive peptide fractionation in 24 fractions and triplicate analyses [26], 3,987 proteins could be identified. This dataset is 23% larger than the one obtained with in solution IEF fractionation, but the latter required a fourth of the analysis time. To evaluate the dynamic range of our analysis, we considered the expression levels of yeast proteins reported by Ghaemmaghami et al. [27] that were later confirmed by mass spectrometry [28].

In our dataset most of the proteins expressed with more than 100,000 copies per cell were present; more interestingly, 85 out of 236 (36%) proteins expressed at less than 250 copies per cell were detected in our analysis. Thus, a dynamic range of 5 orders of magnitude could be achieved with this approach.

The abundance of three proteins reported with <128 copies per cell by Ghaemmaghami et al. [27] (YKR031C, YGL006W, YNR067C) was confirmed by Single Reaction Monitoring (SRM) approach [28]. In our analysis we could detect two of those proteins (YGL006W, YNR067C); interestingly only one unique peptide for YNR067C was identified, while for YGL006W 17 unique peptides were found. Such high coverage of YGL006W despite its low abundance was also observed by Thakur et al. [3]. Furthermore, we could detect 245 out of 1718 proteins that were not detected by Ghaemmaghami et al. [27], showing that urea extraction/in-solution IEF/LC-MS/MS approach can overlap with and complement other techniques.

Functional annotation of identified HEK cells proteins

The functional annotation of large protein dataset is of extreme utility for monitoring the global changes within cellular pathways. Here we used the pathway database of Kyoto Encyclopedia of Genes and Genomes (KEGG) to evaluate the functional information content of our dataset. Notably, only 14 pathways were not represented in our dataset and for 221 pathways at least two proteins were present (supplementary Table 10). In total 42% of the represented pathways have a coverage equal or higher than 50%, including major metabolic processes such as oxidative phosphorylation, tricarboxylic acid (TCA) cycle (supplementary Figures 3 and 4) and purine metabolism, major molecular machineries such as spliceosome (supplementary Figure 5), ribosome and the DNA replication machinery. Also signaling pathways such as mTOR and phosphatydyl-inositol signaling pathways were largely covered (57% and 46% coverage, respectively).

Furthermore, the GeneOntology analysis of the identified proteins did not suggest a major bias toward proteins from different cellular compartments (Figure 4). For example, the proportion of proteins with 'membrane' annotation is above 30%, close to results obtained with detergent-based extraction procedures [25].

Interestingly, the enrichment analysis of GO terms for biological process shows a significant enrichment in proteins involved in various RNA processing activities, while the enrichment analysis of GO terms for molecular function shows enrichment for RNA and nucleotide binding proteins (supplementary Table 9). This could be explained with high efficiency of the urea extraction methods for nuclear and cytosolic proteins.

Conclusions

High proteome coverage is still a challenging task and requires laborious sample preparation, expensive instrumentation, special instrumental setup and long working times. For that reason, we developed a simple, fast and robust workflow for sample preparation and peptide fractionation. Using urea extraction and in solution IEF, we could identify more than 5,800 proteins with an average sequence coverage of 25% in just 48 hours including all experimental procedures.

In-solution IEF evidenced to be an excellent solution for peptide fractionation prior to the LC-MS/MS analysis. It performs robustly, especially when large amounts of peptides are loaded. In addition, the focusing is faster than IPG-IEF and is completed in less than 3 hours.

Furthermore, this technique can be coupled with different preprocessing methods (data not shown) and can be used also as enrichment step for phospho-peptides [17] since it can be applied with several mg of starting material.

The established workflow reduces time and working steps and allows deep proteome analyses with a high depth and sequence coverage using a normal nanoscale liquid chromatography coupled to an Orbitrap Velos system.

The chosen protein extraction can influence the detected fraction of the proteome; urea extraction works excellently with hydrophilic proteins. In our HEK293 cell dataset we could observe enrichment in nuclear, cytosolic and RNA binding proteins. However, the presence of a large proportion of membrane proteins, as well as membrane-associated complexes such as proteins of oxidative phosphorylation, suggests the absence of any major bias against hydrophobic and membrane proteins. With the applied strategy, nearly all the annotated enzymatic pathways could be detected and a large proportion extensively covered.

Comparison of our data with published results from SDS extraction [25] and IPG-IEF [12] showed comparable proteome coverage. Moreover, the overlap of the identified peptides and proteins suggests that a deeper coverage of the proteome can be obtained by combining different techniques. Reduction of sample complexity allowed protein detection over several orders of magnitude, similarly to the results obtained by targeted proteomic approaches.

In our group, this fractionation strategy has been already successfully applied to samples from different organisms, yielding high proteome coverage in short time [18,19,29].

We believe that in solution-IEF will be a valid alternative method for peptide fractionation. It is straightforward to think that several improvements in the future will increase the proteome coverage achievable with this method. The use of longer columns and UPLC systems will enhance the chromatographic resolution and increase proteome coverage (data not shown), while improvement of the MS instrumentation, e.g. Orbitrap analyzer with higher resolution [30] or new instrumentation, as the Q Exactive [31], will allow the detection of a higher number of peptides.

Acknowledgement

We thank Julia Diesbach for her excellent technical assistance. This research was funded by the Federal Ministry for Education and Research (BMBF), the HepatomaSys project and the Senate of Berlin, Berlin, Germany.

Author contributions

G.M. and S.K designed the project. F.B. and C.Z. prepared part of the HEK cells samples, L.A. prepared the yeast sample. G.M. setup the entire protocol, prepared part of the HEK cells samples, performed the LC-MS/MS measurements and analyzed the data. G.M. and S.K. wrote the manuscript. S.K. supervised the project.

References

1. Cox J, Mann M (2007) Is proteomics the new genomics? Cell 130: 395-398.
2. Beck M, Claassen M, Aebersold R (2011) Comprehensive proteomics. Curr Opin Biotechnol 22: 3-8.
3. Thakur SS, Geiger T, Chatterjee B, Bandilla P, Froehlich F, et al. (2011) Deep and highly sensitive proteome coverage by LC-MS/MS without prefractionation. Mol Cell Proteomics 10.
4. Liu H, Sadygov RG, Yates JR III (2004) A model for random sampling and estimation of relative protein abundance in shotgun proteomics. Anal Chem 76: 4193-4201.
5. Michalski A, Cox J, Mann M (2011) More than 100,000 detectable peptide species elute in single shotgun proteomics runs but the majority is inaccessible to data-dependent LC-MS/MS. J Proteome Res 10: 1785-1793.
6. Blagoev B, Ong SE, Kratchmarova I, Mann M (2004) Temporal analysis of phosphotyrosine-dependent signaling networks by quantitative proteomics. Nat Biotechnol 22: 1139-1145.
7. de Godoy LM, Olsen JV, de Souza GA, Li G, Mortensen P, et al. (2006) Status of complete proteome analysis by mass spectrometry: SILAC labeled yeast as a model system. Genome Biol 7: R50.
8. Shevchenko A, Tomas H, Havlis J, Olsen JV, Mann M (2006) In-gel digestion for mass spectrometric characterization of proteins and proteomes. Nat Protoc 1: 2856-2860.
9. Peng J, Elias JE, Thoreen CC, Licklider LJ, Gygi SP (2003) Evaluation of multidimensional chromatography coupled with tandem mass spectrometry (LC/LC-MS/MS) for large-scale protein analysis: the yeast proteome. J Proteome Res 2: 43-50.
10. Washburn MP, Wolters D, Yates JR 3rd (2001) Large-scale analysis of the yeast proteome by multidimensional protein identification technology. Nat Biotechnol 19: 242-247.
11. Horth P, Miller CA, Preckel T, Wenz C (2006) Efficient fractionation and improved protein identification by peptide OFFGEL electrophoresis. Mol Cell Proteomics 5: 1968-1974.
12. Hubner NC, Ren S, Mann M (2008) Peptide separation with immobilized pI strips is an attractive alternative to in-gel protein digestion for proteome analysis. Proteomics 8: 4862-4872.
13. Bier M (1998) Recycling isoelectric focusing and isotachophoresis. Electrophoresis, 19: 1057-1063.
14. Gazzana G, Borlak J (2007) Improved method for proteome mapping of the liver by 2-DE MALDI-TOF MS. J Proteome Res 6: 3143-3151.
15. Thorsell A, Portelius E, Blennow K, Westman-Brinkmalm A (2007) Evaluation of sample fractionation using micro-scale liquid-phase isoelectric focusing on mass spectrometric identification and quantitation of proteins in a SILAC experiment. Rapid Commun Mass Spectrom 21: 771-778.
16. Xiao Z, Conrads TP, Lucas DA, Janini GM, Schaefer CF, et al. (2004) Direct ampholyte-free liquid-phase isoelectric peptide focusing: application to the human serum proteome. Electrophoresis 25: 128-133.
17. Rogers LD, Fang Y, Foster LJ (2010) An integrated global strategy for cell lysis, fractionation, enrichment and mass spectrometric analysis of phosphorylated peptides. Mol Biosyst 6: 822-829.
18. Adamidi C, Wang Y, Gruen D, Mastrobuoni G, You X, et al. (2011) De novo assembly and validation of planaria transcriptome by massive parallel sequencing and shotgun proteomics. Genome research 21: 1193-1200.
19. Jungkamp AC, Stoeckius M, Mecenas D, Grun D, Mastrobuoni G, et al. (2011) In vivo and transcriptome-wide identification of RNA binding protein target sites. Molecular cell 44: 828-840.
20. Bradford MM (1976) A rapid and sensitive method for the quantitation of microgram quantities of protein utilizing the principle of protein-dye binding. Analytical biochemistry 72: 248-254

21. Rappsilber J, Ishihama Y, Mann M (2003) Stop and go extraction tips for matrix-assisted laser desorption/ionization, nanoelectrospray, and LC/MS sample pretreatment in proteomics. Anal Chem 75: 663-670.

22. Cox J, Mann M (2008) MaxQuant enables high peptide identification rates, individualized p.p.b.-range mass accuracies and proteome-wide protein quantification. Nat Biotechnol 26: 1367-1372.

23. Cox J, Neuhauser N, Michalski A, Scheltema RA, Olsen JV, et al. (2011) Andromeda: a peptide search engine integrated into the MaxQuant environment. J Proteome Res 10: 1794-1805.

24. Huang W, Sherman BT, Lempicki RA (2009) Bioinformatics enrichment tools: paths toward the comprehensive functional analysis of large gene lists. Nucleic Acids Res 37: 1-13.

25. Wisniewski JR, Zougman A, Nagaraj N, Mann M (2009) Universal sample preparation method for proteome analysis. Nat Methods 6: 359-U360.

26. de Godoy LM, Olsen JV, Cox J, Nielsen ML, Hubner NC, et al. (2008) Comprehensive mass-spectrometry-based proteome quantification of haploid versus diploid yeast. Nature 455: 1251-1254.

27. Ghaemmaghami S, Huh WK, Bower K, Howson RW, Belle A, et al. (2003) Global analysis of protein expression in yeast. Nature 425: 737-741.

28. Picotti P, Bodenmiller B, Mueller LN, Domon B, Aebersold R (2009): Full dynamic range proteome analysis of S. cerevisiae by targeted proteomics. Cell 138: 795-806.

29. Mastrobuoni G, Irgang S, Pietzke M, Assmus HE, Wenzel M, et al. (2012) Proteome dynamics and early salt stress response of the photosynthetic organism Chlamydomonas reinhardtii. BMC genomics 13: 215.

30. Makarov A, Denisov E, Lange O (2009) Performance evaluation of a high-field Orbitrap mass analyzer. J Am Soc Mass Spectrom 20: 1391-1396.

31. Michalski A, Damoc E, Hauschild JP, Lange O, Wieghaus A, et al. (2011) Mass spectrometry-based proteomics using Q Exactive, a high-performance benchtop quadrupole Orbitrap mass spectrometer. Mol Cell Proteomics 10.

Validation Study of Analysis of 1-Phenyl-2-Propanone in Illicit Methamphetamine Samples by Dynamic Headspace Gas Chromatography Mass Spectrometry

Arnoldi S, Roda G*, Casagni E, Coceanig A, Dell'Acqua L, Farè F, Rusconi C, Tamborini L, Visconti GL and Gambaro V

Dipartimento di Scienze Farmaceutiche, Università degli Studi di Milano, Via Mangiagalli 25, 20133, Milano, Italy

Abstract

A new method based on dynamic headspace sampling (DHS) coupled to GC/MS analysis was developed, optimized and validated for the analysis of 1-phenyl-2-propanone (P2P) in illicit methamphetamine (MAMP) samples. The DHS parameters were investigated to reach the sensitivity suitable for this kind of analysis. The method showed of a good specificity, linearity, accuracy, precision and robustness. The analysis of ten MAMP samples seized by the judicial authority was carried out. P2P was found in all the seizure, confirming that P2P is the starting compound of the synthesis of amphetamines.

Keywords: 1-phenyl-2-propanone; P2P; Methamphetamine; DHS-GC/MS

Introduction

Amphetamine (AMP) and Methamphetamine (MAMP) are widely abused drugs all over the world [1,2]. In fact, the abuse of amphetamine-like compounds has increased continuously in the last years becoming a global problem in recent years [3]. For this reason, it is important to investigate how they are synthesized and where they are sold and diffused [4], analyzing their chemical impurity profiling. This information is useful for characterizing links between different samples originating from the same seizures, but also for determining the synthetic scheme [5,6], which gives details about the origin of the drug.

There are two major raw materials from which AMP and MAMP are prepared: ephedrine and pseudoephedrine and 1-phenyl-2-propanone (P2P). In order to prevent illicit drug manufacture, law enforcement authorities try to control illicit producers of the main drug precursor P2P (Scheme 1) [7,8].

P2P is colorless or slightly yellowish oil, with a density similar to that of water and a characteristic pleasant flavor [9]. P2P is listed in Table 1 of the United Nations Convention against Illicit Trafficking in Narcotic Drugs and Psychotropic Substances of 1988 and it is strictly controlled all over the world. It is prepared in clandestine laboratories starting from phenyl acetic acid [1]. The determination of P2P therefore is of utmost importance for determining the synthetic route by which MAMP is produced and its origin.

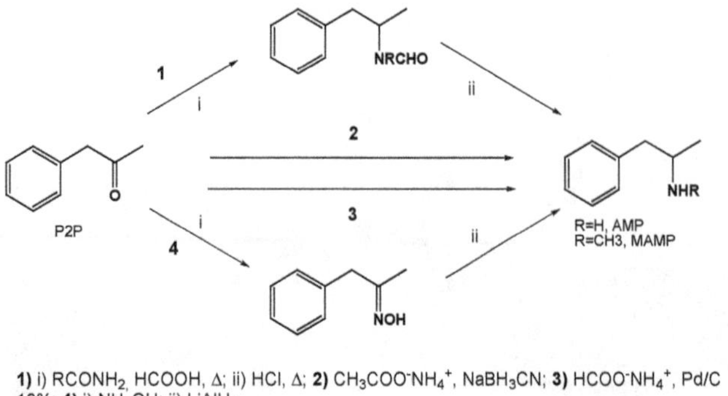

1) i) $RCONH_2$, HCOOH, Δ; ii) HCl, Δ; **2)** $CH_3COO^-NH_4^+$, $NaBH_3CN$; **3)** $HCOO^-NH_4^+$, Pd/C 10%; **4)** i) NH_2OH; ii) $LiAlH_4$

Scheme 1: Synthetic routes to obtain amphetamines from P2P.

In a previous work we reported a method for determining P2P by static headspace gas chromatography (HS-GC/MS), which allowed us to resolve the case of a particular seizure. This finding was made of cornstarch soaked with P2P and it was sold as drug of abuse ("wet amphetamine") to deceive the consumer [10].

In this frame we were interested in applying the same method for the determination of P2P in seized methamphetamine samples, but we realized that it was not sensitive enough for this kind of analysis. So, we developed, optimized and validated a method based on dynamic headspace sampling [11-16], used for the first time in this field, which allowed us to obtain the optimal features for the analysis of impurities in illicit drug samples of MAMP. In this paper we report the results obtained from the analysis of samples containing MAMP seized by the judicial authority and delivered to our laboratory. In all these samples we detected traces of P2P as residual solvent.

Materials and Methods

Reagents and chemicals

All reagents were of analytical grade and were stored as indicated by the supplier. Propylene carbonate and acetophenone, chosen as internal standard (IS), were purchased from Sigma Aldrich (St. Louis, MO, USA). Water (18.2 MΩ·cm^{-1}) was prepared by a Milli-Q System (Millipore, Darmstadt, Germany). Stock solutions of P2P (1 mg/mL) and IS (0.02 mg/mL) were prepared in propylene carbonate.

Synthesis of 1-phenyl-2-propanone: 1-Phenyl-2-propanol (1.36 g, 10 mmol) and a catalytic amount of PCC (5 mg) were added to a solution of periodic acid (2.28 g, 10 mmol) in acetonitrile (50 mL) at 0°C. The mixture was stirred for 30 min and then ethyl acetate (50 mL)

***Corresponding author:** Gabriella Roda, Dipartimento di Scienze Farmaceutiche, Università degli Studi di Milano, Via Mangiagalli 25, 20133, Milano, Italy
E-mail: gabriella.roda@unimi.it

was added. The organic phase was washed with brine (50 mL) and a saturated solution of sodium sulfite (50 mL) and dried with anhydrous sodium sulfate; and the solvent was evaporated under reduced pressure. The residue was purified by flash chromatography (cyclohexane/ethyl acetate, 97/3, v/v). The yield was 90%. The structure of the synthesized compound was confirmed by ^{1}H-nuclear magnetic resonance (NMR) spectroscopy. The NMR spectrum was recorded in DMSO-d_6 at 300 MHz. The chemical shifts (data in δ ppm) were 2.15 (s, 3H); 3.70 (s, 2H); 7.18-7.22 (m, 2H); 7.22-7.38 (m, 3H).

Instruments

GC/MS: Analyses were carried on a HP5890 Series II GC system (Agilent, Santa Clara, CA), with a split-splitless injection system operated in a split mode and an Agilent MSD HP5971 Detector operated in electron impact mode (70 eV). The GC was equipped with a capillary column VF-624MS (60 m, 0.25 mm i.d., film thickness 1.4 µm) (Agilent, Santa Clara, CA) and a Master DHS autosampler (DANI Instruments, Milano, Italy) with 20 mL headspace vials.

The GC/MS system was operated under following conditions: split ratio, 60:1; solvent delay, 4.5 min; injector temperature, 280°C; interface transfer line, 300°C; ion source, 180°C; oven temperature program, from 100°C to 250°C, at 10°C/min, final isotherm, 10 min. Analysis time 25 min (Figure 1).

Helium was used as the carrier gas at a flow rate of 1.0 mL/min. For qualitative analysis the MS detector was operated in SCAN mode, mass range: 10 to 550 m/z. For quantitative determinations, the MS detector was operated in SIM mode. Solvent delay: 4.5 min. From 4.51 to 13.20 min only the m/z=51, 77, 105, 120 ions were acquired for the IS and the m/z=43, 65, 91, 134 ions were acquired for P2P. Quantitative analysis was performed integrating the peak at m/z 105 for the IS and 91 for P2P.

Dynamic headspace analysis: Incubation temperature: 150°C with slow shaking; Incubation time: 30 min; Stripping time: 5 min; Stripping flow: 100 mL/min; Stripping carrier: Nitrogen; Trap: Tenax® GR (70% porous polymers based on 2,6-diphenyl-p-phenylene oxide, 30% graphite carbon, Scientific Instruments Services, Inc Old York RD, Ringoes, NJ); Trap adsorption temp: 0°C; Dry step time: 10 min; Dry Step Flow: 30 mL/min; Dry Step Trap Temp: 40°C; Injection time: 1 min; Trap desorption temperature: 280°C; Dew Stop temperature: 200°C; Transfer line temperature: 280°C; Switching valve temperature: 250°C.

Samples and sample preparation

Ten samples seized by the judicial authority in 2014 were analyzed (Figure 2). The samples were characterized by different content of MAMP (Table 1) and different color and crystal shape.

In a 20 mL headspace vial, 25 mg of powder were weighed and added with 1 mL of IS and 2 mL of propylene carbonate. For the preparation of working Standard Samples (WSS) in a 20 mL headspace vial, different volumes of the stock solution of P2P (see 2.1) were made up to 2 mL with propylene carbonate and 1 mL of IS was added.

Results and Discussion

Applying the conditions previously developed [10], we were able to obtain peaks with a good shape and a good chromatographic response either for P2P or for the IS (acetophenone). On the other hand, we were interested in increasing sensitivity in order to reach a LOD suitable for the analysis of impurities in drugs of abuse. To this end, we studied and optimized the DHS parameters.

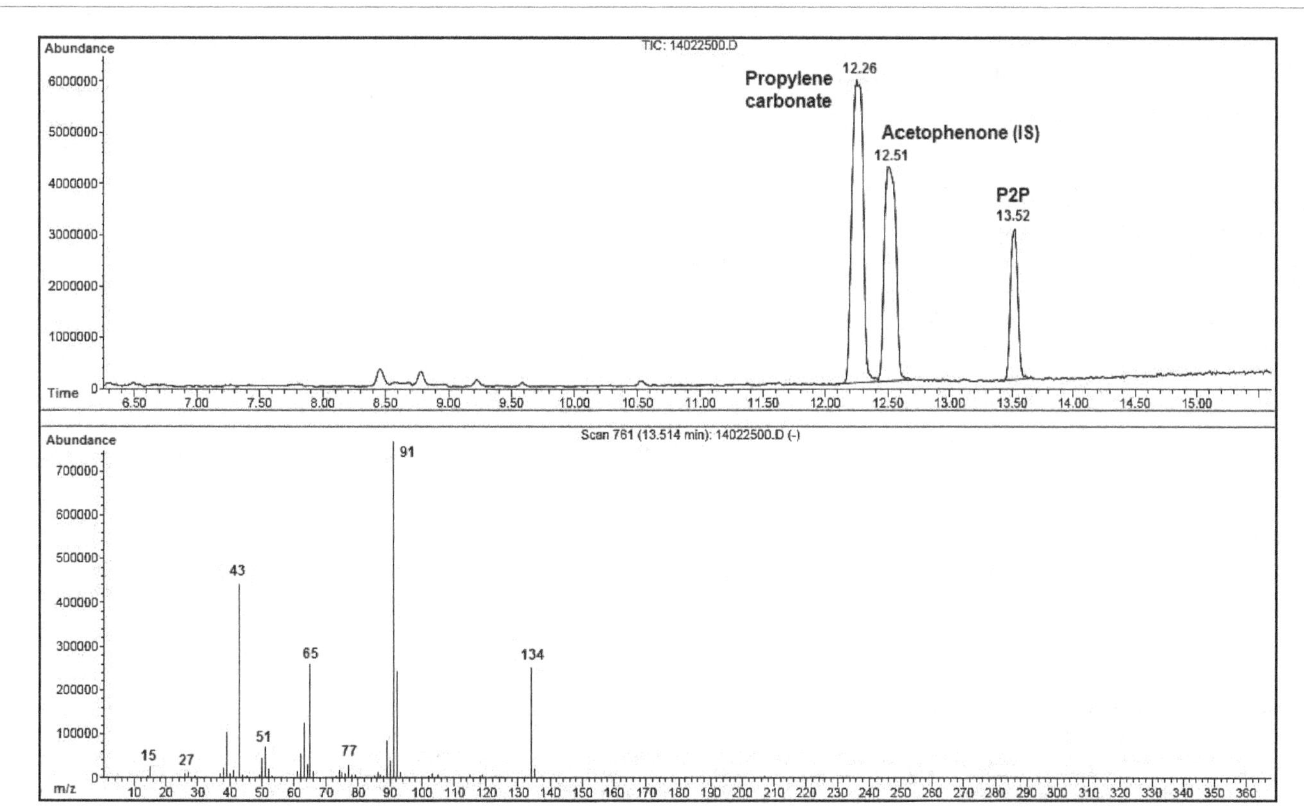

Figure 1: DHS-GC/MS SCAN chromatogram and spectrum of a working standard solution (WSS) of 1-phenyl-2-propanone (P2P).

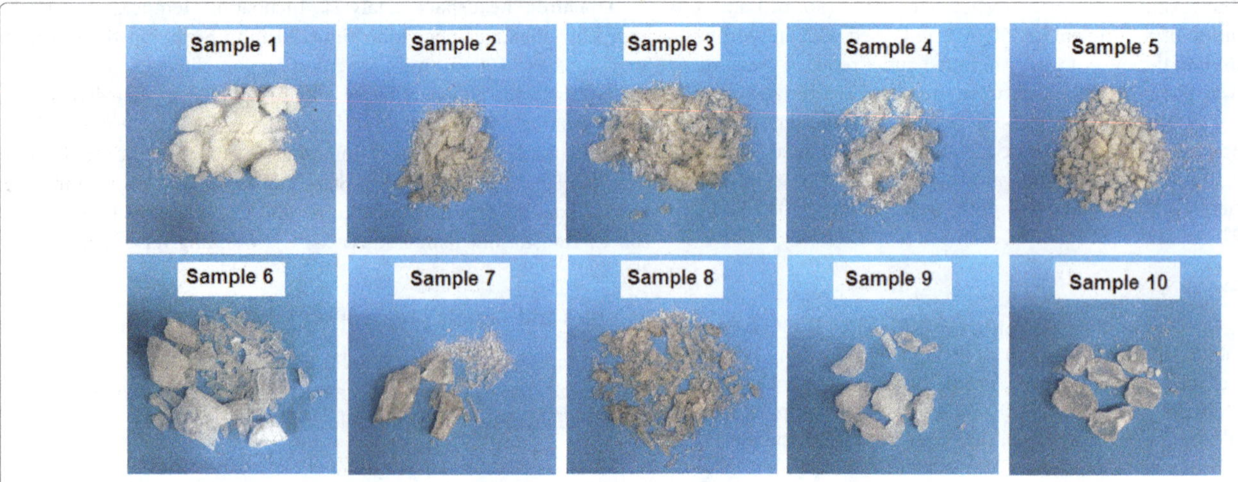

Figure 2: Sample seized by the judicial authority.

Sample	% of MAMP	% P2P
1	83.8	0.287
2	88.2	0.359
3	89.9	0.188
4	90.4	0.426
5	77.6	0.398
6	98.3	0.335
7	91.8	0.277
8	95.8	0.299
9	96.8	0.123
10	95.5	0.619

Table 1: % of methamphetamine and P2P contained in the different seizures.

Optimization of DHS parameters

Incubation time and temperature: In the preliminary analysis [10], the incubation temperature was set at 80°C and the incubation time was 15 min. To favor the transition to the vapor phase of P2P, thus increasing extraction efficiency, we tested an increase of the incubation time (30 min) and an increase of the incubation temperature (100°C). As it is evident from Figure 3, increasing both the temperature and the incubation time led to a marked gain in extraction recovery. When analyzing a real sample of methamphetamine, on the other hand, the peak relative to P2P was not detected, so we decided to further increase the incubation temperature to 150°C. In this way, the chromatographic peak of P2P was evidenced.

Stripping: The duration of the stripping step is very important to transfer all the analyte to the vapor phase. We tested three different stripping times: 1, 5 and 10 min. Going from 1 to 5 min, there is a significant increase in extraction recovery either for P2P or for the IS, while with a stripping time of 10 min, there is a dramatic increase in the response of acetophenone (IS) but the peak related to P2P remains almost unchanged (Figure 4). So, the ideal stripping time resulted to be 5 min.

Trap adsorption temperature: Generally, the lower the trap's temperature, the greater is the adsorption of the analytes. Three different temperatures were tested: -10, 0 and 10°C. The worst result was obtained at 10°C; between 0 and −10°C there is not a big difference, but, at 0°C, the peaks have a sharper shape and propylene carbonate, the solvent, is less adsorbed (Figure 5).

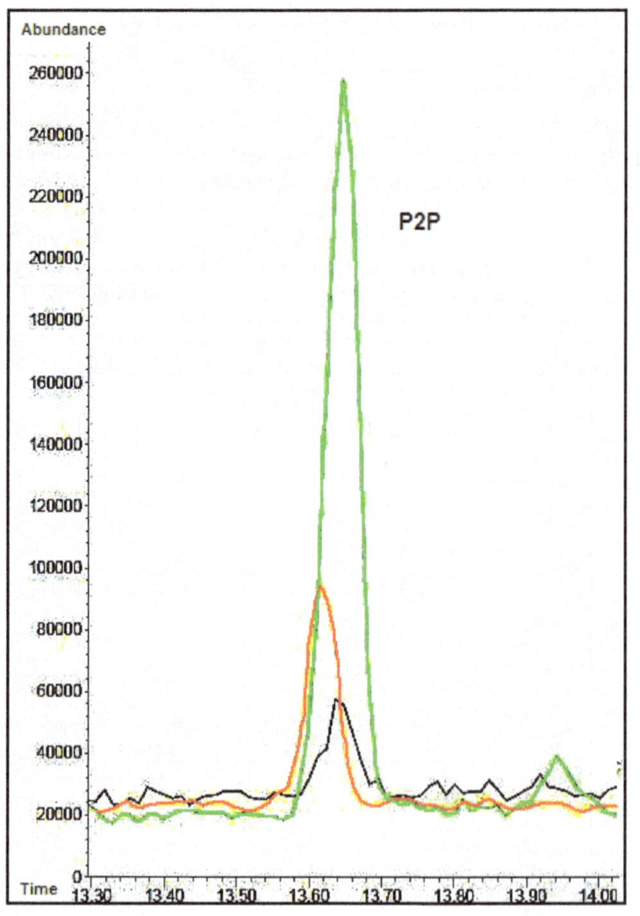

Figure 3: Increase of sensitivity of 1-phenyl-2-propanone (P2P), varying incubation conditions. Black line: 80°C, 15 min; yellow line: 80°C, 30 min; green line 100°C, 30 min.

Injection: The injection of the sample is a crucial step. The analyte has to be desorbed in the shortest time as possible, to obtain sharp peaks and increase sensitivity. We evaluated that the best results were obtained with 1 min of injection time and an inlet and transfer line temperatures of 280°C, in order to ensure the immediate volatilization of the analytes and to avoid the formation of condensate inside the instrument.

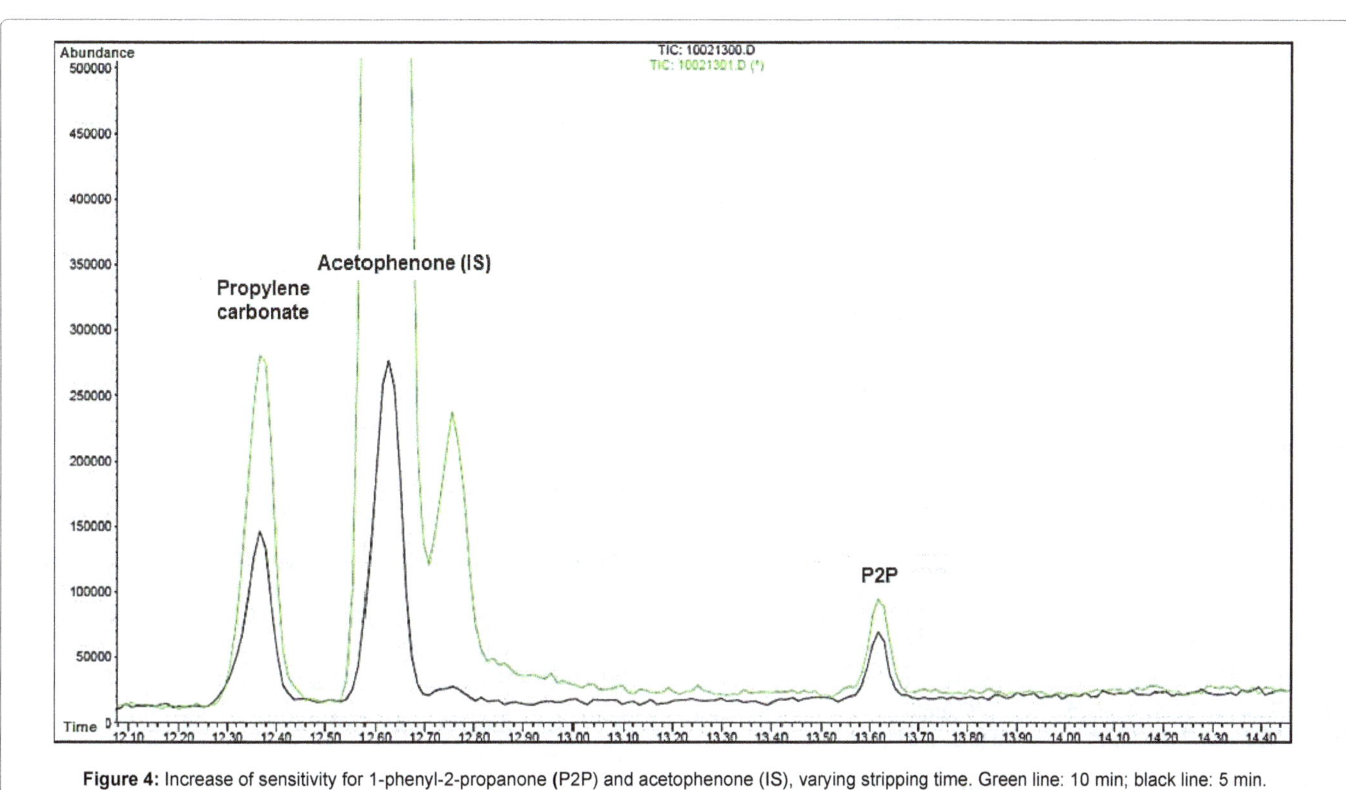

Figure 4: Increase of sensitivity for 1-phenyl-2-propanone (P2P) and acetophenone (IS), varying stripping time. Green line: 10 min; black line: 5 min.

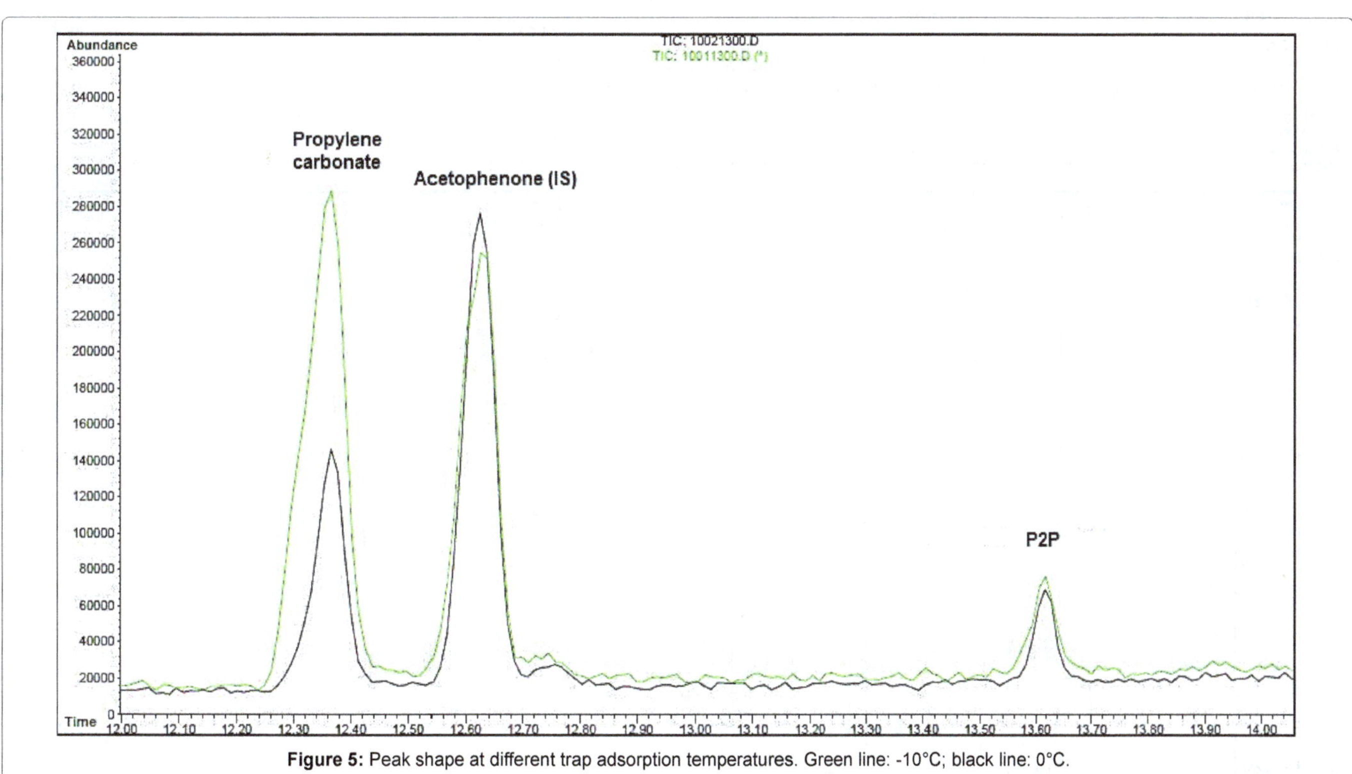

Figure 5: Peak shape at different trap adsorption temperatures. Green line: -10°C; black line: 0°C.

Validation of the DHS-GC/MS method

The DHS-GC/MS method was validated [17] for the identification and quantification of the analytes meeting the requirements of forensic analysis, analyzing WSS specifically prepared.

Specificity: When dealing with seized material synthesized in an illicit way, it is not possible to analyze a "matrix" identical to that of the seizures, which does not contain the analytes. Specificity is assessed if the retention time of standard P2P corresponds perfectly to that of the peak found in unknown samples and the ratio among the four characteristic ions of P2P (m/z 43, 65, 91, 134) is found identical either in standard P2P or in seized MAMP samples.

Linearity: A series of WSS was prepared, made up of seven solutions at the following concentrations of P2P: 10, 50, 100, 150, 200, 250 and 300 μg_{tot}. Each solution was put in a 20 mL headspace vial, starting from the following volumes of the stock solution of P2P (1 mg/mL): 10, 50, 100, 150, 200 250 and 300 μL made up to 2 mL with propylene carbonate, then 1 mL of IS was added. Three series of WSS were analyzed.

The linearity of the method was adequate in all the range. The equation was $y=5.10 \times 10^{-3}-4.39 \times 10^{-2}$ ($R^2=0.9987$).

The calibration curve was built plotting the ratio between the area of the P2P peak and the IS peak against the ratio between the concentration of P2P and the concentration of the IS; the areas were obtained integrating the peaks related to one single ion for each analyte (m/z 105 for IS and 91 for P2P). In this way, it was possible to minimize the noise, increasing accuracy and precision.

LOQ: LOQ was considered as the lowest concentration in which linearity was still satisfied and it corresponded to 10 μg_{tot} of P2P. At this concentration, the signal to noise ratio was 10.83.

LOD: LOD was assessed progressively diluting the solution prepared for the determination of the LOQ, until a signal to noise ratio of 3 was reached. LOD for P2P resulted to be 5 μg_{tot}.

Accuracy: Three different WSS were analyzed in triplicate. In different headspace vials, 50, 100, 150 μL of stock solution of P2P were put and made up to 2 mL with propylene carbonated. Then, 1 mL of IS was added.

Accuracy was evaluated as % recovery according to the following formula:

$$\%REC=(\text{Analytical concentration}/\ \text{real concentration}) \times 100$$

Where the analytical concentration is the concentration of P2P calculated on the basis of the calibration curve and the real concentration is the effective concentration of P2P (μg_{tot}) present in each sample.

% REC ranged from 98.33 to 104.93%, showing a good degree of accuracy of the method.

Precision: In a headspace vial, 100 μL of a P2P stock solution (1 mg/mL) were added to 1 mL of IS and made up to 3 mL with propylene carbonate. Intra-day precision was assessed analyzing six samples in the same day in the same operative conditions.

The mean RA (A_{P2P}/A_{IS}), SD and %RSD were respectively $6.22 \cdot 10^{-2}$, $2.71 \cdot 10^{-3}$ and 4.35%. Inter-day precision was evaluated analyzing other six samples in a different day and combining the results with those of intra-day precision. The mean RA, SD and %RSD resulted respectively $6.08 \cdot 10^{-2}$, $3.05 \cdot 10^{-3}$ and 5.02%, confirming the good precision of the method.

Robustness: Robustness was determined analyzing a solution prepared as described in 3.2.6 with a 10% increase of the carrier gas flow (from 18.6 to 20.5 psi). The carrier gas flow directly affects the desorption of the analyte from the trap. % REC for P2P was 98.82%. The method therefore showed a good degree of robustness.

Quantitative determination of P2P in MAMP samples seized on the illegal market

The validated DHS-GC method was applied for the determination of P2P in ten methamphetamine samples seized by the judicial authority (Figure 2). The results are shown in Table 1. P2P was found in all the samples, demonstrating that P2P is at the moment the most frequently used precursor for the synthesis of MAMP [6,7]. In Figure 6, the chromatogram obtained for Sample 1 is shown. The large peak

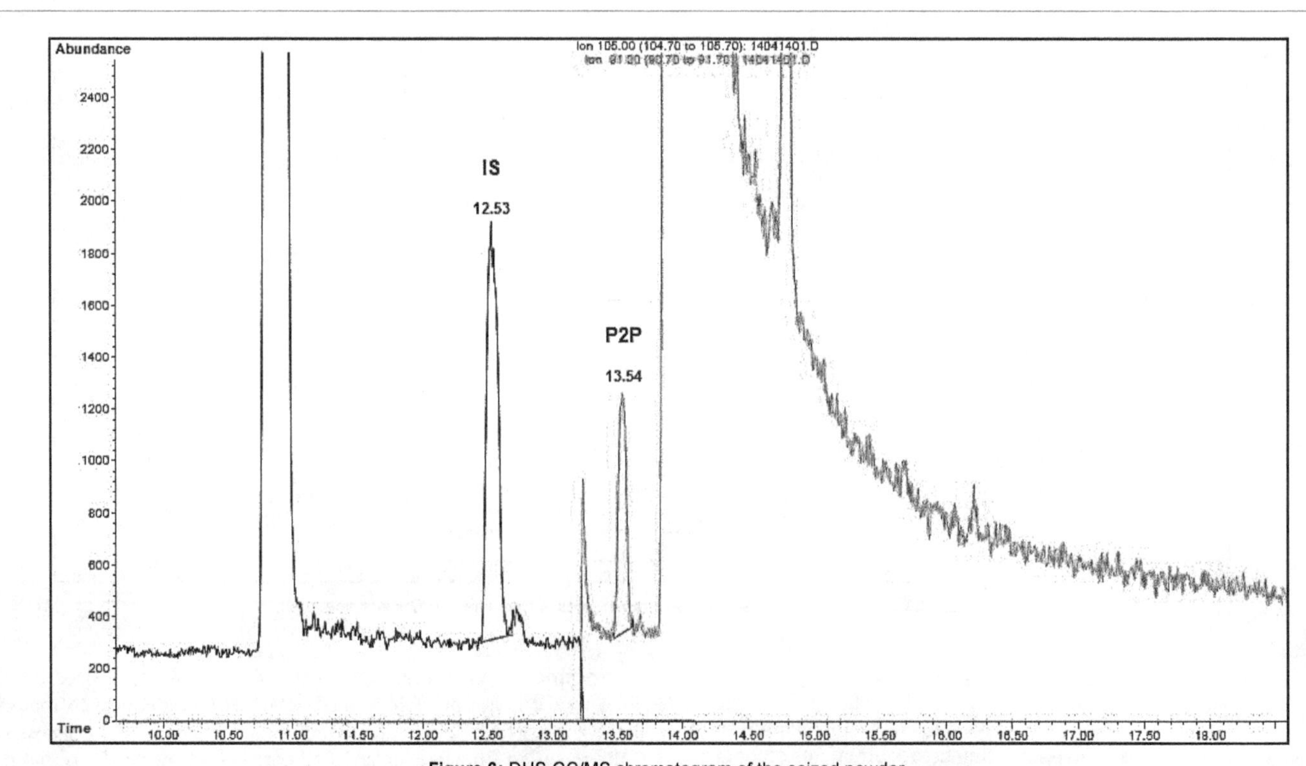

Figure 6: DHS-GC/MS chromatogram of the seized powder.

is due to methamphetamine that in the operative conditions passes to the vapor phase but it does not interfere with the analysis of P2P. It is interesting to note that the % of P2P is not correlated with the % of MAMP. There might have been expected that the more MAMP is pure the less is the concentration of P2P, but in Sample 1, which is less pure respect to Sample 10, the % of P2P instead of being higher resulted lower.

Concluding Remarks

A simple, sensitive and efficient dynamic headspace gas chromatography (DHS-GC/MS) method was optimized and validated for the analysis of P2P in MAMP samples seized by the judicial authority; P2P was found in all the seizure, confirming that P2P is the starting compound of the synthesis of amphetamines (Scheme 1). This method could be used to determine the impurity profile of a MAMP sample, thus establishing similarities among different production batches and laboratories in which the synthesis of the drug of abuse was carried out illegally.

References

1. Tsujikawa K, Kuwayama K, Miyaguchi H, Kanamori T, Iwata YT, et al. (2013) Chemical profiling of seized methamphetamine putatively synthesized from phenylacetic acid derivatives. Forensic Sci Int 227: 42-44.
2. Inoue H, Iwata YT, Kuwayama K (2008) Characterization and profiling of methamphetamine seizures. J Health Sci 54: 615-622.
3. Recommended methods for the identification and analysis of amphetamine, methamphetamine and their ring-substituted analogues in seized materials (2006) United Nations Publication, New York, USA.
4. Lee JS, Chung HS, Kuwayama K, Inoue H, Lee MY, et al. (2008) Determination of impurities in illicit methamphetamine seized in Korea and Japan. Anal Chim Acta 619: 20-25.
5. Tsujikawa K, Kuwayama K, Miyaguchi H, Kanamori T, Iwata YT, et al. (2013) Chemical profiling of seized methamphetamine putatively synthesized from phenylacetic acid derivatives. Forensic Sci Int 227: 42-44.
6. Leuckart R (1885) Uebereineneue Bildungsweise von Tribenzylamin. Ber Dtsch Chem Ges 18: 2341-2344.
7. Di Giovanni S, Varriale A, Marzullo VM, Ruggiero G, Staiano M, et al. (2012) Determination of benzylmethylketone - a commonly used precursor in amphetamine manufacture. Anal Methods 4: 3558-3564.
8. Stojanovska N, Fu S, Tahtouh M, Kelly T, Beavis A, et al. (2013) A review of impurity profiling and synthetic route of manufacture of methylamphetamine, 3,4-methylenedioxymethylamphetamine, amphetamine, dimethylamphetamine and p-methoxyamphetamine. Forensic Sci Int 224: 8-26.
9. Krawczyk W, Kunda T, Perkowska I, Dudek D (2005) Impurity profiling/ comparative analyses of samples of 1-phenyl-2-propanone. Bull Narc 57: 33-62.
10. Arnoldi S, Coceanig A, Dell'Acqua L, Casagni E, Farè F, et al. (2016) Determination of 1-phenyl-2-propanone (P2P) by HS-GC/MS in a material sold as "wet amphetamine". Forensic Toxicol 1-3.
11. Tanaka K, Ohmori T, Inoue T, Seta S (1994) Impurity profiling analysis of illicit methamphetamine by capillary gas chromatography. J Forensic Sci 39: 500-511.
12. Inoue T, Tanaka K, Ohmori T, Togawa Y, Seta S (1994) Impurity profiling analysis of methamphetamine seized in Japan. Forensic Sci Int 69: 97-102.
13. Inoue H, Kanamori T, Iwata YT, Ohmae Y, Tsujikawa K, et al. (2003) Methamphetamine impurity profiling using a 0.32 mm i.d. nonpolar capillary column. Forensic Sci Int 135: 42-47.
14. Kuwayama K, Inoue H, Phorachata J, Kongpatnitiroj K, Puthaviriyakorn V, et al. (2008) Comparison and classification of methamphetamine seized in Japan and Thailand using gas chromatography with liquid-liquid extraction and solid-phase microextraction. Forensic Sci Int 175: 85-92.
15. Cheng H, Qin ZH, Guo XF, Hu XS, Wu JH (2013) Geographical origin identification of propolis using GC-MS and electronic nose combined with principal component analysis. Food Res Int 51: 813-822.
16. Pacioni G, Cerretani L, Procida G, Cichelli A (2014) Composition of commercial truffle flavored oils with GC-MS analysis and discrimination with an electronic nose. Food Chem 146: 30-35.
17. Peters FT, Drummer OH, Musshoff F (2007) Validation of new methods. Forensic Sci Int 165: 216-224.

Permissions

All chapters in this book were first published in JCST, by OMICS International; hereby published with permission under the Creative Commons Attribution License or equivalent. Every chapter published in this book has been scrutinized by our experts. Their significance has been extensively debated. The topics covered herein carry significant findings which will fuel the growth of the discipline. They may even be implemented as practical applications or may be referred to as a beginning point for another development.

The contributors of this book come from diverse backgrounds, making this book a truly international effort. This book will bring forth new frontiers with its revolutionizing research information and detailed analysis of the nascent developments around the world.

We would like to thank all the contributing authors for lending their expertise to make the book truly unique. They have played a crucial role in the development of this book. Without their invaluable contributions this book wouldn't have been possible. They have made vital efforts to compile up to date information on the varied aspects of this subject to make this book a valuable addition to the collection of many professionals and students.

This book was conceptualized with the vision of imparting up-to-date information and advanced data in this field. To ensure the same, a matchless editorial board was set up. Every individual on the board went through rigorous rounds of assessment to prove their worth. After which they invested a large part of their time researching and compiling the most relevant data for our readers.

The editorial board has been involved in producing this book since its inception. They have spent rigorous hours researching and exploring the diverse topics which have resulted in the successful publishing of this book. They have passed on their knowledge of decades through this book. To expedite this challenging task, the publisher supported the team at every step. A small team of assistant editors was also appointed to further simplify the editing procedure and attain best results for the readers.

Apart from the editorial board, the designing team has also invested a significant amount of their time in understanding the subject and creating the most relevant covers. They scrutinized every image to scout for the most suitable representation of the subject and create an appropriate cover for the book.

The publishing team has been an ardent support to the editorial, designing and production team. Their endless efforts to recruit the best for this project, has resulted in the accomplishment of this book. They are a veteran in the field of academics and their pool of knowledge is as vast as their experience in printing. Their expertise and guidance has proved useful at every step. Their uncompromising quality standards have made this book an exceptional effort. Their encouragement from time to time has been an inspiration for everyone.

The publisher and the editorial board hope that this book will prove to be a valuable piece of knowledge for researchers, students, practitioners and scholars across the globe.

List of Contributors

Arae H, Sahoo SK and Watanabe Y
Project for Environmental Dynamics and Radiation Effects, National Institute of Radiological Sciences, 4-9-1 Anagawa, Inage-ku, Chiba 263-8555, Japan

Mishra S
Project for Environmental Dynamics and Radiation Effects, National Institute of Radiological Sciences, 4-9-1 Anagawa, Inage-ku, Chiba 263-8555, Japan
Environmental Monitoring and Assessment Section, Bhabha Atomic Research Centre, Mumbai, Trombay - 400 085, India

Mietelski JW
Department of Nuclear Physical Chemistry, The Henryk Niewodniczanski Institute of Nuclear Physics, Polish Academy of Sciences, Krakow, Radzikowskiego 152, Poland

Afshin Davarpanah
Department of Petroleum, Science and Research Branch, Islamic Azad University, Tehran, Iran

Lara Varden, Britannia Smith and Fadi Bou-Abdallah
Department of Chemistry, State University of New York (SUNY) at Potsdam, 44 Pierrepont Avenue, Potsdam, NY, USA

Justin R. Denton and Thomas Loughlin
Manufacturing Division: Supply Analytical Services, Rahway, NJ, USA

Yun Chen
Research Laboratories: Analytical Research and Development, Rahway, NJ, USA

Akl MA, Mostafa MM and Mohammed SA Bashanaini
Chemistry Department, Faculty of Science, Mansoura University, Mansoura, Egypt

Fariq Fitri MSM, Dzolkhifli Omar and Norhayu Asib
Department of Plant Protection, Faculty of Agriculture, Universiti Putra Malaysia, 43400 Serdang, Malaysia

Halimah Muhamad
Analytical and Quality Development Unit, Malaysian Palm Oil Board, No. 6, Persiaran Institusi, Bandar Baru Bangi, 43000 Selangor, Malaysia China

Anumolu PD, Krishna VL, Rajesh CH, Alekya V, Priyanka B and Sunitha G
Department of Pharmaceutical Analysis, Gokaraju Rangaraju College of Pharmacy, Osmania University, Hyderabad, Telangana, India

George Kuriakose and Saeed Al-Shahrani
SABIC Plastic Application Development Center, King Saud University Campus, Riyadh, KSA 12373, Saudi Arabia

Raghunandana KS
SABIC Technology Center-Jubail, P.O. Box 11669, Al-Jubail, 31961 Saudi Arabia

Heydartaemeh MR
Faculty of Mining, Petroleum and Geophysics, Shahrood University of Technology, Shahrood, Iran

Aslani S and Doulati Ardejani F
College of Mining Engineering, University of Tehran, Tehran, Iran

Rousová J, Ondrušová K, Karlová P and Kubátová A
University of North Dakota, Department of Chemistry, 151 Cornell Street Stop 9024, Grand Forks, ND 58202, USA

Alarfaj NA and El-Tohamy MF
Department of Chemistry, College of Science, King Saud University, PO Box 22452, Riyadh 11495, Saudi Arabia

Fawzia Ibrahim, Nahed El-Enany, Shereen Shalan and Rasha Elsharawy
Department of Analytical Chemistry, Faculty of Pharmacy, University of Mansoura, Mansoura, Egypt

Ghahramani MR, Garibov AA and Agayev TN
Institute of Radiation Problems, Azerbaijan national academy of sciences, Baku, Azerbaijan

Tecleab AG
Department of Pathology and Laboratory Medicine, Staten Island University Hospital, Staten Island, New York 10305, USA

Schofield RC, Ramanathan LV and Dean C Carlow
Department of Laboratory Medicine, Memorial Sloan Kettering Cancer Center, New York, NY 10065, USA

Mustafa GUZEL
Istanbul Medipol University, International School of Medicine, Department of Medical Pharmacology (Chair), Regenerative and Restrorative Medicine Research Center (REMER) (Molecular Discovery and Development Group), Kavacık Campus, Kavacık/ Beykoz-ISTANBUL 34810

Cevdet AKBAY, Yatzka HOYOS, David H. AHLSTROM
Department of Chemistry and Physics, Fayetteville State University, Fayetteville, NC 28301, USA

Mitroshkov AV, Ryan JV, Thomas ML and Neeway JJ
Environmental Systems Group, Pacific Northwest National Laboratory, USA

Naugler DG and Prosser RS
Department of Chemistry, University of Toronto, ON, Canada

Stepfanie NS, Sareh K, Teo SS and Patrick NO
Department of Applied Sciences, UCSI University, 56000 Cheras, Kuala Lumpur, Malaysia

Gabriel AA
Department of Pharmaceutical Science, UCSI University, 56000 Cheras, Kuala Lumpur, Malaysia

Farahnaz A
Department of Anti-aging and Regenerative medicine, UCSI University, 56000 Cheras, Kuala Lumpur, Malaysia

Juan José Berzas Nevado, Carmen Guiberteau-Cabanillas, María Jesus Villasenor Llerena and Virginia Rodríguez-Robledo
Department of Analytical Chemistry and Food Technology, University of Castilla-La Mancha, 13071 Ciudad Real, Spain

Kuvshinova SA, Burmistrov VA, Novikov IV and Alexandriysky VV
Research Institute of Macroheterocyclic Compounds, Ivanovo State University of Chemistry and Technology, Sheremetevskii pr. 7, Ivanovo, 153000 Russia

Koifman OI
Research Institute of Macroheterocyclic Compounds, Ivanovo State University of Chemistry and Technology, Sheremetevskii pr. 7, Ivanovo, 153000 Russia
Institute of Solutions Chemistry, Russian Academy of Sciences, Russia

Vilma del C Salvatierra-Stamp, Norma S Pano-Farias, Jorge Gonzalez, Valentin Ibarra-Galván and Roberto Muñiz-Valencia
Facultad de Ciencias Químicas, Universidad de Colima, Carretera Colima-Coquimatlán, Coquimatlán, Colima, Mexico

Silvia G Ceballos-Magaña
Facultad de Ciencias, Universidad de Colima, c/Bernal Díaz del Castillo 340, Colima, Mexico

Schoeman C, Mashiane M and Dlamini M
Department of Environmental, Water & Earth Sciences, Faculty of Science, Tshwane University of Technology, Rand Water, 2 Barrage Road, Vereeniging, South Africa

Okonkwo OJ
Department of Environmental, Water & Earth Sciences, Faculty of Science, Tshwane University of Technology, 175 Mandela Drive, Pretoria, South Africa

Myron P and Azad SA
Borneo Marine Research Institute, University Malaysia Sabah, 88400 Kota Kinabalu, Sabah, Malaysia

Siddiquee S and Yong YS
Biotechnology Research Institute, University Malaysia Sabah, 88400 Kota Kinabalu, Sabah, Malaysia

Heena and Susheela Rani, Ashok Kumar Malik
Department of Chemistry, Punjabi University, Patiala, Punjab, India

Gaurav
Punjabi University College of Engineering and Management, Rampura Phul, Punjab, India

Abuzar Kabir and Kenneth G Furton
Department of Chemistry and Biochemistry, International Forensic Research Institute, Florida International University, Miami, FL 33193, USA

Sawsan MA, and
Analytical Chemistry department, Faculty of Pharmacy, Cairo University, Kasr-El Aini Street, 11562Cairo, Egypt

Marianne N
Analytical Chemistry department, Faculty of Pharmacy, Cairo University, Kasr-El Aini Street, 11562Cairo, Egypt
Pharmaceutical Analytical Chemistry department, Faculty of Pharmacy & Drug Technology, Heliopolis University, 3 Cairo Belbeis desert road, 2834El- Horria, Cairo, Egypt

Hesham S
Pharmaceutical Analytical Chemistry department, Faculty of Pharmacy, Deraya University, Minia, Egypt

El-Maraghy MC
Analytical Chemistry department, Faculty of Pharmacy, October University for Modern Sciences and Arts (MSA), 11787 6th October city, Egypt

Fabrice Gritti
Waters Corporations, 34 Maple St., Milford, MA, USA

Trivedi MK, Branton A, Trivedi D and Nayak G
Trivedi Global Inc., 10624 S Eastern Avenue Suite A-969, Henderson, NV 89052, USA

Bairwa K and Jana S
Trivedi Science Research Laboratory Pvt. Ltd., Hall-A, Chinar Mega Mall, Chinar Fortune City, Hoshangabad Rd., Bhopal, Madhya Pradesh, India

Suresh Babu VV and Sudhakar V
Natco Pharma Limited, Hyderabad, Andhra Pradesh, India

Murthy TEGK
Department of Pharmaceutical Analysis & Quality Assurance, Bapatla College of Pharmacy, Bapatla, Andhra Pradesh, India

El-Shahawi MS , Hamza A, Alwael H, Bashammakh AS, Al-Sibaai AA and Saigl ZM
Department of Chemistry, Faculty of Science, Damietta University, Damietta, Egypt

Hafez HM
Department of Pharmaceutical Science, Zagazig University, Zagazig, Egypt

Abdullah AE and Abdelaziz LM
Department of Medicinal Chemistry, Faculty of Pharmacy, Zagazig University, Zagazig, Egypt

Kamal MM
Department of Chemistry, Nancy University, EIPICO, Egypt

Zhao Yun
China Waterborne Transport Research Institute, China

Mastrobuoni G, Zasada C, Bindel F, Aeberhard L and Kempa S
Integrative proteomics and metabolomics, Berlin Institute for Medical Systems Biology at the Max Delbrück Center for Molecular Medicine, Robert-Roessle-Strasse 10, 13125 Berlin, Germany

Arnoldi S, Roda G, Casagni E, Coceanig A, Dell'Acqua L, Farè F, Rusconi C, Tamborini L, Visconti GL and Gambaro V
Dipartimento di Scienze Farmaceutiche, Università degli Studi di Milano, Via Mangiagalli 25, 20133, Milano, Italy

Index

www.ingramcontent.com/pod-product-compliance
Lightning Source LLC
Chambersburg PA
CBHW080536200326
41458CB00012B/4450
9781632386243